*Study and Solutions Guide*

# College Algebra:
# A Graphing Approach
### Fourth Edition

Larson/Hostetler/Edwards

# Bruce H. Edwards

University of Florida
Gainesville, Florida

Houghton Mifflin Company    Boston    New York

# Preface

This *Study and Solutions Guide* is a supplement to *College Algebra: A Graphing Approach*, Fourth Edition, by Ron Larson, Robert P. Hostetler, and Bruce H. Edwards.

Solutions to the exercises in the text are given in two parts. Part I contains solutions to odd-numbered Section and Review Exercises; summaries of the chapters; and Practice Tests with solutions. Part II contains solutions to the Chapter and Cumulative Tests from the textbook.

This *Study and Solutions Guide* is the result of the efforts of Larson Texts, Inc., and Meridian Creative Group. If you have any corrections or suggestions for improving this guide, we would appreciate hearing from you.

Bruce H. Edwards
358 Little Hall
University of Florida
Gainesville, FL 32611
Be@math.ufl.edu

Publisher: Jack Shira
Associate Sponsoring Editor: Cathy Cantin
Development Manager: Maureen Ross
Assistant Editor: Lisa Pettinato
Supervising Editor: Karen Carter
Senior Project Editor: Patty Bergin
Editorial Assistant: Allison Seymour
Art and Design Manager: Gary Crespo
Senior Marketing Manager: Danielle Potvin
Marketing Associate: Nicole Mollica
Senior Manufacturing Coordinator: Priscilla Bailey
Composition and Art: Meridian Creative Group

Printed in the United States of America

ISBN: 0-618-39443-5

23456789-POO-08 07 06 05 04

# Contents

# PART I

## CHAPTER P
### Prerequisites

# CHAPTER P
# Prerequisites

## Section P.1    Real Numbers

■ You should know the following sets.

(a) The set of real numbers includes the rational numbers and the irrational numbers.

(b) The set of rational numbers includes all real numbers that can be written as the ratio $p/q$ of two integers, where $q \neq 0$.

(c) The set of irrational numbers includes all real numbers which are not rational.

(d) The set of integers: $\{\ldots, -3, -2, -1, 0, 1, 2, 3, \ldots\}$

(e) The set of whole numbers: $\{0, 1, 2, 3, 4, \ldots\}$

(f) The set of natural numbers: $\{1, 2, 3, 4, \ldots\}$

■ The real number line is used to represent the real numbers.

■ Know the inequality symbols.

(a) $a < b$ means $a$ is less than $b$.

(b) $a \leq b$ means $a$ is less than or equal to $b$.

(c) $a > b$ means $a$ is greater than $b$.

(d) $a \geq b$ means $a$ is greater than or equal to $b$.

■ You should know that

$$|a| = \begin{cases} a, & \text{if } a \geq 0 \\ -a, & \text{if } a < 0. \end{cases}$$

■ Know the properties of absolute value.

(a) $|a| \geq 0$    (b) $|-a| = |a|$    (c) $|ab| = |a|\,|b|$    (d) $\left|\dfrac{a}{b}\right| = \dfrac{|a|}{|b|}$

■ The distance between $a$ and $b$ on the real line is $|b - a| = |a - b|$.

■ You should be able to identify the terms in an algebraic expression.

■ You should know and be able to use the basic rules of algebra.

■ Commutative Property

(a) Addition: $a + b = b + a$    (b) Multiplication: $a \cdot b = b \cdot a$

■ Associative Property

(a) Addition: $(a + b) + c = a + (b + c)$    (b) Multiplication: $(ab)c = a(bc)$

■ Identity Property

(a) Addition: 0 is the identity; $a + 0 = 0 + a = a$

(b) Multiplication: 1 is the identity; $a \cdot 1 = 1 \cdot a = a$

■ Inverse Property

(a) Addition: $-a$ is the inverse of $a$; $a + (-a) = -a + a = 0$

(b) Multiplication: $1/a$ is the inverse of $a$, $a \neq 0$; $a(1/a) = (1/a)a = 1$

■ Distributive Property

(a) Left: $a(b + c) = ab + ac$    (b) Right: $(a + b)c = ac + bc$

—CONTINUED—

**Section P.1 —CONTINUED—**
- Properties of Negatives
  - (a) $(-1)a = -a$
  - (b) $-(-a) = a$
  - (c) $(-a)b = a(-b) = -ab$
  - (d) $(-a)(-b) = ab$
  - (e) $-(a + b) = (-a) + (-b) = -a - b$
- Properties of Zero
  - (a) $a \pm 0 = a$
  - (b) $a \cdot 0 = 0$
  - (c) $0 \div a = 0/a = 0, a \neq 0$
  - (d) If $ab = 0$, then $a = 0$ or $b = 0$.
  - (e) $a/0$ is undefined.
- Properties of Fractions ($b \neq 0, d \neq 0$)
  - (a) Equivalent Fractions: $a/b = c/d$ if and only if $ad = bc$.
  - (b) Rule of Signs: $-a/b = a/(-b) = -(a/b)$ and $-a/(-b) = a/b$
  - (c) Equivalent Fractions: $a/b = ac/bc, c \neq 0$
  - (d) Addition and Subtraction
    - 1. Like Denominators: $(a/b) \pm (c/b) = (a \pm c)/b$
    - 2. Unlike Denominators: $(a/b) \pm (c/d) = (ad \pm bc)/bd$
  - (e) Multiplication: $(a/b) \cdot (c/d) = ac/bd$
  - (f) Division: $(a/b)/(c/d) = (a/b) \cdot (d/c) = ad/bc$, if $c \neq 0$.
- Properties of Equality
  - (a) If $a = b$, then $a + c = b + c$.
  - (b) If $a = b$, then $ac = bc$.
  - (c) If $a + c = b + c$, then $a = b$.
  - (d) If $ac = bc$ and $c \neq 0$, then $a = b$.

**Solutions to Odd-Numbered Exercises**

**1.** $-9, -\frac{7}{2}, 5, \frac{2}{3}, \sqrt{2}, 0, 1, -4, -1$

   (a) Whole numbers: $5, 1, 0$

   (b) Natural numbers: $5, 1$

   (c) Integers: $-9, 5, 0, 1, -4, -1$

   (d) Rational numbers: $-9, -\frac{7}{2}, 5, \frac{2}{3}, 0, 1, -4, -1$

   (e) Irrational numbers: $\sqrt{2}$

**3.** $2.01, 0.666\ldots, -13, 0.010110111\ldots,$
   $1, -10, 20$

   (a) Whole numbers: $1, 20$

   (b) Natural numbers: $1, 20$

   (c) Integers: $-13, 1, -10, 20$

   (d) Rational numbers: $2.01, 0.666\ldots,$
   $\qquad\qquad\qquad -13, 1, -10, 20$

   (e) Irrational numbers: $0.010110111\ldots$

**5.** $-\pi, -\frac{1}{3}, \frac{6}{3}, \frac{1}{2}\sqrt{2}, -7.5, -2, 3, -3$

   (a) Whole numbers: $\frac{6}{3}, 3$

   (b) Natural numbers: $\frac{6}{3}$ (since it equals 2), 3

   (c) Integers: $\frac{6}{3}, -2, 3, -3$

   (d) Rational numbers: $-\frac{1}{3}, \frac{6}{3}, -7.5, -2, 3, -3$

   (e) Irrational numbers: $-\pi, \frac{1}{2}\sqrt{2}$

**7.** $\dfrac{5}{8} = 0.625$

**9.** $\dfrac{41}{333} = 0.\overline{123}$

**11.** $\dfrac{-100}{11} = -9.\overline{09}$

**13.** $4.6 = 4\frac{6}{10} = \frac{46}{10} = \frac{23}{5}$        **15.** $6.5 = 6\frac{5}{10} = 6\frac{1}{2} = \frac{13}{2}$        **17.** $-1 < 2.5$

**19.** $-4 > -8$                **21.** $\frac{3}{2} < 7$                **23.** $\frac{5}{6} > \frac{2}{3}$

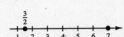

**25.** (a) The inequality $x \le 5$ is the set of all real numbers less than or equal to 5.

(b)

(c) The interval is unbounded.

**27.** (a) The inequality $x < 0$ is the set of all negative real numbers.

(b)

(c) The interval is unbounded.

**29.** (a) The inequality $-2 < x < 2$ is the set of all real numbers greater than $-2$ and less than 2.

(b)

(c) The interval is bounded.

**31.** (a) The inequality $-1 \le x < 0$ is the set of all negative real numbers greater than or equal to $-1$.

(b)

(c) The interval is bounded.

**33.** $x < 0;\ (-\infty, 0)$        **35.** $y \ge 0;\ [0, \infty)$        **37.** $-1 \le p < 9;\ [-1, 9)$

**39.** The interval $(-6, \infty)$ consists of all real numbers greater than $-6$.

**41.** The interval $(-\infty, 2]$ consists of all real numbers less than or equal to 2.

**43.** $|-10| = -(-10) = 10$        **45.** $-3|-3| = -3[-(-3)] = -9$

**47.** If $x + 2 > 0$, then $\dfrac{|x + 2|}{x + 2} = \dfrac{x + 2}{x + 2} = 1$

If $x + 2 < 0$, then $\dfrac{|x + 2|}{x + 2} = \dfrac{-(x + 2)}{x + 2} = -1$

Therefore, $\dfrac{|x + 2|}{x + 2} = \begin{cases} 1 \text{ if } x > -2 \\ -1 \text{ if } x < -2 \end{cases}$

The expression is undefined if $x = -2$.

**49.** $|-3| > -|-3|$ since $3 > -3$.

**51.** $-5 = -|5|$ since $-5 = -5$.        **53.** $-|-2| = -|2|$ since $-2 = -2$.        **55.** $d(126, 75) = |75 - 126| = 51$

**57.** $d\left(-\frac{5}{2}, 0\right) = \left|0 - \left(-\frac{5}{2}\right)\right| = \frac{5}{2}$        **59.** $d\left(\frac{16}{5}, \frac{112}{75}\right) = \left|\frac{112}{75} - \frac{16}{5}\right| = \frac{128}{75}$

**61.** $d(x, 5) = |x - 5|$ and $d(x, 5) \le 3$, thus $|x - 5| \le 3$        **63.** $d(y, 0) = |y - 0| = |y|$ and $d(y, 0) \ge 6$, thus $|y| \ge 6$

**65.** $d(57, 236) = |236 - 57| = 179$ miles

**67.**

| Budgeted Expense, b | Actual Expense, a | $|a - b|$ | $0.05b$ |
|---|---|---|---|
| $112,700 | $113,356 | $656 | $5635 |

The actual expense difference is greater than $500 (but is less than 5% of the budget) so it does not pass the test.

**69.**

| Budgeted Expense, b | Actual Expense, a | $|a - b|$ | $0.05b$ |
|---|---|---|---|
| $37,640 | $37,335 | $305 | $1882 |

Since $305 < $500 and $305 < $1882, it passes the "budget variance test."

**71.** Receipts: 92.5 billion $|92.5 - 92.2| = 0.3$ billion budget surplus for 1960.

**73.** Receipts: 517.1 billion $|517.1 - 590.9| = 73.8$ billion budget deficit for 1980.

**75.** Receipts: 2025.2 billion

$|2025.2 - 1788.8| = 236.4$ billion surplus for 2000.

**77.** Terms:  $7x$, 4

Coefficient of $7x$ is 7

**79.** Terms: $\sqrt{3}x^2, -8x, -11$

Coefficient of $\sqrt{3}x^2$ is $\sqrt{3}$

Coefficient of $-8x$ is $-8$

**81.** Terms: $4x^3, \dfrac{x}{2}, -5$

Coefficient of $4x^3$ is 4

Coefficient of $\dfrac{x}{2}$ is $\dfrac{1}{2}$

**83.** $4x - 6$

(a) $4(-1) - 6 = -4 - 6 = -10$

(b) $4(0) - 6 = 0 - 6 = -6$

**85.** $-x^2 + 5x - 4$

(a) $-(-1)^2 + 5(-1) - 4 = -10$

(b) $-(1)^2 + 5(1) - 4 = 0$

**87.** $x + 9 = 9 + x$

Commutative (addition)

**89.** $\dfrac{1}{(h + 6)}(h + 6) = 1, h \neq -6$

Inverse (multiplication)

**91.** $2(x + 3) = 2x + 6$

Distributive Property

**93.** $x + (y + 10) = (x + y) + 10$

Associative Property of Addition

**95.** $\frac{3}{16} + \frac{5}{16} = \frac{8}{16} = \frac{1}{2}$

**97.** $\dfrac{5}{8} - \dfrac{5}{12} + \dfrac{1}{6} = \dfrac{15}{24} - \dfrac{10}{24} + \dfrac{4}{24} = \dfrac{9}{24} = \dfrac{3}{8}$

**99.** $\dfrac{x}{6} + \dfrac{3x}{4} = \dfrac{2x}{12} + \dfrac{9x}{12} = \dfrac{11x}{12}$

**101.** $\dfrac{12}{x} \div \dfrac{1}{8} = \dfrac{12}{x} \cdot \dfrac{8}{1} = \dfrac{96}{x}$

**103.** $\left(\dfrac{2}{5} \div 4\right) - \left(4 \cdot \dfrac{3}{8}\right) = \left(\dfrac{2}{5} \cdot \dfrac{1}{4}\right) - \dfrac{12}{8} = \dfrac{1}{10} - \dfrac{3}{2}$

$= \dfrac{1}{10} - \dfrac{15}{10} = \dfrac{-14}{10} = -\dfrac{7}{5}$

**105.** $14\left(-3 + \dfrac{3}{7}\right) = -36$

**107.** $\dfrac{11.46 - 5.37}{3.91} \approx 1.56$

**109.** $\dfrac{\frac{2}{3}(-2 - 6)}{-\frac{2}{5}} \approx 13.33$

**111.** (a)

| $n$ | 1 | 0.5 | 0.01 | 0.0001 | 0.000001 |
|-----|---|-----|------|--------|----------|
| $5/n$ | 5 | 10 | 500 | 50,000 | 5,000,000 |

(b) As $n$ approaches 0, $5/n$ approaches infinity ($\infty$). That is, $5/n$ increases without bound.

**113.** This is false. For example, $3 > 2$, but $\frac{1}{3} < \frac{1}{2}$.

**115.** (a) $-A$ is negative, $-A < 0$, because $A > 0$.

(b) $B - A$ is negative, $B - A < 0$, because $B < 0$ and $-A < 0$.

**117.** Consider $|u + v|$ and $|u| + |v|$

(a) No, the values of the expressions are not always equal. For example, if $u = 5$ and $v = -2$, then $|u + v| = |5 - 2| = 3$, whereas $|u| + |v| = |5| + |-2| = 5 + 2 = 7$. In general, the expressions are unequal if $u$ is positive and $v$ is negative, or if $u$ is negative and $v$ is positive.

(b) $|u + v|$ is always less than or equal to $|u| + |v|$: $|u + v| \le |u| + |v|$. The expressions are equal if $u$ and $v$ have the same sign or one or both of $u$ and $v$ is 0. If they differ in sign (one positive and the other negative), then $|u + v| < |u| + |v|$.

**119.** Answers will vary. The set of natural numbers is given by $\{1, 2, 3, \ldots\}$. The whole numbers are the natural numbers together with 0. The integers include the natural numbers, their negatives, and 0: $\{\ldots -3, -2, -1, 0, 1, 2, 3, \ldots\}$. The rational numbers include the integers. The rational numbers consist of all ratios of the form $\frac{p}{q}$, where $p$ and $q$ are integers, $q \neq 0$. The irrational numbers consist of all real numbers that are not rational numbers.

## Section P.2   **Exponents and Radicals**

■ You should know the properties of exponents.

(a) $a^1 = a$

(b) $a^0 = 1, a \neq 0$

(c) $a^m a^n = a^{m+n}$

(d) $a^m/a^n = a^{m-n}, a \neq 0$

(e) $a^{-n} = 1/a^n, a \neq 0$

(f) $(a^m)^n = a^{mn}$

(g) $(ab)^n = a^n b^n$

(h) $(a/b)^n = a^n/b^n, b \neq 0$

(i) $(a/b)^{-n} = (b/a)^n, a \neq 0, b \neq 0$

(j) $|a^2| = |a|^2 = a^2$

■ You should be able to write numbers in scientific notation, $\pm c \times 10^n$, where $1 \le c < 10$ and $n$ is an integer.

■ You should be able to use your calculator to evaluate expressions involving exponents.

■ You should know the properties of radicals.

(a) $\sqrt[n]{a^m} = \left(\sqrt[n]{a}\right)^m = a^{m/n}$

(b) $\sqrt[n]{a} \cdot \sqrt[n]{b} = \sqrt[n]{ab}$

(c) $\dfrac{\sqrt[n]{a}}{\sqrt[n]{b}} = \sqrt[n]{\dfrac{a}{b}}$

(d) $\sqrt[m]{\sqrt[n]{a}} = \sqrt[mn]{a}$

(e) $\left(\sqrt[n]{a}\right)^n = a$

(f) For $n$ even, $\sqrt[n]{a^n} = |a|$

   For $n$ odd, $\sqrt[n]{a^n} = a$

(g) $a^{1/n} = \sqrt[n]{a}$

■ You should be able to simplify radicals.

(a) All possible factors have been removed from the radical sign.

(b) All fractions have radical-free denominators.

(c) The index for the radical has been reduced as far as possible.

■ You should be able to use your calculator to evaluate radicals.

**Solutions to Odd-Numbered Exercises**

**1.** (a) $4^2 \cdot 3 = 16 \cdot 3 = 48$

    (b) $3 \cdot 3^3 = 3^4 = 81$

**3.** (a) $(3^3)^2 = 3^6 = 729$

    (b) $-3^2 = -9$

**5.** (a) $\dfrac{3}{3^{-4}} = 3^{1+4} = 3^5 = 243$

    (b) $24(-2)^{-5} = \dfrac{24}{(-2)^5} = \dfrac{24}{-32} = -\dfrac{3}{4}$

**7.** (a) $2^{-1} + 3^{-1} = \dfrac{1}{2} + \dfrac{1}{3} = \dfrac{5}{6}$

    (b) $(2^{-1})^{-2} = 2^2 = 4$

**9.** When $x = 2$, $7x^{-2} = 7(2^{-2}) = 7\dfrac{1}{2^2} = \dfrac{7}{4}$

**11.** When $x = -3$, $2x^3 = 2(-3)^3 = 2(-27) = -54$

**13.** When $x = -\frac{1}{2}$, $4x^2 = 4\left(-\frac{1}{2}\right)^2 = 4\left(\frac{1}{4}\right) = 1$

**15.** (a) $(-5z)^3 = (-5)^3 z^3 = -125z^3$

    (b) $5x^4(x^2) = 5x^{4+2} = 5x^6$

**17.** (a) $\dfrac{7x^2}{x^3} = 7x^{2-3} = 7x^{-1} = \dfrac{7}{x}$

    (b) $\dfrac{12(x+y)^3}{9(x+y)} = \dfrac{4}{3}(x+y)^{3-1} = \dfrac{4}{3}(x+y)^2,\ x+y \neq 0$

**19.** (a) $\left[(x^2 y^{-2})^{-1}\right]^{-1} = x^2 y^{-2} = \dfrac{x^2}{y^2},\ x \neq 0$

    (b) $\left(\dfrac{a^{-2}}{b^{-2}}\right)\left(\dfrac{b}{a}\right)^3 = \dfrac{b^2}{a^2} \cdot \dfrac{b^3}{a^3} = \dfrac{b^5}{a^5},\ b \neq 0$

**21.** $(-4)^3(5^2) = (-64)(25) = -1600$

**23.** $\dfrac{3^6}{7^3} = \dfrac{729}{343} \approx 2.125$

**25.** $57{,}300{,}000 = 5.73 \times 10^7$ square miles

**27.** $0.0000899 = 8.99 \times 10^{-5}$ gram per cm$^3$

**29.** $5.64 \times 10^8 = 564{,}000{,}000$ servings

**31.** $1.6022 \times 10^{-19} = 0.00000000000000000016022$ coulombs

**33.** $\sqrt{25 \times 10^8} = \sqrt{5^2 \times (10^4)^2} = 5 \times 10^4 = 50{,}000$

**35.** (a) $(9.3 \times 10^6)^3 (6.1 \times 10^{-4}) \approx 4.907 \times 10^{17}$

    (b) $\dfrac{(2.414 \times 10^4)^6}{(1.68 \times 10^5)^5} \approx 1.479$

**37.** (a) $\sqrt{4.5 \times 10^9} \approx 67{,}082.039$

    (b) $\sqrt[3]{6.3 \times 10^4} \approx 39.791$

**39.** $\sqrt{121} = \sqrt{11^2} = 11$

**41.** $-\sqrt[3]{-27} = -(-3) = 3$

**43.** $\left(\sqrt[3]{-125}\right)^3 = -125$

**45.** $32^{-3/5} = \dfrac{1}{32^{3/5}} = \dfrac{1}{\left(\sqrt[5]{32}\right)^3} = \dfrac{1}{(2)^3} = \dfrac{1}{8}$

**47.** $\left(-\frac{1}{64}\right)^{-1/3} = (-64)^{1/3} = \sqrt[3]{-64} = -4$

**49.** $\sqrt[5]{-27^3} = (-27)^{3/5} \approx -7.225$

**51.** $(3.4)^{2.5} \approx 21.316$

**53.** $(1.2^{-2})\sqrt{75} + 3\sqrt{8} \approx 14.499$

**55.** (a) $\left(\sqrt[4]{3}\right)^4 = (3^{1/4})^4 = 3$

(b) $\sqrt[5]{96x^5} = (2^5 \cdot 3 \cdot x^5)^{1/5} = 2x\sqrt[5]{3}$

**57.** (a) $\sqrt{54xy^4} = \sqrt{3^2 \cdot 6 \cdot x(y^2)^2} = 3y^2\sqrt{6x}$

(b) $\sqrt[3]{\dfrac{32a^2}{b^2}} = \left(\dfrac{2^3 \cdot 2^2\, a^2}{b^2}\right)^{1/3} = 2\sqrt[3]{\dfrac{4a^2}{b^2}}$

**59.** (a) $2\sqrt{50} + 12\sqrt{8} = 2\sqrt{25 \cdot 2} + 12\sqrt{4 \cdot 2}$
$$= 2(5\sqrt{2}) + 12(2\sqrt{2}) = 10\sqrt{2} + 24\sqrt{2} = 34\sqrt{2}$$

(b) $10\sqrt{32} - 6\sqrt{18} = 10\sqrt{16 \cdot 2} - 6\sqrt{9 \cdot 2}$
$$= 10(4\sqrt{2}) - 6(3\sqrt{2}) = 40\sqrt{2} - 18\sqrt{2} = 22\sqrt{2}$$

**61.** (a) $3\sqrt{x+1} + 10\sqrt{x+1} = 13\sqrt{x+1}$

(b) $7\sqrt{80x} - 2\sqrt{125x} = 7\sqrt{16 \cdot 5x} - 2\sqrt{25 \cdot 5x}$
$$= 7(4)\sqrt{5x} - 2(5)\sqrt{5x}$$
$$= 28\sqrt{5x} - 10\sqrt{5x} = 18\sqrt{5x}$$

**63.** $\sqrt{5} + \sqrt{3} \approx 3.968.$ and $\sqrt{5+3} = \sqrt{8} \approx 2.828$

Thus, $\sqrt{5} + \sqrt{3} > \sqrt{5+3}$

**65.** $\sqrt{3^2 + 2^2} = \sqrt{9 + 4} = \sqrt{13} \approx 3.606$

Thus, $5 > \sqrt{3^2 + 2^2}$

**67.** (a) $\dfrac{1}{\sqrt{3}} = \dfrac{1}{\sqrt{3}} \cdot \dfrac{\sqrt{3}}{\sqrt{3}} = \dfrac{\sqrt{3}}{3}$

**69.** $\dfrac{5}{\sqrt{14} - 2} = \dfrac{5}{\sqrt{14} - 2} \cdot \dfrac{\sqrt{14} + 2}{\sqrt{14} + 2} = \dfrac{5\left(\sqrt{14} + 2\right)}{14 - 4} = \dfrac{5\left(\sqrt{14} + 2\right)}{10} = \dfrac{\sqrt{14} + 2}{2}$

**71.** $\dfrac{\sqrt{8}}{2} = \dfrac{\sqrt{4 \cdot 2}}{2} = \dfrac{2\sqrt{2}}{2} = \dfrac{\sqrt{2}}{1} \cdot \dfrac{\sqrt{2}}{\sqrt{2}} = \dfrac{2}{\sqrt{2}}$

**73.** $\dfrac{\sqrt{5} + \sqrt{3}}{3} = \dfrac{\sqrt{5} + \sqrt{3}}{3} \cdot \dfrac{\sqrt{5} - \sqrt{3}}{\sqrt{5} - \sqrt{3}} = \dfrac{5 - 3}{3\left(\sqrt{5} - \sqrt{3}\right)} = \dfrac{2}{3\left(\sqrt{5} - \sqrt{3}\right)}$

| | *Radical Form* | *Rational Exponent Form* |
|---|---|---|
| **75.** | $\sqrt[3]{64} = 4$ Given | $64^{1/3} = 4$ Answer |
| **77.** | $\sqrt[5]{32} = 2$ Answer | $32^{1/5} = 2$ Given |
| **79.** | $\sqrt[3]{-216} = -6$ Given | $(-216)^{1/3} = -6$ Answer |
| **81.** | $\sqrt[4]{81^3} = 27$ Given | $81^{3/4} = 27$ Answer |

**83.** $\dfrac{(2x^2)^{3/2}}{2^{1/2}x^4} = \dfrac{2^{3/2}(x^2)^{3/2}}{2^{1/2}x^4} = \dfrac{2^{3/2}x^3}{2^{1/2}x^4} = 2^{3/2 - 1/2}\,x^{3-4} = 2^1 x^{-1} = \dfrac{2}{x}$

**85.** $\dfrac{x^{-3} \cdot x^{1/2}}{x^{3/2} \cdot x^{-1}} = \dfrac{x^{1/2} \cdot x^1}{x^{3/2} \cdot x^3} = x^{1/2 + 1 - 3/2 - 3} = x^{-3} = \dfrac{1}{x^3}, x > 0$

**87.** (a) $\sqrt[4]{3^2} = 3^{2/4} = 3^{1/2} = \sqrt{3}$

(b) $\sqrt[6]{(x+1)^4} = (x+1)^{4/6} = (x+1)^{2/3} = \sqrt[3]{(x+1)^2}$

**89.** (a) $\sqrt{\sqrt{32}} = (32^{1/2})^{1/2} = 32^{1/4} = \sqrt[4]{32} = \sqrt[4]{16 \cdot 2} = 2\sqrt[4]{2}$

(b) $\sqrt{\sqrt[4]{2x}} = ((2x)^{1/4})^{1/2} = (2x)^{1/8} = \sqrt[8]{2x}$

**91.** For $L = 2$, $T = 2\pi\sqrt{\dfrac{2}{32}} = 2\pi\sqrt{\dfrac{1}{16}} = 2\pi\sqrt{\left(\dfrac{1}{4}\right)^2} = 2\pi\left(\dfrac{1}{4}\right)$

$$= \dfrac{\pi}{2} \text{ seconds}$$

$$\approx 1.57 \text{ seconds}$$

**93.** For $v = \dfrac{3}{4}$, size $= 0.03\sqrt{\dfrac{3}{4}} = 0.03\dfrac{\sqrt{3}}{\sqrt{4}} = 0.03\dfrac{\sqrt{3}}{2} \approx 0.026$ inches

**95.** For $x \neq 0$, this is true because $\dfrac{x^{k+1}}{x} = \dfrac{x^{k+1}}{x^1} = x^k$

**97.** For $a \neq 0$, $1 = \dfrac{a}{a} = \dfrac{a^1}{a^1} = a^{1-1} = a^0$. Thus, $a^0 = 1$.

**99.** Consider $x^2 = n$, $x$ a positive integer

| unit digit of $x$ | unit digit of $n = x^2$ |
|:---:|:---:|
| 1 | 1 |
| 2 | 4 |
| 3 | 9 |
| 4 | 6 |
| 5 | 5 |
| 6 | 6 |
| 7 | 9 |
| 8 | 4 |
| 9 | 1 |
| 0 | 0 |

Therefore, the possible digits are 0, 1, 4, 5, 6, and 9 thus $\sqrt{5233}$ is *not* an integer because its unit digit is 3.

## Section P.3    Polynomials and Factoring

- Given a polynomial in $x$, $a_n x^n + a_{n-1}x^{n-1} + \cdots + a_1 x + a_0$, where $a_n \neq 0$, and $n$ is a nonnegative integer, you should be able to identify the following:
  - (a) Degree: $n$
  - (b) Terms: $a_n x^n, a_{n-1}x^{n-1}, \ldots, a_1 x, a_0$
  - (c) Coefficients: $a_n, a_{n-1}, \ldots, a_1, a_0$
  - (d) Leading coefficient: $a_n$
  - (e) Constant term: $a_0$
- You should be able to add and subtract polynomials.
- You should be able to multiply polynomials by either
  - (a) The Distributive Properties
  - (b) The Vertical Method

—CONTINUED—

—CONTINUED—

■ You should know the special binomial products.

(a) $(ax + b)(cx + d) = acx^2 + adx + bcx + bd$    FOIL

$$= acx^2 + (ad + bc)x + bd$$

(b) $(u \pm v)^2 = u^2 \pm 2uv + v^2$

(c) $(u + v)(u - v) = u^2 - v^2$

(d) $(u \pm v)^3 = u^3 \pm 3u^2v + 3uv^2 \pm v^3$

■ You should be able to factor out all common factors, the first step in factoring.

■ You should be able to factor the following special polynomial forms.

(a) $u^2 - v^2 = (u + v)(u - v)$

(b) $u^2 \pm 2uv + v^2 = (u \pm v)^2$

(c) $mx^2 + nx + r = (ax + b)(cx + d)$, where $m = ac, r = bd, n = ad + bc$

**Note:** Not all trinomials can be factored (using real coefficients).

(d) $u^3 \pm v^3 = (u \pm v)(u^2 \mp uv + v^2)$

■ You should be able to factor by grouping.

## Solutions to Odd-Numbered Exercises

**1.** 7 is a polynomial of degree zero. Matches (d)

**3.** $-4x^3 + 1$ is a binomial with leading coefficient $-4$. Matches (b)

**5.** $\frac{3}{4}x^4 + x^2 + 14$ is a trinomial with leading coefficient $\frac{3}{4}$. Matches (f)

**7.** $-2x^3 + 4x$ is one possible answer

**9.** $-4x^4 + 3$ is one possible answer

**11.** $3x + 4x^2 + 2 = 4x^2 + 3x + 2$    Standard form

Degree: 2

Leading coefficient: 4

**13.** $1 + x^7 = x^7 + 1$    Standard form

Degree: 7

Leading coefficient: 1

**15.** $1 - x + 6x^4 - 2x^5 = -2x^5 + 6x^4 - x + 1$    Standard form

Degree: 5

Leading coefficient: $-2$

**17.** This is a polynomial: $-2x^3 + 7x + 10$

**19.** $\sqrt{x^2 - x^4}$ is not a polynomial

**21.** $(6x + 5) - (8x + 15) = 6x + 5 - 8x - 15$

$$= (6x - 8x) + (5 - 15) = -2x - 10$$

**23.** $-(x^3 - 2) + (4x^3 - 2x) = -x^3 + 2 + 4x^3 - 2x$

$$= (4x^3 - x^3) - 2x + 2 = 3x^3 - 2x + 2$$

**25.** $(15x^2 - 6) - (-8.1x^3 - 14.7x^2 - 17) = 15x^2 - 6 + 8.1x^3 + 14.7x^2 + 17$

$$= 8.1x^3 + 29.7x^2 + 11$$

**27.** $3x(x^2 - 2x + 1) = 3x(x^2) + 3x(-2x) + 3x(1) = 3x^3 - 6x^2 + 3x$

**29.** $-5z(3z - 1) = -5z(3z) + (-5z)(-1) = -15z^2 + 5z$

**31.** $(1 - x^3)(4x) = 1(4x) - x^3(4x) = 4x - 4x^4 = -4x^4 + 4x$

**33.** $(2.5x^2 + 5)(-3x) = (2.5x^2)(-3x) + 5(-3x) = -7.5x^3 - 15x$

**35.** $-2x\left(\dfrac{1}{8}x + 3\right) = -2x\left(\dfrac{1}{8}x\right) - 2x(3) = -\dfrac{1}{4}x^2 - 6x$

**37.** $(x + 3)(x + 4) = x^2 + 4x + 3x + 12$  FOIL

$$= x^2 + 7x + 12$$

**39.** $(3x - 5)(2x + 1) = 6x^2 + 3x - 10x - 5$  FOIL

$$= 6x^2 - 7x - 5$$

**41.** $(2x - 5y)^2 = 4x^2 - 2(5y)(2x) + 25y^2$

$$= 4x^2 - 20xy + 25y^2$$

**43.** $(x + 10)(x - 10) = x^2 - 100$

**45.** $(x + 2y)(x - 2y) = x^2 - (2y)^2 = x^2 - 4y^2$

**47.** $(2r^2 - 5)(2r^2 + 5) = (2r^2)^2 - 5^2 = 4r^4 - 25$

**49.** $(x + 1)^3 = x^3 + 3x^2(1) + 3x(1^2) + 1^3$

$$= x^3 + 3x^2 + 3x + 1$$

**51.** $(2x - y)^3 = (2x)^3 - 3(2x)^2y + 3(2x)y^2 - y^3$

$$= 8x^3 - 12x^2y + 6xy^2 - y^3$$

**53.** $\left(\dfrac{1}{2}x - 5\right)^2 = \left(\dfrac{1}{2}x\right)^2 - 2(5)\left(\dfrac{1}{2}x\right) + 5^2 = \dfrac{1}{4}x^2 - 5x + 25$

**55.** $\left(\dfrac{1}{4}x - 3\right)\left(\dfrac{1}{4}x + 3\right) = \left(\dfrac{1}{4}x\right)^2 - 3^2 = \dfrac{1}{16}x^2 - 9$

**57.** $(2.4x + 3)^2 = (2.4x)^2 + 2(3)(2.4x) + 3^2 = 5.76x^2 + 14.4x + 9$

**59.**

$$
\begin{array}{r}
-x^2 + x - 5 \\
3x^2 + 4x + 1 \\
\hline
-x^2 + x - 5 \\
-4x^3 + 4x^2 - 20x \\
-3x^4 + 3x^3 - 15x^2 \\
\hline
-3x^4 - x^3 - 12x^2 - 19x - 5
\end{array}
$$

Answer: $-3x^4 - x^3 - 12x^2 - 19x - 5$

**61.** $[(m - 3) + n][(m - 3) - n] = (m - 3)^2 - n^2$

$$= m^2 - 6m + 9 - n^2$$

$$= m^2 - n^2 - 6m + 9$$

**63.** $[(x - 3) + y]^2 = (x - 3)^2 + 2y(x - 3) + y^2$

$$= x^2 - 6x + 9 + 2xy - 6y + y^2$$

$$= x^2 + 2xy + y^2 - 6x - 6y + 9$$

**65.** $5x(x + 1) - 3x(x + 1) = (5x - 3x)(x + 1)$
$$= 2x(x + 1)$$
$$= 2x^2 + 2x$$

**67.** $(u + 2)(u - 2)(u^2 + 4) = (u^2 - 4)(u^2 + 4)$
$$= u^4 - 16$$

**69.** $2x + 8 = 2(x + 4)$

**71.** $2x^3 - 6x = 2x(x^2 - 3)$

**73.** $3x(x - 5) + 8(x - 5) = (3x + 8)(x - 5)$

**75.** $x^2 - 64 = x^2 - 8^2 = (x + 8)(x - 8)$

**77.** $32y^2 - 18 = 2(16y^2 - 9) = 2[(4y)^2 - 3^2]$
$$= 2(4y + 3)(4y - 3)$$

**79.** $4x^2 - \frac{1}{9} = (2x)^2 - \left(\frac{1}{3}\right)^2 = \left(2x + \frac{1}{3}\right)\left(2x - \frac{1}{3}\right)$

**81.** $(x - 1)^2 - 4 = [(x - 1) + 2][(x - 1) - 2]$
$$= (x + 1)(x - 3)$$

**83.** $x^2 - 4x + 4 = x^2 - 2(2)x + 2^2 = (x - 2)^2$

**85.** $x^2 + x + \frac{1}{4} = x^2 + 2\left(\frac{1}{2}\right)x + \left(\frac{1}{2}\right)^2 = \left(x + \frac{1}{2}\right)^2$

**87.** $4t^2 + 4t + 1 = (2t)^2 + 2(2t)(1) + 1^2$
$$= (2t + 1)^2$$

**89.** $9t^2 + \frac{3}{2}t + \frac{1}{16} = (3t)^2 + 2(3t)\left(\frac{1}{4}\right) + \left(\frac{1}{4}\right)^2 = \left(3t + \frac{1}{4}\right)^2$

**91.** $x^3 - 8 = x^3 - 2^3 = (x - 2)(x^2 + 2x + 4)$

**93.** $y^3 + 216 = y^3 + 6^3 = (y + 6)(y^2 - 6y + 36)$

**95.** $x^3 - \frac{8}{27} = x^3 - \left(\frac{2}{3}\right)^3 = \left(x - \frac{2}{3}\right)\left(x^2 + \frac{2}{3}x + \frac{4}{9}\right)$

**97.** $8x^3 - 1 = (2x)^3 - 1 = (2x - 1)(4x^2 + 2x + 1)$

**99.** $\frac{1}{8}x^3 + 1 = \left(\frac{x}{2}\right)^3 + 1 = \left(\frac{x}{2} + 1\right)\left(\frac{x^2}{4} - \frac{x}{2} + 1\right)$

**101.** $x^2 + x - 2 = (x + 2)(x - 1)$

**103.** $s^2 - 5s + 6 = (s - 3)(s - 2)$

**105.** $20 - y - y^2 = (5 + y)(4 - y)$
or $-(y + 5)(y - 4)$

**107.** $3x^2 - 5x + 2 = (3x - 2)(x - 1)$

**109.** $2x^2 - x - 1 = (2x + 1)(x - 1)$

**111.** $5x^2 + 26x + 5 = (5x + 1)(x + 5)$

**113.** $-5u^2 - 13u + 6 = -(5u^2 + 13u - 6) = -(5u - 2)(u + 3)$ or $(2 - 5u)(u + 3)$

**115.** $x^3 - x^2 + 2x - 2 = x^2(x - 1) + 2(x - 1)$
$$= (x - 1)(x^2 + 2)$$

**117.** $6x^2 + x - 2$

$a = 6, c = -2, ac = -12 = 4(-3)$ where $4 - 3 = 1 = b$. Thus,

$6x^2 + x - 2 = 6x^2 + 4x - 3x - 2$
$$= 2x(3x + 2) - (3x + 2)$$
$$= (2x - 1)(3x + 2)$$

**119.** $x^3 - 16x = x(x^2 - 16) = x(x + 4)(x - 4)$

**121.** $x^3 - x^2 = x^2(x - 1)$

**123.** $x^2 - 2x + 1 = (x - 1)^2$

**125.** $1 - 4x + 4x^2 = (1 - 2x)^2$
$$= (2x - 1)^2$$

**127.** $2x^2 + 4x - 2x^3 = -2x(-x - 2 + x^2)$
$$= -2x(x^2 - x - 2)$$
$$= -2x(x + 1)(x - 2)$$

**129.** $9x^2 + 10x + 1 = (9x + 1)(x + 1)$

**131.** $\frac{1}{8}x^2 - \frac{1}{96}x - \frac{1}{16} = \frac{1}{8}\left(x^2 - \frac{1}{12}x - \frac{1}{2}\right) = \frac{1}{8}\left(x - \frac{3}{4}\right)\left(x + \frac{2}{3}\right) = \frac{1}{96}(4x - 3)(3x + 2)$

**133.** $3x^3 + x^2 + 15x + 5 = x^2(3x + 1) + 5(3x + 1)$
$$= (3x + 1)(x^2 + 5)$$

**135.** $3u - 2u^2 + 6 - u^3 = -u^3 - 2u^2 + 3u + 6$
$$= -u^2(u + 2) + 3(u + 2)$$
$$= (3 - u^2)(u + 2)$$

**137.** $25 - (z + 5)^2 = [5 + (z + 5)][5 - (z + 5)] = -z(z + 10)$

**139.** $(x^2 + 1)^2 - 4x^2 = [(x^2 + 1) + 2x][(x^2 + 1) - 2x]$
$$= (x^2 + 2x + 1)(x^2 - 2x + 1)$$
$$= (x + 1)^2(x - 1)^2$$

**141.** $2t^3 - 16 = 2(t^3 - 8) = 2(t - 2)(t^2 + 2t + 4)$

**143.** $4x(2x - 1) + 2(2x - 1)^2 = 2(2x - 1)(2x + (2x - 1))$
$$= 2(2x - 1)(4x - 1)$$

**145.** $2(x + 1)(x - 3)^2 - 3(x + 1)^2(x - 3) = (x + 1)(x - 3)[2(x - 3) - 3(x + 1)]$
$$= (x + 1)(x - 3)[2x - 6 - 3x - 3]$$
$$= (x + 1)(x - 3)(-x - 9)$$
$$= -(x + 1)(x - 3)(x + 9)$$

**147.** $7x(2)(x^2 + 1)(2x) - (x^2 + 1)^2(7)$
$$= 7(x^2 + 1)[4x^2 - (x^2 + 1)]$$
$$= 7(x^2 + 1)(3x^2 - 1)$$

**149.** $2x(x - 5)^4 - x^2(4)(x - 5)^3$
$$= 2x(x - 5)^3[(x - 5) - 2x]$$
$$= 2x(x - 5)^3(-x - 5)$$
$$= -2x(x - 5)^3(x + 5)$$

**151.** (a)  $500(1 + r)^2 = 500(r + 1)^2 = 500(r^2 + 2r + 1)$
$$= 500r^2 + 1000r + 500$$

(b)

| $r$ | $2\frac{1}{2}\%$ | $3\%$ | $4\%$ | $4\frac{1}{2}\%$ | $5\%$ |
|---|---|---|---|---|---|
| $500(1 + r)^2$ | \$525.31 | \$530.45 | \$540.80 | \$546.01 | \$551.25 |

Remember to write the interest rate in decimal form: $2\frac{1}{2}\% = 0.025$

(c) As $r$ increases, the amount increases.

**153.** Volume = (width)(length)(height)

$$= (15 - 2x)\left(\frac{45 - 3x}{2}\right)(x)$$

$$= \frac{3}{2}x(x - 15)(2x - 15)$$

If $x = 3$, volume = 486 cubic centimeter.

If $x = 5$, volume = 375 cubic centimeter.

If $x = 7$, volume = 84 cubic centimeter.

**155.** (a) $T = R + B = 1.1x + (0.0475x^2 - 0.001x + 0.23)$

$$= 0.0475x^2 + 1.099x + 0.23$$

(b)

| $x$ mi/hr | 30 | 40 | 55 |
|---|---|---|---|
| $T$ feet | 75.95 | 120.19 | 204.36 |

(c) As the speed $x$ increases, the total stopping distance increases.

**157.** $a^2 - b^2 = (a + b)(a - b)$

Matches model (b)

**159.** $a^2 + 2a + 1 = (a + 1)^2$

Matches model (a)

**161.** $3x^2 + 7x + 2 = (3x + 1)(x + 2)$

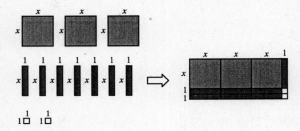

**163.** $2x^2 + 7x + 3 = (2x + 1)(x + 3)$

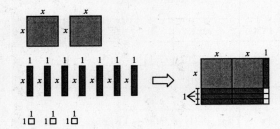

**165.** $A = \pi(r + 2)^2 - \pi r^2 = \pi[(r + 2)^2 - r^2]$

$$= \pi[r^2 + 4r + 4 - r^2] = \pi(4r + 4)$$

$$= 4\pi(r + 1)$$

**167.** $A = 8(18) - 4x^2$

$$= 4(36 - x^2)$$

$$= 4(6 - x)(6 + x)$$

**169.** For $x^2 + bx - 15 = (x + m)(x + n)$ to be factorable, $b$ must equal $m + n$ where $mn = -15$.

| Factors of $-15$ | Sum of factors |
|---|---|
| $(15)(-1)$ | $15 + (-1) = 14$ |
| $(-15)(1)$ | $-15 + 1 = -14$ |
| $(3)(-5)$ | $3 + (-5) = -2$ |
| $(-3)(5)$ | $-3 + 5 = 2$ |

The possible $b$ values are $14, -14, -2,$ or $2$.

**171.** For $x^2 + bx + 50 = (x + m)(x + n)$ to be factorable, $b$ must equal $m + n$ where $mn = 50$.

| Factors of 50 | Sum of factors |
|---|---|
| $(50)(1)$ | 51 |
| $(-50)(-1)$ | $-51$ |
| $(25)(2)$ | 27 |
| $(-25)(-2)$ | $-27$ |
| $(10)(5)$ | 15 |
| $(-10)(-5)$ | $-15$ |

The possible $b$ values are $51, -51, 27, -27, 15$ or $-15$.

**173.** For $2x^2 + 5x + c$ to be factorable, the factors of $2c$ must add up to 5.

| Possible $c$ values | $2c$ | Factors of $2c$ that add up to 5 |
|---|---|---|
| 2 | 4 | $(1)(4) = 4$ and $1 + 4 = 5$ |
| 3 | 6 | $(2)(3) = 6$ and $2 + 3 = 5$ |
| $-3$ | $-6$ | $(6)(-1) = -6$ and $6 + (-1) = 5$ |
| $-7$ | $-14$ | $(7)(-2) = -14$ and $7 + (-2) = 5$ |
| $-12$ | $-24$ | $(8)(-3) = -24$ and $8 + (-3) = 5$ |

These are a few possible $c$ values. There are *many* correct answers.

If $c = 2$: $2x^2 + 5x + 2 = (2x + 1)(x + 2)$

If $c = 3$: $2x^2 + 5x + 3 = (2x + 3)(x + 1)$

If $c = -3$: $2x^2 + 5x - 3 = (2x - 1)(x + 3)$

If $c = -7$: $2x^2 + 5x - 7 = (2x + 7)(x - 1)$

If $c = -12$: $2x^2 + 5x - 12 : (2x - 3)(x + 4)$

**175.** For $3x^2 - 10x + c$ to be factorable, the factors of $3c$ must add up to $-10$.

| Possible $c$ values | $3c$ | Factors of $2c$ that must add up to $-10$ |
|---|---|---|
| 3 | 9 | $(-9)(-1) = 9$ and $-9 - 1 = -10$ |
| 7 | 21 | $(-3)(-7) = 21$ and $-3 - 7 = -10$ |
| 8 | 24 | $(-4)(-6) = 24$ and $-4 - 6 = -10$ |
| $-8$ | $-24$ | $(2)(-12) = -24$ and $-12 + 2 = -10$ |

Other $c$ values are possible. The above values yield the following factorizations. There are many correct answers.

If $c = 3$: $3x^2 - 10x + 3 = (3x - 1)(x - 3)$

If $c = 7$: $3x^2 - 10x + 7 = (3x - 7)(x - 1)$

If $c = 8$: $3x^2 - 10x + 8 = (3x - 4)(x - 2)$

If $c = -8$: $3x^2 - 10x - 8 = (3x + 2)(x - 4)$

**177.** $V = \pi R^2 h - \pi r^2 h$

(a) $V = \pi h[R^2 - r^2] = \pi h(R - r)(R + r)$

(b) The average radius is $\dfrac{R + r}{2}$. The thickness of the shell is $R - r$.

Therefore,

$$V = \pi h(R + r)(R - r) = 2\pi\left(\frac{R + r}{2}\right)(R - r)h = 2\pi(\text{average radius})(\text{thickness})h$$

**179.** False. The product of the two binomials is not always a second-degree polynomial. For instance, $(x^2 + 2)(x^2 - 3) = x^4 - x^2 - 6$ is a fourth-degree polynomial.

**181.** False. For example $3^2 + 4^2 \neq (3 + 4)^2$.

**183.** If $m < n$, then the sum of two polynomials of degree $m$ and $n$ is $n$, the larger degree.

**185.** A polynomial is in factored form if it is written as a product (not a sum).

**187.** $9x^2 - 9x - 54 = 9(x^2 - x - 6) = 9(x + 2)(x - 3)$

The error in the problem in the book was that 3 was factored out of the first binomial but not out of the second binomial.

$(3x + 6)(3x - 9) = 3(x + 2)(3)(x - 3) = 9(x + 2)(x - 3).$

**189.** No, the sum of two second-degree polynomial could be of lesser degree. For example

$(x^2 + 2x) + (-x^2 + x + 1) = 3x + 1$     (degree 1)

## Section P.4   Rational Expressions

- ■   You should be able to find the domain of a fractional expression.
- ■   You should know that a rational expression is the quotient of two polynomials.
- ■   You should be able to simplify rational expressions by reducing them to lowest terms. This may involve factoring both the numerator and the denominator.
- ■   You should be able to add, subtract, multiply, and divide rational expressions.
- ■   You should be able to simplify compound fractions.

**Solutions to Odd-Numbered Exercises**

**1.** The domain of the polynomial $3x^2 - 4x + 7$ is the set of all real numbers.

**3.** The domain of the polynomial $4x^3 + 3, x \geq 0$ is the set of non-negative real numbers, since the polynomial is restricted to that set.

**5.** The domain of $\dfrac{1}{3 - x}$ is the set of all real numbers $x$ such that $x \neq 3$.

**7.** The domain of $\sqrt{x + 7}$ is the set of all real numbers $x$ such that $x \geq -7$.

**9.** $\dfrac{5}{2x} = \dfrac{5(3x)}{(2x)(3x)} = \dfrac{5(3x)}{6x^2}, x \neq 0$

**11.** $\dfrac{15x^2}{10x} = \dfrac{5x(3x)}{5x(2)} = \dfrac{3x}{2}, x \neq 0$

**13.** $\dfrac{3xy}{xy + x} = \dfrac{x(3y)}{x(y + 1)} = \dfrac{3y}{y + 1}, x \neq 0$

**15.** $\dfrac{4y - 8y^2}{10y - 5} = \dfrac{4y(1 - 2y)}{5(2y - 1)} = \dfrac{-4y(2y - 1)}{5(2y - 1)} = \dfrac{-4y}{5}, y \neq \dfrac{1}{2}$

**17.** $\dfrac{x - 5}{10 - 2x} = \dfrac{x - 5}{-2(x - 5)} = -\dfrac{1}{2}, x \neq 5$

**19.** $\dfrac{y^2 - 16}{y + 4} = \dfrac{(y + 4)(y - 4)}{y + 4} = y - 4, y \neq -4$

**21.** $\dfrac{x^3 + 5x^2 + 6x}{x^2 - 4} = \dfrac{x(x + 2)(x + 3)}{(x + 2)(x - 2)} = \dfrac{x(x + 3)}{x - 2}, x \neq -2$

**23.** $\dfrac{y^2 - 7y + 12}{y^2 + 3y - 18} = \dfrac{(y - 3)(y - 4)}{(y + 6)(y - 3)} = \dfrac{y - 4}{y + 6}, y \neq 3$

**25.** $\dfrac{2 - x + 2x^2 - x^3}{x - 2} = \dfrac{(2 - x) + x^2(2 - x)}{-(2 - x)} = \dfrac{(2 - x)(1 + x^2)}{-(2 - x)} = -(1 + x^2), x \neq 2$

**27.** $\dfrac{z^3 - 8}{z^2 + 2z + 4} = \dfrac{(z - 2)(z^2 + 2z + 4)}{z^2 + 2z + 4} = z - 2$

**29.**

| $x$ | 0 | 1 | 2 | 3 | 4 | 5 | 6 |
|---|---|---|---|---|---|---|---|
| $\dfrac{x^2 - 2x - 3}{x - 3}$ | 1 | 2 | 3 | undef. | 5 | 6 | 7 |
| $x + 1$ | 1 | 2 | 3 | 4 | 5 | 6 | 7 |

The expressions are equivalent except at $x = 3$. In fact,

$$\dfrac{x^2 - 2x - 3}{x - 3} = \dfrac{(x - 3)(x + 1)}{x - 3} = x + 1, \; x \neq 3.$$

**31.** $\dfrac{5x^3}{2x^3 + 4} = \dfrac{5x^3}{2(x^3 + 2)}.$

There are no common factors so this expression is in reduced form. In this case factors of terms were incorrectly cancelled.

**33.** $\dfrac{\pi r^2}{(2r)^2} = \dfrac{\pi r^2}{4r^2} = \dfrac{\pi}{4}$

**35.** $\dfrac{5}{x - 1} \cdot \dfrac{x - 1}{25(x - 2)} = \dfrac{1}{5(x - 2)}, x \neq 1$

**37.** $\dfrac{r}{r - 1} \div \dfrac{r^2}{r^2 - 1} = \dfrac{r}{r - 1} \cdot \dfrac{r^2 - 1}{r^2} = \dfrac{r(r + 1)(r - 1)}{(r - 1)r^2}$

$$= \dfrac{r + 1}{r}, \; r \neq 1$$

**39.** $\dfrac{t^2 - t - 6}{t^2 + 6t + 9} \cdot \dfrac{t + 3}{t^2 - 4} = \dfrac{(t - 3)(t + 2)(t + 3)}{(t + 3)^2(t + 2)(t - 2)} = \dfrac{t - 3}{(t + 3)(t - 2)}, t \neq -3, -2$

**41.** $\dfrac{3(x + y)}{4} \div \dfrac{x + y}{2} = \dfrac{3(x + y)}{4} \cdot \dfrac{2}{x + y} = \dfrac{3}{2}, x \neq -y$

**43.** $\dfrac{5}{x - 1} + \dfrac{x}{x - 1} = \dfrac{5 + x}{x - 1} = \dfrac{x + 5}{x - 1}$

**45.** $\dfrac{6}{2x + 1} - \dfrac{x}{x + 3} = \dfrac{6(x + 3) - x(2x + 1)}{(2x + 1)(x + 3)}$

$$= \dfrac{6x + 18 - 2x^2 - x}{(2x + 1)(x + 3)}$$

$$= \dfrac{-2x^2 + 5x + 18}{(2x + 1)(x + 3)}$$

$$= -\dfrac{2x^2 - 5x - 18}{(2x + 1)(x + 3)}$$

**47.** $\dfrac{3}{x - 2} + \dfrac{5}{2 - x} = \dfrac{3}{x - 2} - \dfrac{5}{x - 2} = -\dfrac{2}{x - 2}$

**49.** $\dfrac{1}{x^2 - x - 2} - \dfrac{x}{x^2 - 5x + 6} = \dfrac{1}{(x - 2)(x + 1)} - \dfrac{x}{(x - 2)(x - 3)}$

$$= \dfrac{(x - 3) - x(x + 1)}{(x + 1)(x - 2)(x - 3)}$$

$$= \dfrac{-x^2 - 3}{(x + 1)(x - 2)(x - 3)} = -\dfrac{x^2 + 3}{(x + 1)(x - 2)(x - 3)}$$

**51.** $-\dfrac{1}{x} + \dfrac{2}{x^2 + 1} - \dfrac{1}{x^3 + x} = \dfrac{-(x^2 + 1)}{x(x^2 + 1)} + \dfrac{2x}{x(x^2 + 1)} - \dfrac{1}{x(x^2 + 1)}$

$$= \dfrac{-x^2 - 1 + 2x - 1}{x(x^2 + 1)}$$

$$= -\dfrac{x^2 - 2x + 2}{x(x^2 + 1)}$$

**53.** $\dfrac{\left(\dfrac{x}{2} - 1\right)}{(x - 2)} = \dfrac{\left(\dfrac{x}{2} - \dfrac{2}{2}\right)}{\left(\dfrac{x - 2}{1}\right)} = \dfrac{x - 2}{2} \cdot \dfrac{1}{x - 2} = \dfrac{1}{2}, x \neq 2$

**55.** $\dfrac{\left[\dfrac{x^2}{(x + 1)^2}\right]}{\left[\dfrac{x}{(x + 1)^3}\right]} = \dfrac{x^2}{(x + 1)^2} \cdot \dfrac{(x + 1)^3}{x} = \dfrac{x(x + 1)}{1} = x^2 + x, \ x \neq 0, \ -1$

**57.** $\dfrac{\left[\dfrac{1}{(x + h)^2} - \dfrac{1}{x^2}\right]}{h} = \dfrac{\left[\dfrac{1}{(x + h)^2} - \dfrac{1}{x^2}\right]}{h} \cdot \dfrac{x^2(x + h)^2}{x^2(x + h)^2} = \dfrac{x^2 - (x + h)^2}{hx^2(x + h)^2}$

$$= \dfrac{x^2 - (x^2 + 2xh + h^2)}{hx^2(x + h)^2} = \dfrac{-h(2x + h)}{hx^2(x + h)^2} = -\dfrac{2x + h}{x^2(x + h)^2}, h \neq 0$$

**59.** $\dfrac{\left(\sqrt{x} - \dfrac{1}{2\sqrt{x}}\right)}{\sqrt{x}} = \dfrac{\left(\sqrt{x} - \dfrac{1}{2\sqrt{x}}\right)}{\sqrt{x}} \cdot \dfrac{2\sqrt{x}}{2\sqrt{x}} = \dfrac{2x - 1}{2x}, x > 0$

**61.** $x^5 - 2x^{-2} = x^{-2}(x^7 - 2) = \dfrac{x^7 - 2}{x^2}$

**63.** $x^2(x^2 + 1)^{-5} - (x^2 + 1)^{-4} = (x^2 + 1)^{-5}[x^2 - (x^2 + 1)]$

$$= -\dfrac{1}{(x^2 + 1)^5}$$

**65.** $2x^2(x - 1)^{1/2} - 5(x - 1)^{-1/2} = (x - 1)^{-1/2}(2x^2(x - 1) - 5) = \dfrac{2x^3 - 2x^2 - 5}{(x - 1)^{1/2}}$

**67.** $\dfrac{2x^{3/2} - x^{-1/2}}{x^2} = \dfrac{x^{-1/2}(2x^2 - 1)}{x^2} = \dfrac{2x^2 - 1}{x^{5/2}}$

**69.** $\dfrac{\sqrt{x+2}-\sqrt{x}}{2} = \dfrac{\sqrt{x+2}-\sqrt{x}}{2} \cdot \dfrac{\sqrt{x+2}+\sqrt{x}}{\sqrt{x+2}+\sqrt{x}}$

$\qquad = \dfrac{(x+2)-x}{2(\sqrt{x+2}+\sqrt{x})} = \dfrac{2}{2(\sqrt{x+2}+\sqrt{x})}$

$\qquad = \dfrac{1}{\sqrt{x+2}+\sqrt{x}}$

**71.** (a) $\dfrac{1}{16}$ minute to copy one page

    (b) $x\dfrac{1}{16} = \dfrac{x}{16}$ minutes to copy $x$ pages

    (c) $\dfrac{60}{16} = \dfrac{15}{4}$ minutes to copy 60 pages

**73.** Probability $= \dfrac{\text{Area shaded rectangle}}{\text{Area large rectangle}}$

$\qquad = \dfrac{x\left(\dfrac{x}{2}\right)}{x(2x+1)} = \dfrac{\dfrac{x}{2}}{2x+1} \cdot \dfrac{2}{2} = \dfrac{x}{2(2x+1)}$

**75.** (a)

| $t$ | 0 | 2 | 4 | 6 | 8 | 10 | 12 | 14 | 16 | 18 | 20 | 22 |
|---|---|---|---|---|---|---|---|---|---|---|---|---|
| $T$ | 75° | 55.9 | 48.3 | 45 | 43.3 | 42.3 | 41.7 | 41.3 | 41.1 | 40.9 | 40.7 | 40.6 |

    (b) $T = 10\left(\dfrac{4t^2 + 16t + 75}{t^2 + 4t + 10}\right)$ appears to be approaching 40.

**77.** False. For $n$ odd, the domain of $(x^{2n} - 1)/(x^n - 1)$ is all $x \neq 1$, unlike the domain of the right-hand side.

**79.** Completely factor the numerator and denominator to determine if they have any common factors.

# Section P.5    The Cartesian Plane

■    You should be able to plot points.

■    You should know that the distance between $(x_1, y_1)$ and $(x_2, y_2)$ in the plane is

$\qquad d = \sqrt{(x_2 - x_1)^2 + (y_2 - y_1)^2}$.

■    You should know that the midpoint of the line segment joining $(x_1, y_1)$ and $(x_2, y_2)$ is

$\qquad \left(\dfrac{x_1 + x_2}{2}, \dfrac{y_1 + y_2}{2}\right)$.

■    You should know the equation of a circle: $(x - h)^2 + (y - k)^2 = r^2$.

■    You should be able to translate points in the plane.

**Solutions to Odd-Numbered Exercises**

**1.** $A$: $(2, 6)$, $B$: $(-6, -2)$, $C$: $(4, -4)$, $D$: $(-3, 2)$

**3.** Plot $(-4, 2), (-3, -6), (0, 5), (1, -4)$

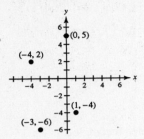

**5.**

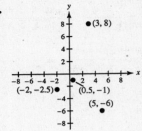

**7.** The coordinates are $(-5, 4)$

**9.** The coordinates are $(-6, -6)$

**11.** $x > 0 \implies$ The point lies in Quadrant I or in Quadrant IV.

$y < 0 \implies$ The point lies in Quadrant III or in Quadrant IV.

$x > 0$ and $y < 0 \implies (x, y)$ lies in Quadrant IV.

**13.** $x = -4 \implies x$ is negative $\implies$ The point lies in Quadrant II or in Quadrant III.

$y > 0 \implies$ The point lies in Quadrant I or Quadrant II.

$x = -4$ and $y > 0 \implies (x, y)$ lies in Quadrant II.

**15.** $y < -5 \implies y$ is negative $\implies$ The point lies in either Quadrant III or Quadrant IV.

**17.** Since $(x, -y)$ is in Quadrant II, we know that $x < 0$ and $-y > 0$. If $-y > 0$, then $y < 0$.

$x < 0 \implies$ The point lies in Quadrant II or in Quadrant III.

$y < 0 \implies$ The point lies in Quadrant III or in Quadrant IV.

$x < 0$ and $y < 0 \implies (x, y)$ lies in Quadrant III.

**19.** If $xy > 0$, then either $x$ and $y$ are both positive, or both negative. Hence, $(x, y)$ lies in either Quadrant I or Quadrant III.

**21.**

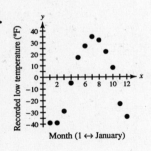

Month (1 ↔ January)

**23.** $(6, -3), (6, 5)$

$$d = \sqrt{(6 - 6)^2 + (5 - (-3))^2} = \sqrt{64} = 8$$

**25.** $(-3, -1), (2, -1)$

$$d = \sqrt{(2 - (-3))^2 + (-1 - (-1))^2} = \sqrt{25} = 5$$

**27.** $d = \sqrt{(3 - (-2))^2 + (-6 - 6)^2} = \sqrt{5^2 + (-12)^2}$
$$= \sqrt{25 + 144} = \sqrt{169} = 13.$$

**29.** $\left(\frac{1}{2}, \frac{4}{3}\right), (2, -1)$

$$d = \sqrt{\left(\frac{1}{2} - 2\right)^2 + \left(\frac{4}{3} + 1\right)^2}$$
$$= \sqrt{\frac{9}{4} + \frac{49}{9}}$$
$$= \sqrt{\frac{277}{36}} = \sqrt{\frac{277}{6}} \approx 2.77$$

**31.** $(-4.2, 3.1), (-12.5, 4.8)$

$$d = \sqrt{(-4.2 + 12.5)^2 + (3.1 - 4.8)^2}$$
$$= \sqrt{68.89 + 2.89}$$
$$= \sqrt{71.78} \approx 8.47$$

**33.** (a) The distance between $(0, 2)$ and $(4, 2)$ is 4.

The distance between $(4, 2)$ and $(4, 5)$ is 3.

The distance between $(0, 2)$ and $(4, 5)$ is

$$\sqrt{(4 - 0)^2 + (5 - 2)^2} = \sqrt{16 + 9} = \sqrt{25} = 5.$$

(b) $4^2 + 3^2 = 16 + 9 = 25 = 5^2$

**35.** (a) The distance between $(-1, 1)$ and $(9, 1)$ is 10.

The distance between $(9, 1)$ and $(9, 4)$ is 3.

The distance between $(-1, 1)$ and $(9, 4)$ is

$$\sqrt{(9 - (-1))^2 + (4 - 1)^2} = \sqrt{100 + 9} = \sqrt{109}.$$

(b) $10^2 + 3^2 = 109 = \left(\sqrt{109}\right)^2$

**37.** Find distances between pairs of points.

$$d_1 = \sqrt{(4 - 2)^2 + (0 - 1)^2} = \sqrt{5}$$

$$d_2 = \sqrt{(4 + 1)^2 + (0 + 5)^2} = \sqrt{50}$$

$$d_3 = \sqrt{(2 + 1)^2 + (1 + 5)^2} = \sqrt{45}$$

$$\left(\sqrt{5}\right)^2 + \left(\sqrt{45}\right)^2 = \left(\sqrt{50}\right)^2$$

Because $d_1{}^2 + d_3{}^2 = d_2{}^2$, the triangle is a right triangle.

**39.** Find distances between pairs of points.

$$d_1 = \sqrt{(0 - 2)^2 + (9 - 5)^2} = \sqrt{4 + 16} = \sqrt{20} = 2\sqrt{5}$$

$$d_2 = \sqrt{(-2 - 0)^2 + (0 - 9)^2} = \sqrt{4 + 81} = \sqrt{85}$$

$$d_3 = \sqrt{(0 - (-2))^2 + (-4 - 0)^2} = \sqrt{4 + 16} = \sqrt{20} = 2\sqrt{5}$$

$$d_4 = \sqrt{(0 - 2)^2 + (-4 - 5)^2} = \sqrt{4 + 81} = \sqrt{85}$$

Opposite sides have equal lengths of $2\sqrt{5}$ and $\sqrt{85}$, so the figure is a parallelogram.

**41.**

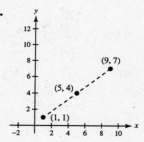

(b) $d = \sqrt{(9 - 1)^2 + (7 - 1)^2}$

$\quad = \sqrt{64 + 36} = 10$

(c) $\left(\dfrac{9 + 1}{2}, \dfrac{7 + 1}{2}\right) = (5, 4)$

**43.**

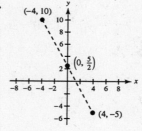

(b) $d = \sqrt{(4 + 4)^2 + (-5 - 10)^2}$

$\quad = \sqrt{64 + 225} = 17$

(c) $\left(\dfrac{4 - 4}{2}, \dfrac{-5 + 10}{2}\right) = \left(0, \dfrac{5}{2}\right)$

**45.**

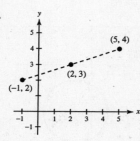

(b) $d = \sqrt{(5 + 1)^2 + (4 - 2)^2}$

$\quad = \sqrt{36 + 4} = \sqrt{40} = 2\sqrt{10}$

(c) $\left(\dfrac{-1 + 5}{2}, \dfrac{2 + 4}{2}\right) = (2, 3)$

**47.**

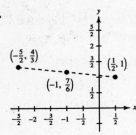

(b) $d = \sqrt{\left(\dfrac{1}{2} + \dfrac{5}{2}\right)^2 + \left(1 - \dfrac{4}{3}\right)^2}$

$d = \sqrt{9 + \dfrac{1}{9}} = \dfrac{\sqrt{82}}{3}$

(c) $\left(\dfrac{-\frac{5}{2} + \frac{1}{2}}{2}, \dfrac{\frac{4}{3} + 1}{2}\right) = \left(-1, \dfrac{7}{6}\right)$

**49.**

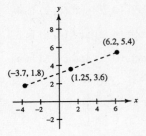

(b) $d = \sqrt{(6.2 + 3.7)^2 + (5.4 - 1.8)^2}$

$= \sqrt{98.01 + 12.96} = \sqrt{110.97}$

(c) $\left(\dfrac{6.2 - 3.7}{2}, \dfrac{5.4 + 1.8}{2}\right) = (1.25, 3.6)$

**51.** $\left(\dfrac{1998 + 2002}{2}, \dfrac{839.6 + 1480.0}{2}\right) = (2000, 1159.8)$

The sales in 2000 are \$1159.8 million.

**53.** Since $x_m = \dfrac{x_1 + x_2}{2}$ and $y_m = \dfrac{y_1 + y_2}{2}$ we have:

$2x_m = x_1 + x_2 \qquad 2y_m = y_1 + y_2$

$2x_m - x_1 = x_2 \qquad 2y_m - y_1 = y_2$

So, $(x_2, y_2) = (2x_m - x_1, 2y_m - y_1)$.

(a) $(x_2, y_2) = (2x_m - x_1, 2y_m - y_1) = (2(4) - 1, 2(-1) - (-2)) = (7, 0)$

(b) $(x_2, y_2) = (2x_m - x_1, 2y_m - y_1) = (2(2) - (-5), 2(4) - 11) = (9, -3)$

**55.** $(x - 0)^2 + (y - 0)^2 = 3^2$

$\phantom{(x-0)^2 + (y-0)^2}x^2 + y^2 = 9$

**57.** $(x - 2)^2 + (y + 1)^2 = 4^2$

$(x - 2)^2 + (y + 1)^2 = 16$

**59.** $(x + 1)^2 + (y - 2)^2 = r^2$

$(0 + 1)^2 + (0 - 2)^2 = r^2 \implies r^2 = 5$

$(x + 1)^2 + (y - 2)^2 = 5$

**61.** $r = \dfrac{1}{2}\sqrt{(6 - 0)^2 + (8 - 0)^2} = \dfrac{1}{2}\sqrt{100} = 5$

$\text{Center}\left(\dfrac{0 + 6}{2}, \dfrac{0 + 8}{2}\right) = (3, 4)$

$(x - 3)^2 + (y - 4)^2 = 25$

**63.** $x^2 + y^2 = 25$

Center: $(0, 0)$

Radius $= 5$

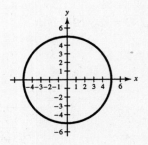

**65.** Center: $(1, -3)$

Radius $= 2$

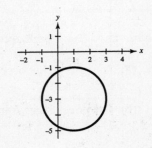

**67.** Center: $\left(\frac{1}{2}, \frac{1}{2}\right)$

Radius $= \frac{3}{2}$

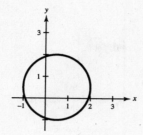

**69.** The *x*-coordinates are increased by 2, and the *y*-coordinates are increased by 5

| Old vertex | Shifted vertex |
|---|---|
| $(-1, -1)$ | $(1, 4)$ |
| $(-2, -4)$ | $(0, 1)$ |
| $(2, -3)$ | $(4, 2)$ |

**71.**

| Old vertex | Shifted vertex |
|---|---|
| $(0, 2)$ | $(-1, 5)$ |
| $(3, 5)$ | $(2, 8)$ |
| $(5, 2)$ | $(4, 5)$ |
| $(2, -1)$ | $(1, 2)$ |

**73.** The point $(65, 83)$ represents an entrance exam score of 65.

**75.** (a) It appears that the number of artists elected alternates between 6 and 8 per year from 1990 to 2002. If the trend continues, 6, 7 or 8 would be elected in 2005.

(b) Since 1986 and 1987 were the first two years that artists were elected, there was a larger number elected.

**77.**
$$d = \sqrt{(45 - 10)^2 + (40 - 15)^2}$$
$$= \sqrt{35^2 + 25^2}$$
$$= \sqrt{1850}$$
$$= 5\sqrt{74}$$
$$\approx 43 \text{ yards}$$

**79.** False. It would be sufficient to use the midpoint formula 15 times.

**81.** False. The polygon could be a rhombus. For example, consider the points $(4, 0)$, $(0, 6)$, $(-4, 0)$ and $(0, -6)$

**83.** No, the scales can be different. The scales depend on the magnitude of the coordinates. See Figure P.13.

# Section P.6    Exploring Data: Representing Data Graphically

- You should be able to construct line plots.
- You should be able to construct histograms or frequency distributions.
- You should be able to construct bar graphs.
- You should be able to construct line graphs.

### Solutions to Odd-Numbered Exercises

**1.** (a) The price 1.709 occurred with the greatest frequency (6).

(b) The prices range from 1.599 to 1.789. The range is $1.789 - 1.599 = 0.19$

**3.**

```
        ×
        ×       × × ×
      × × ×     × × × ×
      × × ×     × × × ×
×   × × × ×   × × × × ×   ×
+--+--+--+--+--+--+--+--+--+
10  12  14  16  18  20  22  24
        Quiz Scores
```

The score of 15 occurred with the greatest frequency.

**5.**   *Interval*          *Tally*

[800, 1000)          ||

[1000, 1200)         ℍℍ ℍℍ ||||

[1200, 1400)         ℍℍ ℍℍ ℍℍ |

[1400, 1600)         ℍℍ ℍℍ

[1600, 1800)         ℍℍ ||

[1800, 2000)         |

[2000, 2200)         |

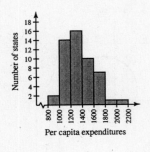

**7.**   1997 to 2000: $\dfrac{943 - 753}{753} \approx 0.252$ or 25.2%

2000 to 2001: $\dfrac{882 - 943}{943} \approx -0.065$ or 6.5%

decrease

**9.**

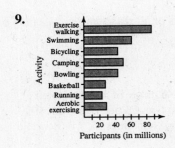

**11.**   The highest price was approximately $2.14 per pound in 2000.

**13.**   1996 to 2002: $\dfrac{1900 - 1100}{1100} \approx 0.727$ or 72.7%

**15.**

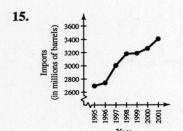

**17.**

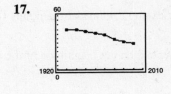

**19.**

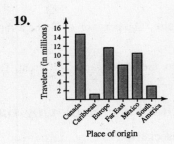

The graph indicates that oil imports increased yearly from 1995 to 2001.

**21.**   A histogram has a portion of the real number line as its horizontal axis, and the bars are not separated by spaces. A bar graph can be either horizontal or vertical. The labels are not necessarily numbers, and the bars are usually separated by spaces.

**23.**   The second graph is misleading because the vertical scale is too small which makes small changes look large. Answers will vary.

## Review Exercises for Chapter P

**Solutions to Odd-Numbered Exercises**

**1.** $\{11, -14, -\frac{8}{9}, \frac{5}{2}, \sqrt{6}, 0.4\}$

    (a) Whole numbers: 11

    (b) Natural numbers: 11

    (c) Integers: $11, -14$

    (d) Rational numbers: $11, -14, -\frac{8}{9}, \frac{5}{2}, 0.4$

    (e) Irrational numbers: $\sqrt{6}$

**3.** $\frac{5}{6} = 0.8\overline{3}$

    $\frac{7}{8} = 0.875$

    $\frac{5}{6} < \frac{7}{8}$

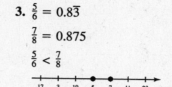

**5.** $x \le 7$, The set consists of all real numbers less than or equal to 7.

**7.** $d(a, b) = |b - a| = |48 - (-74)| = |48 + 74| = 122$

**9.** $d(x, 8) = |x - 8|$ and $d(x, 8) \ge 3$, so $|x - 8| \ge 3$.

**11.** $d(y, -30) = |y - (-30)| = |y + 30|$ and $d(y, -30) < 5$, so $|y + 30| < 5$.

**13.** $10x - 3$

    (a) $x = -1$: $10(-1) - 3 = -10 - 3 = -13$

    (b) $x = 3$: $10(3) - 3 = 30 - 3 = 27$

**15.** $-2x^2 - x + 3$

    (a) $x = 3$: $-2(3)^2 - 3 + 3 = -2(9) = -18$

    (b) $x = -3$: $-2(-3)^2 - (-3) + 3 = -18 + 6 = -12$

**17.** $2x + (3x - 10) = (2x + 3x) - 10$

Associative Property of Addition

**19.** $(t + 4)(2t) = (2t)(t + 4)$

Commutative Property of Multiplication

**21.** $\frac{2}{3} + \frac{8}{9} = \frac{2 \cdot 3}{3 \cdot 3} + \frac{8}{9} = \frac{6 + 8}{9} = \frac{14}{9}$

**23.** $\frac{3}{16} \div \frac{9}{2} = \frac{3}{16} \cdot \frac{2}{9} = \frac{3 \cdot 2}{2 \cdot 8 \cdot 3 \cdot 3} = \frac{1}{8 \cdot 3} = \frac{1}{24}$

**25.** $\frac{x}{5} + \frac{7x}{12} = \frac{12(x) + 7x(5)}{60} = \frac{12x + 35x}{60} = \frac{47x}{60}$

**27.** (a) $(-2z)^3 = (-2)^3 z^3 = -8z^3$

    (b) $(a^2 b^4)(3ab^{-2}) = 3a^{2+1}b^{4-2} = 3a^3 b^2$

**29.** (a) $\frac{6^2 u^3 v^{-3}}{12u^{-2}v} = \frac{36u^{3-(-2)}v^{-3-1}}{12} = 3u^5 v^{-4} = \frac{3u^5}{v^4}$

    (b) $\frac{3^{-4}m^{-1}n^{-3}}{9^{-2}mn^{-3}} = \frac{9^2 n^3}{3^4 mmn^3} = \frac{81}{81m^2} = \frac{1}{m^2} = m^{-2}$

**31.** $43,800,000,000 = 4.38 \times 10^{10}$

**33.** $4.836 \times 10^8 = 483,600,000$

**35.** $\left(\sqrt[4]{78}\right)^4 = (78^{1/4})^4 = 78^1 = 78$

**37.** $\sqrt{4x^4} = 2x^2$

**39.** $\sqrt{\dfrac{81}{144}} = \sqrt{\dfrac{9 \cdot 9}{12 \cdot 12}} = \dfrac{9}{12} = \dfrac{3}{4}$

**41.** $\sqrt[3]{\dfrac{2x^3}{27}} = \sqrt[3]{\dfrac{2x^3}{3^3}} = \dfrac{x}{3}\sqrt[3]{2} = \dfrac{2^{1/3}x}{3}$

**43.** $\sqrt{50} - \sqrt{18} = \sqrt{25 \cdot 2} - \sqrt{9 \cdot 2} = 5\sqrt{2} - 3\sqrt{2} = 2\sqrt{2}$

**45.** $8\sqrt{3x} - 5\sqrt{3x} = 3\sqrt{3x}$

**47.** $\sqrt{8x^3} + \sqrt{2x} = \sqrt{2 \cdot 2^2 \cdot x \cdot x^2} + \sqrt{2x} = 2x\sqrt{2x} + \sqrt{2x}$
$$= (2x + 1)\sqrt{2x}$$

**49.** $A = wh = \left(12\sqrt{2}\right)\sqrt{24^2 - \left(12\sqrt{2}\right)^2}$
$$= 12\sqrt{2}\sqrt{288}$$
$$= 12\sqrt{2} \cdot 12\sqrt{2}$$
$$= 288 \text{ square inches}$$

The shape is a square because $w = h = 12\sqrt{2}$

**51.** $\dfrac{1}{2 - \sqrt{3}} = \dfrac{1}{2 - \sqrt{3}} \cdot \dfrac{2 + \sqrt{3}}{2 + \sqrt{3}} = \dfrac{2 + \sqrt{3}}{4 - 3}$
$$= \dfrac{2 + \sqrt{3}}{1} = 2 + \sqrt{3}$$

**53.** $\dfrac{\sqrt{20}}{4} = \dfrac{2\sqrt{5}}{4} = \dfrac{\sqrt{5}}{2} = \dfrac{\sqrt{5}}{2} \cdot \dfrac{\sqrt{5}}{\sqrt{5}} = \dfrac{5}{2\sqrt{5}}$

**55.** $81^{3/2} = (81^{1/2})^3 = 9^3 = 729$

**57.** $(-3x^{2/5})(-2x^{1/2}) = 6x^{2/5 + 1/2}$
$$= 6x^{9/10}$$

**59.** $-2x^5 - x^4 + 3x^3 + 15x^2 + 5$
Degree: 5
Leading coefficient: $-2$

**61.** $-(3x^2 + 2x) + (1 - 5x) = -3x^2 - 2x + 1 - 5x$
$$= -3x^2 - 7x + 1$$

**63.** $(2x^3 - 5x^2 + 10x - 7) + (4x^2 - 7x - 2)$
$$= 2x^3 - 5x^2 + 4x^2 + 10x - 7x - 7 - 2$$
$$= 2x^3 - x^2 + 3x - 9$$

**65.** $(x^2 - 2x + 1)(x^3 - 1) = (x^2 - 2x + 1)x^3 - (x^2 - 2x + 1)$
$$= x^5 - 2x^4 + x^3 - x^2 + 2x - 1$$

**67.** $(y^2 - y)(y^2 + 1)(y^2 + y + 1) = (y^4 - y^3 + y^2 - y)(y^2 + y + 1)$
$$= (y^4 - y^3 + y^2 - y)y^2 + (y^4 - y^3 + y^2 - y)y + (y^4 - y^3 + y^2 - y)$$
$$= y^6 - y^5 + y^4 - y^3 + y^5 - y^4 + y^3 - y^2 + y^4 - y^3 + y^2 - y$$
$$= y^6 + y^4 - y^3 - y.$$

**69.** $(x + 8)(x - 8) = x^2 - 8^2 = x^2 - 64$

**71.** $(x - 4)^3 = x^3 - 12x^2 + 48x - 64$

**73.** $(m - 4 + n)(m - 4 - n) = [(m - 4) + n][(m - 4) - n]$

$$= (m - 4)^2 - n^2$$
$$= m^2 - 8m + 16 - n^2$$
$$= m^2 - n^2 - 8m + 16$$

**75.** $(x + 3)(x + 5) = x(x + 5) + 3(x + 5)$

Distributive Property

**77.** $7x + 35 = 7(x + 5)$

**79.** $x^3 - x = x(x^2 - 1)$

**81.** $2x^3 + 18x^2 - 4x = 2x(x^2 + 9x - 2)$

**83.** (a)

The surface is the sum of the area of the side, $2\pi rh$, and the areas of the top and bottom which are each $\pi r^2$.

$$S = 2\pi rh + \pi r^2 + \pi r^2 = 2\pi rh + 2\pi r^2.$$

(b) $S = 2\pi rh + 2\pi r^2 = 2\pi r(r + h)$

**85.** $x^2 - 169 = (x + 13)(x - 13)$

**87.** $x^3 + 216 = (x + 6)(x^2 - 6x + 36)$

**89.** $x^2 - 6x - 27 = (x - 9)(x + 3)$

**91.** $2x^2 + 21x + 10 = (2x + 1)(x + 10)$

**93.** $x^3 - 4x^2 - 3x + 12 = x^2(x - 4) - 3(x - 4)$

$$= (x - 4)(x^2 - 3)$$

**95.** $2x^2 - x - 15$

$a = 2, c = -15, ac = -30 = (-6)5$ and $-6 + 5 = -1 = b$

So, $2x^2 - x - 15 = 2x^2 - 6x + 5x - 15$

$$= 2x(x - 3) + 5(x - 3)$$
$$= (2x + 5)(x - 3)$$

**97.** Domain: all $x$

**99.** Domain: all $x \neq \dfrac{3}{2}$

**101.** $\dfrac{4x^2}{4x^3 + 28x} = \dfrac{x}{x^2 + 7}, x \neq 0$

**103.** $\dfrac{x^2 - x - 30}{x^2 - 25} = \dfrac{(x - 6)(x + 5)}{(x + 5)(x - 5)} = \dfrac{x - 6}{x - 5}, x \neq -5$

**105.** $\dfrac{x^2 - 4}{x^4 - 2x^2 - 8} \cdot \dfrac{x^2 + 2}{x^2} = \dfrac{(x - 2)(x + 2)}{(x^2 - 4)(x^2 + 2)} \cdot \dfrac{x^2 + 2}{x^2}$

$$= \dfrac{(x - 2)(x + 2)}{(x - 2)(x + 2)} \cdot \dfrac{1}{x^2} = \dfrac{1}{x^2}, x \neq \pm 2$

**107.** $\dfrac{x^2(5x - 6)}{2x + 3} \div \dfrac{5x}{2x + 3} = \dfrac{x^2(5x - 6)}{2x + 3} \cdot \dfrac{2x + 3}{5x} = \dfrac{x(5x - 6)}{5}, x \neq 0, -\dfrac{3}{2}$

**109.** $x - 1 + \dfrac{1}{x + 2} + \dfrac{1}{x - 1} = \dfrac{(x - 1)^2(x + 2) + (x - 1) + (x + 2)}{(x + 2)(x - 1)}$

$= \dfrac{(x^2 - 2x + 1)(x + 2) + 2x + 1}{(x + 2)(x - 1)}$

$= \dfrac{x^3 - 2x^2 + x + 2x^2 - 4x + 2 + (2x + 1)}{(x + 2)(x - 1)}$

$= \dfrac{x^3 - x + 3}{(x + 2)(x - 1)}$

**111.** $\dfrac{1}{x} - \dfrac{x - 1}{x^2 + 1} = \dfrac{1(x^2 + 1) - x(x - 1)}{x(x^2 + 1)}$

$= \dfrac{x^2 + 1 - x^2 + x}{x(x^2 + 1)}$

$= \dfrac{x + 1}{x(x^2 + 1)}$

**113.** $\dfrac{\dfrac{1}{x} - \dfrac{1}{y}}{(x^2 - y^2)} = \dfrac{y - x}{xy} \cdot \dfrac{1}{(x - y)(x + y)}$

$= \dfrac{-1}{xy(x + y)}, x \neq y$

**115.**

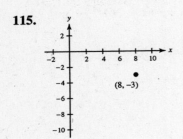

Quadrant IV

**117.**

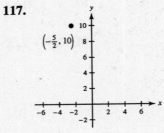

Quadrant II

**119.** $x > 0, y = -2 \rightarrow$

Quadrant IV

**121.**

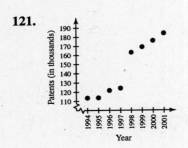

**123.** $(-3, 8), (1, 5)$

$d = \sqrt{(1 - (-3))^2 + (5 - 8)^2}$

$= \sqrt{4^2 + 3^2} = \sqrt{25} = 5$

**125.**

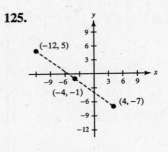

Midpoint:

$\left( \dfrac{-12 + 4}{2}, \dfrac{5 - 7}{2} \right) = (-4, -1)$

**127.** Radius:

$$\sqrt{(3 - (-5))^2 + (-1 - 1)^2} = \sqrt{64 + 4} = \sqrt{68}$$

$$(x - 3)^2 + (y + 1)^2 = 68$$

**129.**

| *Original vertices* | *Shifted vertices* |
|---|---|
| (4, 8) | (4 − 2, 8 − 3) = (2, 5) |
| (6, 8) | (6 − 2, 8 − 3) = (4, 5) |
| (4, 3) | (4 − 2, 3 − 3) = (2, 0) |
| (6, 3) | (6 − 2, 3 − 3) = (4, 0) |

**131.**

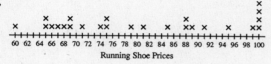

Running Shoe Prices

The price of 100 occurs with the greatest frequency (4)

**133.**

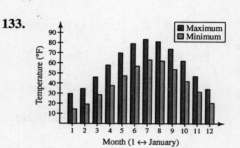

**135.** False. Not true for $x = 1$

**137.** $(2x)^4 = 2^4 x^4 = 16x^4$

The exponent is to be applied to the whole quantity inside the parentheses.

**139.** The expression $\sqrt{5u} + \sqrt{3u}$ does not equal $2\sqrt{2u}$ because radicals cannot be combined unless the index and the radicand are the same. In this case, radicands are not the same.

# C H A P T E R 1
# Functions and Their Graphs

# C H A P T E R   1
# Functions and Their Graphs

## Section 1.1    Graphs of Equations

---

■  You should be able to use the point-plotting method of graphing.

■  You should be able to find $x$- and $y$-intercepts.

    (a) To find the $x$-intercepts, let $y = 0$ and solve for $x$.

    (b) To find the $y$-intercepts, let $x = 0$ and solve for $y$.

■  You should know how to graph an equation with a graphing utility. You should be able to determine an appropriate viewing rectangle.

■  You should be able to use the zoom and trace features of a graphing utility.

---

**Solutions to Odd-Numbered Exercises**

**1.** $y = \sqrt{x + 4}$

    (a) $(0, 2)$:  $2 \overset{?}{=} \sqrt{0 + 4}$

                $2 = 2$ ✓

    Yes, the point *is* on the graph.

    (b) $(5, 3)$:  $3 \overset{?}{=} \sqrt{5 + 4}$

                $3 = \sqrt{9}$ ✓

    Yes, the point *is* on the graph.

**3.** $y = 4 - |x - 2|$

    (a) $(1, 5)$:  $5 \overset{?}{=} 4 - |1 - 2|$

                $5 \neq 4 - 1$

    No, the point *is not* on the graph.

    (b) $(1.2, 3.2)$: $3.2 \overset{?}{=} 4 - |1.2 - 2|$

                    $3.2 \overset{?}{=} 4 - |-.8|$

                    $3.2 \overset{?}{=} 4 - .8$

                    $3.2 \overset{?}{=} 3.2$ ✓

    Yes, the point *is* on the graph.

**5.** $x^2 + y^2 = 20$

    (a) $(3, -2)$:  $3^2 + (-2)^2 \overset{?}{=} 20$

                 $9 + 4 \overset{?}{=} 20$

                  $13 \neq 20$

    No, the point *is not* on the graph.

    (b) $(-4, 2)$:  $(-4)^2 + 2^2 \overset{?}{=} 20$

                 $16 + 4 \overset{?}{=} 20$

                 $20 = 20$

    Yes, the point *is* on the graph.

**7.** $y = -2x + 3$

| $x$ | $-1$ | $0$ | $1$ | $\frac{3}{2}$ | $2$ |
|---|---|---|---|---|---|
| $y$ | $5$ | $3$ | $1$ | $0$ | $-1$ |
| Solution point | $(-1, 5)$ | $(0, 3)$ | $(1, 1)$ | $\left(\frac{3}{2}, 0\right)$ | $(2, -1)$ |

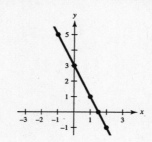

**9.** $y = x^2 - 2x$

| $x$ | $-1$ | $0$ | $1$ | $2$ | $3$ |
|---|---|---|---|---|---|
| $y$ | $3$ | $0$ | $-1$ | $0$ | $3$ |
| Solution point | $(-1, 3)$ | $(0, 0)$ | $(1, -1)$ | $(2, 0)$ | $(3, 3)$ |

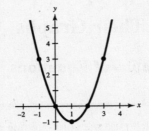

**11. (a)** $y = \frac{1}{4}x - 3$

| $x$ | $-2$ | $-1$ | $0$ | $1$ | $2$ |
|---|---|---|---|---|---|
| $y$ | $-\frac{7}{2}$ | $-\frac{13}{4}$ | $-3$ | $-\frac{11}{4}$ | $-\frac{5}{2}$ |

**(b)**

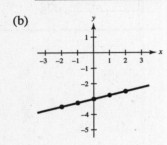

**(c)** $y = -\frac{1}{4}x - 3$

| $x$ | $-2$ | $-1$ | $0$ | $1$ | $2$ |
|---|---|---|---|---|---|
| $y$ | $-\frac{5}{2}$ | $-\frac{11}{4}$ | $-3$ | $-\frac{13}{4}$ | $-\frac{7}{2}$ |

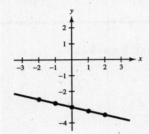

Both graphs are lines. The first graph rises to the right, whereas the second falls. Both pass through $(0, -3)$.

**13.** $y = 1 - x$ has intercepts $(1, 0)$ and $(0, 1)$.

Matches graph (d).

**15.** $y = \sqrt{9 - x^2}$ has intercepts $(\pm 3, 0)$ and $(0, 3)$.

Matches graph (f).

**17.** $y = x^3 - x + 1$ has a $y$-intercept of $(0, 1)$ and the points $(1, 1)$ and $(-2, -5)$ are on the graph.

Matches (a).

**19.** $y = -4x + 1$

Slope: $-4$

$y$-intercept: $(0, 1)$

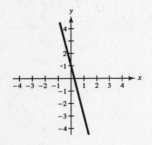

**21.** $y = 2 - x^2$

Intercepts: $(0, 2), \left(\sqrt{2}, 0\right), \left(-\sqrt{2}, 0\right)$

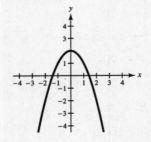

**23.** $y = x^2 - 3x$

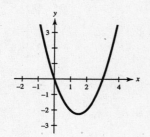

**25.** $y = x^3 + 2$

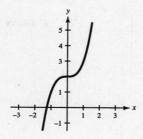

**27.** $y = \sqrt{x - 3}$

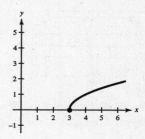

**29.** $y = |x - 2|$

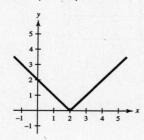

**31.** $x = y^2 - 1$

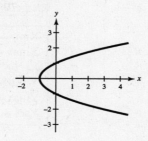

**33.** $y = x - 7$

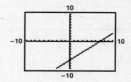

Intercepts: $(0, -7)$, $(7, 0)$

**35.** $y = 3 - \frac{1}{2}x$

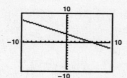

Intercepts: $(6, 0)$, $(0, 3)$

**37.** $y = x^2 - 4x + 3$

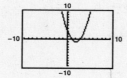

Intercepts: $(3, 0)$, $(1, 0)$, $(0, 3)$

**39.** $y = x(x - 2)^2$

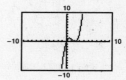

Intercepts: $(0, 0)$, $(2, 0)$

**41.** $y = \dfrac{2x}{x - 1}$

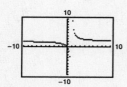

Intercepts: $(0, 0)$

**43.** $y = x\sqrt{x + 3}$

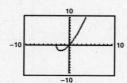

Intercepts: $(0, 0)$, $(-3, 0)$

**45.** $y = \sqrt[3]{x}$

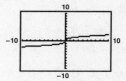

Intercepts: $(0, 0)$

**47.** $y = \frac{5}{2}x + 5$

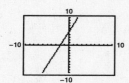

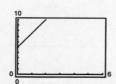

The first setting shows the line and its intercepts. The first setting is better.

The second setting does not show the $x$-intercept $(-2, 0)$.

**49.** $y = -x^2 + 10x - 5$

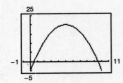

The second viewing window is better because it shows more of the essential features of the function.

**51.** $y = 4x^2 - 25$

Range/Window

```
Xmin = -5
Xmax = 5
Xscl = 1
Ymin = -30
Ymax = 10
Yscl = 5
```

**53.** $y = |x| + |x - 10|$

Range/Window

```
Xmin = -30
Xmax = 30
Xscl = 5
Ymin = -10
Ymax = 50
Yscl = 5
```

**55.** $x^2 + y^2 = 64$

$$y = \pm\sqrt{64 - x^2}$$

Use: $y_1 = \sqrt{64 - x^2}$

$$y_2 = -\sqrt{64 - x^2}$$

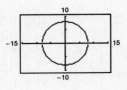

**57.** $(x - 1)^2 + (y - 2)^2 = 16$

$$(y - 2)^2 = 16 - (x - 1)^2$$

$$y - 2 = \pm\sqrt{16 - (x - 1)^2}$$

$$y = 2 \pm \sqrt{16 - (x - 1)^2}$$

Use $y_1 = 2 + \sqrt{16 - (x - 1)^2}$

$$y_2 = 2 - \sqrt{16 - (x - 1)^2}$$

**59.** $y_1 = \frac{1}{4}(x^2 - 8)$

$y_2 = \frac{1}{4}x^2 - 2$

The graphs are identical.

The Distributive Property is illustrated.

**61.** $y_1 = \frac{1}{5}[10(x^2 - 1)]$

$y_2 = 2(x^2 - 1)$

The graphs are identical.

The Associative Property of Multiplication is illustrated.

**63.** $y = \sqrt{5 - x}$

(a) $(2, y) \approx (2, 1.73)$

(b) $(x, 3) = (-4, 3)$

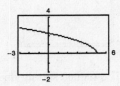

**65.** $y = x^5 - 5x$

   (a) $(-0.5, y) \approx (-0.5, 2.47)$

   (b) $(x, -4) = (1, -4)$ or $(x, -4) \approx (-1.65, -4)$

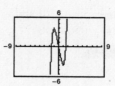

**67.** (a) $y = 225,000 - 20,000t$ , $0 \le t \le 8$

    <u>Window</u>

    $X_{min} = 0$

    $X_{max} = 8$

    $X_{scl} = 1$

    $Y_{min} = 60,000$

    $Y_{max} = 230,000$

    $Y_{scl} = 10,000$

   (b)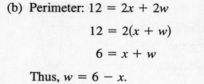

   (c) When $t = 5.8$, $y = 109,000$. Algebraically, $225,000 - 20,000(5.8) = \$109,000$.

   (d) When $t = 2.35$, $y = 178,000$. Algebraically, $225,000 - 20,000(2.35) = \$178,000$.

**69.** (a)

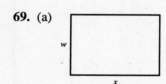

   (b) Perimeter: $12 = 2x + 2w$

                 $12 = 2(x + w)$

                   $6 = x + w$

     Thus, $w = 6 - x$.

     Area: $xw = x(6 - x) \implies A = x(6 - x)$

   (c)

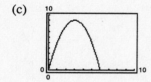

   (d) When $w = 4.9$, $x = 1.1$ and Area $= 5.39$ square meters.

     Algebraically,
     Area $= xw = (1.1)(4.9) = 5.39$ square meters.

   (e) The maximum area corresponds to the highest point on the graph, which appears to be $(3, 9)$.
     Thus, $x = 3$ and $w = 3$, and the rectangle is a square.

**71.** (a) The $y$-intercept $(0, 59.97)$ indicates the model's estimate of the life expectancy in 1930. $(t = 0)$

   (b) $y = 73.2$ when $t \approx 53.3$, which corresponds to 1983. Algebraically,

$$\frac{59.97 + 0.98t}{1 + 0.01t} = 73.2$$

$$59.97 + 0.98t = 73.2 + 0.732t$$

$$0.248t = 13.23$$

$$t = \frac{13.23}{0.248} \approx 53.3 \text{ or } 1983$$

   (c) 1948 corresponds to $t = 18$. Graphically, $y \approx 65.8$ when $t = 18$. Algebraically,

$$\frac{59.97 + 0.98(18)}{1 + 0.01(18)} = \frac{77.61}{1.18} \approx 65.8 \text{ years}$$

   (d) 2010 corresponds to $t = 80$:

$$\frac{59.97 + 0.98(80)}{1 + 0.01(80)} = \frac{138.37}{1.8} \approx 76.9 \text{ years}$$

**73.** False. $y = 1 - x^2$ has two $x$-intercepts, $(1, 0)$ and $(-1, 0)$. Also, $y = x^2 + 1$ has no $x$-intercepts.

**75.** Answers will vary.

**77.** $7\sqrt{72} - 5\sqrt{18} = 7\sqrt{2(6^2)} - 5\sqrt{2(3^2)}$
$$= 42\sqrt{2} - 15\sqrt{2} = 27\sqrt{2}$$

**79.** $7^{3/2} \cdot 7^{11/2} = 7^{(3/2 + 11/2)} = 7^7 = 823{,}543$

**81.** $(9x - 4) + (2x^2 - x + 15) = 2x^2 + 8x + 11$

# Section 1.2     Lines in the Plane

---

You should know the following important facts about lines.

■  The graph of $y = mx + b$ is a straight line. It is called a linear equation.

■  The slope of the line through $(x_1, y_1)$ and $(x_2, y_2)$ is

$$m = \frac{y_2 - y_1}{x_2 - x_1}.$$

■  (a) If $m > 0$ the line rises from left to right.

(b) If $m = 0$, the line is horizontal.

(c) If $m < 0$,, the line falls from left to right.

(d) If $m$ is undefined, the line is vertical.

■  Equations of Lines

(a) Slope-Intercept:  $y = mx + b$

(b) Point-Slope:  $y - y_1 = m(x - x_1)$

(c) Two-Point:  $y - y_1 = \dfrac{y_2 - y_1}{x_2 - x_1}(x - x_1)$

(d) General:  $Ax + By + c = 0$

(e) Vertical:  $x = a$

(f) Horizontal:  $y = b$

■  Given two distinct nonvertical lines

$$L_1: y = m_1x + b_1 \quad \text{and} \quad L_2: y = m_2x + b_2$$

(a) $L_1$ is parallel to $L_2$ if and only if $m_1 = m_2$ and $b_1 \neq b_2$.

(b) $L_1$ is perpendicular to $L_2$ if and only if $m_1 = -1/m_2$.

---

**Solutions to Odd-Numbered Exercises**

**1.** (a) $m = \frac{2}{3}$. Since the slope is positive, the line rises. Matches $L_2$.

(b) $m$ is undefined. The line is vertical. Matches $L_3$.

(c) $m = -2$. The line falls. Matches $L_1$.

**3.**

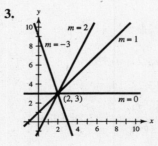

**5.** Slope $= \dfrac{\text{rise}}{\text{run}} = \dfrac{3}{2}$

**7.** slope $= \dfrac{0 - (-10)}{-4 - 0} = \dfrac{10}{-4} = -\dfrac{5}{2}$

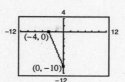

**9.**

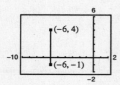

Slope is undefined.

**11.** Since $m = 0$, $y$ does not change. Three points are $(0, 1)$, $(3, 1)$, and $(-1, 1)$.

**13.** Since $m$ is undefined, $x$ does not change and the line is vertical. Three points are $(1, 1)$, $(1, 2)$, and $(1, 3)$.

**15.** Since $m = -2$, $y$ decreases 2 for every unit increase in $x$. Three points are $(1, -11)$, $(2, -13)$, and $(3, -15)$.

**17.** Since $m = \frac{1}{2}$, $y$ increases 1 for every increase of 2 in $x$. Three points are $(9, -1)$, $(11, 0)$, $(13, 1)$.

**19.** $5x - y + 3 = 0$

$\qquad y = 5x + 3$

(a) Slope: $m = 5$

$\quad y$-intercept: $(0, 3)$

(b)

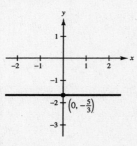

**21.** $5x - 2 = 0$

$\qquad x = \frac{2}{5}$

(a) Slope: undefined

$\quad$ No $y$-intercept

(b)

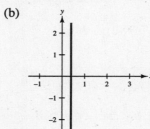

**23.** $3y + 5 = 0$

(a) $y = -\frac{5}{3}$

$\quad$ Slope: $m = 0$

$\quad y$-intercept: $\left(0, -\frac{5}{3}\right)$

(b)

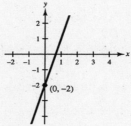

**25.** $y + 2 = 3(x - 0)$

$\qquad y = 3x - 2 \implies 3x - y - 2 = 0$

**27.** $y - 0 = 4(x - 0)$

$y = 4x$

$4x - y = 0$

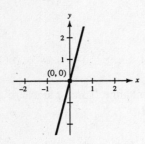

**29.** $x = 6$

$x - 6 = 0$

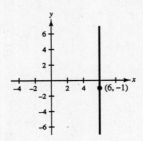

**31.** $y - \frac{3}{2} = 0\left(x + \frac{1}{2}\right)$

$y - \frac{3}{2} = 0$ horizontal line

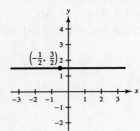

**33.** $y + 1 = \dfrac{5 + 1}{-5 - 5}(x - 5)$

$y = -\frac{3}{5}(x - 5) - 1$

$y = -\frac{3}{5}x + 2$

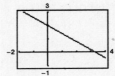

**35.** Since both points have $x = -8$, the slope is undefined.

$x = -8$

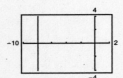

**37.** $y - \frac{1}{2} = \dfrac{\frac{5}{4} - \frac{1}{2}}{\frac{1}{2} - 2}(x - 2)$

$-y = -\frac{1}{2}(x - 2) + \frac{1}{2}$

$y = -\frac{1}{2}x + \frac{3}{2}$

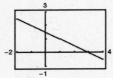

**39.** $y + \frac{3}{5} = \dfrac{-\frac{9}{5} + \frac{3}{5}}{\frac{9}{10} + \frac{1}{10}}\left(x + \frac{1}{10}\right)$

$y + \frac{3}{5} = -\frac{6}{5}\left(x + \frac{1}{10}\right)$

$y = -\frac{6}{5}x - \frac{18}{25}$

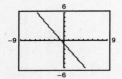

**41.** $y - 0.6 = \dfrac{-0.6 - 0.6}{-2 - 1}(x - 1)$

$y = 0.4(x - 1) + 0.6$

$y = 0.4x + 0.2$

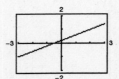

**43.** Using the points $(2000, 28500)$ and $(2002, 32900)$ we have

$$m = \frac{32900 - 28500}{2002 - 2000} = \frac{4400}{2} = 2200$$

$$S - 28500 = 2200(t - 2000)$$

$$S = 2200t - 4,371,500$$

When $t = 2006$,
$$S = 2200(2006) - 4,371,500 = \$41,700$$

**45.** $x - 2y = 4$

$$-2y = -x + 4$$

$$y = \tfrac{1}{2}x - 2$$

Slope: $\tfrac{1}{2}$

$y$-intercept: $(0, -2)$

The graph passes through $(0, -2)$ and rises 1 unit for each horizontal increase of 2.

**47.** $x = -6$

slope is undefined

no $y$-intercept

The line is vertical and passes through $(-6, 0)$.

**49.** $y = 0.5x - 3$

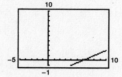 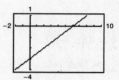

The second setting shows the $x$- and $y$-intercepts more clearly.

**51.** $m_{L_1} = \dfrac{9 + 1}{5 - 0} = 2$

$$m_{L_2} = \frac{1 - 3}{4 - 0} = -\frac{1}{2} = -\frac{1}{m_{L_1}}$$

$L_1$ and $L_2$ are perpendicular.

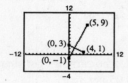

**53.** $m_{L_1} = \dfrac{0 - 6}{-6 - 3} = \dfrac{2}{3}$

$$m_{L_2} = \frac{\tfrac{7}{3} + 1}{5 - 0} = \frac{2}{3} = m_{L_1}$$

$L_1$ and $L_2$ are parallel.

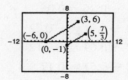

**55.** $4x - 2y = 3$

$$y = 2x - \tfrac{3}{2}$$

Slope: $m = 2$

(a) $y - 1 = 2(x - 2)$

$$y = 2x - 3$$

(b) $y - 1 = -\tfrac{1}{2}(x - 2)$

$$y = -\tfrac{1}{2}x + 2$$

**57.** $3x + 4y = 7$

$$y = -\tfrac{3}{4}x + \tfrac{7}{4}$$

Slope: $m = -\tfrac{3}{4}$

(a) $y - \tfrac{7}{8} = -\tfrac{3}{4}\left(x + \tfrac{2}{3}\right)$

$$y = -\tfrac{3}{4}x + \tfrac{3}{8}$$

(b) $y - \tfrac{7}{8} = \tfrac{4}{3}\left(x + \tfrac{2}{3}\right)$

$$y = \tfrac{4}{3}x + \tfrac{127}{72}$$

**59.** $x - 4 = 0$   vertical line

slope not defined

(a) $x - 3 = 0$ passes through $(3, -2)$

(b) $y + 2 = 0$ passes through $(3, -2)$ and is horizontal

**61.** (a) $y = 2x$   (b) $y = -2x$   (c) $y = \frac{1}{2}x$

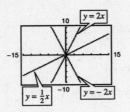

(b) and (c) are perpendicular.

**63.** (a) $y = -\frac{1}{2}x$   (b) $y = -\frac{1}{2}x + 3$

(c) $y = 2x - 4$

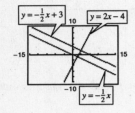

(a) and (b) are parallel.

(c) is perpendicular to (a) and (b).

**65.** (a) $m = 135$. The sales are increasing 135 units per year.

(b) $m = 0$. There is no change in sales.

(c) $m = -40$. The sales are decreasing 40 units per year.

**67.** (a)

| Years | Slope |
|-------|-------|
| 1992–1993 | $0.68 - 0.58 = 0.1$ |
| 1993–1994 | $0.86 - 0.68 = 0.18$ |
| 1994–1995 | $0.91 - 0.86 = 0.05$ |
| 1995–1996 | $0.69 - 0.91 = -0.22$ |
| 1996–1997 | $0.57 - 0.69 = -0.12$ |
| 1997–1998 | $0.74 - 0.57 = 0.17$ |
| 1998–1999 | $1.60 - 0.74 = 0.86$ |
| 1999–2000 | $0.82 - 1.60 = -0.78$ |
| 2000–2001 | $0.92 - 0.82 = 0.1$ |
| 2001–2002 | $0.08 - 0.92 = -0.84$ |

Greatest increase: 1998–1999   $(0.86)$

Greatest decrease: 2001–2002   $(-0.84)$

(b) $(2, 0.58)$, $(12, 0.08)$:

$$y - 0.58 = \frac{0.08 - 0.58}{12 - 2}(x - 2)$$

$$y = -0.05(x - 2) + 0.58$$

$$y = -0.05x + 0.68$$

(c) Between 1992 and 2002, the earnings per share decreased at a rate of 0.05 per year.

(d) For 2006, $x = 16$ and
$y = -0.05(16) + 0.68 = -0.12$, which is probably reasonable.

**69.** $\dfrac{\text{rise}}{\text{run}} = \dfrac{3}{4} = \dfrac{x}{\frac{1}{2}(32)}$

$$\frac{3}{4} = \frac{x}{16}$$

$$4x = 48$$

$$x = 12$$

The maximum height in the attic is 12 feet.

**71.** $(4, 2540)$, $m = 125$

$$V - 2540 = 125(t - 4)$$

$$V - 2540 = 125t - 500$$

$$V = 125t + 2040$$

**73.** $(4, 20400), m = -2000$

$$V - 20{,}400 = -2000(t - 4)$$

$$V - 20{,}400 = -2000t + 8000$$

$$V = -2000t + 28{,}400$$

**75.** The slope is $m = -10$. This represents the decrease in the amount of the loan each week.

Matches graph (b).

**77.** The slope is $m = 0.35$ This represents the increase in travel cost for each mile driven.

Matches graph (a).

**79.** Using the points $(0, 32)$ and $(100, 212)$, we have

$$m = \frac{212 - 32}{100 - 0} = \frac{180}{100} = \frac{9}{5}$$

$$F - 32 = \frac{9}{5}(C - 0)$$

$$F = \frac{9}{5}C + 32.$$

**81.** (a) Using the points $(0, 875)$ and $(5, 0)$, where the first coordinate represents the year $t$ and the second coordinate represents the value $V$, we have

$$m = \frac{0 - 875}{5 - 0} = -175$$

$$V = -175t + 875, \ 0 \le t \le 5.$$

(c) $t = 0: V = -175(0) + 875 = 875$

$t = 1: V = -175(1) + 875 = 700$

$t = 2: V = -175(2) + 875 = 525$

$t = 3: V = -175(3) + 875 = 350$

$t = 4: V = -175(4) + 875 = 175$

$t = 5: V = -175(5) + 875 = 0$

(b)

| $t$ | 0 | 1 | 2 | 3 | 4 | 5 |
|---|---|---|---|---|---|---|
| $V$ | 875 | 700 | 525 | 350 | 175 | 0 |

**83.** (a) $C = 36{,}500 + 5.25t + 11.50t$

$= 16.75t + 36{,}500$

(c) $P = R - C$

$= 27t - (16.75t + 36{,}500)$

$= 10.25t - 36{,}500$

(b) $R = 27t$

(d) $\quad 0 = 10.25t - 36{,}500$

$36{,}500 = 10.25t$

$t \approx 3561$ hours

**85.** (a) $\dfrac{83{,}038 - 75{,}365}{2002 - 1990} = \dfrac{7673}{12} \approx 639$ student/year

(b) 1984: $75{,}365 - 6(639) \approx 71{,}531$ students

1997: $75{,}365 + 7(639) \approx 79{,}838$ students

2000: $75{,}365 + 10(639) \approx 81{,}755$ students

(Answer could vary slightly)

(c) Let $t = 0$ represent 1990.

$(0, 75{,}365), (12, 83{,}038)$

$$y - 75{,}365 = \frac{83{,}038 - 75{,}365}{12 - 0} \ t - 0$$

$$y = \frac{7673}{12}t + 75{,}365 \approx 639t + 75{,}365$$

The slope is the annual increase in students. It is positive, indicating that Penn State University increased its students from 1990 to 2002.

**87.** False. The slopes are different:

$$\frac{4-2}{-1+8} = \frac{2}{7}$$

$$\frac{7+4}{-7-0} = -\frac{11}{7}$$

**89.**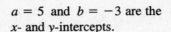

$$\frac{x}{5} + \frac{y}{-3} = 1$$

$$-3x + 5y + 15 = 0$$

$a = 5$ and $b = -3$ are the $x$- and $y$-intercepts.

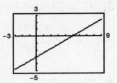

**91.**    $$\frac{x}{2} + \frac{y}{3} = 1$$

$$3x + 2y - 6 = 0$$

**93.**    $$\frac{x}{-\frac{1}{6}} + \frac{y}{-\frac{2}{3}} = 1$$

$$-6x - \frac{3}{2}y = 1$$

$$12x + 3y + 2 = 0$$

**95.** The line with slope $-3$ is steeper.

**97.** One way is to calculate the lengths of the sides.

$$d(A, B) = \sqrt{(2-2)^2 + (9-3)^2} = 6$$

$$d(B, C) = \sqrt{(7-2)^2 + (3-9)^2} = \sqrt{25+36} = \sqrt{61}$$

$$d(A, C) = \sqrt{(7-2)^2 + (3-3)^2} = 5$$

Then, $[d(A, B)]^2 + [d(A, C)]^2 = [d(B, C)]^2$, and the triangle is a right triangle.

Another way is to calculate the slopes of the lines joining $AB$ and $AC$.

slope of $AB$: undefined    (vertical line)

slope of $AC$: 0        (horizontal line)

**99.** Yes. $x + 20$

**101.** No. The term $x^{-1} = \frac{1}{x}$ causes the expression to not be a polynomial.

**103.** No. This expression is not defined for $x = \pm 3$.

**105.** $x^2 - 6x - 27 = (x - 9)(x + 3)$

**107.** $2x^2 + 11x - 40 = (2x - 5)(x + 8)$

# Section 1.3    Functions

■    Given a set or an equation, you should be able to determine if it represents a function.
■    Given a function, you should be able to do the following.
    (a) Find the domain.
    (b) Evaluate it at specific values.

**Solutions to Odd-Numbered Exercises**

**1.** Yes, it does represent a function. Each domain value is matched with only one range value.

**3.** No, it does not represent a function. The domain values are each matched with three range values.

**5.** Yes, it does represent a function. Each input value is matched with only one output value.

**7.** No, it does not represent a function. The input values of 10 and 7 are each matched with two output values.

**9.** (a) Each element of $A$ is matched with exactly one element of $B$, so it does represent a function.

   (b) The element 1 in $A$ is matched with two elements, $-2$ and 1 of $B$, so it does not represent a function.

   (c) Each element of $A$ is matched with exactly one element of $B$, so it does represent a function.

   (d) The element 2 of $A$ is not matched to any element of $B$, so it does not represent a function.

**11.** Each are functions. For each year there corresponds one and only one circulation.

**13.** $x^2 + y^2 = 4 \implies y = \pm\sqrt{4 - x^2}$

   Thus, $y$ *is not* a function of $x$. For instance, the values $y = 2$ and $-2$ both correspond to $x = 0$.

**15.** $x^2 + y = -1$

   $$y = -x^2 - 1$$

   Thus, $y$ *is* a function of $x$.

**17.** $2x + 3y = 4 \implies y = \frac{1}{3}(4 - 2x)$

   Thus, $y$ *is* a function of $x$.

**19.** $y^2 = x^2 - 1 \implies y = \pm\sqrt{x^2 - 1}$

   Thus, $y$ *is not* a function of $x$. For instance, the values $y = \sqrt{3}$ and $-\sqrt{3}$ both correspond to $x = 2$.

**21.** $y = |4 - x|$

   This is a function of $x$.

**23.** $x = -7$ does not represent $y$ as a function of $x$. All values of $y$ correspond to $x = -7$.

**25.** $f(x) = \dfrac{1}{x + 1}$

   (a) $f(4) = \dfrac{1}{(4) + 1} = \dfrac{1}{5}$

   (b) $f(0) = \dfrac{1}{(0) + 1} = 1$

   (c) $f(4t) = \dfrac{1}{(4t) + 1} = \dfrac{1}{4t + 1}$

   (d) $f(x + c) = \dfrac{1}{(x + c) + 1} = \dfrac{1}{x + c + 1}$

**27.** $f(x) = 2x - 3$

   (a) $f(1) = 2(1) - 3 = -1$

   (b) $f(-3) = 2(-3) - 3 = -9$

   (c) $f(x - 1) = 2(x - 1) - 3 = 2x - 5$

**29.** $h(t) = t^2 - 2t$

   (a) $h(2) = 2^2 - 2(2) = 0$

   (b) $h(1.5) = (1.5)^2 - 2(1.5) = -0.75$

   (c) $h(x + 2) = (x + 2)^2 - 2(x + 2) = x^2 + 2x$

**31.** $f(y) = 3 - \sqrt{y}$

   (a) $f(4) = 3 - \sqrt{4} = 1$

   (b) $f(0.25) = 3 - \sqrt{0.25} = 2.5$

   (c) $f(4x^2) = 3 - \sqrt{4x^2} = 3 - 2|x|$

**33.** $q(x) = \dfrac{1}{x^2 - 9}$

   (a) $q(0) = \dfrac{1}{0^2 - 9} = -\dfrac{1}{9}$

   (b) $q(3) = \dfrac{1}{3^2 - 9}$ is undefined.

   (c) $q(y + 3) = \dfrac{1}{(y + 3)^2 - 9} = \dfrac{1}{y^2 + 6y}$

**35.** $f(x) = \dfrac{|x|}{x}$

(a) $f(2) = \dfrac{|2|}{2} = 1$

(b) $f(-2) = \dfrac{|-2|}{-2} = -1$

(c) $f(x^2) = \dfrac{|x^2|}{x^2} = 1, \ x \neq 0$

**37.** $f(x) = \begin{cases} 2x + 1, & x < 0 \\ 2x + 2, & x \geq 0 \end{cases}$

(a) $f(-1) = 2(-1) + 1 = -1$

(b) $f(0) = 2(0) + 2 = 2$

(c) $f(2) = 2(2) + 2 = 6$

**39.** $h(t) = \frac{1}{2}|t + 3|$

| $t$ | $-5$ | $-4$ | $-3$ | $-2$ | $-1$ |
|---|---|---|---|---|---|
| $h(t)$ | 1 | $\frac{1}{2}$ | 0 | $\frac{1}{2}$ | 1 |

**41.** $f(x) = \begin{cases} -\frac{1}{2}x + 4, & x \leq 0 \\ (x - 2)^2, & x > 0 \end{cases}$

| $x$ | $-2$ | $-1$ | 0 | 1 | 2 |
|---|---|---|---|---|---|
| $f(x)$ | 5 | $\frac{9}{2}$ | 4 | 1 | 0 |

**43.** $f(x) = 15 - 3x = 0$

$\qquad 3x = 15$

$\qquad\ \ x = 5$

**45.** $f(x) = \dfrac{3x - 4}{5} = 0$

$\qquad 3x - 4 = 0$

$\qquad\quad\ 3x = 4$

$\qquad\quad\ \ x = \frac{4}{3}$

**47.** $\qquad\qquad f(x) = g(x)$

$\qquad\qquad\quad x^2 = x + 2$

$\qquad\quad x^2 - x - 2 = 0$

$\qquad (x + 1)(x - 2) = 0$

$\qquad x = -1 \ \text{ or } \ x = 2$

**49.** $f(x) = 5x^2 + 2x - 1$

Since $f(x)$ is a polynomial, the domain is all real numbers $x$.

**51.** $h(t) = \dfrac{4}{t}$

Domain: All real numbers except $t = 0$

**53.** $f(x) = \sqrt[3]{x - 4}$

Domain: all real numbers

**55.** $g(x) = \dfrac{1}{x} - \dfrac{3}{x + 2}$

Domain: All real numbers except $x = 0, \ x = -2$

**57.** $g(y) = \dfrac{y + 2}{\sqrt{y - 10}}$

$\quad y - 10 > 0$

$\qquad\quad y > 10$

Domain: all $y > 10$.

**59.** $f(x) = \sqrt{4 - x^2}$

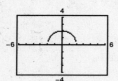

Domain: $[-2, 2]$

Range: $[0, 2]$

**61.** $g(x) = |2x + 3|$

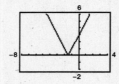

Domain: $(-\infty, \infty)$

Range: $[0, \infty)$

**63.** $f(x) = x^2$

$\{(-2, 4), (-1, 1), (0, 0), (1, 1), (2, 4)\}$

**65.** $f(x) = |x| + 2$

$\{(-2, 4), (-1, 3), (0, 2), (1, 3), (2, 4)\}$

**67.** $A = \pi r^2, \quad C = 2\pi r$

$r = \dfrac{C}{2\pi}$

$A = \pi\left(\dfrac{C}{2\pi}\right)^2 = \dfrac{C^2}{4\pi}$

**69.** (a) According to the table, the maximum profit is 3375 for $x = 150$

(b) 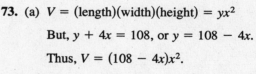    Yes, $P$ is a function of $x$.

(c) Profit = Revenue − Cost

= (price per unit)(number of units) − (cost)(number of units)

= $[90 - (x - 100)(0.15)]x - 60x$

= $(105 - 0.15x)x - 60x$

= $45x - 0.15x^2, \quad x > 100$

$P = \begin{cases} 30x, & x \le 100 \\ 45x - 0.15x^2, & x > 100 \end{cases}$

**71.** $A = \frac{1}{2}(\text{base})(\text{height}) = \frac{1}{2}xy.$

Since $(0, y), (2, 1)$ and $(x, 0)$ all lie on the same line, the slopes between any pair of points are equal.

$\dfrac{1 - y}{2 - 0} = \dfrac{1 - 0}{2 - x}$

$1 - y = \dfrac{2}{2 - x}$

$y = 1 - \dfrac{2}{2 - x} = \dfrac{x}{x - 2}$

Therefore, $A = \dfrac{1}{2}xy = \dfrac{1}{2}x\left(\dfrac{x}{x - 2}\right) = \dfrac{x^2}{2x - 4}$

The domain is $x > 2$, since $A > 0$.

**73.** (a) $V = (\text{length})(\text{width})(\text{height}) = yx^2$

But, $y + 4x = 108$, or $y = 108 - 4x$.

Thus, $V = (108 - 4x)x^2$.

(b) Since $y = 108 - 4x > 0$

$4x < 108$

$x < 27$

Domain: $0 < x < 27$

(c)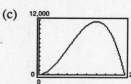

(d) The highest point on the graph occurs at $x = 18$. The dimensions that maximize the volume are $18 \times 18 \times 36$ inches.

**75.** The domain of $-1.97x + 26.3$ is $7 \le x \le 12$.

The domain of $0.505x^2 - 1.47x + 6.3$ is $1 \le x \le 6$.

You can tell by comparing the models to the given data. The models fit the data well on the domains above.

**77.** $f(11) = -1.97(11) + 26.3 = 4.63$

$4,630 in monthly revenue for November.

**79.** $n(t) = \begin{cases} -9.2t^2 + 84.5t + 575, & 0 \le t \le 4 \\ 26.8t + 657, & 5 \le t \le 10 \end{cases}$

$t = 0$ corresponds to 1990

| $t$ | 0 | 1 | 2 | 3 | 4 | 5 | 6 | 7 | 8 | 9 | 10 |
|---|---|---|---|---|---|---|---|---|---|---|---|
| $n(t)$ (in billions) | 575 | 650 | 707 | 746 | 766 | 791 | 818 | 845 | 871 | 898 | 925 |

**81.** (a) $F(y) = 149.76\sqrt{10}\,y^{5/2}$

| $y$ | 5 | 10 | 20 | 30 | 40 |
|---|---|---|---|---|---|
| $F(y)$ | $2.65 \times 10^4$ | $1.50 \times 10^5$ | $8.47 \times 10^5$ | $2.33 \times 10^6$ | $4.79 \times 10^6$ |

(Answers will vary.)

$F$ increases very rapidly as $y$ increases.

(b)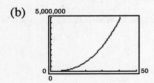

(c) From the table, $y \approx 22$ ft (slightly above 20). You could obtain a better approximation by completing the table for values of $y$ between 20 and 30.

(d) By graphing $F(y)$ together with the horizontal line $y_2 = 1,000,000$, you obtain $y \approx 21.37$ feet.

**83.** $f(x) = 2x$

$$\frac{f(x + c) - f(x)}{c} = \frac{2(x + c) - 2x}{c}$$

$$= \frac{2c}{c} = 2, \quad c \neq 0$$

**85.** $f(x) = x^2 - x + 1, \quad f(2) = 3$

$$\frac{f(2 + h) - f(2)}{h} = \frac{(2 + h)^2 - (2 + h) + 1 - 3}{h}$$

$$= \frac{4 + 4h + h^2 - 2 - h + 1 - 3}{h}$$

$$= \frac{h^2 + 3h}{h} = h + 3, \quad h \neq 0$$

**87.** $f(t) = \dfrac{1}{t}, \quad f(1) = 1$

$$\frac{f(t) - f(1)}{t - 1} = \frac{\dfrac{1}{t} - 1}{t - 1} = \frac{1 - t}{t(t - 1)} = \frac{-1}{t}, \quad t \neq 1$$

**89.** False. The range of $f(x)$ is $[-1, \infty)$.

**91.** Since the function is undefined at 0, we have $r(x) = \dfrac{c}{x}$. Since $(-4, -8)$ is on the graph, we have

$-8 = \dfrac{c}{-4} \implies c = 32$. Thus, $r(x) = \dfrac{32}{x}$.

**93.** The domain is the set of inputs of the function and the range is the set of corresponding outputs.

**95.** $12 - \dfrac{4}{x+2} = \dfrac{12(x+2) - 4}{x+2} = \dfrac{12x + 20}{x+2}$

**97.** $\dfrac{2x^3 + 11x^2 - 6x}{5x} \cdot \dfrac{x+10}{2x^2 + 5x - 3} = \dfrac{x(2x^2 + 11x - 6)(x+10)}{5x(2x-1)(x+3)}$

$\qquad\qquad\qquad\qquad\qquad = \dfrac{(2x-1)(x+6)(x+10)}{5(2x-1)(x+3)}$

$\qquad\qquad\qquad\qquad\qquad = \dfrac{(x+6)(x+10)}{5(x+3)}, x \neq 0, \dfrac{1}{2}$

# Section 1.4    Graphs of Functions

- You should be able to determine the domain and range of a function from its graph.
- You should be able to use the vertical line test for functions.
- You should be able to determine when a function is constant, increasing, or decreasing.
- You should be able to find relative maximum and minimum values of a function.
- You should know that $f$ is
  (a) Odd if $f(-x) = -f(x)$.
  (b) Even if $f(-x) = f(x)$.

**Solutions to Odd-Numbered Exercises**

**1.** $f(x) = 1 - x^2$

Domain:  All real numbers

Range:  $(-\infty, 1]$

$f(0) = 1$

**3.** $f(x) = \sqrt{16 - x^2}$

Domain:  $[-4, 4]$

Range:  $[0, 4]$

$f(0) = 4$

**5.** $f(x) = 2x^2 + 3$

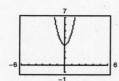

Domain: All real numbers

Range:  $[3, \infty)$

**7.** $f(x) = \sqrt{x - 1}$

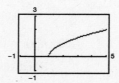

Domain: $x - 1 \geq 0 \implies x \geq 1$
or $[1, \infty)$

Range: $[0, \infty)$

**9.** $f(x) = |x + 3|$

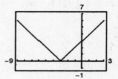

Domain:  All real numbers

Range:  $[0, \infty)$

**11.** $y = \frac{1}{2}x^2$

A vertical line intersects the graph just once, so $y$ is a function of $x$.

**13.** $x - y^2 = 1 \implies y = \pm\sqrt{x - 1}$

$y$ is not a function of $x$. Graph

$y_1 = \sqrt{x - 1}$ and $y_2 = -\sqrt{x - 1}$.

**15.** $x^2 = 2xy - 1$

A vertical line intersects the graph just once, so $y$ is a function of $x$. Solve for $y$ and graph

$$y = \frac{x^2 + 1}{2x}.$$

**17.** $f(x) = \frac{3}{2}x$

$f$ is increasing on $(-\infty, \infty)$.

**19.** $f(x) = x^3 - 3x^2 + 2$

$f$ is increasing on $(-\infty, 0)$ and $(2, \infty)$.

$f$ is decreasing on $(0, 2)$.

**21.** $f(x) = 3$

(a)

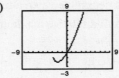

(b) $f$ is constant on $(-\infty, \infty)$

**23.** $f(x) = x^{2/3}$

(a)

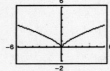

(b)  Increasing on $(0, \infty)$

Decreasing on $(-\infty, 0)$

**25.** $f(x) = x\sqrt{x + 3}$

(a)

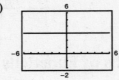

(b)  Increasing on $(-2, \infty)$

Decreasing on $(-3, -2)$

**27.** $f(x) = |x + 1| + |x - 1|$

(a)

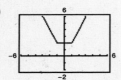

(b)  Increasing on $(1, \infty)$, constant on $(-1, 1)$, decreasing on $(-\infty, -1)$

**29.** $f(x) = x^2 - 6x$

Relative minimum: $(3, -9)$

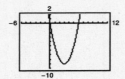

**31.** $y = 2x^3 + 3x^2 - 12x$

Relative minimum:  $(1, -7)$

Relative maximum:  $(-2, 20)$

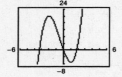

**33.** $h(x) = (x - 1)\sqrt{x}$

Relative minimum:  $(0.33, -0.38)$

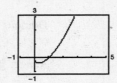

$(0, 0)$ is not a relative maximum because it occurs at the endpoint of the domain $[0, \infty)$.)

**35.** $f(x) = x^2 - 4x - 5$

(a)

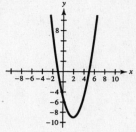

Minimum: $(2, -9)$

(b)

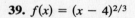

Minimum: $(2, -9)$

(c) Answers are the same

**37.** $f(x) = x^3 - 8x$

(a)

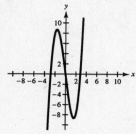

Maximum: Approximately $(-2, 9)$

Minimum: Approximately $(2, -9)$

(b)

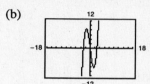

Maximum: $(-1.63, 8.71)$

Minimum: $(1.63, -8.71)$

(c) The answers are similar.

**39.** $f(x) = (x - 4)^{2/3}$

(a)

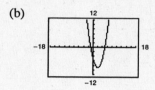

Minimum: $(4, 0)$

(b)

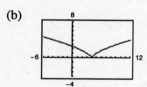

Minimum: $(4, 0)$

(c) The answers are the same.

**41.** $f(x) = \begin{cases} 2x + 3, & x < 0 \\ 3 - x, & x \geq 0 \end{cases}$

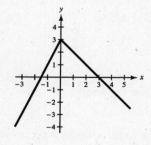

**43.** $f(x) = \begin{cases} \sqrt{x + 4}, & x < 0 \\ \sqrt{4 - x}, & x \geq 0 \end{cases}$

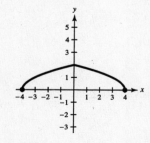

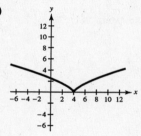

**45.** $f(x) = \begin{cases} x + 3, & x \le 0 \\ 3, & 0 < x \le 2 \\ 2x - 1, & x > 2 \end{cases}$

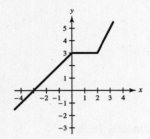

**47.** $f(x) = \begin{cases} 2x + 1, & x \le -1 \\ x^2 - 2, & x > -1 \end{cases}$

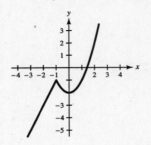

**49.** $f(-t) = (-t)^2 + 2(-t) - 3$

$= t^2 - 2t - 3$

$\ne f(t) \ne -f(t)$

$f$ is neither even nor odd.

**51.** $g(-x) = (-x)^3 - 5(-x)$

$= -x^3 + 5x$

$= -g(x)$

$g$ is odd.

**53.** $f(-x) = (-x)\sqrt{1 - (-x)^2} = -x\sqrt{1 - x^2} = -f(x)$

The function is odd.

**55.** $g(-s) = 4(-s)^{2/3} = 4s^{2/3} = g(s)$

The function is even.

**57.** $\left(-\frac{3}{2}, 4\right)$

(a) If $f$ is even, another point is $\left(\frac{3}{2}, 4\right)$.

(b) If $f$ is odd, another point is $\left(\frac{3}{2}, -4\right)$.

**59.** $(4, 9)$

(a) If $f$ is even, another point is $(-4, 9)$.

(b) If $f$ is odd, another point is $(-4, -9)$.

**61.** $(x, -y)$

(a) If $f$ is even, another point is $(-x, -y)$.

(b) If $f$ is odd, another point is $(-x, y)$.

**63.** $f(x) = 5$, even

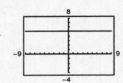

**65.** $f(x) = 3x - 2$, neither even nor odd

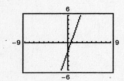

**67.** $h(x) = x^2 - 4$, even

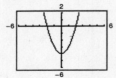

**69.** $f(x) = \sqrt{1 - x}$, neither even nor odd

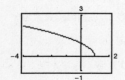

**71.** $f(x) = |x + 2|$, neither even nor odd

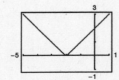

**73.** $f(x) = 4 - x \geq 0$

$\qquad 4 \geq x$

$(-\infty, 4]$

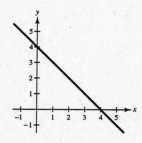

**75.** $f(x) = x^2 - 9 \geq 0$

$\qquad x^2 \geq 9$

$\qquad x \geq 3 \quad \text{or} \quad x \leq -3$

$\qquad [3, \infty) \quad \text{or} \quad (-\infty, -3]$

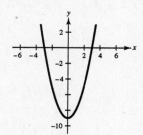

**77.** $s(x) = 2\left(\frac{1}{4}x - \left[\!\left[\frac{1}{4}x\right]\!\right]\right)$

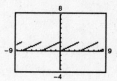

Domain: $(-\infty, \infty)$

Range: $[0, 2)$

Sawtooth pattern

**79.** (a) Let $x$ and $y$ be the length and width of the rectangle. Then $100 = 2x + 2y$ or $y = 50 - x$. Thus, the area is $A = xy = x(50 - x)$.

(b)
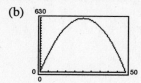

(c) The maximum area is 625 m² when $x = y = 25$ m. That is, the rectangle is a square.

**81.** (a) The second model is correct. For instance,

$$C_2\left(\tfrac{1}{2}\right) = 1.05 - 0.38\left[\!\left[-\left(\tfrac{1}{2} - 1\right)\right]\!\right]$$

$$= 1.05 - 0.38\left[\!\left[\tfrac{1}{2}\right]\!\right] = 1.05.$$

(b)
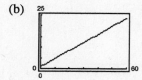

The cost of an 18-minute 45-second call is

$$C_2\left(18\tfrac{45}{60}\right)$$

$$= C_2(18.75) = 1.05 - 0.38[[-(18.75 - 1)]]$$

$$= 1.05 - 0.38[[-17.75]] = 1.05 - 0.38(-18)$$

$$= 1.05 + 0.38(18) = \$7.89$$

**83.** $h = \text{top} - \text{bottom}$

$\qquad = (-x^2 + 4x - 1) - 2$

$\qquad = -x^2 + 4x - 3, \ 1 \leq x \leq 3$

**85.** $L = \text{right} - \text{left}$

$\qquad = \tfrac{1}{2}y^2 - 0$

$\qquad = \tfrac{1}{2}y^2, \ 0 \leq y \leq 4$

**87.** $P = -0.76t^2 + 9.9t + 618, \quad 5 \leq t \leq 11$ where $t = 5$ corresponds to 1995

(a)
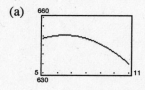

(b) Increasing on $(5, 6.5)$ or 1995 to middle of 1996

Decreases on $(6.5, 11)$ or middle of 1996 to 2001

(c) Maximum population was 650,200 for $t \approx 6.5$.

**89.** False. The domain of $f(x) = \sqrt{x^2}$ is the set of all real numbers.

**91.** $f(x) = a_{2n+1}x^{2n+1} + a_{2n-1}x^{2n-1} + \cdots + a_3x^3 + a_1x$

$f(-x) = a_{2n+1}(-x)^{2n+1} + a_{2n-1}(-x)^{2n-1} + \cdots + a_3(-x)^3 + a_1(-x)$

$\qquad = -a_{2n+1}x^{2n+1} - a_{2n-1}x^{2n-1} - \cdots - a_3x^3 - a_1x = -f(x)$

Therefore, $f(x)$ is odd.

**93.** $f$ is an even function.

(a) $g(x) = -f(x)$ is even because
$g(-x) = -f(-x) = -f(x) = g(x)$.

(b) $g(x) = f(-x)$ is even because
$g(-x) = f(-(-x)) = f(x) = f(-x) = g(x)$.

(c) $g(x) = f(x) - 2$ is even because
$g(-x) = f(-x) - 2 = f(x) - 2 = g(x)$.

(d) $g(x) = -f(x - 2)$ is neither even nor odd because
$g(-x) = -f(-x - 2) = -f(x + 2) \neq g(x)$ nor $-g(x)$.

**95.** No, $x^2 + y^2 = 25$ does not represent $x$ as a function of $y$. For instance, $(-3, 4)$ and $(3, 4)$ both lie on the graph.

**97.** $-2x^2 + 8x$

Terms: $-2x^2, 8x$

Coefficients: $-2, 8$

**99.** $\dfrac{x}{3} - 5x^2 + x^3$

Terms: $\dfrac{x}{3}, -5x^2, x^3$

Coefficients: $\dfrac{1}{3}, -5, 1$

**101.** (a) $d = \sqrt{(6 - (-2))^2 + (3 - 7)^2}$

$\qquad = \sqrt{64 + 16} = \sqrt{80} = 4\sqrt{5}$

(b) midpoint $= \left(\dfrac{-2 + 6}{2}, \dfrac{7 + 3}{2}\right) = (2, 5)$

**103.** (a) $d = \sqrt{\left(-\dfrac{3}{2} - \dfrac{5}{2}\right)^2 + (4 - (-1))^2}$

$\qquad = \sqrt{16 + 25} = \sqrt{41}$

(b) midpoint $= \left(\dfrac{\dfrac{5}{2} - \dfrac{3}{2}}{2}, \dfrac{-1 + 4}{2}\right) = \left(\dfrac{1}{2}, \dfrac{3}{2}\right)$

**105.** $f(x) = 5x - 1$

(a) $f(6) = 5(6) - 1 = 29$

(b) $f(-1) = 5(-1) - 1 = -6$

(c) $f(x - 3) = 5(x - 3) - 1 = 5x - 16$

**107.** $f(x) = x\sqrt{x - 3}$

(a) $f(3) = 3\sqrt{3 - 3} = 0$

(b) $f(12) = 12\sqrt{12 - 3}$

$\qquad = 12\sqrt{9} = 12(3) = 36$

(c) $f(6) = 6\sqrt{6 - 3} = 6\sqrt{3}$

**109.** $f(x) = x^2 - 2x + 9$

$f(3 + h) = (3 + h)^2 - 2(3 + h) + 9 = 9 + 6h + h^2 - 6 - 2h + 9$

$\qquad\qquad = h^2 + 4h + 12$

$f(3) = 3^2 - 2(3) + 9 = 12$

$\dfrac{f(3 + h) - f(3)}{h} = \dfrac{(h^2 + 4h + 12) - 12}{h} = \dfrac{h(h + 4)}{h} = h + 4, h \neq 0$

# Section 1.5    Shifting, Reflecting, and Stretching Graphs

■ You should know the graphs of the most commonly used functions in algebra, and be able to reproduce them on your graphing utility.

(a) Constant function: $f(x) = c$

(b) Identity function: $f(x) = x$

(c) Absolute value function: $f(x) = |x|$

(d) Square root function: $f(x) = \sqrt{x}$

(e) Squaring function: $f(x) = x^2$

(f) Cubing function: $f(x) = x^3$

■ You should know how the graph of a function is changed by vertical and horizontal shifts.

■ You should know how the graph of a function is changed by reflection.

■ You should know how the graph of a function is changed by nonrigid transformations, like stretches and shrinks.

■ You should know how the graph of a function is changed by a sequence of transformations.

## Solutions to Odd-Numbered Exercises

**1.**

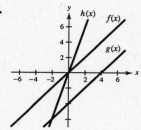

**3.**

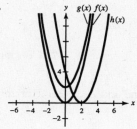

**5.**

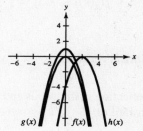

**7.**

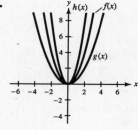

**9.**

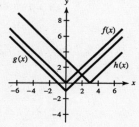

**11.**

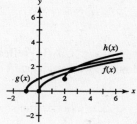

**13.** (a) $y = f(x) + 2$

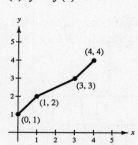

(b) $y = -f(x)$

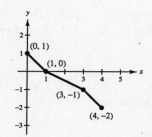

(c) $y = f(x - 2)$

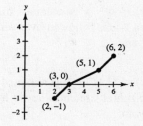

**—CONTINUED—**

**13. —CONTINUED—**

(d) $y = f(x + 3)$

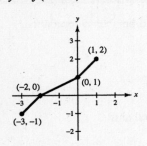

(e) $y = 2f(x)$

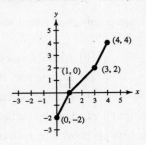

(f) $y = f(-x)$

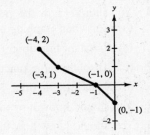

(g) Let $g(x)$ and $f\left(\frac{1}{2}x\right)$. Then from the graph,

$$g(0) = f\left(\tfrac{1}{2}(0)\right) = f(0) = -1$$

$$g(2) = f\left(\tfrac{1}{2}(2)\right) = f(1) = 0$$

$$g(6) = f\left(\tfrac{1}{2}(6)\right) = f(3) = 1$$

$$g(8) = f\left(\tfrac{1}{2}(8)\right) = f(4) = 2$$

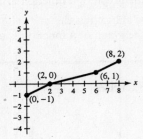

**15.** Vertical shrink of $y = x$ : $y = \frac{1}{2}x$

**17.** Constant function: $y = 7$

**19.** Reflection in the $x$-axis and a vertical shift one unit upward of $y = \sqrt{x}$: $y = 1 - \sqrt{x}$

**21.** Horizontal shift of $y = |x|$ : $y = |x + 2|$

**23.** Vertical shift one unit downward of $y = x^2$

$$y = x^2 - 1$$

**25.** Reflection in the $x$-axis and a vertical shift one unit upward of $y = x^3$: $y = 1 - x^3$

**27.** $y = -\sqrt{x} - 1$ is $f(x)$ reflected in the $x$-axis, followed by a vertical shift 1 unit downward.

**29.** $y = \sqrt{x - 2}$ is $f(x)$ shifted right two units.

**31.** $y = \sqrt{2x}$ is a horizontal shrink of $f(x)$ by 2.

**33.** $y = |x + 5|$ is $f(x)$ shifted left five units.

**35.** $y = -|x|$ is $f(x)$ reflected in the $x$-axis.

**37.** $y = 4|x|$ is a vertical stretch of $f(x)$.

**39.** $g(x) = 4 - x^3$ is obtained from $f(x)$ by a reflection in the $x$-axis followed by a vertical shift upward of four units.

**41.** $h(x) = \frac{1}{4}(x + 2)^3$ is obtained from $f(x)$ by a left shift of two units and a vertical shrink by a factor of $\frac{1}{4}$.

**43.** $p(x) = \left(\frac{1}{3}x\right)^3 + 2$ is obtained from $f(x)$ by a horizontal stretch followed by a vertical shift 2 units upward.

**45.** $f(x) = x^3 - 3x^2$

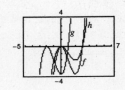

$g(x) = f(x + 2) = (x + 2)^3 - 3(x + 2)^2$ is a horizontal shift 2 units to left

$h(x) = \frac{1}{2}f(x) = \frac{1}{2}(x^3 - 3x^2)$ is a vertical shrink.

**47.** $f(x) = x^3 - 3x^2$

$g(x) = -\frac{1}{3}f(x) = -\frac{1}{3}(x^3 - 3x^2)$    reflection in the *x*-axis and vertical shrink

$h(x) = f(-x) = (-x)^3 - 3(-x)^2$    reflection in the *y*-axis

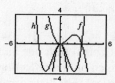

**49.** The graph of *g* is obtained from that of *f* by first negating *f*, and then shifting vertically one unit upward: $g(x) = -x^3 + 3x^2 + 1$.

**51.** (a) $f(x) = x^2$

(b) $g(x) = 2 - (x + 5)^2$ is obtained from *f* by a horizontal shift to the left 5 units, a reflection in the *x*-axis, and a vertical shift upward 2 units.

(c)

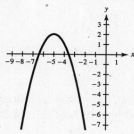

(d) $g(x) = 2 - f(x + 5)$

**53.** (a) $f(x) = x^2$

(b) $g(x) = 3 + 2(x - 4)^2$ is obtained from *f* by a horizontal shift 4 units to the right, a vertical stretch of 2, and a vertical shift upward 3 units.

(c)

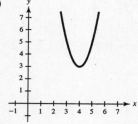

(d) $g(x) = 3 + 2f(x - 4)$

**55.** (a) $f(x) = x^3$

(b) $g(x) = 3(x - 2)^3$ is obtained from *f* by a horizontal shift 2 units to the right followed by a vertical stretch of 3.

(c)

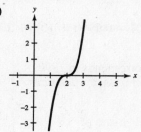

(d) $g(x) = 3f(x - 2)$

**57.** (a) $f(x) = x^3$

(b) $g(x) = (x - 1)^3 + 2$ is obtained from *f* by a horizontal shift 1 unit to the right, and a vertical shift upward 2 units.

(c)

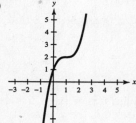

(d) $g(x) = f(x - 1) + 2$

**59.** (a) $f(x) = |x|$

(b) $g(x) = |x + 4| + 8$ is obtained from *f* by a horizontal shift 4 units to the left, followed by a vertical shift 8 units upward.

(c)

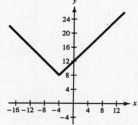

(d) $g(x) = f(x + 4) + 8$

**61.** (a) $f(x) = |x|$

(b) $g(x) = -2|x - 1| - 4$ is obtained from $f$ by a horizontal shift one unit to the right, a vertical stretch of 2, a reflection in the $x$-axis, and a vertical shift downward 4 units.

(c)

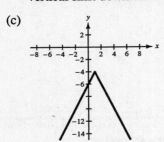

(d) $g(x) = -2f(x - 1) - 4$

**63.** (a) $f(x) = \sqrt{x}$

(b) $g(x) = -\frac{1}{2}\sqrt{x + 3} - 1$ is obtained from $f$ by a horizontal shift 3 units to the left, a vertical shrink, a reflection in the $x$-axis, and a vertical shift 1 unit downward.

(c)

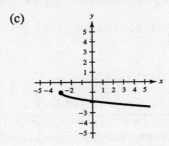

(d) $g(x) = -\frac{1}{2}f(x + 3) - 1$

**65.** (a) $P(x) = 80 + 20x - 0.5x^2, \quad 0 \le x \le 20$

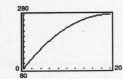

(b) $P(x)$ is shifted downward by a vertical shift of 25.

$P(x) = 55 + 20x - 0.5x^2, \quad 0 \le x \le 20$

(c) $P(x)$ is changed by a horizontal stretch.

$$P(x) = 80 + 20\left(\frac{x}{100}\right) - 0.5\left(\frac{x}{100}\right)^2$$

$$= 80 + 0.2x - 0.00005x^2$$

**67.** $F(t) = 0.036t^2 + 20.1, \quad 0 \le t \le 20$

$t = 0$ corresponds to 1980.

(a) $F$ is obtained from $f(t) = t^2$ by a vertical shrink of 0.036 followed by a vertical shift of 20.1 units upward

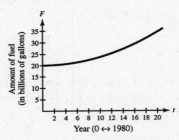

(b) $G(t) = F(t + 10) = 0.036(t + 10)^2 + 20.1$
$-10 \le t \le 10$

$G(0) = F(10)$ corresponds to 1990.

**69.** True. $|x| = |-x|$ implies $f(x) = |x| - 5 = |-x| - 5 = g(x)$

**71.** (a)

(b)

(c)

(d)

(e)

(f)

All the graphs pass through the origin. The graphs of the odd powers of $x$ are symmetric to the origin and the graphs of the even powers are symmetric to the $y$-axis. As the powers increase, the graphs become flatter in the interval $-1 < x < 1$.

**73.** slope $L_1$: $\dfrac{10 + 2}{2 + 2} = 3$

slope $L_2$: $\dfrac{9 - 3}{3 + 1} = \dfrac{3}{2}$

Neither parallel nor perpendicular

**77.** Domain: $100 - x^2 \geq 0 \implies x^2 \leq 100 \implies -10 \leq x \leq 10$

**75.** Domain: All $x \neq 9$

# Section 1.6    Combinations of Functions

> ■ Given two functions, $f$ and $g$, you should be able to form the following functions (if defined):
> 1. Sum: $(f + g)(x) = f(x) + g(x)$
> 2. Difference: $(f - g)(x) = f(x) - g(x)$
> 3. Product: $(fg)(x) = f(x)g(x)$
> 4. Quotient: $(f/g)(x) = f(x)/g(x), \, g(x) \neq 0$
> 5. Composition of $f$ with $g$: $(f \circ g)(x) = f(g(x))$
> 6. Composition of $g$ with $f$: $(g \circ f)(x) = g(f(x))$

**Solutions to Odd-Numbered Exercises**

**1.**

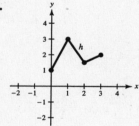

**3.**

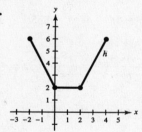

**5.** $f(x) = x + 3, \quad g(x) = x - 3$

   (a) $(f + g)(x) = f(x) + g(x) = (x + 3) + (x - 3) = 2x$

   (b) $(f - g)(x) = f(x) - g(x) = (x + 3) - (x - 3) = 6$

   (c) $(fg)(x) = f(x)g(x) = (x + 3)(x - 3) = x^2 - 9$

   (d) $\left(\dfrac{f}{g}\right)(x) = \dfrac{f(x)}{g(x)} = \dfrac{x + 3}{x - 3}, \quad x \neq 3$

      Domain: all $x \neq 3$

**7.** $f(x) = x^2, g(x) = 1 - x$

   (a) $(f + g)(x) = f(x) + g(x) = x^2 + (1 - x) = x^2 - x + 1$

   (b) $(f - g)(x) = f(x) - g(x) = x^2 - (1 - x) = x^2 + x - 1$

   (c) $(fg)(x) = f(x) \cdot g(x) = x^2(1 - x) = x^2 - x^3$

   (d) $\left(\dfrac{f}{g}\right)(x) = \dfrac{f(x)}{g(x)} = \dfrac{x^2}{1 - x}, x \neq 1$

      Domain: all $x \neq 1$.

**9.** $f(x) = x^2 + 5, g(x) = \sqrt{1 - x}$

(a) $(f + g)(x) = f(x) + g(x) = (x^2 + 5) + \sqrt{1 - x}$

(b) $(f - g)(x) = f(x) - g(x) = (x^2 + 5) - \sqrt{1 - x}$

(c) $(fg)(x) = f(x) \cdot g(x) = (x^2 + 5)\sqrt{1 - x}$

(d) $\left(\dfrac{f}{g}\right)(x) = \dfrac{f(x)}{g(x)} = \dfrac{x^2 + 5}{\sqrt{1 - x}}, x < 1$

Domain: $x < 1$.

**11.** $f(x) = \dfrac{1}{x}, g(x) = \dfrac{1}{x^2}$

(a) $(f + g)(x) = f(x) + g(x) = \dfrac{1}{x} + \dfrac{1}{x^2} = \dfrac{x + 1}{x^2}$

(b) $(f - g)(x) = f(x) - g(x) = \dfrac{1}{x} - \dfrac{1}{x^2} = \dfrac{x - 1}{x^2}$

(c) $(fg)(x) = f(x) \cdot g(x) = \dfrac{1}{x}\left(\dfrac{1}{x^2}\right) = \dfrac{1}{x^3}$

(d) $\left(\dfrac{f}{g}\right)(x) = \dfrac{f(x)}{g(x)} = \dfrac{1/x}{1/x^2} = \dfrac{x^2}{x} = x, x \neq 0$

Domain: $x \neq 0$.

**13.** $(f + g)(3) = f(3) + g(3) = (3^2 + 1) + (3 - 4) = 9$

**15.** $(f - g)(0) = f(0) - g(0) = [0^2 + 1] - (0 - 4) = 5$

**17.** $(fg)(4) = f(4)g(4) = (4^2 + 1)(4 - 4) = 0$

**19.** $\left(\dfrac{f}{g}\right)(-5) = \dfrac{f(-5)}{g(-5)} = \dfrac{(-5)^2 + 1}{(-5 - 4)} = \dfrac{26}{-9} = -\dfrac{26}{9}$

**21.** $(f - g)(2t) = f(2t) - g(2t) = [(2t)^2 + 1] - (2t - 4) = 4t^2 - 2t + 5$

**23.** $(fg)(-5t) = f(-5t)g(-5t) = [(-5t)^2 + 1][(-5t) - 4]$

$= (25t^2 + 1)(-5t - 4) = -125t^3 - 100t^2 - 5t - 4$

**25.** $\left(\dfrac{f}{g}\right)(-t) = \dfrac{f(-t)}{g(-t)} = \dfrac{(-t)^2 + 1}{-t - 4} = \dfrac{t^2 + 1}{-t - 4}, t \neq -4$

**27.** $f(x) = \frac{1}{2}x, g(x) = x - 1, (f + g)(x) = \frac{3}{2}x - 1$

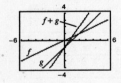

**29.** $f(x) = x^2, g(x) = -2x, (f + g)(x) = x^2 - 2x$

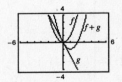

**31.** $f(x) = 3x, g(x) = -\dfrac{x^3}{10}, (f + g)(x) = 3x - \dfrac{x^3}{10}$

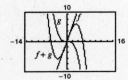

For $0 \leq x \leq 2, f(x)$ contributes more to the magnitude.

For $x > 6, g(x)$ contributes more to the magnitude.

**33.** $f(x) = 3x + 2, g(x) = -\sqrt{x + 5},$

$(f + g)(x) = 3x + 2 - \sqrt{x + 5}$

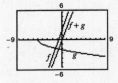

$f(x) = 3x + 2$ contributes more to the magnitude in both intervals.

**35.** $f(x) = x^2, g(x) = x - 1$

    (a) $(f \circ g)(x) = f(g(x)) = f(x - 1) = (x - 1)^2$         (c) $(f \circ g)(0) = (0 - 1)^2 = 1$

    (b) $(g \circ f)(x) = g(f(x)) = g(x^2) = x^2 - 1$

**37.** $f(x) = 3x + 5, g(x) = 5 - x$

    (a) $(f \circ g)(x) = f(g(x)) = f(5 - x) = 3(5 - x) + 5 = 20 - 3x$

    (b) $(g \circ f)(x) = g(f(x)) = g(3x + 5) = 5 - (3x + 5) = -3x$

    (c) $(f \circ g)(0) = 20$

**39.** (a) $(f \circ g)(x) = f(g(x)) = f(x^2) = \sqrt{x^2 + 4}$     (b) 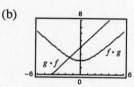     They are not equal.

        $(g \circ f)(x) = g(f(x)) = g\left(\sqrt{x + 4}\right) = \left(\sqrt{x + 4}\right)^2$

                $= x + 4, \; x \geq -4$

**41.** (a) $(f \circ g)(x) = f(g(x)) = f(3x + 1)$     (b) 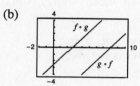     They are not equal.

            $= \frac{1}{3}(3x + 1) - 3 = x - \frac{8}{3}$

    (b) $(g \circ f)(x) = g(f(x)) = g\left(\frac{1}{3}x - 3\right) \circ$

            $= 3\left(\frac{1}{3}x - 3\right) + 1 = x - 8$

**43.** (a) $(f \circ g)(x) = f(g(x)) = f(x^6) = (x^6)^{2/3} = x^4$     (b) 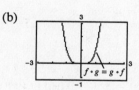     They are equal.

        $(g \circ f)(x) = g(f(x)) = g(x^{2/3}) = (x^{2/3})^6 = x^4$

**45.** (a) $(f \circ g)(x) = f(g(x)) = f(4 - x) = 5(4 - x) + 4 = 24 - 5x$

        $(g \circ f)(x) = g(f(x)) = g(5x + 4) = 4 - (5x + 4) = -5x$

    (b) No, $(f \circ g)(x) \neq (g \circ f)(x)$ because $24 - 5x \neq -5x$.

(c)

| $x$ | $f(g(x))$ | $g(f(x))$ |
|-----|-----------|-----------|
| 0   | 24        | 0         |
| 1   | 19        | -5        |
| 2   | 14        | -10       |
| 3   | 9         | -15       |

**47.** (a) $(f \circ g)(x) = f(g(x)) = f(x^2 - 5) = \sqrt{(x^2 - 5) + 6} = \sqrt{x^2 + 1}$

        $(g \circ f)(x) = g(f(x)) = g\left(\sqrt{x + 6}\right) = \left(\sqrt{x + 6}\right)^2 - 5 = (x + 6) - 5 = x + 1, \; x \geq -6$

    (b) No, $(f \circ g)(x) \neq (g \circ f)(x)$ because $\sqrt{x^2 + 1} \neq x + 1$.

    (c)

| $x$  | $f(g(x))$    | $g(f(x))$ |
|------|--------------|-----------|
| 0    | 1            | 1         |
| -2   | $\sqrt{5}$   | -1        |
| 3    | $\sqrt{10}$  | 4         |

**49.** (a) $(f \circ g)(x) = f(g(x)) = f(2x - 1) = |(2x - 1) + 3| = |2x + 2| = 2|x + 1|$

       $(g \circ f)(x) = g(f(x)) = g(|x + 3|) = 2|x + 3| - 1$

   (b) No, $(f \circ g)(x) \neq (g \circ f)(x)$ because $2|x + 1| \neq 2|x + 3| - 1$.

   (c)

| $x$ | $f(g(x))$ | $g(f(x))$ |
|---|---|---|
| $-1$ | 0 | 3 |
| 0 | 2 | 5 |
| 1 | 4 | 7 |

**51.** (a) $(f + g)(3) = f(3) + g(3) = 2 + 1 = 3$

   (b) $\left(\dfrac{f}{g}\right)(2) = \dfrac{f(2)}{g(2)} = \dfrac{0}{2} = 0$

**53.** (a) $(f \circ g)(2) = f(g(2)) = f(2) = 0$

   (b) $(g \circ f)(2) = g(f(2)) = g(0) = 4$

**55.** Let $f(x) = x^2$ and $g(x) = 2x + 1$, then $(f \circ g)(x) = h(x)$. This is not a unique solution. For example, if $f(x) = (x + 1)^2$ and $g(x) = 2x$, then $(f \circ g)(x) = h(x)$ as well.

**57.** Let $f(x) = \sqrt[3]{x}$ and $g(x) = x^2 - 4$, then $(f \circ g)(x) = h(x)$. This answer is not unique. Other possibilities may be:

     $f(x) = \sqrt[3]{x - 4}$ and $g(x) = x^2$ or

     $f(x) = \sqrt[3]{-x}$ and $g(x) = 4 - x^2$ or

     $f(x) = \sqrt[9]{x}$ and $g(x) = (x^2 - 4)^3$

**59.** Let $f(x) = 1/x$ and $g(x) = x + 2$, then $(f \circ g)(x) = h(x)$. Again, this is not a unique solution. Other possibilities may be:

     $f(x) = \dfrac{1}{x + 2}$ and $g(x) = x$

     or $f(x) = \dfrac{1}{x + 1}$ and $g(x) = x + 1$

**61.** Let $f(x) = x^2 + 2x$ and $g(x) = x + 4$. Then $(f \circ g)(x) = h(x)$. (Answer is not unique.)

**63.** (a) The domain of $f(x) = \sqrt{x + 4}$ is $x + 4 \geq 0$ or $x \geq -4$

   (b) The domain of $g(x) = x^2$ is all real numbers.

   (c) $(f \circ g)(x) = f(g(x)) = f(x^2) = \sqrt{x^2 + 4}$.

     The domain of $(f \circ g)$ is all real numbers.

**65.** (a) The domain of $f(x) = x^2 + 1$ is all real numbers.

   (b) The domain of $g(x) = \sqrt{x}$ is all $x \geq 0$

   (c) $(f \circ g)(x) = f(g(x)) = f(\sqrt{x})$

               $= (\sqrt{x})^2 + 1 = x + 1, \quad x \geq 0$

     The domain of $f \circ g$ is $x \geq 0$

**67.** (a) The domain of $f(x) = \dfrac{1}{x}$ is all $x \neq 0$.

   (b) The domain of $g(x) = x + 3$ is all real numbers.

   (c) The domain of $(f \circ g)(x) = f(x + 3) = \dfrac{1}{x + 3}$ is all $x \neq -3$.

**69.** (a) The domain of $f(x) = |x - 4|$ is all real numbers.

   (b) The domain of $g(x) = 3 - x$ is all real numbers.

   (c) $(f \circ g)(x) = f(g(x)) = f(3 - x) = |(3 - x) - 4| = |-x - 1| = |x + 1|$

     Domain: all real numbers

**71.** (a) The domain of $f(x) = x + 2$ is all real numbers.

(b) The domain of $g(x) = \dfrac{1}{x^2 - 4}$ is all $x \neq \pm 2$

(c) $(f \circ g)(x) = f(g(x)) = f\left(\dfrac{1}{x^2 - 4}\right) = \dfrac{1}{x^2 - 4} + 2$

Domain: $x \neq \pm 2$

**73.** (a) $T(x) = R(x) + B(x) = \frac{3}{4}x + \frac{1}{15}x^2$

(b)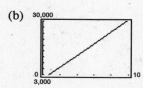

(c) $B(x)$ contributes more to $T(x)$ at higher speeds.

**75.** Let $t = 4$ represent 1994

$y_1 = 8.93t + 103.0$

$y_2 = 1.886t^2 - 5.24t + 305.7$

$y_3 = -0.361t^2 + 7.97t + 14.2$

| Year | 1994 | 1995 | 1996 | 1997 | 1998 | 1999 | 2000 |
|------|------|------|------|------|------|------|------|
| $y_1$ | 138.7 | 147.7 | 156.6 | 165.5 | 174.4 | 183.4 | 192.3 |
| $y_2$ | 314.9 | 326.7 | 342.2 | 361.4 | 384.5 | 411.3 | 441.9 |
| $y_3$ | 40.3 | 45.0 | 49.0 | 52.3 | 54.9 | 56.7 | 57.8 |

**77.** $(A \circ r)(t)$ gives the area of the circle as a function of time.

$(A \circ r)(t) = A(r(t))$

$= A(0.6t)$

$= \pi(0.6t)^2 = 0.36\pi t^2$

**79.** (a) $(C \circ x)(t) = C(x(t))$

$= 60(50t) + 750$

$= 3000t + 750$

$(C \circ x)(t)$ represents the cost after $t$ production hours.

(b) 30,000 / 0 / 3,000 / 10

*decrease y-axis scale to obtain greater accuracy in t*

The cost increases to \$15,000 when $t = 4.75$ hours.

**81.** $g(f(x)) = g(x - 500{,}000) = 0.03(x - 500{,}000)$ represents 3 percent of the amount over \$500,000.

**83.** False. $(f \circ g)(x) = f(6x) = 6x + 1$, but $(g \circ f)(x) = g(x + 1) = 6(x + 1)$

**85.** Let $f(x)$ and $g(x)$ be odd functions, and define $h(x) = f(x)g(x)$. Then,

$h(-x) = f(-x)g(-x)$

$\quad = [-f(x)][-g(x)]$ since $f$ and $g$ are both odd

$\quad = f(x)g(x) = h(x).$

Thus, $h$ is even.

Let $f(x)$ and $g(x)$ be even functions, and define $h(x) = f(x)g(x)$. Then,

$h(-x) = f(-x)g(-x)$

$\quad = f(x)g(x)$ since $f$ and $g$ are both even

$\quad = h(x).$

Thus, $h$ is even.

**87.** $g(-x) = \frac{1}{2}[f(-x) + f(-(-x))] = \frac{1}{2}[f(-x) + f(x)] = g(x),$

which shows that $g$ is even.

$h(-x) = \frac{1}{2}[f(-x) - f(-(-x))] = \frac{1}{2}[f(-x) - f(x)]$

$\qquad = -\frac{1}{2}[f(x) - f(-x)] = -h(x),$

which shows that $h$ is odd.

**89.** $(0, -5), (1, -5), (2, -7)$

(other answers possible)

**91.** $\left(\sqrt{24}, 0\right), \left(-\sqrt{24}, 0\right), \left(0, \sqrt{24}\right)$

(other answers possible)

**93.** $y - (-2) = \dfrac{8 - (-2)}{-3 - (-4)}(x - (-4))$

$y + 2 = 10(x + 4)$

$y - 10x - 38 = 0$

**95.** $y - (-1) = \dfrac{4 - (-1)}{-\frac{1}{3} - \frac{3}{2}}\left(x - \frac{3}{2}\right)$

$y + 1 = \dfrac{5}{-\frac{11}{6}}\left(x - \frac{3}{2}\right) = -\frac{30}{11}\left(x - \frac{3}{2}\right)$

$11y + 11 = -30x + 45$

$30x + 11y - 34 = 0$

**97.** Figure shifts 4 units to the right

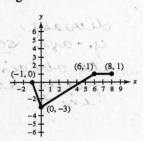

**99.** Figure shifts 4 units upward

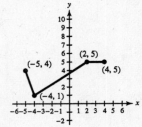

**101.** Vertical stretch by 2

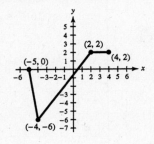

## Section 1.7    Inverse Functions

- Two functions $f$ and $g$ are inverses of each other if $f(g(x)) = x$ for every $x$ in the domain of $g$ and $g(f(x)) = x$ for every $x$ in the domain of $f$.

- Be able to find the inverse of a function, if it exists.

  1. Replace $f(x)$ with $y$.

  2. Interchange $x$ and $y$.

  3. Solve for $y$. If this equation represents $y$ as a function of $x$, then you have found $f^{-1}(x)$. If this equation does not represent $y$ as a function of $x$, then $f$ does not have an inverse function.

- A function $f$ has an inverse function if and only if no **horizontal** line crosses the graph of $f$ at more than one point.

- A function $f$ has an inverse function if and only if $f$ is one-to-one.

**Solutions to Odd-Numbered Exercises**

**1.**    $f(x) = 6x$

$f^{-1}(x) = \frac{1}{6}x$

$f(f^{-1}(x)) = f(\frac{1}{6}x) = 6(\frac{1}{6}x) = x$

$f^{-1}(f(x)) = f^{-1}(6x) = \frac{1}{6}(6x) = x$

**3.**    $f(x) = x + 7$

$f^{-1}(x) = x - 7$

$f(f^{-1}(x)) = f(x - 7) = (x - 7) + 7 = x$

$f^{-1}(f(x)) = f^{-1}(x + 7) = (x + 7) - 7 = x$

**5.**  $f^{-1}(x) = \dfrac{x - 1}{2}$

$f(f^{-1}(x)) = f\left(\dfrac{x - 1}{2}\right) = 2\left(\dfrac{x - 1}{2}\right) + 1 = (x - 1) + 1 = x$

$f^{-1}(f(x)) = f^{-1}(2x + 1) = \dfrac{(2x + 1) - 1}{2} = \dfrac{2x}{2} = x$

**7.** $f^{-1}(x) = x^3$

$f(f^{-1}(x)) = f(x^3) = \sqrt[3]{x^3} = x$

$f^{-1}(f(x)) = f^{-1}(\sqrt[3]{x}) = (\sqrt[3]{x})^3 = x$

**9.** (a) $f(g(x)) = f\left(-\dfrac{2x + 6}{7}\right) = -\dfrac{7}{2}\left(-\dfrac{2x + 6}{7}\right) - 3 = \dfrac{2x + 6}{2} - 3 = (x + 3) - 3 = x$

$g(f(x)) = g\left(-\dfrac{7}{2}x - 3\right) = -\dfrac{2(-\frac{7}{2}x - 3) + 6}{7} = -\dfrac{-7x - 6 + 6}{7} = \dfrac{7x}{7} = x$

(b)

| $x$ | 2 | 0 | $-2$ | $-4$ | $-6$ |
|---|---|---|---|---|---|
| $f(x)$ | $-10$ | $-3$ | 4 | 11 | 18 |

| $x$ | $-10$ | $-3$ | 4 | 11 | 18 |
|---|---|---|---|---|---|
| $g(x)$ | 2 | 0 | $-2$ | $-4$ | $-6$ |

Note that the entries in the tables are the same except that the rows are interchanged.

**11.** (a) $f(g(x)) = f(\sqrt[3]{x - 5}) = \left[\sqrt[3]{x - 5}\right]^3 + 5 = (x - 5) + 5 = x$

$g(f(x)) = g(x^3 + 5) = \sqrt[3]{(x^3 + 5) - 5} = \sqrt[3]{x^3} = x$

(b)

| $x$ | $-3$ | $-2$ | $-1$ | 0 | 1 |
|---|---|---|---|---|---|
| $f(x)$ | $-22$ | $-3$ | 4 | 5 | 6 |

| $x$ | $-22$ | $-3$ | 4 | 5 | 6 |
|---|---|---|---|---|---|
| $g(x)$ | $-3$ | $-2$ | $-1$ | 0 | 1 |

Note that the entries in the tables are the same except that the rows are interchanged.

**13.** (a) $f(g(x)) = f(8 + x^2) = -\sqrt{(8 + x^2) - 8} = -\sqrt{x^2} = -(-x) = x$   $x \le 0$

    [Since $x \le 0$, $\sqrt{x^2} = -x$]    *Restrict domain*

    $g(f(x)) = g(-\sqrt{x - 8}) = 8 + [(-)\sqrt{x - 8}]^2 = 8 + ((x) - 8) = x$

(b)

| $x$ | 8 | 9 | 12 | 17 | 24 |
|---|---|---|---|---|---|
| $f(x)$ | 0 | $-1$ | $-2$ | $-3$ | $-4$ |

| $x$ | 0 | $-1$ | $-2$ | $-3$ | $-4$ |
|---|---|---|---|---|---|
| $g(x)$ | 8 | 9 | 12 | 17 | 24 |

Note that the entries in the tables are the same except that the rows are interchanged.

**15.** (a) $f(g(x)) = f(\sqrt[3]{x}) = (\sqrt[3]{x})^3 = x$

    $g(f(x)) = g(x^3) = \sqrt[3]{x^3} = x$

(b)

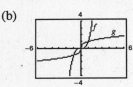

Reflections in the line $y = x$

**17.** (a) $f(g(x)) = f(x^2 + 4)$, $x \ge 0$

           $= \sqrt{(x^2 + 4) - 4} = x$

    $g(f(x)) = g(\sqrt{x - 4})$

           $= (\sqrt{x - 4})^2 + 4 = x$

(b)

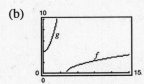

Reflections in the line $y = x$

**19.** (a) $f(g(x)) = f(\sqrt[3]{1 - x}) = 1 - (\sqrt[3]{1 - x}))^3 = 1 - (1 - x) = x$

    $g(f(x)) = g(1 - x^3) = \sqrt[3]{1 - (1 - x^3)} = \sqrt[3]{x^3} = x$

(b)              Reflections in the line $y = x$

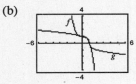

**21.** The inverse is a line through $(-1, 0)$.

    Matches graph (c).

**23.** The inverse is half a parabola starting at $(1, 0)$.

    Matches graph (a).

**25.** $f(x) = 2x$,   $g(x) = \dfrac{x}{2}$.

(a)

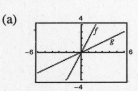

Reflection in the line $y = x$

(b)

| $x$ | $-2$ | $-1$ | 0 | 1 | 2 |
|---|---|---|---|---|---|
| $f(x)$ | $-4$ | $-2$ | 0 | 2 | 4 |

| $x$ | $-4$ | $-2$ | 0 | 2 | 4 |
|---|---|---|---|---|---|
| $g(x)$ | $-2$ | $-1$ | 0 | 1 | 2 |

The entries in the tables are the same, except that the rows are interchanged.

**27.** $f(x) = \dfrac{x-1}{x+5},$   $g(x) = -\dfrac{5x+1}{x-1} = \dfrac{5x+1}{1-x}$

(a)

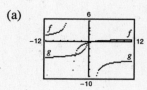

Reflection in the line $y = x$

(b)

| $x$ | $-2$ | $-1$ | $0$ | $3$ | $5$ |
|-----|------|------|-----|-----|-----|
| $f(x)$ | $-1$ | $-\frac{1}{2}$ | $-\frac{1}{5}$ | $\frac{1}{4}$ | $\frac{2}{5}$ |

| $x$ | $-1$ | $-\frac{1}{2}$ | $-\frac{1}{5}$ | $\frac{1}{4}$ | $\frac{2}{5}$ |
|-----|------|------|-----|-----|-----|
| $g(x)$ | $-2$ | $-1$ | $0$ | $3$ | $5$ |

The entries in the tables are the same, except that the rows are interchanged.

**29.** $f(x) = 3 - \dfrac{1}{2}x$

$f$ is one-to-one because a horizontal line will intersect the graph at most once.

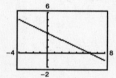

**31.** $h(x) = \dfrac{x^2}{x^2 + 1}$

$h$ is not one-to-one because some horizontal lines intersect the graph twice.

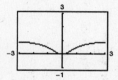

**33.** $h(x) = \sqrt{16 - x^2}$

$h$ is not one-to-one because some horizontal lines intersect the graph twice.

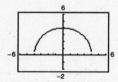

**35.** $f(x) = 10$

$f$ is not one-to-one because the horizontal line $y = 10$ intersects the graph at every point on the graph.

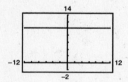

**37.** $g(x) = (x + 5)^3$

$g$ is one-to-one because a horizontal line will intersect the graph at most once.

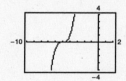

**39.** $h(x) = |x + 4| - |x - 4|$

$h$ is not one-to-one because some horizontal lines intersect the graph more than once.

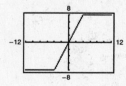

**41.** $f(x) = x^4$

$y = x^4$

$x = y^4$

$y = \pm\sqrt[4]{x}$

$f$ is not one-to-one.

This does not represent $y$ as a function of $x$. $f$ does not have an inverse.

**43.** $f(x) = \dfrac{3x + 4}{5}$

$y = \dfrac{3x + 4}{5}$

$x = \dfrac{3y + 4}{5}$

$5x = 3y + 4$

$5x - 4 = 3y$

$(5x - 4)/3 = y$

$f^{-1}(x) = \dfrac{5x - 4}{3}$

$f$ is one-to-one and has an inverse.

**45.** $f(x) = \dfrac{1}{x^2}$ is not one-to-one, and does not have an inverse. For example, $f(1) = f(-1) = 1$.

**47.** $f(x) = (x + 3)^2,\ x \geq -3,\ y \geq 0$

$y = (x + 3)^2,\ x \geq -3,\ y \geq 0$

$x = (y + 3)^2,\ y \geq -3,\ x \geq 0$

$\sqrt{x} = y + 3,\ y \geq -3,\ x \geq 0$

$y = \sqrt{x} - 3,\ x \geq 0,\ y \geq -3$

$f$ is one-to-one.

This is a function of $x$, so $f$ has an inverse.

$f^{-1}(x) = \sqrt{x} - 3,\ x \geq 0$

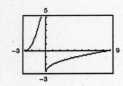

**49.** $f(x) = \sqrt{2x + 3} \implies x \geq -\dfrac{3}{2},\ y \geq 0$

$y = \sqrt{2x + 3},\ x \geq -\dfrac{3}{2},\ y \geq 0$

$x = \sqrt{2y + 3},\ y \geq -\dfrac{3}{2},\ x \geq 0$

$x^2 = 2y + 3,\ x \geq 0,\ y \geq -\dfrac{3}{2}$

$y = \dfrac{x^2 - 3}{2},\ x \geq 0,\ y \geq -\dfrac{3}{2}$

$f$ is one to one.

This is a function of $x$, so $f$ has an inverse.

$f^{-1}(x) = \dfrac{x^2 - 3}{2},\ x \geq 0$

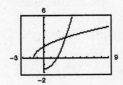

**51.** $f(x) = |x - 2|,\ x \leq 2,\ y \geq 0$

$y = |x - 2|$

$x = |y - 2| \quad y \leq 2,\ x \geq 0$

$x = -(y - 2)$ since $y - 2 \leq 0$

$x = -y + 2$

$y = -x + 2,\ x \geq 0,\ y \leq 2$

$f^{-1}(x) = -x + 2,\ x \geq 0$

**53.**  $f(x) = 2x - 3$

$y = 2x - 3$

$x = 2y - 3$

$y = \dfrac{x + 3}{2}$

$f^{-1}(x) = \dfrac{x + 3}{2}$

Reflections in the line $y = x$

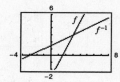

**55.**  $f(x) = x^5$

$y = x^5$

$x = y^5$

$y = \sqrt[5]{x}$

$f^{-1}(x) = \sqrt[5]{x}$

Reflections in the line $y = x$

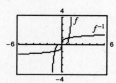

**57.**  $f(x) = x^{3/5}$

$y = x^{3/5}$

$x = y^{3/5}$

$y = x^{5/3}$

$f^{-1}(x) = x^{5/3}$

Reflections in the line $y = x$

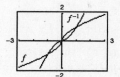

**59.** $f(x) = \sqrt{4 - x^2}, 0 \le x \le 2$

$y = \sqrt{4 - x^2}$

$x = \sqrt{4 - y^2}$

$x^2 = 4 - y^2$

$y^2 = 4 - x^2$

$y = \sqrt{4 - x^2}$

$f^{-1}(x) = \sqrt{4 - x^2}, 0 \le x \le 2$

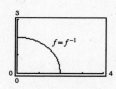

Reflections in the line $y = x$

**61.** $f(x) = \dfrac{4}{x}$

$y = \dfrac{4}{x}$

$x = \dfrac{4}{y}$

$xy = 4$

$y = \dfrac{4}{x}$

$f^{-1}(x) = \dfrac{4}{x}$

Reflections in the line $y = x$

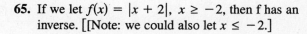

**63.** If we let $f(x) = (x - 2)^2$, $x \ge 2$, then $f$ has an inverse. [Note: we could also let $x \le 2$.]

$f(x) = (x - 2)^2$, $x \ge 2$, $y \ge 0$

$y = (x - 2)^2$, $x \ge 2$, $y \ge 0$

$x = (y - 2)^2$, $x \ge 0$, $y \ge 2$

$\sqrt{x} = y - 2$, $x \ge 0$, $y \ge 2$

$\sqrt{x} + 2 = y$, $x \ge 0$, $y \ge 2$

Thus, $f^{-1}(x) = \sqrt{x} + 2$, $x \ge 0$.

**65.** If we let $f(x) = |x + 2|$, $x \ge -2$, then $f$ has an inverse. [[Note: we could also let $x \le -2$.]

$f(x) = |x + 2|$, $x \ge -2$

$f(x) = x + 2$ when $x \ge -2$

$y = x + 2$, $x \ge -2$, $y \ge 0$

$x = y + 2$, $x \ge 0$, $y \ge -2$

$x - 2 = y$, $x \ge 0$, $y \ge -2$

Thus, $f^{-1}(x) = x - 2$, $x \ge 0$.

**67.**

| $x$ | $f(x)$ |
|-----|--------|
| $-2$ | $-4$ |
| $-1$ | $-2$ |
| $1$ | $2$ |
| $3$ | $3$ |

| $x$ | $f^{-1}(x)$ |
|-----|-------------|
| $-4$ | $-2$ |
| $-2$ | $-1$ |
| $2$ | $1$ |
| $3$ | $3$ |

**69.** $f(x) = x^3 + x + 1$

The graph of the inverse relation is an inverse function since it satisfies the vertical line test.

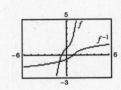

**71.** $g(x) = \dfrac{3x^2}{x^2 + 1}$

The graph of the inverse relation is not an inverse function since it does not satisfy the vertical line test.

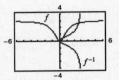

In Exercises 73, 75, and 77, $f(x) = \frac{1}{8}x - 3$, $f^{-1}(x) = 8(x + 3)$, $g(x) = x^3$, $g^{-1}(x) = \sqrt[3]{x}$.

**73.** $(f^{-1} \circ g^{-1})(1) = f^{-1}(g^{-1}(1)) = f^{-1}(\sqrt[3]{1}) = 8(\sqrt[3]{1} + 3) = 8(1 + 3) = 32$

**75.** $(f^{-1} \circ f^{-1})(6) = f^{-1}(f^{-1}(6)) = f^{-1}(8[6 + 3]) = f^{-1}(72) = 8(72 + 3) = 600$

**77.**   $(f \circ g)(x) = f(g(x)) = f(x^3) = \frac{1}{8}x^3 - 3$. Now find the inverse of $(f \circ g)(x) = \frac{1}{8}x^3 - 3$:

$$y = \tfrac{1}{8}x^3 - 3$$
$$x = \tfrac{1}{8}y^3 - 3$$
$$x + 3 = \tfrac{1}{8}y^3$$
$$8(x + 3) = y^3$$
$$\sqrt[3]{8(x + 3)} = y$$
$$(f \circ g)^{-1}(x) = 2\sqrt[3]{x + 3}$$

Note: $(f \circ g)^{-1} = g^{-1} \circ f^{-1}$

In Exercises 79 and 81, $f(x) = x + 4$, $f^{-1}(x) = x - 4$, $g(x) = 2x - 5$, $g^{-1}(x) = \dfrac{x + 5}{2}$.

**79.** $(g^{-1} \circ f^{-1})(x) = g^{-1}(f^{-1}(x)) = g^{-1}(x - 4) = \dfrac{(x - 4) + 5}{2} = \dfrac{x + 1}{2}$

**81.** $(f \circ g)(x) = f(g(x)) = f(2x - 5) = (2x - 5) + 4 = 2x - 1$. Now find the inverse of $(f \circ g)(x) = 2x - 1$:

$$y = 2x - 1$$
$$x = 2y - 1$$
$$x + 1 = 2y$$
$$y = \dfrac{x + 1}{2}$$

$$(f \circ g)^{-1}(x) = \dfrac{x + 1}{2}$$

Note that $(f \circ g)^{-1}(x) = (g^{-1} \circ f^{-1})(x)$; see Exercise 87.

**83.** (a) Yes, $f$ is one-to-one

(b) $f^{-1}$ gives the year corresponding to the 7 values in the second column

(c) $f^{-1}(650.3) = 10$ because $f(10) = 650.3$

(d) No, because $f(8) = f(12) = 546.3$.

**85.** False. $f(x) = x^2$ is even, but $f^{-1}$ does not exist

**87.** We will show that $(f \circ g)^{-1}(x) = (g^{-1} \circ f^{-1})(x)$ for all $x$ in their domains.
Let $y = (f \circ g)^{-1}(x) \implies (f \circ g)(y) = x$ then $f(g(y)) = x \implies f^{-1}(x) = g(y)$.

Hence, $(g^{-1} \circ f^{-1})(x) = g^{-1}(f^{-1}(x)) = g^{-1}(g(y)) = y = (f \circ g)^{-1}(x)$.

Thus, $g^{-1} \circ f^{-1} = (f \circ g)^{-1}$

**89.** $\dfrac{27x^3}{3x^2} = 9x,\ x \neq 0$

**91.** $\dfrac{x^2 - 36}{6 - x} = \dfrac{(x-6)(x+6)}{-(x-6)} = \dfrac{x+6}{-1} = -x - 6,\ x \neq 6$

**93.** $4x - y = 3$

$y = 4x - 3$

Yes, $y$ is a function of $x$.

**95.** $x^2 + y^2 = 9$

$y = \pm\sqrt{9 - x^2}$

No, $y$ is not a function of $x$.

**97.** $y = \sqrt{x + 2}$

Yes, $y$ is a function of $x$

# Review Exercises for Chapter 1

**Solutions to Odd-Numbered Exercises**

**1.** $y = -\frac{1}{2}x + 2$

| $x$ | $-2$ | $0$ | $2$ | $3$ | $4$ |
|---|---|---|---|---|---|
| $y$ | $3$ | $2$ | $1$ | $\frac{1}{2}$ | $0$ |
| Solution point | $(-2, 3)$ | $(0, 2)$ | $(2, 1)$ | $(3, \frac{1}{2})$ | $(4, 0)$ |

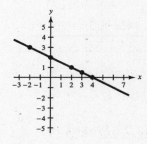

**3.** $y = 4 - x^2$

| $x$ | $-2$ | $-1$ | $0$ | $1$ | $2$ |
|---|---|---|---|---|---|
| $y$ | $0$ | $3$ | $4$ | $3$ | $0$ |
| Solution point | $(-2, 0)$ | $(-1, 3)$ | $(0, 4)$ | $(1, 3)$ | $(2, 0)$ |

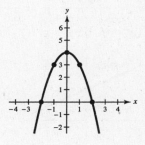

**5.** $y = \frac{1}{4}(x + 1)^3$

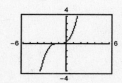

Intercepts: $(-1, 0),\ \left(0, \frac{1}{4}\right)$

**7.** $y = \frac{1}{4}x^4 - 2x^2$

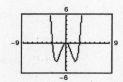

Intercepts: $(0, 0),\ \left(\pm 2\sqrt{2}, 0\right) \approx (\pm 2.83, 0)$

**9.** $y = x\sqrt{9 - x^2}$

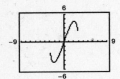

Intercepts: $(0, 0), (\pm 3, 0)$

**11.** $y = |x - 4| - 4$

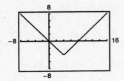

Intercepts: $(0, 0), (8, 0)$

**13.** The setting is:

Xmin = -20
Xmax = 50
Xscl = 10
Ymin = -2
Ymax = 1
Yscl = 0.5

**15.** $y = 13{,}500 - 1100t, \quad 0 \leq t \leq 6$

(a) $0 \leq x \leq 6, \quad 6900 \leq y \leq 13{,}500$

(Answers will vary.)

(b)

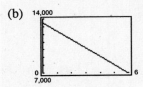

(c) When $y = 9100, \quad t = 4$

**17.** $m = \dfrac{2 - 2}{8 - (-3)} = \dfrac{0}{11} = 0$

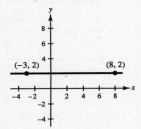

**19.** $m = \dfrac{5/2 - 1}{5 - 3/2} = \dfrac{3/2}{7/2} = \dfrac{3}{7}$

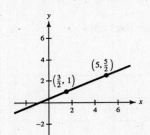

**21.** $(-4.5, 6), (2.1, 3)$

$m = \dfrac{3 - 6}{2.1 - (-4.5)} = \dfrac{-3}{6.6} = -\dfrac{30}{66} = -\dfrac{5}{11}$

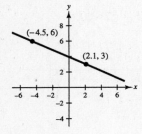

**23.** (a) $\qquad y + 1 = \frac{1}{4}(x - 2)$

$\qquad\qquad 4y + 4 = x - 2$

$\qquad -x + 4y + 6 = 0$

(b) Three additional points:

$(2 + 4, -1 + 1) = (6, 0)$

$(6 + 4, 0 + 1) = (10, 1)$

$(10 + 4, 1 + 1) = (14, 2)$

(other answers possible)

**25.** (a) $\qquad y + 5 = \frac{3}{2}(x - 0)$

$\qquad\qquad 2y + 10 = 3x$

$\qquad -3x + 2y + 10 = 0$

(b) Three additional points:

$(0 + 2, -5 + 3) = (2, -2)$

$(2 + 2, -2 + 3) = (4, 1)$

$(4 + 2, 1 + 3) = (6, 4)$

(other answers possible)

**27.** (a) $y + 5 = -1\left(x - \frac{1}{5}\right)$

$y + 5 = -x + \frac{1}{5}$

$5y + 25 = -5x + 1$

$5x + 5y + 24 = 0$

(b) Three additional points:

$\left(\frac{1}{5} + 1, -5 - 1\right) = \left(\frac{6}{5}, -6\right)$

$\left(\frac{6}{5} + 1, -6 - 1\right) = \left(\frac{11}{5}, -7\right)$

$\left(\frac{11}{5} + 1, -7 - 1\right) = \left(\frac{16}{5}, -8\right)$

(other answers possible)

**29.** (a) $y - 6 = 0(x + 2)$

$y - 6 = 0$

(b) Three additional points:

$(0, 6), (1, 6), (2, 6)$

(other answers possible)

**31.** (a) $m$ is undefined means that the line is vertical.

$x - 10 = 0$

(b) Three additional points: $(10, 0), (10, 1), (10, 2)$

(other answers possible)

**33.** (a) $y + 1 = \dfrac{-1 + 1}{4 - 2}(x - 2) = 0(x - 2) = 0 \implies y = -1$ (slope = 0)

(b)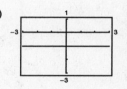

**35.** (a) $y - 0 = \dfrac{2 - 0}{6 - (-1)}(x + 1) = \frac{2}{7}(x + 1) = \frac{2}{7}x + \frac{2}{7} \implies y = \frac{2}{7}x + \frac{2}{7}$

(b)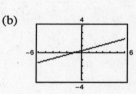

**37.** $t = 5$ corresponds to 2005.

Point: $(5, 12500)$, slope: 850

$y - 12{,}500 = 850(t - 5)$

$y = 850t + 8250$

**39.** $(2, 160{,}000), \; (3, 185{,}000)$

$m = \dfrac{185{,}000 - 160{,}000}{3 - 2} = 25{,}000$

$S - 160{,}000 = 25{,}000(t - 2)$

$S = 25{,}000t + 110{,}000$

For the fourth quarter let $t = 4$. Then we have

$S = 25{,}000(4) + 110{,}000 = \$210{,}000.$

**41.** $5x - 4y = 8 \implies y = \frac{5}{4}x - 2$ and $m = \frac{5}{4}$

(a) Parallel slope: $m = \frac{5}{4}$

$y - (-2) = \frac{5}{4}(x - 3)$

$4y + 8 = 5x - 15$

$0 = 5x - 4y - 23$

$y = \frac{5}{4}x - \frac{23}{4}$

(b) Perpendicular slope: $m = -\frac{4}{5}$

$y - (-2) = -\frac{4}{5}(x - 3)$

$5y + 10 = -4x + 12$

$4x + 5y - 2 = 0$

$y = -\frac{4}{5}x + \frac{2}{5}$

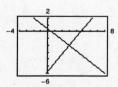

**43.** $x = 4$ is a vertical line; the slope is not defined.

   (a) Parallel line: $x = -6$

   (b) Perpendicular slope: $m = 0$

      Perpendicular line: $y - 2 = 0(x + 6) = 0 \implies y = 2$

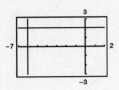

**45.** (a) Not a function. 20 is assigned two different values.

   (b) Function

   (c) Function

   (d) Not a function. No value is assigned to 30.

**47.**    $16x - y^4 = 0$

$$y^4 = 16x$$

$$y = \pm 2\sqrt[4]{x}$$

$y$ is *not* a function of $x$. Some $x$-values correspond to two $y$-values.

For example, $x = 1$ corresponds to $y = 2$ and $y = -2$.

**49.** $y = \sqrt{1 - x}$

Each $x$ value, $x \leq 1$, corresponds to only one $y$ value so y is a function of $x$.

**51.** $f(x) = x^2 + 1$

   (a) $f(2) = 2^2 + 1 = 5$

   (b) $f(-4) = (-4)^2 + 1 = 17$

   (c) $f(t^2) = (t^2)^2 + 1 = t^4 + 1$

   (d) $-f(x) = -(x^2 + 1) = -x^2 - 1$

**53.** $h(x) = \begin{cases} 2x + 1, & x \leq -1 \\ x^2 + 2, & x > -1 \end{cases}$

   (a) $h(-2) = 2(-2) + 1 = -3$

   (b) $h(-1) = 2(-1) + 1 = -1$

   (c) $h(0) = 0^2 + 2 = 2$

   (d) $h(2) = 2^2 + 2 = 6$

**55.** $f(x) = (x - 1)(x + 2)$ is defined for all real numbers.

Domain: $(-\infty, \infty)$

**57.** $f(x) = \sqrt{25 - x^2}$

Domain:       $25 - x^2 \geq 0$

             $(5 + x)(5 - x) \geq 0$

Domain: $[-5, 5]$

**59.** $g(s) = \dfrac{5}{3s - 9} = \dfrac{5}{3(s - 3)}$

Domain: All real numbers except $s = 3$

**61.** (a) $C(x) = 16{,}000 + 5.35x$

   (b) $P(x) = R(x) - C(x)$

           $= 8.20x - (16{,}000 + 5.35x)$

           $= 2.85x - 16{,}000$

**63.**

$$f(x) = 2x^2 + 3x - 1$$

$$f(x + h) = 2(x + h)^2 + 3(x + h) - 1$$

$$= 2x^2 + 4xh + 2h^2 + 3x + 3h - 1$$

$$\frac{f(x + h) - f(x)}{h} = \frac{(2x^2 + 4xh + 2h^2 + 3x + 3h - 1) - (2x^2 + 3x - 1)}{h}$$

$$= \frac{4xh + 2h^2 + 3h}{h}$$

$$= 4x + 2h + 3, \quad h \neq 0$$

**65.** Domain: All real numbers

Range: $y \leq 3$

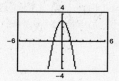

**67.** Domain: $36 - x^2 \geq 0 \implies x^2 \leq 36 \implies -6 \leq x \leq 6$

Range: $0 \leq y \leq 6$

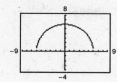

**69.** (a) $y = \dfrac{x^2 + 3x}{6}$

(b) $y$ is a function of $x$.

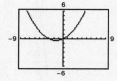

**71.** (a) $3x + y^2 = 2$

$$y^2 = 2 - 3x$$

$$y = \pm\sqrt{2 - 3x}$$

(b) $y$ is not a function of $x$.

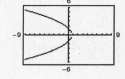

**73.** $f(x) = x^3 - 3x$

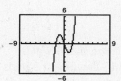

Increasing on $(-\infty, -1)$ and $(1, \infty)$. Decreasing on $(-1, 1)$.

**75.** $f(x) = x\sqrt{x - 6}$

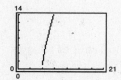

Increasing on $(6, \infty)$

**77.** $f(x) = (x^2 - 4)^2$. Relative minimums at $(-2, 0)$ and $(2, 0)$. Relative maximum at $(0, 16)$.

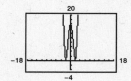

**79.** $h(x) = 4x^3 - x^4$. Relative maximum $(3, 27)$

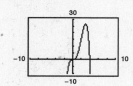

**81.** $f(x) = \begin{cases} 3x + 5, & x < 0 \\ x - 4, & x \geq 0 \end{cases}$

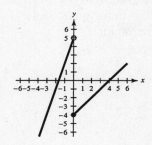

**83.** $f(-x) = ((-x)^2 - 8)^2 = (x^2 - 8)^2 = f(x)$.

$f$ is even.

**85.** $f(x) = -2$ is a constant function.

**87.** $g(x) = -x^3 - 2$ is obtained from $f(x) = x^3$ by a reflection in the $y$-axis, followed by a vertical shift 2 units downward. $g(x) = -f(x) - 2$.

**89.** $h(x) = x^2 - 6$

(a) $f(x) = x^2$

(b) The graph of $h$ is a vertical shift of $f$ 6 units downward.

(c)

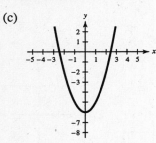

(d) $h(x) = x^2 - 6 = f(x) - 6$

**91.** $h(x) = (x - 1)^3 + 7$

(a) $f(x) = x^3$

(b) The graph of $h$ is obtained from $f$ by a horizontal shift 1 unit to the right, followed by a vertical shift 7 units upward.

(c)

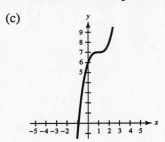

(d) $h(x) = (x - 1)^3 + 7 = f(x - 1) + 7$

**93.** $h(x) = \sqrt{x} - 5$

(a) $f(x) = \sqrt{x}$

(b) The graph of $h$ is a vertical shift of $f$ 5 units downward.

(c)

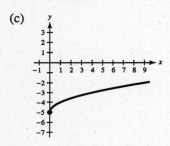

(d) $h(x) = \sqrt{x} - 5 = f(x) - 5$

**95.** $h(x) = -x^2 - 3$

(a) $f(x) = x^2$

(b) The graph of $h$ is a refelction in the $x$-axis, followed by a vertical shift downward 3 units of the graph of $f$.

(c)

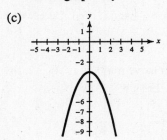

(d) $h(x) = -x^2 - 3 = -f(x) - 3$

**97.** $h(x) = -2x^2 + 3$

(a) $f(x) = x^2$

(b) The graph of $h$ is obtained from $f$ by a vertical stretch of 2, a reflection in the $x$-axis, and a vertical shift 3 units upward.

(c)

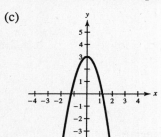

(d) $h(x) = -2x^2 + 3 = -2f(x) + 3$

**99.** $h(x) = -\frac{1}{2}|x| + 9$

(a) $f(x) = |x|$

(b) The graph of $h$ is obtained from $f$ by a vertical shrink of $\frac{1}{2}$, a reflection in the $x$-axis, and a vertical shift 9 units upward.

(c)

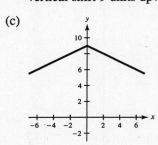

(d) $h(x) = -\frac{1}{2}|x| + 9 = -\frac{1}{2}f(x) + 9$

**101.** $(f - g)(4) = f(4) - g(4)$
$$= [3 - 2(4)] - \sqrt{4}$$
$$= -5 - 2$$
$$= -7$$

**103.** $(f + g)(25) = f(25) + g(25) = -47 + 5 = -42$

**105.** $(fh)(1) = f(1)h(1) = (3 - 2(1))(3(1)^2 + 2)$
$$= (1)(5) = 5$$

**107.** $(h \circ g)(7) = h(g(7))$
$$= h\left(\sqrt{7}\right)$$
$$= 3\left(\sqrt{7}\right)^2 + 2$$
$$= 23$$

**109.** $(f \circ h)(-4) = f(h(-4)) = f(50) = -97$

**111.** Let $t = 6$ represents 1996
$$y_1 = -2.75t^2 + 86.8t + 659$$
$$y_2 = -1.88t^2 + 62.4t + 616$$

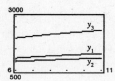

**113.** $f(x) = 6x \implies f^{-1}(x) = \frac{1}{6}x$
$$f(f^{-1}(x)) = f\left(\tfrac{1}{6}x\right) = 6\left(\tfrac{1}{6}x\right) = x$$
$$f^{-1}(f(x)) = f^{-1}(6x) = \tfrac{1}{6}(6x) = x$$

**115.** (a)

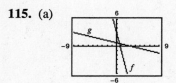

Reflection in the line $y = x$

(b)
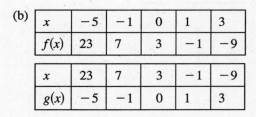

| $x$ | $-5$ | $-1$ | 0 | 1 | 3 |
|-----|------|------|---|---|---|
| $f(x)$ | 23 | 7 | 3 | $-1$ | $-9$ |

| $x$ | 23 | 7 | 3 | $-1$ | $-9$ |
|-----|----|---|---|------|------|
| $g(x)$ | $-5$ | $-1$ | 0 | 1 | 3 |

The entries in the table are the same except that their rows are interchanged.

**117.**

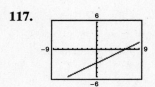

$f(x) = \frac{1}{2}x - 3$ passes the Horizontal Line Test, and hence is one-to-one and has an inverse $(f^{-1}(x) = 2(x + 3))$.

**119.**

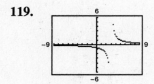

$h(t) = \dfrac{2}{t - 3}$ passes the Horizontal Line Test, and hence is one-to-one.

**121.** $f(x) = \dfrac{x}{12}$

$y = \dfrac{x}{12}$

$x = \dfrac{y}{12}$

$12x = y$

$f^{-1}(x) = 12x$

**123.** $f(x) = 4x^3 - 3$

$y = 4x^3 - 3$

$x = 4y^3 - 3$

$x + 3 = 4y^3$

$\dfrac{x + 3}{4} = y^3$

$f^{-1}(x) = \sqrt[3]{\dfrac{x + 3}{4}}$

**125.** $f(x) = \sqrt{x + 10}$

$y = \sqrt{x + 10}, x \geq -10, y \geq 0$

$x = \sqrt{y + 10}, y \geq -10, x \geq 0$

$x^2 = y + 10$

$x^2 - 10 = y$

$f^{-1}(x) = x^2 - 10, x \geq 0$

**127.** False. $g(x) = -[(x - 6)^2 + 3] = -(x - 6)^2 - 3$
and $g(-1) = -52 \neq 28$

**129.** False. $f(x) = \dfrac{1}{x}$ or $f(x) = x$ satisfy $f = f^{-1}$.

# Chapter 1    Practice Test

1. Use a graphing utility to graph the equation $y = 4/x^2 - 5$. Approximate any $x$-intercepts of the graph.

2. Use a graphing utility to graph the equation $y = |x - 3| + 2$. Approximate any $x$-intercepts of the graph.

3. Graph $3x - 5y = 15$ by hand.

4. Graph $y = \sqrt{9 - x}$ by hand.

5. Solve $5x + 4 = 7x - 8$.

6. Solve $\dfrac{x}{3} - 5 = \dfrac{x}{5} + 1$.

7. Solve $\dfrac{3x + 1}{6x - 7} = \dfrac{2}{5}$ graphically and analytically.

8. Solve $(x - 3)^2 + 4 = (x + 1)^2$ graphically and analytically.

9. Find an equation for the line passing through the points $(3, -2)$ and $(4, -5)$. Use a graphing utility to sketch a graph of the line.

10. Find an equation of the line that passes through the point $(-1, 5)$ and has slope $-3$. Use a graphing utility to sketch a graph of the line.

11. Does the equation $x^4 + y^4 = 16$ represent $y$ as a function of $x$?

12. Evaluate the function $f(x) = |x - 2|/(x - 2)$ at the points $x = 0$, $x = 2$, and $x = 4$.

13. Find the domain of the function $f(x) = 5/(x^2 - 16)$.    14. Find the domain of the function $g(t) = \sqrt{4 - t}$.

15. Use a graphing utility to sketch the graph of the function $f(x) = 3 - x^6$ and determine if the function is even, odd, or neither.

16. Use a graphing utility to approximate any relative minimum or maximum values of the function $y = 4 - x + x^3$.

17. Compare the graph of $f(x) = x^3 - 3$ with the graph of $y = x^3$.

18. Compare the graph of $f(x) = \sqrt{x - 6}$ with the graph of $y = \sqrt{x}$.

19. Find $g \circ f$ if $f(x) = \sqrt{x}$ and $g(x) = x^2 - 2$. What is the domain of $g \circ f$?

20. Find $f/g$ if $f(x) = 3x^2$ and $g(x) = 16 - x^4$. What is the domain of $f/g$?

21. Show that $f(x) = 3x + 1$ and $g(x) = \dfrac{x - 1}{3}$ are inverse functions algebraically and graphically.

22. Find the inverse of $f(x) = \sqrt{9 - x^2}$, $0 \le x \le 3$. Graph $f$ and $f^{-1}$ in the same viewing rectangle.

# CHAPTER 2
## Solving Equations and Inequalities

---

# CHAPTER 2
## Solving Equations and Inequalities

### Section 2.1    Linear Equations and Problem Solving

■ You should know how to solve linear equations: $ax + b = 0$.

■ An identity is an equation whose solution consists of every real number in its domain.

■ To solve an equation you can:
   (a) Add or subtract the same quantity from both sides.
   (b) Multiply or divide both sides by the same nonzero quantity.

■ To solve an equation that can be simplified to a linear equation:
   (a) Remove all symbols of grouping and all fractions.    (b) Combine like terms.
   (c) Solve by algebra.    (d) Check the answer.

■ A "solution" that does not satisfy the original equation is called an extraneous solution.

■ You should be able to set up mathematical models to solve problems.

■ You should be able to translate key words and phrases.

   (a) Equality:    (b) Addition:
   Equals, Equal to, is, are, was,    Sum, plus, greater, increased by, more than,
   will be, represents    exceeds, total of

   (c) Subtraction:    (d) Multiplication:
   Difference, minus, less than,    Product, multiplied by, twice, times, percent of
   decreased by, subtracted from,
   reduced by, the remainder

   (e) Division:    (f) Consecutive:
   Quotient, divided by, ratio, per    Next, subsequent

■ You should know the following formulas:

   (a) Perimeter:    (b) Area:
   1. Square: $P = 4s$    1. Square: $A = s^2$
   2. Rectangle: $P = 2L + 2W$    2. Rectangle: $A = LW$
   3. Circle: $C = 2\pi r$    3. Circle: $A = \pi r^2$

   4. Triangle: $A = \left(\dfrac{1}{2}\right)bh$

   (c) Volume    (d) Simple Interest: $I = Prt$
   1. Cube: $V = s^3$
   2. Rectangular solid: $V = LWH$
   3. Cylinder: $V = \pi r^2 h$
   4. Sphere: $V = \left(\dfrac{4}{3}\right)\pi r^3$

   (e) Compound Interest: $A = P\left(1 + \dfrac{r}{n}\right)^{nt}$    (f) Distance: $D = r \cdot t$

   (g) Temperature: $F = \dfrac{9}{5}C + 32$

■ You should be able to solve word problems. Study the examples in the text carefully.

**Solutions to Odd-Numbered Exercises**

**1.** $\dfrac{5}{2x} - \dfrac{4}{x} = 3$

(a) $\dfrac{5}{2(-1/2)} - \dfrac{4}{(-1/2)} \overset{?}{=} 3$

$$3 = 3$$

$x = -\frac{1}{2}$ *is* a solution.

(c) $\dfrac{5}{2(0)} - \dfrac{4}{0}$ is undefined.

$x = 0$ *is not* a solution.

(b) $\dfrac{5}{2(4)} - \dfrac{4}{4} \overset{?}{=} 3$

$$-\dfrac{3}{8} \neq 3$$

$x = 4$ *is not* a solution.

(d) $\dfrac{5}{2(1/4)} - \dfrac{4}{1/4} \overset{?}{=} 3$

$$-6 \neq 3$$

$x = \frac{1}{4}$ *is not* a solution.

**3.** $3 + \dfrac{1}{x + 2} = 4$

(a) $3 + \dfrac{1}{(-1) + 2} \overset{?}{=} 4$

$$4 = 4$$

$x = -1$ *is* a solution

(b) $3 + \dfrac{1}{(-2) + 2} = 3 + \dfrac{1}{0}$ is undefined

$x = -2$ *is not* a solution

(c) $3 + \dfrac{1}{0 + 2} \overset{?}{=} 4$

$$\dfrac{7}{2} \neq 4$$

$x = 0$ *is not* a solution

(d) $3 + \dfrac{1}{5 + 2} \overset{?}{=} 4$

$$\dfrac{22}{7} = 4$$

$x = 5$ *is not* a solution

**5.** $\dfrac{\sqrt{x + 4}}{6} + 3 = 4$

(a) $\dfrac{\sqrt{-3 + 4}}{6} + 3 \overset{?}{=} 4$

$$\dfrac{19}{6} \neq 4$$

$x = -3$ *is not* a solution

(b) $\dfrac{\sqrt{0 + 4}}{6} + 3 \overset{?}{=} 4$

$$\dfrac{10}{3} \neq 4$$

$x = 0$ *is not* a solution

(c) $\dfrac{\sqrt{21 + 4}}{6} + 3 \overset{?}{=} 4$

$$\dfrac{23}{6} \neq 4$$

$x = 21$ *is not* a solution

(d) $\dfrac{\sqrt{32 + 4}}{6} + 3 \overset{?}{=} 4$

$$4 = 4$$

$x = 32$ *is* a solution

**7.** $2(x - 1) = 2x - 2$ is an *identity* by the Distributive Property. It is true for all real values of $x$.

**9.** $x^2 - 8x + 5 = (x - 4)^2 - 11$ is an *identity* since $(x - 4)^2 - 11 = x^2 - 8x + 16 - 11 = x^2 - 8x + 5$.

**11.** $3 + \dfrac{1}{x + 1} = \dfrac{4x}{x + 1}$ is *conditional*. There are real values of $x$ for which the equation is not true.

**13.** Method 1:    $\dfrac{3x}{8} - \dfrac{4x}{3} = 4$

$$\dfrac{9x - 32x}{24} = 4$$

$$-23x = 96$$

$$x = -\dfrac{96}{23}$$

Method 2: Graph $y_1 = \dfrac{3x}{8} - \dfrac{4x}{3}$ and $y_2 = 4$ in the same viewing window. These lines intersect at $x \approx -4.1739 \approx -\dfrac{96}{23}$

**15.**    $\dfrac{x}{5} - \dfrac{x}{2} = 3$

$$\dfrac{2x - 5x}{10} = 3$$

$$-3x = 30$$

$$x = -10$$

**17.**    $\dfrac{3}{2}(z + 5) - \dfrac{1}{4}(z + 24) = 0$

$$4\left(\dfrac{3}{2}\right)(z + 5) - 4\left(\dfrac{1}{4}\right)(z + 24) = 4(0)$$

$$6(z + 5) - (z + 24) = 0$$

$$6z + 30 - z - 24 = 0$$

$$5z = -6$$

$$z = -\dfrac{6}{5}$$

**19.**    $\dfrac{100 - 4u}{3} = \dfrac{5u + 6}{4} + 6$

$$12\left(\dfrac{100 - 4u}{3}\right) = 12\left(\dfrac{5u + 6}{4}\right) + 12(6)$$

$$4(100 - 4u) = 3(5u + 6) + 72$$

$$400 - 16u = 15u + 18 + 72$$

$$-31u = -310$$

$$u = 10$$

**21.**    $\dfrac{5x - 4}{5x + 4} = \dfrac{2}{3}$

$$3(5x - 4) = 2(5x + 4)$$

$$15x - 12 = 10x + 8$$

$$5x = 20$$

$$x = 4$$

**23.**    $\dfrac{1}{x - 3} + \dfrac{1}{x + 3} = \dfrac{10}{x^2 - 9}$

$$\dfrac{(x + 3) + (x - 3)}{x^2 - 9} = \dfrac{10}{x^2 - 9}$$

$$2x = 10$$

$$x = 5$$

**25.**    $\dfrac{7}{2x + 1} - \dfrac{8x}{2x - 1} = -4$

$$7(2x - 1) - 8x(2x + 1) = -4(2x + 1)(2x - 1)$$

$$14x - 7 - 16x^2 - 8x = -16x^2 + 4$$

$$6x = 11$$

$$x = \dfrac{11}{6}$$

**27.**    $\dfrac{1}{x} + \dfrac{2}{x - 5} = 0$

$$1(x - 5) + 2x = 0$$

$$3x - 5 = 0$$

$$3x = 5$$

$$x = \dfrac{5}{3}$$

**29.** $\dfrac{3}{x(x-3)} + \dfrac{4}{x} = \dfrac{1}{x-3}$

$3 + 4(x-3) = x$

$3 + 4x - 12 = x$

$3x = 9$

$x = 3$

A check reveals that $x = 3$ is an extraneous solution, so there is no solution.

**31.** $A = \dfrac{1}{2}bh$

$2A = bh$

$\dfrac{2A}{b} = h$

**33.** $A = P\left(1 + \dfrac{r}{n}\right)^{nt}$

$P = \dfrac{A}{\left(1 + \dfrac{r}{n}\right)^{nt}}$

$P = A\left(1 + \dfrac{r}{n}\right)^{-nt}$

**35.** $S = \dfrac{rL - a}{r - 1}$

$Sr - S = rL - a$

$Sr - rL = S - a$

$r(S - L) = S - a$

$r = \dfrac{S - a}{S - L}$

**37.** $V = \dfrac{4}{3}\pi a^2 b$

$\dfrac{3V}{4\pi a^2} = b$

**39.** $16 = 0.432x - 10.44$

$26.44 = 0.432x$

$\dfrac{26.44}{0.432} = x$

$\approx 61.2$ inches

**41.** (a)

(b) $\ell = 1.5w$

$P = 2\ell + 2w = 2(1.5w) + 2w = 5w$

(c) $25 = 5w \implies w = 5$ meters and $\ell = (1.5)(5) = 7.5$ meters

**43.** (a) test average $= (\text{test } 1 + \text{test } 2 + \text{test } 3 + \text{test } 4)/4$

(b) 4 tests at 100 points each $\implies$ 400 points. $(0.90)(400) = 360$ points needed for an A. So far, $87 + 92 + 84 = 263$ points. $360 - 263 = 97$ points needed on fourth test.

**45.** Rate $= \dfrac{\text{Distance}}{\text{Time}} = \dfrac{50 \text{ kilometers}}{\frac{1}{2} \text{ hours}} = 100$ Kilometers/hour

Total time $= \dfrac{\text{Total distance}}{\text{Rate}} = \dfrac{300 \text{ kilometers}}{100 \text{ kilometers/hour}} = 3$ hours

**47.** *Model:* $(\text{Distance}) = (\text{rate})(\text{time}_1 + \text{time}_2)$

*Labels:* Distance $= 2 \cdot 200 = 400$ miles, rate $= r$,

$\text{time}_1 = \dfrac{\text{distance}}{\text{rate}_1} = \dfrac{200}{55}$ hours, $\text{time}_2 = \dfrac{\text{distance}}{\text{rate}_2} = \dfrac{200}{40}$ hours

*Equation:* $400 = r\left(\dfrac{200}{55} + \dfrac{200}{40}\right)$

$400 = r\left(\dfrac{1600}{440} + \dfrac{2200}{440}\right) = \dfrac{3800}{440}r$

$46.3 \approx r$ The average speed for the round trip was approximately 46.3 miles per hour.

**49.** *Model:* (Distance) = (rate)(time)

    *Labels:* Distance = $1.5 \times 10^{11}$ meters, rate = $3.0 \times 10^{8}$ meters per second, time = $t$

    *Equation:* $1.5 \times 10^{11} = (3.0 \times 10^{8})t$

$$500 = t$$

Light from the sun travels to the earth in 500 seconds or approximately 8.33 minutes.

**51.** Let $h$ = height of the building in feet

(a)

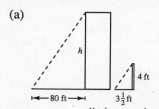

                                                   4 ft

    ←— 80 ft —→    $3\frac{1}{2}$ ft

             *Not drawn to scale*

(b) $\dfrac{h \text{ feet}}{80 \text{ feet}} = \dfrac{4 \text{ feet}}{3.5 \text{ feet}}$

$$\frac{h}{80} = \frac{4}{3.5}$$

$$3.5h = 320$$

$$h \approx 91.4 \text{ feet}$$

**53.** Let $x$ = amount in $4\frac{1}{2}$% fund. Then $12{,}000 - x$ = amount in 5% fund.

$$560 = .045x + 0.05(12{,}000 - x)$$

$$560 = .045x + 600 - 0.05x$$

$$.005x = 40$$

$$x = 8000$$

You must invest \$8000 in the $4\frac{1}{2}$% fund, and $12{,}000 - 8000 = 4000$ in the 5% fund.

**55.** Let $x$ = number of pounds of \$2.49 nuts. Then $100 - x$ = number of pounds of \$3.89 nuts.

$$2.49x + 3.89(100 - x) = 3.19(100)$$

$$2.49x + 389 - 3.89x = 319$$

$$-1.40x = -70$$

$$x = \frac{-70}{-1.40} = 50$$

$$x = 50 \text{ lbs of \$2.49 nuts}$$

$$100 - x = 50 \text{ lbs of \$3.89 nuts}$$

Use 50 pounds of each kind.

**57.** $A = \dfrac{1}{2}bh$

$$h = \frac{2A}{b} = \frac{2(182.25)}{13.5} = 27 \text{ feet}$$

**59.** $V = \dfrac{4}{3}\pi r^{3} = 47{,}712.94$

$$r = \sqrt[3]{\frac{3V}{4\pi}} \approx 22.5 \text{ cm}$$

**61.** $W_1 x = W_2(L - x)$

$$50x = 75(10 - x)$$

$$50x = 750 - 75x$$

$$125x = 750$$

$$x = 6 \text{ feet from the 50-pound child.}$$

**63.** False. $x(3 - x) = 10$ is a quadratic equation

$$-x^2 + 3x = 10 \text{ or } x^2 - 3x + 10 = 0$$

**65.** You need $a(-3) + b = c(-3)$ or $b = (c - a)(-3) = 3(a - c)$. One answer is $a = 2$, $c = 1$ and $b = 3$ ($2x + 3 = x$) and another is $a = 6$, $c = 1$, and $b = 15$ ($6x + 15 = x$).

**67.** Equivalent equations are derived from the substitution principle and simplification techniques. They have the same solution(s).

$2x + 3 = 8$ and $2x = 5$ are equivalent equations.

**69.** $y = \frac{5}{8}x - 2$

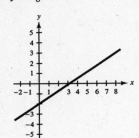

**71.** $y = (x - 3)^2 + 7$

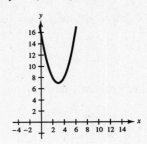

**73.** $y = -\frac{1}{2}|x + 4| - 1$

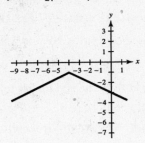

**75.** $(f + g)(-3) = f(-3) + g(-3) = [-3^2 + 4] + [6(-3) - 5] = -5 - 23 = -28$

**77.** $(fg)(8) = f(8)g(8) = [-64 + 4][48 - 5] = -2580$     **79.** $(f \circ g)(4) = f(g(4)) = f(19) = -357$

# Section 2.2    Solving Equations Graphically

- You should be able to find the intercepts of the graph of an equation.
- You should be able to find the zeros of a function $y = f(x)$ by solving the equation $f(x) = 0$.
- You should be able to find the solutions of an equation graphically using a graphing utility.
- You should be able to use the zoom and trace features to find solutions to any desired accuracy.
- You should be able to find the points of intersection of two graphs.

**Solutions to Odd-Numbered Exercises**

**1.** $y = x - 5$

Let $y = 0$: $0 = x - 5 \implies x = 5 \implies (5, 0)$   $x$-intercept

Let $x = 0$: $y = 0 - 5 \implies y = -5 \implies (0, -5)$   $y$-intercept

**3.** $y = x^2 + x - 2$

Let $y = 0$: $(x^2 + x - 2) = (x + 2)(x - 1) = 0 \implies x = -2, 1 \implies (-2, 0), (1, 0)$   $x$-intercepts

Let $x = 0$: $y = 0^2 + 0 - 2 = -2 \implies (0, -2)$   $y$-intercept

**5.** $y = x\sqrt{x + 2}$

Let $y = 0$: $0 = x\sqrt{x + 2} \implies x = 0, -2 \implies (0, 0), (-2, 0)$   $x$-intercepts

Let $x = 0$: $y = 0\sqrt{0 + 2} = 0 \implies (0, 0)$   $y$-intercept

**7.** $xy = 4$

If $x = 0$, then $0y = 0 = 4$, which is impossible. Similarly, $y = 0$ is impossible.
Hence there are no intercepts.

**9.** $y = |x - 2| - 4$

Let $y = 0$: $|x - 2| - 4 = 0 \Rightarrow |x - 2| = 4 \Rightarrow x = -2, 6 \Rightarrow (-2, 0), (6, 0)$   $x$-intercepts

Let $x = 0$: $|0 - 2| - 4 = |-2| - 4 = 2 - 4 = -2 = y \Rightarrow (0, -2)$   $y$-intercept

**11.** $xy - 2y - x + 1 = 0$

Let $y = 0$: $-x + 1 = 0 \Rightarrow x = 1 \Rightarrow (1, 0)$   $x$-intercept

Let $x = 0$: $-2y + 1 = 0 \Rightarrow y = \frac{1}{2} \Rightarrow \left(0, \frac{1}{2}\right)$   $y$-intercept

**13.**     $f(x) = 5(4 - x)$

$5(4 - x) = 0$

$4 - x = 0$

$x = 4$

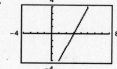

$x = 4$

**15.** $f(x) = x^3 - 6x^2 + 5x$

$x^3 - 6x^2 + 5x = 0$

$x(x^2 - 6x + 5) = 0$

$x(x - 5)(x - 1) = 0$

$x = 0, 5, 1$

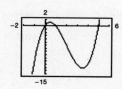

$x = 0, 1, 5$

**17.** $f(x) = \dfrac{x + 2}{3} - \dfrac{x - 1}{5} - 1$

$\dfrac{x + 2}{3} - \dfrac{x - 1}{5} - 1 = 0$

$5(x + 2) - 3(x - 1) - 15 = 0$

$2x = 2$

$x = 1$

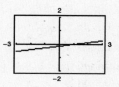

**19.**

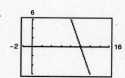     $(3, 0)$

$y = 0 = 2(x - 1) - 4 = 2x - 2 - 4 = 2x - 6 \Rightarrow 2x = 6 \Rightarrow x = 3$

**21.**

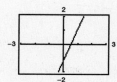     $(10, 0)$

$y = 0 = 20 - (3x - 10) = 20 - 3x + 10 = 30 - 3x \Rightarrow 3x = 30 \Rightarrow x = 10$

**23.** $2.7x - 0.4x = 1.2$

$2.3x = 1.2$

$x = \dfrac{1.2}{2.3} \approx 0.522$

$f(x) = 2.7x - 0.4x - 1.2 = 0$

$x \approx 0.522$

**25.**   $25(x - 3) = 12(x + 2) - 10$

$25x - 75 = 12x + 24 - 10$

$13x - 89 = 0$

$x = \frac{89}{13}$

$f(x) = 25(x - 3) - 12(x + 2) + 10 = 0$

$x = 6.846$

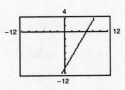

**27.**   $\frac{3x}{2} + \frac{1}{4}(x - 2) = 10$

$\frac{6x}{4} + \frac{x}{4} = 10 + \frac{1}{2}$

$\frac{7x}{4} = \frac{21}{2}$

$x = 6$

$f(x) = \frac{3x}{2} + \frac{1}{4}(x - 2) - 10 = 0$

$x = 6.0$

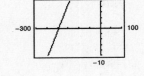

**29.**   $0.60x + 0.40(100 - x) = 1.2$

$0.60x + 40 - 0.40x = 1.2$

$0.20x = -38.8$

$x = -194$

$f(x) = 0.60x + 0.40(100 - x) - 1.2 = 0$

$x = -194$

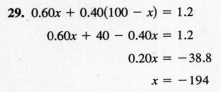

**31.**   $\frac{2x}{3} = 10 - \frac{24}{x}$

$\frac{2x}{3}(3x) = 10(3x) - \frac{24}{x}(3x)$

$2x^2 = 30x - 72$

$2x^2 - 30x + 72 = 0$

$x^2 - 15x + 36 = 0$

$(x - 3)(x - 12) = 0$

$x = 3, 12$

$f(x) = \frac{2x}{3} - 10 + \frac{24}{x}$

$x = 3, 12$

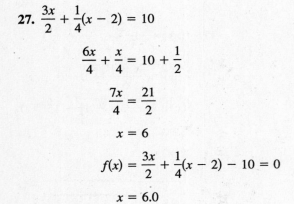

**33.**   $\frac{3}{x + 2} - \frac{4}{x - 2} = 5$

$3(x - 2) - 4(x + 2) = 5(x + 2)(x - 2)$

$3x - 6 - 4x - 8 = 5(x^2 - 4)$

$0 = 5x^2 + x - 6$

$0 = (x - 1)(5x + 6)$

$x = 1, -\frac{6}{5}$

$f(x) = \frac{3}{x + 2} - \frac{4}{x - 2} - 5 = 0$

$x = 1.0, -1.2$

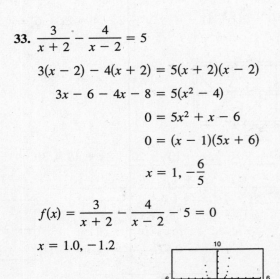

**35.**      $(x + 2)^2 = x^2 - 6x + 1$

$x^2 + 4x + 4 = x^2 - 6x + 1$

$10x = -3$

$x = -\frac{3}{10}$

$f(x) = (x + 2)^2 - x^2 + 6x - 1$

$x = -\frac{3}{10}$

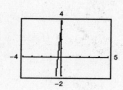

**37.** $\frac{1}{4}(x^2 - 10x + 17) = 0$

$x = 2.172, 7.828$

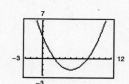

**39.** $x^3 + x + 4 = 0$

$x = -1.379$

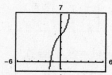

**41.** $2x^3 - x^2 - 18x + 9 = 0$

$x = -3.0, \ 0.5, \ 3.0$

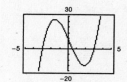

**43.**        $x^4 = 2x^3 + 1$

$x^4 - 2x^3 - 1 = 0$

$x \approx -0.717, \ 2.107$

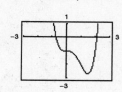

**45.**      $\frac{2}{x + 2} = 3$

$\frac{2}{x + 2} - 3 = 0$

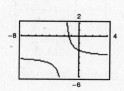

$x = -\frac{4}{3}$

**47.**      $|x - 3| = 4$

$|x - 3| - 4 = 0$

$x = -1, 7$

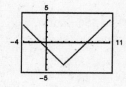

**49.**      $\sqrt{x - 2} = 3$

$\sqrt{x - 2} - 3 = 0$

$x = 11$

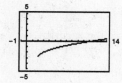

**51.** (a)

| $x$ | $-1$ | 0 | 1 | 2 | 3 | 4 |
|---|---|---|---|---|---|---|
| $3.2x - 5.8$ | $-9$ | $-5.8$ | $-2.6$ | 0.6 | 3.8 | 7.0 |

(b) $1 < x < 2$ because of the sign change.

(c)

| $x$ | 1.5 | 1.6 | 1.7 | 1.8 | 1.9 | 2 |
|---|---|---|---|---|---|---|
| $3.2x - 5.8$ | $-1$ | $-0.68$ | $-0.36$ | $-0.04$ | 0.28 | 0.6 |

(d) $1.8 < x < 1.9$. To improve accuracy, evaluate the expression in this interval and determine where the sign changes.

(e) $x = 1.8125$

**53.** $y = 2 - x$

$y = 2x - 1$

$2 - x = 2x - 1$

$3 = 3x$

$x = 1, y = 2 - 1 = 1$

$(x, y) = (1, 1)$

**55.** $2x + y = 6 \Rightarrow y = 6 - 2x$

$-x + y = 0 \Rightarrow y = x$

$6 - 2x = x$

$6 = 3x$

$x = 2, y = x = 2$

$(x, y) = (2, 2)$

**57.** $x - y = -4 \Rightarrow y = x + 4$

$x^2 - y = -2 \Rightarrow y = x^2 + 2$

$x^2 + 2 = x + 4$

$x^2 - x - 2 = 0$

$(x - 2)(x + 1) = 0$

$x = 2, y = 6$

$x = -1, y = 3$

$(2, 6), (-1, 3)$

**59.** $y = 9 - 2x$

$y = x - 3$

$(4, 1)$

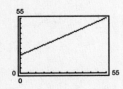

**61.** $y = 4 - x^2$

$y = 2x - 1$

$(x, y) = (1.449, 1.898),$

$(-3.449, -7.899)$

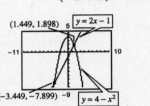

**63.** $y = 2x^2$

$y = x^4 - 2x^2$

$(x, y) = (0, 0), \ (2, 8), \ (-2, 8)$

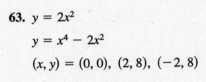

**65.** (a) $\dfrac{1 + 0.73205}{1 - 0.73205} = \dfrac{1.73205}{0.26795}$

$= 6.464079 = 6.46$

(b) $\dfrac{1 + 0.73205}{1 - 0.73205} = \dfrac{1.73205}{0.26795}$

$= \dfrac{1.73}{0.27}$

$= 6.407407 = 6.41$

The second method decreases the accuracy.

**67.** (a) $t = \dfrac{x}{63} + \dfrac{(280 - x)}{54}$

(b) Domain: $0 \le x \le 280$

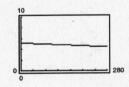

(c) If the time was 4 hours and 45 minutes, then $t = 4\frac{3}{4}$ and $x = 164.5$ miles.

**69.** (a) $A = x + 0.33(55 - x)$

(b) Domain: $0 \le x \le 55$

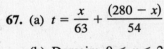

(c) If the final mixture is 60% concentrate,
   then $A = 0.6(55) = 33$ and $x = 22.2$ gallons.

**71.** (a) Area $= A(x) = 4x + 8x = 12x$

(b)

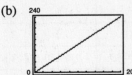

(c) $A(x) = 12x = 200 \Rightarrow x \approx 16.67$ units

**73.** (a) $V = \frac{1}{2}(5)(40)(20) + 4(40)(20) = 5200$ ft³

(b) The line has slope $\frac{5}{40} = \frac{1}{8}$ and passes through $(0, 0)$: $y = \frac{1}{8}x$.

(c) For $0 \le d \le 5$, by similar triangles, $\frac{d}{5} = \frac{b}{40} \implies b = 8d$.

So, $V = \frac{1}{2}bd(20) = \frac{1}{2}(8d)d(20)$

$= 80d^2, 0 \le d \le 5$

For $5 < d \le 9$,

$V = \frac{1}{2}(5)(40)(20) + (d - 5)(40)(20)$

$= 2000 + 800d - 4000$

$= 800d - 2000, 5 < d \le 9$

(d)

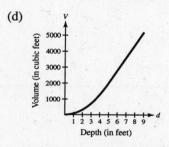

(e)

| $d$ | 3 | 5 | 7 | 9 |
|---|---|---|---|---|
| $V$ | 720 | 2000 | 3600 | 5200 |

(f) If $V = 4800$, $800d - 2000 = 4800 \implies 800d = 6800 \implies d = 8.5$ feet.

(g) Number of gallons $= 5200(7.48) = 38,896$ gallons

**75.** $y = 0.91t + 56.4 = 60$

$0.91t = 3.6$

$t \approx 3.96$

That is, near the end of 1993.

Graphically, graph $y = 0.91t + 56.4$ and $y = 60$ and note that they intersect at $t \approx 3.96$.

**77.** True

**79.** False. Two linear equations could have an infinite number of points of intersection.
For example, $x + y = 1$ and $2x + 2y = 2$.

**81.** $\frac{12}{5\sqrt{3}} \cdot \frac{\sqrt{3}}{\sqrt{3}} = \frac{12\sqrt{3}}{5(3)} = \frac{4\sqrt{3}}{5}$

**83.** $\frac{3}{8 + \sqrt{11}} \cdot \frac{8 - \sqrt{11}}{8 - \sqrt{11}} = \frac{3(8 - \sqrt{11})}{64 - 11} = \frac{3(8 - \sqrt{11})}{53}$

**85.** $(x + 6)(3x - 5) = 3x^2 + 13x - 30$

**87.** $(2x - 9)(2x + 9) = 4x^2 - 81$

# Section 2.3 Complex Numbers

■ You should know how to work with complex numbers.

■ Operations on complex numbers

(a) Addition: $(a + bi) + (c + di) = (a + c) + (b + d)i$

(b) Subtraction: $(a + bi) - (c + di) = (a - c) + (b - d)i$

(c) Multiplication: $(a + bi)(c + di) = (ac - bd) + (ad + bc)i$

(d) Division: $\dfrac{a + bi}{c + di} = \dfrac{a + bi}{c + di} \cdot \dfrac{c - di}{c - di} = \dfrac{ac + bd}{c^2 + d^2} + \dfrac{bc - ad}{c^2 + d^2}i$

■ The complex conjugate of $a + bi$ is $a - bi$:

$(a + bi)(a - bi) = a^2 + b^2$

■ The additive inverse of $a + bi$ is $-a - bi$.

■ The multiplicative inverse of $a + bi$ is

$\dfrac{a - bi}{a^2 + b^2}.$

■ $\sqrt{-a} = \sqrt{a}\,i$ for $a > 0$.

## Solutions to Odd-Numbered Exercises

**1.** $a + bi = -9 + 4i$

$a = -9$

$b = 4$

**3.** $(a - 1) + (b + 3)i = 5 + 8i$

$a - 1 = 5 \implies a = 6$

$b + 3 = 8 \implies b = 5$

**5.** $4 + \sqrt{-25} = 4 + 5i$

**7.** $7 = 7 + 0i$

**9.** $-5i + i^2 = -5i - 1 = -1 - 5i$

**11.** $\left(\sqrt{-75}\right)^2 = -75$

**13.** $\sqrt{-0.09} = \sqrt{0.09}\,i = 0.3i$

**15.** $(4 + i) + (7 - 2i) = 11 - i$

**17.** $\left(-1 + \sqrt{-8}\right) + \left(8 - \sqrt{-50}\right) = 7 + 2\sqrt{2}i - 5\sqrt{2}i = 7 - 3\sqrt{2}i$

**19.** $13i - (14 - 7i) = 13i - 14 + 7i = -14 + 20i$

**21.** $\left(\dfrac{3}{2} + \dfrac{5}{2}i\right) + \left(\dfrac{5}{3} + \dfrac{11}{3}i\right) = \left(\dfrac{3}{2} + \dfrac{5}{3}\right) + \left(\dfrac{5}{2} + \dfrac{11}{3}\right)i$

$= \dfrac{9 + 10}{6} + \dfrac{15 + 22}{6}i$

$= \dfrac{19}{6} + \dfrac{37}{6}i$

**23.** $(1.6 + 3.2i) + (-5.8 + 4.3i) = -4.2 + 7.5i$

**25.** $\sqrt{-6} \cdot \sqrt{-2} = \left(\sqrt{6}i\right)\left(\sqrt{2}i\right) = \sqrt{12}i^2 = (2\sqrt{3})(-1) = -2\sqrt{3}$

**27.** $\left(\sqrt{-10}\right)^2 = \left(\sqrt{10}\,i\right)^2 = 10i^2 = -10$

**29.** $(1 + i)(3 - 2i) = 3 - 2i + 3i - 2i^2$
$$= 3 + i + 2$$
$$= 5 + i$$

**31.** $4i(8 + 5i) = 32i + 20i^2 = 32i + 20(-1) = -20 + 32i$

**33.** $\left(\sqrt{14} + \sqrt{10}\,i\right)\left(\sqrt{14} - \sqrt{10}\,i\right) = 14 - 10i^2 = 14 + 10 = 24$

**35.** $(4 + 5i)^2 - (4 - 5i)^2 = [(4 + 5i) + (4 - 5i)][(4 + 5i) - (4 - 5i)] = 8(10i) = 80i$

**37.** $4 - 3i$ is the complex conjugate of $4 + 3i$
$(4 + 3i)(4 - 3i) = 16 + 9 = 25$

**39.** $-6 + \sqrt{5}\,i$ is the complex conjugate of $-6 - \sqrt{5}\,i$
$\left(-6 - \sqrt{5}\,i\right)\left(-6 + \sqrt{5}\,i\right) = 36 + 5 = 41$

**41.** $-\sqrt{20}\,i$ is the complex conjugate of
$\sqrt{-20} = \sqrt{20}\,i$
$\left(\sqrt{20}\,i\right)\left(-\sqrt{20}\,i\right) = 20$

**43.** $3 + \sqrt{2}\,i$ is the complex conjugate of
$3 - \sqrt{-2} = 3 - \sqrt{2}\,i$
$\left(3 - \sqrt{2}\,i\right)\left(3 + \sqrt{2}\,i\right) = 9 + 2 = 11$

**45.** $\dfrac{6}{i} = \dfrac{6}{i} \cdot \dfrac{-i}{-i} = \dfrac{-6i}{-i^2} = \dfrac{-6i}{1} = -6i$

**47.** $\dfrac{2}{4 - 5i} = \dfrac{2}{4 - 5i} \cdot \dfrac{4 + 5i}{4 + 5i} = \dfrac{8 + 10i}{16 + 25} = \dfrac{8}{41} + \dfrac{10}{41}i$

**49.** $\dfrac{2 + i}{2 - i} = \dfrac{2 + i}{2 - i} \cdot \dfrac{2 + i}{2 + i}$
$$= \dfrac{4 + 4i + i^2}{4 + 1}$$
$$= \dfrac{3 + 4i}{5}$$
$$= \dfrac{3}{5} + \dfrac{4}{5}i$$

**51.** $\dfrac{i}{(4 - 5i)^2} = \dfrac{i}{16 - 25 - 40i}$
$$= \dfrac{i}{-9 - 40i} \cdot \dfrac{-9 + 40i}{-9 + 40i}$$
$$= \dfrac{-40 - 9i}{81 + 40^2}$$
$$= \dfrac{-40}{1681} - \dfrac{9}{1681}i$$

**53.** $\dfrac{2}{1 + i} - \dfrac{3}{1 - i} = \dfrac{2(1 - i) - 3(1 + i)}{(1 + i)(1 - i)}$
$$= \dfrac{2 - 2i - 3 - 3i}{1 + 1}$$
$$= \dfrac{-1 - 5i}{2}$$
$$= -\dfrac{1}{2} - \dfrac{5}{2}i$$

**55.** $\dfrac{i}{3 - 2i} + \dfrac{2i}{3 + 8i} = \dfrac{3i + 8i^2 + 6i - 4i^2}{(3 - 2i)(3 + 8i)}$
$$= \dfrac{-4 + 9i}{9 + 18i + 16}$$
$$= \dfrac{-4 + 9i}{25 + 18i} \cdot \dfrac{25 - 18i}{25 - 18i}$$
$$= \dfrac{-100 + 72i + 225i + 162}{25^2 + 18^2}$$
$$= \dfrac{62 + 297i}{949}$$
$$= \dfrac{62}{949} + \dfrac{297}{949}i$$

**57.** $-6i^3 + i^2 = -6i^2i + i^2$

$\qquad = -6(-1)i + (-1)$

$\qquad = 6i - 1$

$\qquad = -1 + 6i$

**59.** $\left(\sqrt{-75}\right)^3 = \left(5\sqrt{3}i\right)^3 = 5^3\left(\sqrt{3}\right)^3 i^3$

$\qquad = 125\left(3\sqrt{3}\right)(-i)$

$\qquad = -375\sqrt{3}i$

**61.** $\dfrac{1}{i^3} = \dfrac{1}{i^3} \cdot \dfrac{i}{i} = \dfrac{i}{i^4} = \dfrac{i}{1} = i$

**63.** $(2)^3 = 8$

$\left(-1 + \sqrt{3}i\right)^3 = (-1)^3 + 3(-1)^2\left(\sqrt{3}i\right) + 3(-1)\left(\sqrt{3}i\right)^2 + \left(\sqrt{3}i\right)^3$

$\qquad = -1 + 3\sqrt{3}i - 9i^2 + 3\sqrt{3}i^3$

$\qquad = -1 + 3\sqrt{3}i + 9 - 3\sqrt{3}i$

$\qquad = 8$

$\left(-1 - \sqrt{3}i\right)^3 = (-1)^3 + 3(-1)^2\left(-\sqrt{3}i\right) + 3(-1)\left(-\sqrt{3}i\right)^2 + \left(-\sqrt{3}i\right)^3$

$\qquad = -1 - 3\sqrt{3}i - 9i^2 - 3\sqrt{3}i^3$

$\qquad = -1 - 3\sqrt{3}i + 9 + 3\sqrt{3}i$

$\qquad = 8$

The three numbers are cube roots of 8.

**65.** $4 + 3i$

**67.** $4 - 5i$

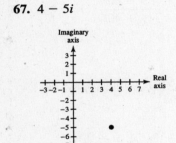

**69.** $3i$

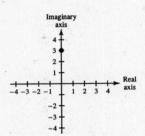

**71.** 1

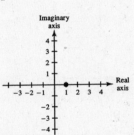

**73.** The complex number $\frac{1}{2}i$, is in the Mandelbrot Set since for $c = \frac{1}{2}i$, the corresponding Mandelbrot sequence is

$$\frac{1}{2}i, -\frac{1}{4} + \frac{1}{2}i, -\frac{3}{16} + \frac{1}{4}i, -\frac{7}{256} + \frac{13}{32}i, -\frac{10{,}767}{65{,}536} + \frac{1957}{4096}i, -\frac{864{,}513{,}055}{4{,}294{,}967{,}296} + \frac{46{,}037{,}845}{134{,}217{,}728}i$$

which is bounded. Or in decimal form

$0.5i, -0.25 + 0.5i, -0.1875 + 0.25i, -0.02734 + 0.40625i,$

$-0.164291 + 0.477783i, -0.201285 + 0.343009i.$

**75.** The complex number 1 is not in the Mandelbrot Set since for $c = 1$, the corresponding Mandelbrot sequence is 1, 2, 5, 26, 677, 458,330 which is unbounded.

**77.** $z_1 = 5 + 2i$

$z_2 = 3 - 4i$

$$\frac{1}{z} = \frac{1}{z_1} + \frac{1}{z_2} = \frac{1}{5 + 2i} + \frac{1}{3 - 4i} = \frac{(3 - 4i) + (5 + 2i)}{(5 + 2i)(3 - 4i)} = \frac{8 - 2i}{23 - 14i}$$

$$z = \frac{23 - 14i}{8 - 2i}\left(\frac{8 + 2i}{8 + 2i}\right) = \frac{212 - 66i}{68} \approx 3.118 - 0.971i$$

**79.** False. A real number $a + 0i = a$ is equal to its conjugate.

**81.** $(4x - 5)(4x + 5) = 16x^2 - 20x + 20x - 25$

$= 16x^2 - 25$

**83.** $\left(3x - \frac{1}{2}\right)(x + 4) = 3x^2 - \frac{1}{2}x + 12x - 2 = 3x^2 + \frac{23}{2}x - 2$

# Section 2.4    Solving Equations Algebraically

> - ■ You should be able to solve a quadratic equation by factoring, if possible.
> - ■ You should be able to solve a quadratic equation of the form $u^2 = d$ by extracting square roots.
> - ■ You should be able to solve a quadratic equation by completing the square.
> - ■ You should know and be able to use the Quadratic Formula: For $ax^2 + bx + c = 0, a \neq 0$,
>
> $$x = \frac{-b \pm \sqrt{b^2 - 4ac}}{2a}.$$
>
> - ■ You should be able to determine the types of solutions of a quadratic equation by checking the discriminant $b^2 - 4ac$.
>   - (a) If $b^2 - 4ac > 0$, there are two distinct real solutions.
>   - (b) If $b^2 - 4ac = 0$, there is one repeated real solution.
>   - (c) If $b^2 - 4ac < 0$, there is no real solution.
> - ■ You should be able to solve certain types of nonlinear or nonquadratic equations.
> - ■ For equations involving radicals or fractional powers, raise both sides to the same power.
> - ■ For equations that are of the quadratic type, $au^2 + bu + c = 0, a \neq 0$, use either factoring or the quadratic equation.
> - ■ For equations with fractions, multiply both sides by the least common denominator to clear the fractions.
> - ■ For equations involving absolute value, remember that the expression inside the absolute value can be positive or negative.
> - ■ Always check for extraneous solutions.

### Solutions to Odd-Numbered Exercises

**1.** $2x^2 = 3 - 5x$

Standard form: $2x^2 + 5x - 3 = 0$

**3.** $\frac{1}{5}(3x^2 - 10) = 12x$

$3x^2 - 10 = 60x$

Standard form: $3x^2 - 60x - 10 = 0$

**5.**   $6x^2 + 3x = 0$

$3x(2x + 1) = 0$

$3x = 0$ or $2x + 1 = 0$

$x = 0$ or $x = -\frac{1}{2}$

**7.**   $x^2 - 2x - 8 = 0$

$(x - 4)(x + 2) = 0$

$x - 4 = 0$ or $x + 2 = 0$

$x = 4$ or $x = -2$

**9.**   $3 + 5x - 2x^2 = 0$

$(3 - x)(1 + 2x) = 0$

$3 - x = 0$ or $1 + 2x = 0$

$x = 3$ or $x = -\frac{1}{2}$

**11.**   $x^2 + 4x = 12$

$x^2 + 4x - 12 = 0$

$(x + 6)(x - 2) = 0$

$x + 6 = 0$ or $x - 2 = 0$

$x = -6$ or $x = 2$

**13.**   $(x + a)^2 - b^2 = 0$

$[(x + a) + b][(x + a) - b] = 0$

$x + a + b = 0 \implies x = -a - b$

$x + a - b = 0 \implies x = -a + b$

**15.**   $x^2 = 49$

$x = \pm\sqrt{49} = \pm 7$

**17.**   $(x - 12)^2 = 16$

$x - 12 = \pm\sqrt{16} = \pm 4$

$x = 12 \pm 4$

$x = 16, 8$

**19.**   $(2x - 1)^2 = 12$

$2x - 1 = \pm\sqrt{12} = \pm 2\sqrt{3}$

$2x = 1 \pm 2\sqrt{3}$

$x = \frac{1}{2} \pm \sqrt{3}$

$(x = 2.23, -1.23)$

**21.**   $(x - 7)^2 = (x + 3)^2$

$x - 7 = \pm(x + 3)$

$x - 7 = x + 3$ impossible

$x - 7 = -(x + 3) \implies 2x = 4$

$\implies x = 2$

**23.**   $x^2 + 4x = 32$

$x^2 + 4x + 4 = 32 + 4$

$(x + 2)^2 = 36$

$x + 2 = \pm 6$

$x = -2 \pm 6$

$x = -8, 4$

**25.**   $x^2 + 6x + 2 = 0$

$x^2 + 6x = -2$

$x^2 + 6x + 3^2 = -2 + 3^2$

$(x + 3)^2 = 7$

$x + 3 = \pm\sqrt{7}$

$x = -3 \pm\sqrt{7}$

**27.** $9x^2 - 18x + 3 = 0$

$$x^2 - 2x + \frac{1}{3} = 0$$

$$x^2 - 2x = -\frac{1}{3}$$

$$x^2 - 2x + 1^2 = -\frac{1}{3} + 1^2$$

$$(x - 1)^2 = \frac{2}{3}$$

$$x - 1 = \pm\sqrt{\frac{2}{3}}$$

$$x = 1 \pm \sqrt{\frac{2}{3}}$$

$$x = 1 \pm \frac{\sqrt{6}}{3}$$

**29.** $8 + 4x - x^2 = 0$

$$x^2 - 4x = 8$$

$$x^2 - 4x + 4 = 8 + 4$$

$$(x - 2)^2 = 12$$

$$x - 2 = \pm\sqrt{12}$$

$$= \pm 2\sqrt{3}$$

$$x = 2 \pm 2\sqrt{3}$$

**31.** $2x^2 + 5x - 8 = 0$

$$x^2 + \frac{5}{2}x - 4 = 0$$

$$x^2 + \frac{5}{2}x + \frac{25}{16} = 4 + \frac{25}{16}$$

$$\left(x + \frac{5}{4}\right)^2 = \frac{89}{16}$$

$$x + \frac{5}{4} = \pm\frac{\sqrt{89}}{4}$$

$$x = \frac{-5}{4} \pm \frac{\sqrt{89}}{4}$$

**33.** $y = (x + 3)^2 - 4$

(a)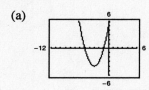

(b) The x-intercepts are $(-1, 0)$ and $(-5, 0)$.

(c)
$$0 = (x + 3)^2 - 4$$
$$4 = (x + 3)^2$$
$$\pm\sqrt{4} = x + 3$$
$$-3 \pm 2 = x$$
$$x = -1 \text{ or } x = -5$$

(d) The *x*-intercepts of the graph are solutions to the equation.

**35.** $y = -4x^2 + 4x + 3$

(a)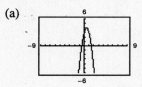

(b) The *x*-intercepts are $\left(-\frac{1}{2}, 0\right)$ and $\left(\frac{3}{2}, 0\right)$.

(c)
$$0 = -4x^2 + 4x + 3$$
$$4x^2 - 4x = 3$$
$$4(x^2 - x) = 3$$
$$x^2 - x = \frac{3}{4}$$
$$x^2 - x + \left(\frac{1}{2}\right)^2 = \frac{3}{4} + \left(\frac{1}{2}\right)^2$$
$$\left(x - \frac{1}{2}\right)^2 = 1$$
$$x - \frac{1}{2} = \pm\sqrt{1}$$
$$x = \frac{1}{2} \pm 1$$
$$x = \frac{3}{2} \text{ or } x = -\frac{1}{2}$$

(d) The *x*-intercepts of the graph are solutions to the equation.

**37.** $y = \frac{1}{4}(4x^2 - 20x + 25)$

(a)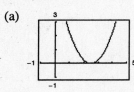

(b) The *x*-intercept is $\left(\frac{5}{2}, 0\right)$.

(c) $\frac{1}{4}(4x^2 - 20x + 25) = 0$

$$4x^2 - 20x + 25 = 0$$
$$(2x - 5)^2 = 0$$
$$2x - 5 = 0$$
$$x = \frac{5}{2}$$

(d) The *x*-intercepts of the graph are solutions to the equation.

**39.** $2x^2 - 5x + 5 = 0$

The graph does not have any $x$-intercepts and thus the equation has no real solution.

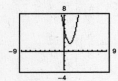

**41.** $\frac{4}{7}x^2 - 8x + 28 = 0$

The graph has one $x$-intercept $(7, 0)$ and hence the equation has one real solution

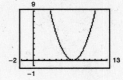

**43.** $-0.2x^2 + 1.2x - 8 = 0$

The graph does not have any $x$-intercepts and hence the equation has no real solution

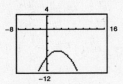

**45.** $-x^2 + 2x + 2 = 0$

$$x = \frac{-b \pm \sqrt{b^2 - 4ac}}{2a}$$

$$= \frac{-2 \pm \sqrt{2^2 - 4(-1)(2)}}{2(-1)}$$

$$= \frac{-2 \pm 2\sqrt{3}}{-2} = 1 \pm \sqrt{3}$$

**47.** $x^2 + 8x - 4 = 0$

$$x = \frac{-b \pm \sqrt{b^2 - 4ac}}{2a}$$

$$= \frac{-8 \pm \sqrt{8^2 - 4(1)(-4)}}{2(1)}$$

$$= \frac{-8 \pm 4\sqrt{5}}{2}$$

$$= -4 \pm 2\sqrt{5}$$

**49.**    $28x - 49x^2 = 4$

$-49x^2 + 28x - 4 = 0$

$$x = \frac{-b \pm \sqrt{b^2 - 4ac}}{2a}$$

$$= \frac{-28 \pm \sqrt{28^2 - 4(-49)(-4)}}{2(-49)}$$

$$= \frac{-28 \pm 0}{-98} = \frac{2}{7}$$

**51.** $4x^2 + 16x + 17 = 0$

$$x = \frac{-b \pm \sqrt{b^2 - 4ac}}{2a}$$

$$= \frac{-16 \pm \sqrt{16^2 - 4(4)(17)}}{2(4)}$$

$$= \frac{-16 \pm \sqrt{-16}}{8}$$

$$= \frac{-16 \pm 4i}{8}$$

$$= -2 \pm \frac{1}{2}i$$

**53.** $x^2 - 2x - 1 = 0$

$x^2 - 2x = 1$

$x^2 - 2x + 1^2 = 1 + 1^2$

$(x - 1)^2 = 2$

$x - 1 = \pm\sqrt{2}$

$x = 1 \pm \sqrt{2}$

**55.** $(x + 3)^2 = 81$

$x + 3 = \pm 9.$

$x + 3 = 9$ or $x + 3 = -9$

$x = 6$ or    $x = -12$

**57.** $x^2 - x - \frac{11}{4} = 0$

$x^2 - x + \frac{1}{4} = \frac{11}{4} + \frac{1}{4}$

$\left(x - \frac{1}{2}\right)^2 = 3$

$x - \frac{1}{2} = \pm\sqrt{3}$

$x = \frac{1}{2} \pm \sqrt{3}$

$x = \frac{1}{2} + \sqrt{3}, \frac{1}{2} - \sqrt{3}$

**59.**
$$(x + 1)^2 = x^2$$
$$(x + 1)^2 - x^2 = 0$$
$$(x + 1 - x)(x + 1 + x) = 0$$
$$2x + 1 = 0$$
$$2x = -1$$
$$x = -\tfrac{1}{2}$$

**61.**   $4x^4 - 18x^2 = 0$
$$2x^2(2x^2 - 9) = 0$$
$$2x^2 = 0 \implies x = 0$$
$$2x^2 - 9 = 0 \implies x = \pm\frac{3\sqrt{2}}{2}$$

**63.**
$$x^4 - 4x^2 + 3 = 0$$
$$(x^2 - 3)(x^2 - 1) = 0$$
$$\left(x + \sqrt{3}\right)\left(x - \sqrt{3}\right)(x + 1)(x - 1) = 0$$
$$x + \sqrt{3} = 0 \implies x = -\sqrt{3}$$
$$x - \sqrt{3} = 0 \implies x = \sqrt{3}$$
$$x + 1 = 0 \implies x = -1$$
$$x - 1 = 0 \implies x = 1$$

**65.** $5x^3 + 30x^2 + 45x = 0$
$$5x(x^2 + 6x + 9) = 0$$
$$5x(x + 3)^2 = 0$$
$$5x = 0 \implies x = 0$$
$$x + 3 = 0 \implies x = -3$$

**67.**   $x^3 - 3x^2 - x + 3 = 0$
$$x^2(x - 3) - (x - 3) = 0$$
$$(x - 3)(x^2 - 1) = 0$$
$$(x - 3)(x + 1)(x - 1) = 0$$
$$x - 3 = 0 \implies x = 3$$
$$x + 1 = 0 \implies x = -1$$
$$x - 1 = 0 \implies x = 1$$

**69.**
$$4x^4 - 65x^2 + 16 = 0$$
$$(4x^2 - 1)(x^2 - 16) = 0$$
$$(2x + 1)(2x - 1)(x + 4)(x - 4) = 0$$
$$2x + 1 = 0 \implies x = -\tfrac{1}{2}$$
$$2x - 1 = 0 \implies x = \tfrac{1}{2}$$
$$x + 4 = 0 \implies x = -4$$
$$x - 4 = 0 \implies x = 4$$

**71.**   $\dfrac{1}{t^2} + \dfrac{8}{t} + 15 = 0$
$$1 + 8t + 15t^2 = 0$$
$$(1 + 3t)(1 + 5t) = 0$$
$$1 + 3t = 0 \implies t = -\frac{1}{3}$$
$$1 + 5t = 0 \implies t = -\frac{1}{5}$$

**73.** $6\left(\dfrac{s}{s + 1}\right)^2 + 5\left(\dfrac{s}{s + 1}\right) - 6 = 0$

Let $u = s/(s + 1)$.
$$6u^2 + 5u - 6 = 0$$
$$(3u - 2)(2u + 3) = 0$$
$$3u - 2 = 0 \implies u = \frac{2}{3}$$
$$2u + 3 = 0 \implies u = -\frac{3}{2}$$
$$\frac{s}{s + 1} = \frac{2}{3} \implies s = 2$$
$$\frac{s}{s + 1} = -\frac{3}{2} \implies s = -\frac{3}{5}$$

**75.**
$$2x + 9\sqrt{x} - 5 = 0$$
$$(2\sqrt{x} - 1)(\sqrt{x} + 5) = 0$$
$$\sqrt{x} = \tfrac{1}{2} \implies x = \tfrac{1}{4}$$
$(\sqrt{x} = -5$ is not possible.)

Note: You can see graphically that there is only one solution.

**77.**
$$3x^{1/3} + 2x^{2/3} = 5$$
$$2x^{2/3} + 3x^{1/3} - 5 = 0$$
$$(2x^{1/3} + 5)(x^{1/3} - 1) = 0$$
$$x^{1/3} = -\frac{5}{2} \implies x = \frac{-125}{8}$$
$$x^{1/3} = 1 \implies x = 1$$

**79.** $y = x^3 - 2x^2 - 3x$

(a)
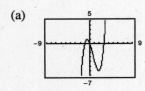

(b) $x$-intercepts: $(-1, 0), (0, 0), (3, 0)$

(c) $0 = x^3 - 2x^2 - 3x$

$0 = x(x + 1)(x - 3)$

$x = 0$

$x + 1 = 0 \implies x = -1$

$x - 3 = 0 \implies x = 3$

(d) The $x$-intercepts are the same as the solutions.

**81.** $y = x^4 - 10x^2 + 9$

(a)
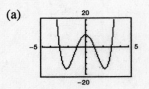

(b) $x$-intercepts: $(\pm 1, 0),\ (\pm 3, 0)$

(c) $0 = x^4 - 10x^2 + 9$

$0 = (x^2 - 1)(x^2 - 9)$

$0 = (x + 1)(x - 1)(x + 3)(x - 3)$

$x + 1 = 0 \implies x = -1$

$x - 1 = 0 \implies x = 1$

$x + 3 = 0 \implies x = -3$

$x - 3 = 0 \implies x = 3$

(d) The $x$-intercepts are the same as the solutions.

**83.**
$$\sqrt{x - 10} - 4 = 0$$
$$\sqrt{x - 10} = 4$$
$$x - 10 = 16$$
$$x = 26$$

**85.**
$$\sqrt{x + 1} - 3x = 1$$
$$\sqrt{x + 1} = 3x + 1$$
$$x + 1 = 9x^2 + 6x + 1$$
$$0 = 9x^2 + 5x$$
$$0 = x(9x + 5)$$
$$x = 0$$
$$9x + 5 = 0 \implies x = -\tfrac{5}{9}, \text{ extraneous}$$

**87.**
$$\sqrt[3]{2x + 1} + 8 = 0$$
$$\sqrt[3]{2x + 1} = -8$$
$$2x + 1 = -512$$
$$2x = -513$$
$$x = -\frac{513}{2} = -256.5$$

**89.**
$$\sqrt{x} - \sqrt{x - 5} = 1$$
$$\sqrt{x} = 1 + \sqrt{x - 5}$$
$$\left(\sqrt{x}\right)^2 = \left(1 + \sqrt{x - 5}\right)^2$$
$$x = 1 + 2\sqrt{x - 5} + x - 5$$
$$4 = 2\sqrt{x - 5}$$
$$2 = \sqrt{x - 5}$$
$$4 = x - 5$$
$$9 = x$$

**91.** $(x - 5)^{2/3} = 16$

$$x - 5 = \pm 16^{3/2}$$

$$x - 5 = \pm 64$$

$$x = 69, \; -59$$

**93.** $3x(x - 1)^{1/2} + 2(x - 1)^{3/2} = 0$

$$(x - 1)^{1/2}[3x + 2(x - 1)] = 0$$

$$(x - 1)^{1/2}(5x - 2) = 0$$

$$(x - 1)^{1/2} = 0 \implies x - 1 = 0 \implies x = 1$$

$$5x - 2 = 0 \implies x = \tfrac{2}{5} \text{ which is extraneous.}$$

**95.** $y = \sqrt{11x - 30} - x$

(a)

(b) $x$-intercepts: $(5, 0), (6, 0)$

(c) $0 = \sqrt{11x - 30} - x$

$$x = \sqrt{11x - 30}$$

$$x^2 = 11x - 30$$

$$x^2 - 11x + 30 = 0$$

$$(x - 5)(x - 6) = 0$$

$$x - 5 = 0 \implies x = 5$$

$$x - 6 = 0 \implies x = 6$$

(d) The $x$-intercepts and the solutions are the same.

**97.** $y = \sqrt{7x + 36} - \sqrt{5x + 16} - 2$

(a)

(b) $x$-intercepts:  $(0, 0), (4, 0)$

(c) $0 = \sqrt{7x + 36} - \sqrt{5x + 16} - 2$

$$\sqrt{7x + 36} = 2 + \sqrt{5x + 16}$$

$$\left(\sqrt{7x + 36}\right)^2 = \left(2 + \sqrt{5x + 16}\right)^2$$

$$7x + 36 = 4 + 4\sqrt{5x + 16} + 5x + 16$$

$$7x + 36 = 5x + 20 + 4\sqrt{5x + 16}$$

$$2x + 16 = 4\sqrt{5x + 16}$$

$$x + 8 = 2\sqrt{5x + 16}$$

$$x^2 + 16x + 64 = 4(5x + 16)$$

$$x^2 + 16x + 64 = 20x + 64$$

$$x^2 - 4x = 0$$

$$x(x - 4) = 0$$

$$x = 0$$

$$x - 4 = 0 \implies x = 4$$

(d) The $x$-intercepts and the solutions are the same.

**99.** $\dfrac{20 - x}{x} = x$

$$20 - x = x^2$$

$$0 = x^2 + x - 20$$

$$0 = (x + 5)(x - 4)$$

$$x + 5 = 0 \implies x = -5$$

$$x - 4 = 0 \implies x = 4$$

**101.**
$$\frac{1}{x} - \frac{1}{x+1} = 3$$

$$x(x+1)\frac{1}{x} - x(x+1)\frac{1}{x+1} = x(x+1)(3)$$

$$x + 1 - x = 3x(x+1)$$

$$1 = 3x^2 + 3x$$

$$0 = 3x^2 + 3x - 1; \quad a = 3, \quad b = 3, \quad c = -1$$

$$x = \frac{-3 \pm \sqrt{(3)^2 - 4(3)(-1)}}{2(3)} = \frac{-3 \pm \sqrt{21}}{6}$$

**103.**
$$x = \frac{3}{x} + \frac{1}{2}$$

$$(2x)(x) = (2x)\left(\frac{3}{x}\right) + (2x)\left(\frac{1}{2}\right)$$

$$2x^2 = 6 + x$$

$$2x^2 - x - 6 = 0$$

$$(2x + 3)(x - 2) = 0$$

$$2x + 3 = 0 \implies x = -\frac{3}{2}$$

$$x - 2 = 0 \implies x = 2$$

**105.** $\quad |2x - 1| = 5$

$$2x - 1 = 5 \implies x = 3$$

$$-(2x - 1) = 5 \implies x = -2$$

**107.**
$$|x| = x^2 + x - 3$$

$$x = x^2 + x - 3 \quad \text{OR} \quad -x = x^2 + x - 3$$

$$x^2 - 3 = 0 \qquad\qquad x^2 + 2x - 3 = 0$$

$$x = \pm\sqrt{3} \qquad\qquad (x - 1)(x + 3) = 0$$

$$x - 1 = 0 \implies x = 1$$

$$x + 3 = 0 \implies x = -3$$

Only $x = \sqrt{3}$, and $x = -3$ are solutions to the original equation. $x = -\sqrt{3}$ and $x = 1$ are extraneous. Note that the graph of $y = x^2 + x - 3 - |x|$ has two $x$-intercepts.

**109.** $y = \dfrac{1}{x} - \dfrac{4}{x-1} - 1$

(a)

(b) $x$-intercept: $(-1, 0)$

(c) $0 = \dfrac{1}{x} - \dfrac{4}{x-1} - 1$

$$0 = (x - 1) - 4x - x(x - 1)$$

$$0 = x - 1 - 4x - x^2 + x$$

$$0 = -x^2 - 2x - 1$$

$$0 = x^2 + 2x + 1$$

$$x + 1 = 0 \implies x = -1$$

(d) The $x$-intercepts and the solutions are the same.

**111.** $y = |x + 1| - 2$

(a)
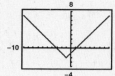

(b) $x$-intercept: $(1, 0), (-3, 0)$

(c) $0 = |x + 1| - 2$

$2 = |x + 1|$

$x + 1 = 2$ or $-(x + 1) = 2$

$x = 1$ or $\quad -x - 1 = 2$

$-x = 3$

$x = -3$

(d) The $x$-intercepts and the solutions are the same.

**113.** $(x + 6)(x - 5) = 0$

$x^2 + x - 30 = 0$

(other answers possible)

**115.** $\left(x - \sqrt{2}\right)\left(x + \sqrt{2}\right)(x - 4) = 0$

$(x^2 - 2)(x - 4) = 0$

$x^3 - 4x^2 - 2x + 8 = 0$

(other answers possible)

**117.** $(x + 2)(x - 2)(x + i)(x - i) = 0$

$(x^2 - 4)(x^2 + 1) = 0$

$x^4 - 3x^2 - 4 = 0$

(other answers possible)

**119.** The distance between $(1, 2)$ and $(x, -10)$ is 13.

$\sqrt{(x - 1)^2 + (-10 - 2)^2} = 13$

$(x - 1)^2 + (-12)^2 = 13^2$

$x^2 - 2x + 1 + 144 = 169$

$x^2 - 2x - 24 = 0$

$(x + 4)(x - 6) = 0$

$x + 4 = 0 \implies x = -4$

$x - 6 = 0 \implies x = \quad 6$

**121.** (a)

w

$w + 14$

(b) $w(w + 14) = 1632$

$w^2 + 14w - 1632 = 0$

(c) $(w + 48)(w - 34) = 0$

$w = 34$, length $= w + 14 = 48$

width 34 feet, length 48 feet

**123.** Let $x$ be the length of the square base. Then $2x^2 = 200 \implies x^2 = 100 \implies x = 10$. Thus, the original piece of material is of length $x + 4 = 14$ cm. Size: $14 \times 14$ centimeters.

**125.** (a) $s_0 = 1815$ and $s_0 = 0$

$s = -16t^2 + 1815$

(b)

| $t$ | 0 | 2 | 4 | 6 | 8 | 10 | 12 |
|---|---|---|---|---|---|---|---|
| $s$ | 1815 | 1751 | 1559 | 1239 | 791 | 215 | $-489$ |

(c) The object reaches the ground between 10 and 12 seconds, $[10, 12]$. In fact, $t \approx 10.651$ seconds.

**127.** (a) $s = -16t^2 + s_0 t + s_0$

$s_0 = 45, s_0 = 5.5$

$s = -16t^2 + 45t + 5.5$

(b) $s\left(\frac{1}{2}\right) = -16\left(\frac{1}{2}\right)^2 + 45\left(\frac{1}{2}\right) + 5.5 = 24$ feet

(c) $-16t^2 + 45t + 5.5 = 6$

$16t^2 - 45t + 0.5 = 0$

Using the quadratic formula, $t \approx 2.801$ sec.

(d)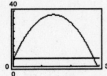

The curves $y = -16t^2 + 45t + 5.5$ and $y = 6$ intersect at $t \approx 2.801$.

**129.** $S = 0.032t^2 - 0.87t + 12.6, 5 \le t \le 12$

(a)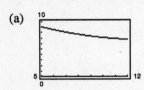

(b) Yes, the parabola eventually turns up. $S = 8$ for $t = 20$, or 2010.

**131.** (a)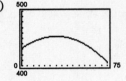

(b) The greatest distance is $d = 453.552$ miles corresponding to $s = 30.313$ mph. The trip will take $\dfrac{d}{s} = \dfrac{453.552}{30.313} = 14.962$ hours.

(c) The distance traveled in 8 hours is $8s$. The point of intersection of $d$ and $y_2 = 8s$ is $s = 54.883$ mph, corresponding to a distance of 439.063 miles.

**133.** $d = \sqrt{100^2 + h^2}$

(a)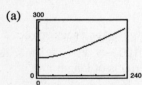

If $d = 200, h \approx 173$.

(b)

| $h$ | 160 | 165 | 170 | 175 | 180 | 185 |
|---|---|---|---|---|---|---|
| $d$ | 188.68 | 192.94 | 197.23 | 201.56 | 205.91 | 210.3 |

$d = 200 \implies h \approx 174$

(c) $\sqrt{100^2 + h^2} = 200$

$100^2 + h^2 = 200^2$

$h^2 = 200^2 - 100^2$

$h \approx 173.205$

(d) Solving graphically or numerically yields an approximate solution. An exact solution is obtained algebraically.

**135.** Let $u$ be the speed of the eastbound plane. Then $u + 50 =$ speed of northbound plane.

$$[u(3)]^2 + [(u + 50)3]^2 = 2440^2$$

$$9u^2 + 9(u + 50)^2 = 2440^2$$

$$18u^2 + 900u + 22,500 = 2440^2$$

$$18u^2 + 900u - 5,931,100 = 0$$

Using a graphing utility, $u \approx 549.57$ mph and $u + 50 = 599.57$ mph.

**137.** False. The solutions are complex numbers.

**139.** False. An equation can have any number of extraneous solutions. For example, $|x| = x^2 + x + 3$

**141.** (a) $ax^2 + bx = 0$

$x(ax + b) = 0$

$x = 0$

$x = -b/a$

(b) $ax^2 - ax = 0$

$ax(x - 1) = 0$

$x = 0$

$x = 1$

**143.** $x^5 - 27x^2 = x^2(x^3 - 27)$

$\qquad = x^2(x - 3)(x^2 + 3x + 9)$

**145.** $x^3 + 5x^2 - 2x - 10 = x^2(x + 5) - 2(x + 5)$

$\qquad = (x^2 - 2)(x + 5)$

$\qquad = \left(x + \sqrt{2}\right)\left(x - \sqrt{2}\right)(x + 5)$

**147.** Yes, $y$ is a function of $x$:
$y = \frac{1}{8}(-1 - 5x)$

**149.** No, $y$ is not a function of $x$:
$y = \pm\sqrt{10 - x}$

**151.** Yes, $y$ is a function of $x$:
$y = |x - 3|$.

# Section 2.5    Solving Inequalities Algebraically and Graphically

- ■ You should know the properties of inequalities.
  - (a) Transitive: $a < b$ and $b < c$ implies $a < c$.
  - (b) Addition: $a < b$ and $c < d$ implies $a + c < b + d$.
  - (c) Adding or Subtracting a Constant: $a \pm c < b \pm c$ if $a < b$.
  - (d) Multiplying or Dividing by a Constant: For $a < b$,

    1. If $a > 0$, then $ac < bc$ and $\dfrac{a}{c} < \dfrac{b}{c}$.

    2. If $c < 0$, then $ac > bc$ and $\dfrac{a}{c} > \dfrac{b}{c}$.

- ■ You should know that

$$|x| = \begin{cases} x & \text{if } x \geq 0 \\ -x & \text{if } x < 0 \end{cases}.$$

**—CONTINUED—**

---

**—CONTINUED—**

■ You should be able to solve absolute value inequalities.

(a) $|x| < a$ if and only if $-a < x < a$.

(b) $|x| > a$ if and only if $x < -a$ or $x > a$.

■ You should be able to solve polynomial inequalities.

(a) Find the critical numbers.

    1. Values that make the expression zero

    2. Values that make the expression undefined

(b) Test one value in each interval on the real number line resulting from the critical numbers.

(c) Determine the solution intervals.

■ You should be able to solve rational and other types of inequalities.

---

**Solutions to Odd-Numbered Exercises**

**1.** $x < 3$

    Matches (d).

**3.** $-3 < x \le 4$

    Matches (c).

**5.** (a) $x = 3$

$$5(3) - 12 \overset{?}{>} 0$$

$$3 > 0$$

    Yes, $x = 3$ is a solution.

(c) $x = \frac{5}{2}$

$$5\left(\tfrac{5}{2}\right) - 12 \overset{?}{>} 0$$

$$\tfrac{1}{2} > 0$$

    Yes, $x = \frac{5}{2}$ is a solution.

(b) $x = -3$

$$5(-3) - 12 \overset{?}{>} 0$$

$$-27 \not> 0$$

    No, $x = -3$ is not a solution.

(d) $x = \frac{3}{2}$

$$5\left(\tfrac{3}{2}\right) - 12 \overset{?}{>} 0$$

$$-\tfrac{9}{2} \not> 0$$

    No, $x = \frac{3}{2}$ is not a solution.

**7.** $-1 < \dfrac{3 - x}{2} \le 1$

(a) $x = 0$

$$-1 \overset{?}{<} \frac{3 - 0}{2} \overset{?}{\le} 1$$

$$-1 \overset{?}{<} \frac{3}{2} \overset{?}{\le} 1$$

    No, $x = 0$ is not a solution.

(c) $x = 1$

$$-1 \overset{?}{<} \frac{3 - 1}{2} \overset{?}{\le} 1$$

$$-1 \overset{?}{<} 1 \overset{?}{\le} 1$$

    Yes, $x = 1$ is a solution.

(b) $x = \sqrt{5}$

$$-1 \overset{?}{<} \frac{3 - \sqrt{5}}{2} \overset{?}{\le} 1$$

$$-1 \overset{?}{<} 0.382 \overset{?}{\le} 1$$

    Yes, $x = \sqrt{5}$ is a solution.

(d) $x = 5$

$$-1 \overset{?}{<} \frac{3 - 5}{2} \le 1$$

$$-1 \overset{?}{<} -1 \overset{?}{\le} 1$$

    No, $x = 5$ is not a solution.

**9.**     $-10x < 40$

$-\frac{1}{10}(-10x) > -\frac{1}{10}(40)$

$x > -4$

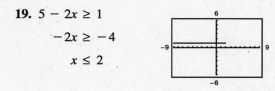

**11.** $4(x + 1) < 2x + 3$

$4x + 4 < 2x + 3$

$2x < -1$

$x < -\frac{1}{2}$

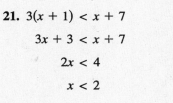

**13.** $\frac{3}{4}x - 6 \le x - 7$

$1 \le \frac{1}{4}x$

$4 \le x$

$x \ge 4$

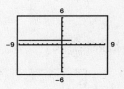

**15.** $-8 \le 1 - 3(x - 2) < 13$

$-8 \le 1 - 3x + 6 < 13$

$-8 \le -3x + 7 < 13$

$-15 \le -3x < 6$

$5 \ge x > -2 \implies -2 < x \le 5$

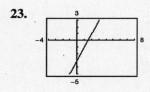

**17.** $-4 < \dfrac{2x - 3}{3} < 4$

$-12 < 2x - 3 < 12$

$-9 < 2x < 15$

$-\dfrac{9}{2} < x < \dfrac{15}{2}$

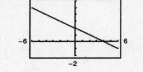

**19.** $5 - 2x \ge 1$

$-2x \ge -4$

$x \le 2$

**21.** $3(x + 1) < x + 7$

$3x + 3 < x + 7$

$2x < 4$

$x < 2$

**23.**

Using the graph, (a) $y \ge 1$ for $x \ge 2$ and (b) $y \le 0$ for $x \le \frac{3}{2}$.

Algebraically, (a)         $y \ge 1$         (b)         $y \le 0$

$2x - 3 \ge 1$                $2x - 3 \le 0$

$2x \ge 4$                      $2x \le 3$

$x \ge 2$                        $x \le \frac{3}{2}$

**25.**

Using the graph, (a) $0 \le y \le 3$ for $-2 \le x \le 4$ and (b) $y \ge 0$ for $x \le 4$

Algebraically, (a)     $0 \le y \le 3$                        (b)         $y \ge 0$

$0 \le -\frac{1}{2}x + 2 \le 3$                $-\frac{1}{2}x + 2 \ge 0$

$-2 \le -\frac{1}{2}x \le 1$                      $2 \ge \frac{1}{2}x$

$4 \ge x \ge -2$                          $4 \ge x$

**27.** $|5x| > 10$

$5x < -10$ or $5x > 10$

$x < -2$   or   $x > 2$

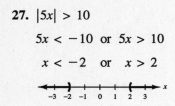

**31.** $|x + 14| + 3 > 17$

$|x + 14| > 14$

$x + 14 < -14$ or $x + 14 > 14$

$x < -28$ or        $x > 0$

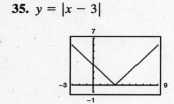

**35.** $y = |x - 3|$

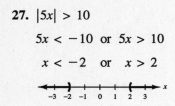

(a) Graphically, $y \le 2$ for $1 \le x \le 5$ and

(b) $y \ge 4$ for $x \le -1$ or $x \ge 7$

**37.** The midpoint of the interval $[-3, 3]$ is 0. The interval represents all real numbers $x$ no more than 3 units from 0.

$|x - 0| \le 3$

$|x| \le 3$

**41.** All real numbers within 10 units of 7

$|x - 7| \le 10$

**29.** $|x - 7| < 6$

$-6 < x - 7 < 6$

$1 < x < 13$

**33.** $10|1 - 2x| < 5$

$|1 - 2x| < \frac{1}{2}$

$-\frac{1}{2} < 1 - 2x < \frac{1}{2}$

$-\frac{3}{2} < -2x < -\frac{1}{2}$

$\frac{3}{4} > x > \frac{1}{4}$

$\frac{1}{4} < x < \frac{3}{4}$

Algebraically,

(a)        $y \le 2$

$|x - 3| \le 2$

$-2 \le x - 3 \le 2$

$1 \le x \le 5$

(b)        $y \ge 4$

$|x - 3| \ge 4$

$x - 3 \le -4$ or   $x - 3 \ge 4$

$x \le -1$ or        $x \ge 7$

**39.** The midpoint of the interval $[-3, 3]$ is 0. The two intervals represent all numbers $x$ more than 3 units from 0.

$|x - 0| > 3$

$|x| > 3$

**43.** $x^2 - 4x - 5 > 0$

$(x - 5)(x + 1) > 0$

Critical numbers: $-1, 5$

Testing the intervals $(-\infty, -1)$, $(-1, 5)$ and $(5, \infty)$, we have $x^2 - 4x - 5 > 0$ on $(-\infty, -1)$ and $(5, \infty)$

Similarly, $x^2 - 4x - 5 < 0$ on $(-1, 5)$

**45.**          $(x + 2)^2 < 25$

$x^2 + 4x + 4 < 25$

$x^2 + 4x - 21 < 0$

$(x + 7)(x - 3) < 0$

Critical numbers: $x = -7, x = 3$

Test intervals: $(-\infty, -7), (-7, 3), (3, \infty)$

Test:  Is $(x + 7)(x - 3) < 0$?

Solution set: $(-7, 3)$

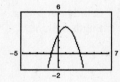

**47.**    $x^2 + 4x + 4 \geq 9$

$x^2 + 4x - 5 \geq 0$

$(x + 5)(x - 1) \geq 0$

Critical numbers: $x = -5, x = 1$

Test intervals: $(-\infty, -5), (-5, 1), (1, \infty)$

Test:  Is $(x + 5)(x - 1) \geq 0$?

Solution set: $(-\infty, -5] \cup [1, \infty)$

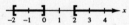

**49.**          $x^3 - 4x \geq 0$

$x(x + 2)(x - 2) \geq 0$

Critical number: $x = 0, x = \pm 2$

Test intervals: $(-\infty, -2), (-2, 0), (0, 2), (2, \infty)$

Test:  Is $x(x + 2)(x - 2) \geq 0$?

Solution set: $[-2, 0] \cup [2, \infty)$

**51.** $y = -x^2 + 2x + 3$

(a) $y \leq 0$ when $x \leq -1$ or $x \geq 3$

(b) $y \geq 3$ when $0 \leq x \leq 2$

Algebraically,

$-x^2 + 2x + 3 \leq 0$

$x^2 - 2x - 3 \geq 0$

$(x - 3)(x + 1) \geq 0$

Critical numbers: $x = -1, x = 3$

Testing the intervals $(-\infty, -1), (-1, 3),$
$(3, \infty)$ you obtain $x \leq -1$ or $x \geq 3$.

$-x^2 + 2x + 3 \geq 3$

$-x^2 + 2x \geq 0$

$x^2 - 2x \leq 0$

$x(x - 2) \leq 0$

Critical numbers: $x = 0, x = 2$

Testing the intervals $(-\infty, 0), (0, 2), (2, \infty)$
you obtain $0 \leq x \leq 2$.

**53.**  $\dfrac{1}{x} - x > 0$

$\dfrac{1 - x^2}{x} > 0$

Critical numbers: $x = 0, x = \pm 1$

Test intervals: $(-\infty, -1), (-1, 0), (0, 1), (1, \infty)$

Test:  Is $\dfrac{1 - x^2}{x} > 0$?

Solution set: $(-\infty, -1) \cup (0, 1)$

**55.**    $\dfrac{x+6}{x+1} - 2 < 0$

$\dfrac{x+6-2(x+1)}{x+1} < 0$

$\dfrac{4-x}{x+1} < 0$

Critical numbers: $x = -1, x = 4$

Test intervals: $(-\infty, -1), (-1, 4), (4, \infty)$

Test: Is $\dfrac{4-x}{x+1} < 0$?

Solution set: $(-\infty, -1) \cup (4, \infty)$

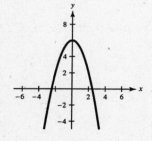

**57.** $y = \dfrac{3x}{x-2}$

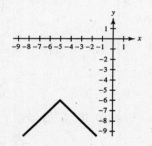

(a) $y \le 0$ when $0 \le x < 2$.

(b) $y \ge 6$ when $2 < x \le 4$.

**59.** $\sqrt{x-5}$

Need $x - 5 \ge 0$

$x \ge 5$

Domain: $[5, \infty)$

**61.** $\sqrt[3]{6-x}$

Domain: all real $x$

**63.** $\sqrt{x^2 - 4}$

Need $x^2 - 4 \ge 0$

$(x+2)(x-2) \ge 0$

$x \le -2$ or $x \ge 2$

Domain: $(-\infty, -2] \cup [2, \infty)$

**65.** (a), (b)

(c) For $y \ge 200$, $x \ge 181.5$ pounds

(d) The model is not accurate. The data is not linear. Other factors include muscle strength, height, physical condition, etc.

**67.** When $t = 2$, $v \approx 333$ vibrations per second

**69.** When $200 \le v \le 400$, $1.2 < t < 2.4$

**71.** False. If $-10 \le x \le 8$, then $10 \ge -x$ and $-x \ge -8$.

**73.** The polynomial $f(x) = (x - a)(x - b)$ is zero at $x = a$ and $x = b$.

**75.** $d = \sqrt{(1 - (-4))^2 + (12 - 2)^2} = \sqrt{25 + 100}$

$= \sqrt{125} = 5\sqrt{5}$

Midpoint: $\left(\dfrac{-4+1}{2}, \dfrac{2+12}{2}\right) = \left(-\dfrac{3}{2}, 7\right)$

**77.** $d = \sqrt{(3 - (-5))^2 + (6 - (-8))^2} = \sqrt{64 + 196}$

$= \sqrt{260} = 2\sqrt{65}$

Midpoint: $\left(\dfrac{3-5}{2}, \dfrac{6-8}{2}\right) = (-1, -1)$

**79.** $f(x) = -x^2 + 6$

**81.** $f(x) = -|x + 5| - 6$

**83.**  $y = 12x$

$x = 12y$

$\dfrac{x}{12} = y$

$f^{-1}(x) = \dfrac{x}{12}$

**85.**  $y = x^3 + 7$

$x = y^3 + 7$

$x - 7 = y^3$

$\sqrt[3]{x - 7} = y$

$f^{-1}(x) = \sqrt[3]{x - 7}$

## Section 2.6   Exploring Data:  Linear Models and Scatter Plots

**Solutions to Odd-Numbered Exercises**

**1.** (a)

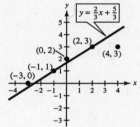

(b)  Yes, the data appears somewhat linear. The more experience, $x$, corresponds to higher sales, $y$.

**3.** Negative correlation—$y$ decreases as $x$ increases.

**5.** No correlation.

**7.** (a)

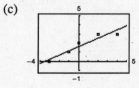

(b)  $y = 0.46x + 1.62$

(c)

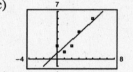

(d)  Yes, the model appears valid.

**9.** (a)

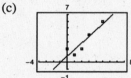

(b)  $y = 0.95x + 0.92$

(c)

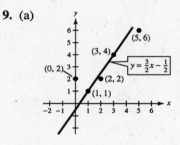

(d)  Yes, the model appears valid.

**11.** (a)

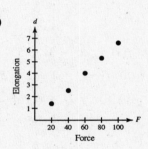

(b)  $d = 0.07F - 0.3$

(c)  $d = 0.066F$  or

$F = 15.13d + 0.096$

(d)  If $F = 55$,

$d = 0.066(55) \approx 3.63$ cm

**13.** (a)  $S = 0.2t - 0.14$

(b)

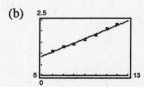

(c)  The slope 0.2 is the average annual increase of salary in millions of dollars.

(d)  For 2006, $t = 16$ and $S \approx 3.06$ or about $\$3.1$ million.

**15.** (a) $S = 0.183t + 7.013$

(b)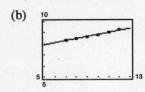

(c) The slope 0.183 is the average annual increase in monthly spending.

(d) For 2008, $t = 18$ and $S \approx \$10.31$

(e)

| Year | 1997 | 1998 | 1999 | 2000 | 2001 | 2002 |
|-------|------|------|------|------|------|------|
| $S$ | 8.30 | 8.50 | 8.65 | 8.80 | 9.00 | 9.25 |
| Model | 8.29 | 8.48 | 8.66 | 8.84 | 9.03 | 9.21 |

The model fits well

**17.** (a) $y = -0.024x + 5.06$

(b) The negative slope indicates that the times are decreasing

(c)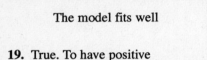

(d) The model is reasonably close

(e) Answers will vary

**19.** True. To have positive correlation, the $y$-values tend to increase as $x$ increases

**21.** Answers will vary

**23.** $P \leq 2, (-\infty, 2]$

**25.** $-3 \leq z \leq 10, [-3, 10]$

**27.** $\dfrac{x^2 - 4}{\left(\dfrac{x+2}{5}\right)} = \dfrac{(x+2)(x-2)5}{x+2} = 5x - 10, x \neq -2$

**29.** $f(x) = 2x^2 - 3x + 5$

(a) $f(-1) = 2 + 3 + 5 = 10$

(b) $f(w + 2) = 2(w+2)^2 - 3(w+2) + 5$
$= 2w^2 + 5w + 7$

**31.** $h(x) = \begin{cases} 1 - x^2, & x \leq 0 \\ 2x + 3, & x > 0 \end{cases}$

(a) $h(1) = 2(1) + 3 = 5$

(b) $h(0) = 1 - 0 = 1$

**33.** $6x + 1 = -9x - 8$

$15x = -9$

$x = -\dfrac{9}{15} = -\dfrac{3}{5}$

**35.** $8x^2 - 10x - 3 = 0$

$(4x + 1)(2x - 3) = 0$

$x = -\dfrac{1}{4}, \dfrac{3}{2}$

**37.** $2x^2 - 7x + 4 = 0$

$x = \dfrac{7 \pm \sqrt{49 - 4(4)(2)}}{4}$

$= \dfrac{7 \pm \sqrt{17}}{4}$

## Review Exercises for Chapter 2

**Solutions to Odd-Numbered Exercises**

**1.** $6 + \dfrac{3}{x-4} = 5$

(a) $x = 5$     $6 + \dfrac{3}{5-4} \overset{?}{=} 5$     No, $x = 5$ is not a solution.

$6 + 3 \overset{?}{=} 5$

$9 \neq 5$

(c) $x = -2$     $6 + \dfrac{3}{-2-4} \overset{?}{=} 5$     No, $x = -2$ is not a solution.

$6 - \dfrac{1}{2} \overset{?}{=} 5$

$5.5 \neq 5$

(b) $x = 0$     $6 + \dfrac{3}{0-4} \overset{?}{=} 5$     No, $x = 0$ is not a solution.

$6 - \dfrac{3}{4} \overset{?}{=} 5$

$5.25 \neq 5$

(d) $x = 1$     $6 + \dfrac{3}{1-4} \overset{?}{=} 5$     Yes, $x = 1$ is a solution.

$6 - 1 \overset{?}{=} 5$

$5 = 5$

**3.** $\dfrac{18}{x} = \dfrac{10}{x-4}$

$18(x-4) = 10x$

$18x - 72 = 10x$

$8x = 72$

$x = 9$

**5.** $14 + \dfrac{2}{x-1} = 10$

$\dfrac{2}{x-1} = -4$

$2 = -4(x-1)$

$2 = -4x + 4$

$4x = 2$

$x = \dfrac{1}{2}$

**7.** $\dfrac{9x}{3x-1} - \dfrac{4}{3x+1} = 3$

$9x(3x+1) - 4(3x-1) = 3(3x-1)(3x+1)$

$27x^2 + 9x - 12x + 4 = 3(9x^2 - 1)$

$27x^2 - 3x + 4 = 27x^2 - 3$

$-3x = -7$

$x = \dfrac{7}{3}$

**9.** September's profit + October's profit = 689,000

Let $x$ = September's profit.

Then $x + 0.12x$ = October's profit

$x + (x + 0.12x) = 689,000$

$2.12x = 689,000$

$x = 325,000$

$x + 0.12x = 364,000$

September: $325,000

October: $364,000

**11.** Let $x$ = the number of quarts of pure antifreeze.

$30\%$ of $(10 - x) + 100\%$ of $x = 50\%$ of $10$

$$0.30(10 - x) + 1.00x = 0.50(10)$$

$$3 - 0.30x + 1.00x = 5$$

$$0.70x = 2$$

$$x = \frac{2}{0.70}$$

$$= \frac{20}{7}$$

$$= 2\frac{6}{7} \text{ liters}$$

**13.** (a)

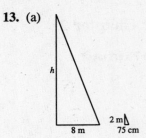

(b) $\dfrac{h}{8} = \dfrac{2}{\frac{3}{4}} = \dfrac{8}{3}$

$h = \dfrac{64}{3} = 21\dfrac{1}{3}$ meters high

**15.**     $F = \frac{9}{5}C + 32$

$25.7 = \frac{9}{5}C + 32$

$-6.3 = \frac{9}{5}C$

$C = -3.5$

$-3.5°$ Celsius

**17.** $-x + y = 3$

Let $x = 0$: $y = 3$.   $y$-intercept: $(0, 3)$

Let $y = 0$: $x = -3$. $x$-intercept: $(-3, 0)$

**19.** $y = x^2 - 9x + 8 = (x - 8)(x - 1)$

Let $x = 0$: $y = 8$.   $y$-intercept: $(0, 8)$

Let $y = 0$: $x = 1, 8$.   $x$-intercepts: $(1, 0), (8, 0)$

**21.** $y = -|x + 5| - 2$

$y$-intercept: $(0, -7)$

No $x$-intercepts

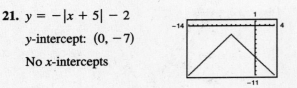

**23.** $5(x - 2) - 1 = 0$

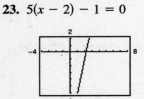

Solution: $x = 2.2$

**25.** $3x^3 - 2x + 4 = 0$

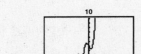

Solution: $x = -1.301$

**27.** $x^4 - 3x + 1 = 0$

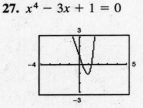

Solutions: $x = 1.307, x = 0.338$

**29.**   $3x + 5y = -7$

$-x - 2y = 3$

From second equation, $x = -2y - 3$. Then

$3(-2y - 3) + 5y = 7$

$-y - 9 = -7$

$y = -2$ and $x = -2(-2) - 3 = 1$

Intersection point $(1, -2)$

**31.** $x^2 + 2y = 14$

$3x + 4y = 1$

From equation 2, $y = \frac{1}{4}(1 - 3x)$. Then

$x^2 + 2\left(\frac{1}{4}\right)(1 - 3x) = 14$

$x^2 + \frac{1}{2} - \frac{3}{2}x = 14$

$2x^2 - 3x - 27 = 0$

$(2x - 9)(x + 3) = 0$

$x = \frac{9}{2} \implies y = \frac{1}{4}\left(1 - 3\left(\frac{9}{2}\right)\right) = -\frac{25}{8}$

$x = -3 \implies y = \frac{1}{4}(1 - 3(-3)) = \frac{5}{2}$

Intersection points: $\left(-3, \frac{5}{2}\right), \left(\frac{9}{2}, -\frac{25}{8}\right)$

**33.** $6 + \sqrt{-25} = 6 + 5i$

**35.** $-2i^2 + 7i = 2 + 7i$

**37.** $(7 + 5i) + (-4 + 2i) = (7 - 4) + (5i + 2i)$
$$= 3 + 7i$$

**39.** $5i(13 - 8i) = 65i - 40i^2 = 40 + 65i$

**41.** $(10 - 8i)(2 - 3i) = 20 - 30i - 16i + 24i^2 = -4 - 46i$

**43.** $(3 + 7i)^2 + (3 - 7i)^2 = (9 + 42i - 49) + (9 - 42i - 49)$
$$= -80$$

**45.** $\dfrac{6 + i}{i} = \dfrac{6 + i}{i} \cdot \dfrac{-i}{-i} = \dfrac{-6i - i^2}{-i^2}$
$$= \dfrac{-6i + 1}{1} = 1 - 6i$$

**47.** $\dfrac{3 + 2i}{5 + i} \cdot \dfrac{5 - i}{5 - i} = \dfrac{15 + 10i - 3i + 2}{25 + 1}$
$$= \dfrac{17}{26} + \dfrac{7}{26}i$$

**49.** $2 - 5i$

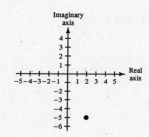

**51.** $-6i$

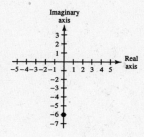

**53.** $3$

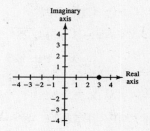

**55.** $6x = 3x^2$
$$0 = 3x^2 - 6x$$
$$0 = 3x(x - 2)$$
$$3x = 0 \implies x = 0$$
$$x - 2 = 0 \implies x = 2$$

**57.** $(x + 4)^2 = 18$
$$x + 4 = \pm\sqrt{18}$$
$$x = -4 \pm 3\sqrt{2}$$

**59.** $x^2 - 12x + 30 = 0$
$$x^2 - 12x = -30$$
$$x^2 - 12x + 36 = -30 + 36$$
$$(x - 6)^2 = 6$$
$$x - 6 = \pm\sqrt{6}$$
$$x = 6 \pm \sqrt{6}$$

**61.** $2x^2 + 9x - 5 = 0$
$$(2x - 1)(x + 5) = 0$$
$$x = \frac{1}{2}, -5$$

**63.** $x^2 + 4x + 10 = 0$
$$x = \frac{-4 \pm \sqrt{16 - 40}}{2}$$
$$= \frac{-4 \pm \sqrt{-24}}{2}$$
$$= -2 \pm \sqrt{6}i$$

**65.** $3x^3 - 26x^2 + 16x = 0$
$$x(3x^2 - 26x + 16) = 0$$
$$x(3x - 2)(x - 8) = 0$$
$$x = 0, \frac{2}{3}, 8$$

**67.** $5x^4 - 12x^3 = 0$
$$x^3(5x - 12) = 0$$
$$x^3 = 0 \quad \text{or} \quad 5x - 12 = 0$$
$$x = 0 \quad \text{or} \quad x = \tfrac{12}{5}$$

**69.** $\sqrt{x + 4} = 3$
$$\left(\sqrt{x + 4}\right)^2 = (3)^2$$
$$x + 4 = 9$$
$$x = 5$$

**71.** $2\sqrt{x} - 5 = 0$

$\qquad 2\sqrt{x} = 5$

$\qquad 4x = 25$

$\qquad x = \frac{25}{4}$

**73.** $\sqrt{2x + 3} + \sqrt{x - 2} = 2$

$\qquad \left(\sqrt{2x + 3}\right)^2 = \left(2 - \sqrt{x - 2}\right)^2$

$\qquad 2x + 3 = 4 - 4\sqrt{x - 2} + x - 2$

$\qquad x + 1 = -4\sqrt{x - 2}$

$\qquad (x + 1)^2 = \left(-4\sqrt{x - 2}\right)^2$

$\qquad x^2 + 2x + 1 = 16(x - 2)$

$\qquad x^2 - 14x + 33 = 0$

$\qquad (x - 3)(x - 11) = 0$

$x = 3$, extraneous or $x = 11$, extraneous

No solution. (You can verify that the graph of
$y = \sqrt{2x + 3} + \sqrt{x - 2} - 2$ lies above the
$x$-axis.)

**75.** $(x - 1)^{2/3} - 25 = 0$

$\qquad (x - 1)^{2/3} = 25$

$\qquad (x - 1)^2 = 25^3$

$\qquad x - 1 = \pm\sqrt{25^3}$

$\qquad x = 1 \pm 125$

$\qquad x = 126 \quad \text{or} \quad x = -124$

**77.** $(x + 4)^{1/2} + 5x(x + 4)^{3/2} = 0$

$(x + 4)^{1/2}[1 + 5x(x + 4)] = 0$

$(x + 4)^{1/2}(5x^2 + 20x + 1) = 0$

$\qquad (x + 4)^{1/2} = 0 \quad \text{or} \quad 5x^2 + 20x + 1 = 0$

$\qquad\qquad x = -4$

$$x = \frac{-20 \pm \sqrt{400 - 20}}{10}$$

$$x = \frac{-20 \pm 2\sqrt{95}}{10}$$

$$x = -2 \pm \frac{\sqrt{95}}{5}$$

**79.** $3\left(1 - \dfrac{1}{5t}\right) = 0$

$\qquad 1 - \dfrac{1}{5t} = 0$

$\qquad 1 = \dfrac{1}{5t}$

$\qquad 5t = 1$

$\qquad t = \dfrac{1}{5}$

**81.** $\dfrac{4}{(x - 4)^2} = 1$

$\qquad 4 = (x - 4)^2$

$\qquad \pm 2 = x - 4$

$\qquad 4 \pm 2 = x$

$\qquad x = 6 \quad \text{or} \quad x = 2$

**83.** $|x - 5| = 10$

$x - 5 = -10$  or  $x - 5 = 10$

$x = -5$            $x = 15$

**85.** $|x^2 - 3| = 2x$

$x^2 - 3 = 2x$    or    $x^2 - 3 = -2x$

$x^2 - 2x - 3 = 0$        $x^2 + 2x - 3 = 0$

$(x - 3)(x + 1) = 0$      $(x + 3)(x - 1) = 0$

$x = 3$  or  $x = -1$    $x = -3$  or  $x = 1$

The only solutions to the original equation are $x = 3$ or $x = 1$. ($x = -3$ and $x = -1$ are extraneous.)

**87.** $P = 0.11t^2 + 1.5t + 728, 5 \leq t \leq 11$

$t = 5$ corresponds to 1995.

(a) $0.11t^2 + 1.5t + 728 = 750$

$0.11t^2 + 1.5t - 22 = 0$

$t = \dfrac{-1.5 \pm \sqrt{1.5^2 - 4(0.11)(-22)}}{2(0.11)}$

$= \dfrac{-1.5 \pm \sqrt{11.93}}{0.22}$

$\approx 8.88$  (use positive root)

Some time in late 1998

(b)

| t | 8.5 | 8.6 | 8.7 | 8.8 | 8.9 | 9.0 |
|---|-----|-----|-----|-----|-----|-----|
| P | 748.7 | 749.0 | 749.4 | 749.7 | 750.1 | 750.4 |

(c)

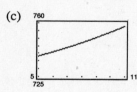

(d) $P = 740$ when $t \approx 5.65$ or between 1995 and 1996

(e) $0.11t^2 + 1.5t + 728 = 740$

$0.11t^2 + 1.5t - 12 = 0$

$t = \dfrac{-1.5 \pm \sqrt{1.5^2 - 4(0.11)(-12)}}{2(0.11)}$

$= \dfrac{-1.5 \pm \sqrt{7.53}}{0.22}$

$\approx 5.64$  (use positive root)

**89.** $8x - 3 < 6x + 15$

$2x < 18$

$x < 9$

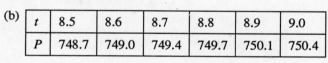

**91.** $\frac{1}{2}(3 - x) > \frac{1}{3}(2 - 3x)$

$3(3 - x) > 2(2 - 3x)$

$9 - 3x > 4 - 6x$

$3x > -5$

$x > -\frac{5}{3}, \left(-\frac{5}{3}, \infty\right)$

**93.** $-2 < -x + 7 \leq 10$

$-9 < -x \leq 3$

$9 > x \geq -3$

$-3 \leq x < 9$

**95.** $|x - 2| < 1$

$-1 < x - 2 < 1$

$1 < x < 3$

which can be written as $(1, 3)$.

**97.** $\left|x - \frac{3}{2}\right| \geq \frac{3}{2}$

$x - \frac{3}{2} \leq -\frac{3}{2}$ or $x - \frac{3}{2} \geq \frac{3}{2}$

$x \leq 0$ or $x \geq 3$

which can be written as $(-\infty, 0] \cup [3, \infty)$.

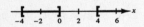

**99.** $4|3 - 2x| \leq 16$

$|3 - 2x| \leq 4$

$-4 \leq 3 - 2x \leq 4$

$-7 \leq -2x \leq 1$

$\frac{7}{2} \geq x \geq -\frac{1}{2}$

$-\frac{1}{2} \leq x \leq \frac{7}{2}$

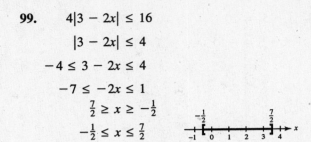

**101.** $x^2 - 2x \geq 3$

$x^2 - 2x - 3 \geq 0$

$(x - 3)(x + 1) \geq 0$

Test intervals: $(-\infty, -1), (-1, 3), (3, \infty)$

$x \geq 3$ or $x \leq -1$

$(-\infty, -1] \cup [3, \infty)$

**103.** $4x^2 - 23x \leq 6$

$4x^2 - 23x - 6 \leq 0$

$(x - 6)(4x + 1) \leq 0$

Critical numbers: $6, -\frac{1}{4}$. Testing the three intervals, we obtain $-\frac{1}{4} \leq x \leq 6$

**105.** $x^3 - 16x \geq 0$

$x(x - 4)(x + 4) \geq 0$

Critical numbers: $0, 4, -4$. Testing the four intervals, we obtain $-4 \leq x \leq 0$ or $x \geq 4$

**107.** $\frac{x - 5}{3 - x} < 0$

Critical numbers: $x = 5, x = 3$

Test intervals: $(-\infty, 3), (3, 5), (5, \infty)$

Test: Is $\frac{x - 5}{3 - x} < 0$?

Solution set: $(-\infty, 3) \cup (5, \infty)$

**109.** $\frac{3x + 8}{x - 3} - 4 \leq 0$

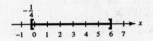

$\frac{3x + 8 - 4(x - 3)}{x - 3} \leq 0$

$\frac{20 - x}{x - 3} \leq 0$

$\frac{x - 20}{x - 3} \geq 0$

Critical numbers: $x = 3, 20$. Testing the three intervals, we obtain $x \geq 20$ or $x < 3$.

**111.** $\left(20.8 - \frac{1}{16}\right)^2 \leq \text{Area} \leq \left(20.8 + \frac{1}{16}\right)^2$

$430.044 \leq \text{Area} \leq 435.244$ square inches

**113.** (a)

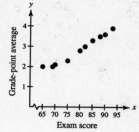

(b) Yes, the relationship is approximately linear. Higher entrance exam scores $x$ are associated with higher grade-point averages, $y$.

**117.** (a) $y = 95.174x - 458.423$

(b)

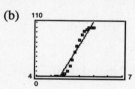

**119.** False. A function can have only one $y$-intercept (Vertical line test)

**123.** They are the same. A point $(a, 0)$ is an $x$-intercept if it is a solution point of the equation. In other words, $a$ is a zero of the function.

**127.** (a) $i^{40} = (i^4)^{10} = 1^{10} = 1$

(b) $i^{25} = i(i^{24}) = i(1) = i$

(c) $i^{50} = i^2(i^{48}) = (-1)(1) = -1$

(d) $i^{67} = i^3(i^{64}) = -i(1) = -i$

**115.** (a)

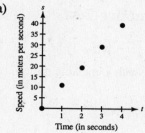

(b) $S \approx 10t$ (approximations will vary)

(c) $S = 9.7t + 0.4$

(d) For $t = 2.5$, $S \approx 24.7$ m/sec

(c) The model does not fit well

(d) No. The data stops at (6.00, 100.0)

**121.** False. The slope can be positive, negative or 0.

**125.** $\sqrt{-6}\sqrt{-6} \neq \sqrt{(-6)(-6)}$

In fact, $\sqrt{-6}\sqrt{-6} = \sqrt{6}i\sqrt{6}i = -6$

# Chapter 2    Practice Test

**1.** Solve the equation $\frac{1}{2}x - \frac{1}{3}(x - 1) = 10$.

Verify your answer with a graphing utility.

**2.** Solve the equation $(x + 1)^2 - 6 = x^2 + 3x$ and verify your answer with a graphing utility.

**3.** Solve $A = \frac{1}{2}(a + b)h$ for $a$.

**4.** 301 is what percent of 4300?

**5.** Cindy has \$6.05 in quarters and nickels. How many of each coin does she have if there are 53 coins in all?

**6.** Ed has \$15,000 invested in two funds paying $9\frac{1}{2}\%$ and $11\%$ simple interest, respectively. How much is invested in each if the yearly interest is \$1582.50?

**7.** Use a graphing utility to approximate any points of intersection of $y = 3x^2 - 4$ and $y = 2 - x$.

**8.** Use a graphing utility to approximate any points of intersection of $y = 2x^2 + 3$ and $y = 5 + \sqrt{x}$.

**9.** Write $\dfrac{2}{1 + i}$ in standard form.

**10.** Write $\dfrac{3 + i}{2} - \dfrac{i + 1}{4}$ in standard form.

**11.** Solve $28 + 5x - 3x^2 = 0$ by factoring.

**12.** Solve $(x - 2)^2 = 24$ by taking the square root of both sides.

**13.** Solve $x^2 - 4x - 9 = 0$ by completing the square.

**14.** Solve $x^2 + 5x - 1 = 0$ by the Quadratic Formula.

**15.** Solve $3x^2 - 2x + 4 = 0$ by the Quadratic Formula.

**16.** The perimeter of a rectangle is 1100 feet. Find the dimension so that the enclosed area will be 60,000 square feet.

**17.** Find two consecutive even positive integers whose product is 624.

**18.** Solve $x^3 - 10x^2 + 24x = 0$ by factoring.

**19.** Solve $\sqrt[3]{6 - x} = 4$.

**20.** Solve $(x^2 - 8)^{2/5} = 4$.

**21.** Solve $x^4 - x^2 - 12 = 0$.

**22.** Solve $4 - 3x > 16$.

**23.** Solve $\left|\dfrac{x - 3}{2}\right| < 5$.

**24.** Solve $\dfrac{x + 1}{x - 3} < 2$.

**25.** Solve $|3x - 4| \geq 9$.

**26.** Use a graphing utility to find the least squares regression line for the points $(-1, 0)$, $(0, 1)$, $(3, 3)$ and $(4, 5)$. Graph the points and the line.

# CHAPTER 3
# Polynomial and Rational Functions

# CHAPTER 3
## Polynomial and Rational Functions

### Section 3.1  Quadratic Functions

---

You should know the following facts about parabolas.

- $f(x) = ax^2 + bx + c$, $a \neq 0$, is a quadratic function, and its graph is a parabola.
- If $a > 0$, the parabola opens upward and the vertex is the minimum point. If $a < 0$, the parabola opens downward and the vertex is the maximum point.
- The vertex is $(-b/2a, f(-b/2a))$.
- To find the $x$-intercepts (if any), solve
  $$ax^2 + bx + c = 0.$$
- The standard form of the equation of a parabola is
  $$f(x) = a(x - h)^2 + k$$
  where $a \neq 0$.
  - (a) The vertex is $(h, k)$.
  - (b) The axis is the vertical line $x = h$.

---

### Solutions to Odd-Numbered Exercises

**1.** $f(x) = (x - 2)^2$ opens upward and has vertex $(2, 0)$.

Matches graph (g).

**3.** $f(x) = x^2 - 2$ opens upward and has vertex $(0, -2)$.

Matches graph (b).

**5.** $f(x) = 4 - (x - 2)^2 = -(x - 2)^2 + 4$ opens downward and has vertex $(2, 4)$.

Matches graph (f).

**7.** $f(x) = x^2 + 3$ opens upward and has vertex $(0, 3)$.

Matches graph (e)

**9.** (a) $y = \frac{1}{2}x^2$ — vertical shrink

    (b) $y = \frac{1}{2}x^2 - 1$ — vertical shrink and vertical shift 1 unit downward

    (c) $y = \frac{1}{2}(x + 3)^2$ — vertical shrink and horizontal shift 3 units to the left

    (d) $y = -\frac{1}{2}(x + 3)^2 - 1$ — horizontal shift 3 units to the left, vertical shrink, reflection in $x$-axis, and vertical shift 1 unit downward

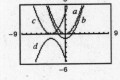

**11.** (a) $y = -2x^2$ — vertical stretch and reflection in the $x$-axis

    (b) $y = -2x^2 - 1$ — vertical stretch, reflection in the $x$-axis, and vertical shift 1 unit downward

    (c) $y = -2(x - 3)^2$ — horizontal shift 3 units to the right, vertical stretch, and reflection in the $x$-axis

    (d) $y = 2(x - 3)^2 - 1$ — horizontal shift 3 units to the right, vertical stretch, and vertical shift 1 unit downward

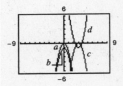

120

**13.** $f(x) = 25 - x^2$

Vertex: $(0, 25)$

$x$-Intercepts: $(-5, 0), (5, 0)$

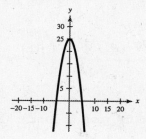

**15.** $f(x) = \frac{1}{2}x^2 - 4$

Vertex: $(0, -4)$

$x$-Intercepts: $(\pm 2\sqrt{2}, 0)$

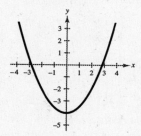

**17.** $f(x) = (x + 4)^2 - 3$

Vertex: $(-4, -3)$

$x$-Intercepts: $(-4 \pm \sqrt{3}, 0)$

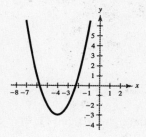

**19.** $h(x) = x^2 - 8x + 16 = (x - 4)^2$

Vertex: $(4, 0)$

$x$-Intercepts: $(4, 0)$

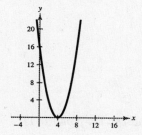

**21.** $f(x) = x^2 - x + \frac{5}{4} = \left(x - \frac{1}{2}\right)^2 + 1$

Vertex: $\left(\frac{1}{2}, 1\right)$

$x$-Intercepts: None

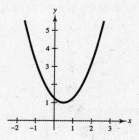

**23.** $f(x) = -x^2 + 2x + 5 = -(x - 1)^2 + 6$

Vertex: $(1, 6)$

$x$-Intercepts: $\left(1 - \sqrt{6}, 0\right), \left(1 + \sqrt{6}, 0\right)$

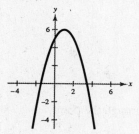

**25.** $h(x) = 4x^2 - 4x + 21 = 4\left(x - \frac{1}{2}\right)^2 + 20$

Vertex: $\left(\frac{1}{2}, 20\right)$

$x$-Intercept: None

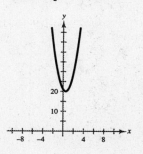

**27.** $f(x) = -(x^2 + 2x - 3) = -(x + 1)^2 + 4$

Vertex: $(-1, 4)$

$x$-Intercepts: $(-3, 0), (1, 0)$

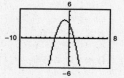

**29.** $g(x) = x^2 + 8x + 11 = (x + 4)^2 - 5$

Vertex: $(-4, -5)$

$x$-Intercepts: $\left(-4 \pm \sqrt{5}, 0\right)$

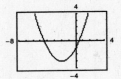

**31.** $f(x) = -2x^2 + 16x - 31$

$\quad = -2\left(x^2 - 8x + \frac{31}{2}\right)$

$\quad = -2\left(x^2 - 8x + 16 - \frac{1}{2}\right)$

$\quad = -2(x - 4)^2 + 1$

Vertex: $(4, 1)$

$x$-Intercept: $\left(4 \pm \frac{1}{2}\sqrt{2}, 0\right)$

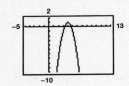

**33.** $g(x) = \frac{1}{2}(x^2 + 4x - 2) = \frac{1}{2}(x^2 + 4x + 4 - 6)$

$\qquad\qquad\qquad\qquad = \frac{1}{2}(x + 2)^2 - 3$

Vertex: $(-2, -3)$

$x$-Intercepts: $\left(-2 \pm \sqrt{6}, 0\right)$

**35.** $(1, 0)$ is the vertex.

$\quad f(x) = a(x - 1)^2 + 0 = a(x - 1)^2$

Since the graph passes through the point $(0, 1)$ we have:

$\quad 1 = a(0 - 1)^2$

$\quad 1 = a$

$\quad f(x) = 1(x - 1)^2 = (x - 1)^2$

**37.** $(-1, 4)$ is the vertex.

$\quad f(x) = a(x + 1)^2 + 4$

Since the graph passes through the point $(1, 0)$ we have

$\quad 0 = a(1 + 1)^2 + 4$

$\quad 0 = 4a + 4$

$\quad -1 = a$

Thus, $f(x) = -(x + 1)^2 + 4$. Note that $(-3, 0)$ is on the parabola.

**39.** $(-2, 5)$ is the vertex.

$\quad f(x) = a(x + 2)^2 + 5$

Since the graph passes through the point $(0, 9)$, we have:

$\quad 9 = a(0 + 2)^2 + 5$

$\quad 4 = 4a$

$\quad 1 = a$

$\quad f(x) = 1(x + 2)^2 + 5 = (x + 2)^2 + 5$

**41.** $\left(\frac{5}{2}, -\frac{3}{4}\right)$ is the vertex.

$\quad f(x) = a\left(x - \frac{5}{2}\right)^2 - \frac{3}{4}$

Since the graph passes through $(-2, 4)$,

$\quad 4 = a\left(-2 - \frac{5}{2}\right)^2 - \frac{3}{4}$

$\quad \frac{19}{4} = a\left(-\frac{9}{2}\right)^2$

$\quad 19 = 81a$

$\quad a = \frac{19}{81}$

Thus, $f(x) = \frac{19}{81}\left(x - \frac{5}{2}\right)^2 - \frac{3}{4}$

**43.** $y = x^2 - 4x - 5 \qquad\quad 0 = x^2 - 4x - 5$

$\quad x$-intercepts: $(5, 0), (-1, 0) \quad 0 = (x - 5)(x + 1)$

$\qquad\qquad\qquad\qquad\qquad\qquad x = 5 \text{ or } x = -1$

**45.** $y = x^2 + 8x + 16$

$x$-intercept: $(-4, 0)$

$0 = x^2 + 8x + 16$

$0 = (x + 4)^2$

$x = -4$

$x$-intercept: $(-4, 0)$

**47.** $y = x^2 - 4x$

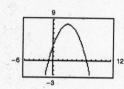

$0 = x^2 - 4x$

$0 = x(x - 4)$

$x = 0$ or $x = 4$

$x$-intercepts: $(0, 0)$, $(4, 0)$

**49.** $y = 2x^2 - 7x - 30$

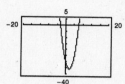

$0 = 2x^2 - 7x - 30$

$0 = (2x + 5)(x - 6)$

$x = -\frac{5}{2}$ or $x = 6$

$x$-intercepts: $\left(-\frac{5}{2}, 0\right)$, $(6, 0)$

**51.** $y = -\frac{1}{2}(x^2 - 6x - 7)$

$0 = -\frac{1}{2}(x^2 - 6x - 7)$

$0 = x^2 - 6x - 7$

$0 = (x + 1)(x - 7)$

$x = -1, 7$

$x$-intercepts: $(-1, 0)$, $(7, 0)$

**53.** $f(x) = [x - (-1)](x - 3)$   opens upward

$= (x + 1)(x - 3)$

$= x^2 - 2x - 3$

$g(x) = -[x - (-1)](x - 3)$   opens downward

$= -(x + 1)(x - 3)$

$= -(x^2 - 2x - 3)$

$= -x^2 + 2x + 3$

Note: $f(x) = a(x + 1)(x - 3)$ has $x$-intercepts $(-1, 0)$ and $(3, 0)$ for all real numbers $a \neq 0$.

**55.** $f(x) = [x - (-3)]\left[x - \left(-\frac{1}{2}\right)\right](2)$   opens upward

$= (x + 3)\left(x + \frac{1}{2}\right)(2)$

$= (x + 3)(2x + 1)$

$= 2x^2 + 7x + 3$

$g(x) = -(2x^2 + 7x + 3)$   opens downward

$= -2x^2 - 7x - 3$

Note:  $f(x) = a(x + 3)(2x + 1)$ has $x$-intercepts $(-3, 0)$ and $\left(-\frac{1}{2}, 0\right)$ for all real numbers $a \neq 0$.

**57.** Let $x =$ the first number and $y =$ the second number. Then the sum is

$$x + y = 110 \implies y = 110 - x.$$

The product is

$$P(x) = xy = x(110 - x) = 110x - x^2.$$

$$P(x) = -x^2 + 110x$$

$$= -(x^2 - 110x + 3025 - 3025)$$

$$= -[(x - 55)^2 - 3025]$$

$$= -(x - 55)^2 + 3025$$

The maximum value of the product occurs at the vertex of $P(x)$ and is 3025. This happens when $x = y = 55$.

**59.** Let $x$ be the first number and $y$ be the second number.

Then $x + 2y = 24 \implies x = 24 - 2y$.

The product is $P = xy = (24 - 2y)y = 24y - 2y^2$.

Completing the square,

$$P = -2y^2 + 24y$$

$$= -2(y^2 - 12y + 36) + 72$$

$$= -2(y - 6)^2 + 72.$$

The maximum value of the product $P$ occurs at the vertex of the parabola and equals 72. This happens when $y = 6$ and $x = 24 - 2(6) = 12$.

**61. (a)**

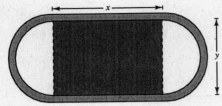

**(b)** Radius of semicircular ends of track: $r = \frac{1}{2}y$ distance around two semicircular parts of track:

$$d = 2\pi r = 2\pi\left(\frac{1}{2}y\right) = \pi y$$

**(c)** Distance traveled around track in one lap:

$$d = \pi y + 2x = 200$$

$$\pi y = 200 - 2x$$

$$y = \frac{200 - 2x}{\pi}$$

**(d)** Area of rectangular region:

$$A = xy = x\left(\frac{200 - 2x}{\pi}\right)$$

$$= \frac{1}{\pi}(200x - 2x^2)$$

$$= -\frac{2}{\pi}(x^2 - 100x)$$

$$= -\frac{2}{\pi}(x^2 - 100x + 2500 - 2500)$$

$$= -\frac{2}{\pi}(x - 50)^2 + \frac{5000}{\pi}$$

The area is maximum when $x = 50$ and

$$y = \frac{200 - 2(50)}{\pi} = \frac{100}{\pi}.$$

**(e)**

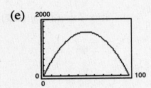

The area is maximum when $x = 50$ and

$$y = \frac{200 - 2(50)}{\pi} = \frac{100}{\pi}.$$

**63.** $y = -\dfrac{1}{12}x^2 + 2x + 4$

**(a)**

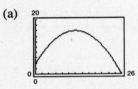

**(b)** When $x = 0$, $y = 4$ feet.

**(c)** The vertex occurs at

$$x = -\frac{b}{2a} = -\frac{2}{2(-1/12)} = 12.$$

The maximum height is

$$y = -\frac{1}{12}(12)^2 + 2(12) + 4$$

$$= 16 \text{ feet.}$$

**(d)** You can solve this part graphically by finding the $x$-intercept of the graph:

$$x \approx 25.856.$$

Algebraically,

$$0 = -\frac{1}{12}x^2 + 2x + 4$$

$$0 = x^2 - 24x - 48 \quad \text{(Multiply both sides by } -12.\text{)}$$

$$x = \frac{-(-24) \pm \sqrt{(-24)^2 - 4(1)(-48)}}{2(1)}$$

$$= \frac{24 \pm \sqrt{768}}{2} = \frac{24 \pm 16\sqrt{3}}{2} = 12 \pm 8\sqrt{3}$$

Using the positive value for $x$, we have

$$x = 12 + 8\sqrt{3} \approx 25.86 \text{ feet.}$$

**65.** $C = 800 - 10x + 0.25x^2$

| $x$ | 10 | 15 | 20 | 25 | 30 |
|---|---|---|---|---|---|
| $C$ | 725 | 706.25 | 700 | 706.25 | 725 |

From the table, the minimum cost seems to be at $x = 20$.

The minimum cost occurs at the vertex.

$$x = -\frac{b}{2a} = -\frac{(-10)}{2(0.25)} = \frac{10}{.5} = 20$$

$C(20) = 700$ is the minimum cost.

Graphically, you could graph
$C = 800 - 10x + 0.25x^2$ in the window
$[0, 40] \times [0, 1000]$ and find the vertex $(20, 700)$.

**67.** (a) $C = 4274 + 3.4t - 1.52t^2$    $0 \le t \le 41$

($t = 0$ corresponds to 1960)

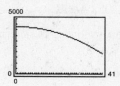

(b) Using a graphing utility, the maximum is about 4276 for $t = 1.12$, or 1961.

Yes the per capita consumption is decreasing. (Answers will vary.)

(c) For 2000 ($t = 40$), $C = 1978$ cigarettes per person. The annual consumption per smoker was

$$\frac{1978(209,128,000)}{48,300,00} \approx 8564 \text{ per smoker per year.}$$

The daily consumption was

$$\frac{8564}{365} \approx 23.5 \text{ cigarettes per smoker per day.}$$

**69.** True

$$-12x^2 - 1 = 0$$

$$12x^2 = -1 \quad \text{impossible}$$

**71.** Model (a) is preferable. $a > 0$ means the parabola opens upward and profits are increasing for $t$ to the right of the vertex,

$$t \ge -\frac{b}{(2a)}.$$

**73.** $x + y = 8 \implies y = 8 - x$.

Then $-\frac{2}{3}x + y = -\frac{2}{3}x + (8 - x) = 6 \implies -\frac{5}{3}x = -2 \implies x = \frac{6}{5}$ and $y = 8 - \frac{6}{5} = \frac{34}{5}$

$(1.2, 6.8)$

**75.** $y = x + 3 = 9 - x^2$

$$x^2 + x - 6 = 0$$

$$(x + 3)(x - 2) = 0$$

$$x = -3, x = 2$$

Thus, $(-3, 0)$ and $(2, 5)$ are the points of intersection.

**77.** $(6 - i) - (2i + 11) = 6 - 11 - i - 2i$

$$= -5 - 3i$$

**79.** $(3i + 7)(-4i + 1) = -12i^2 + 3i - 28i + 7$

$$= 19 - 25i$$

# Section 3.2    Polynomial Functions of Higher Degree

■ You should know the following basic principles about polynomials.

■ $f(x) = a_n x^n + a_{n-1} x^{n-1} + \cdots + a_2 x^2 + a_1 x + a_0, a_n \neq 0$, is a polynomial function of degree $n$.

■ If $f$ is of odd degree and

(a) $a_n > 0$, then

1. $f(x) \to \infty$ as $x \to \infty$.

2. $f(x) \to -\infty$ as $x \to -\infty$.

(b) $a_n < 0$, then

1. $f(x) \to -\infty$ as $x \to \infty$.

2. $f(x) \to \infty$ as $x \to -\infty$.

■ If $f$ is of even degree and

(a) $a_n > 0$, then

1. $f(x) \to \infty$ as $x \to \infty$.

2. $f(x) \to \infty$ as $x \to -\infty$.

(b) $a_n < 0$, then

1. $f(x) \to -\infty$ as $x \to \infty$.

2. $f(x) \to -\infty$ as $x \to -\infty$.

■ The following are equivalent for a polynomial function.

(a) $x = a$ is a zero of a function.

(b) $x = a$ is a solution of the polynomial equation $f(x) = 0$.

(c) $(x - a)$ is a factor of the polynomial.

(d) $(a, 0)$ is an $x$-intercept of the graph of $f$.

■ A polynomial of degree $n$ has at most $n$ distinct zeros.

■ If $f$ is a polynomial function such that $a < b$ and $f(a) \neq f(b)$, then $f$ takes on every value between $f(a)$ and $f(b)$ in the interval $[a, b]$.

■ If you can find a value where a polynomial is positive and another value where it is negative, then there is at least one real zero between the values.

## Solutions to Odd-Numbered Exercises

**1.** $f(x) = -2x + 3$ is a line with $y$-intercept $(0, 3)$. Matches graph (f).

**3.** $f(x) = -2x^2 - 5x$ is a parabola with $x$-intercepts $(0, 0)$ and $\left(-\frac{5}{2}, 0\right)$ and opens downward. Matches graph (c).

**5.** $f(x) = -\frac{1}{4}x^4 + 3x^2$ has intercepts $(0, 0)$ and $\left(\pm 2\sqrt{3}, 0\right)$. Matches graph (e).

**7.** $f(x) = x^4 + 2x^3$ has intercepts $(0, 0)$ and $(-2, 0)$. Matches graph (g).

**9.** $y = x^3$

(a) $f(x) = (x - 2)^3$

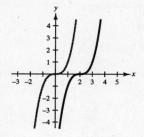

Horizontal shift two units to the right

(b) $f(x) = x^3 - 2$

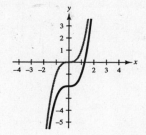

Vertical shift two units downward

(c) $f(x) = -\frac{1}{2}x^3$

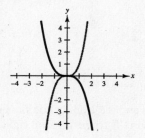

Reflection in the x-axis and a vertical shrink

(d) $f(x) = (x - 2)^3 - 2$

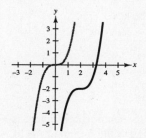

Horizontal shift two units to the right and a vertical shift two units downward

**11.** $y = x^4$

(a) $f(x) = (x + 5)^4$

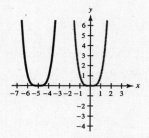

Horizontal shift five units to the left

(b) $f(x) = x^4 - 5$

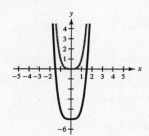

Vertical shift five units downward

(c) $f(x) = 4 - x^4$

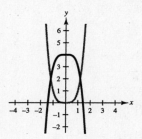

Reflection in the x-axis and then a vertical shift four units upward

(d) $f(x) = \frac{1}{2}(x - 1)^4$

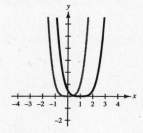

Horizontal shift one unit to the right and a vertical shrink

**13.** $f(x) = 3x^3 - 9x + 1$;   $g(x) = 3x^3$

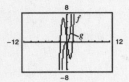

**15.** $f(x) = -(x^4 - 4x^3 + 16x)$;   $g(x) = -x^4$

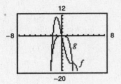

**17.** $f(x) = 2x^2 - 3x + 1$

Degree:  2

Leading coefficient:  2

The degree is even and the leading coefficient is positive. The graph rises to the left and right.

**19.** $g(x) = 5 - \frac{7}{2}x - 3x^2$

Degree:  2

Leading coefficient:  $-3$

The degree is even and the leading coefficient is negative. The graph falls to the left and right.

**21.** $f(x) = \dfrac{6 - 2x + 4x^2 - 5x^3}{3}$

Degree:  3

Leading coefficient:  $-\frac{5}{3}$

The degree is odd and the leading coefficient is negative. The graph rises to the left and falls to the right.

**23.** $h(t) = -\frac{2}{3}(t^2 - 5t + 3)$

Degree:  2

Leading coefficient:  $-\frac{2}{3}$

The degree is even and the leading coefficient is negative. The graph falls to the left and right.

**25.** $f(x) = x^2 - 25$

$\qquad = (x + 5)(x - 5)$

$\qquad x = \pm 5$

**27.** $h(t) = t^2 - 6t + 9$

$\qquad = (t - 3)^2$

$\qquad t = 3$    (multiplicity 2)

**29.** $f(x) = x^2 + x - 2$

$\qquad = (x + 2)(x - 1)$

$\qquad x = -2, 1$

**31.** $f(t) = t^3 - 4t^2 + 4t$

$\qquad = t(t - 2)^2$

$\qquad t = 0, 2$    (multiplicity 2)

**33.** $f(x) = \dfrac{1}{2}x^2 + \dfrac{5}{2}x - \dfrac{3}{2}$

$\qquad = \dfrac{1}{2}(x^2 + 5x - 3)$

$\quad x = \dfrac{-5 \pm \sqrt{25 - 4(-3)}}{2} = -\dfrac{5}{2} \pm \dfrac{\sqrt{37}}{2}$

$\qquad \approx 0.5414, -5.5414$

**35.** (a)

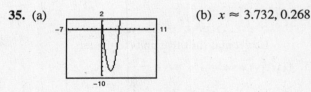

(b) $x \approx 3.732, 0.268$

(c) $f(x) = 3x^2 - 12x + 3$

$\qquad = 3(x^2 - 4x + 1)$

$\quad x = \dfrac{4 \pm \sqrt{16 - 4}}{2} = 2 \pm \sqrt{3}$

**37.** (a)

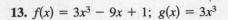

(b) $t = \pm 1$

(c) $g(t) = \frac{1}{2}t^4 - \frac{1}{2}$

$\qquad = \frac{1}{2}(t + 1)(t - 1)(t^2 + 1)$

$\qquad t = \pm 1$

**39.** (a)

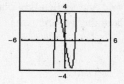

(b) $x = 0, 1.414, -1.414$

(c) $f(x) = x^5 + x^3 - 6x$

$= x(x^4 + x^2 - 6)$

$= x(x^2 + 3)(x^2 - 2)$

$x = 0, \pm\sqrt{2}$

**41.** (a)

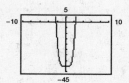

(b) $2.236, -2.236$

(c) $f(x) = 2x^4 - 2x^2 - 40$

$= 2(x^2 + 4)(x + \sqrt{5})(x - \sqrt{5})$

$x = \pm\sqrt{5}$

**43.** (a)

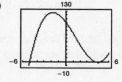

(b) $x = 4, 5, -5$

(c) $f(x) = x^3 - 4x^2 - 25x + 100$

$= x^2(x - 4) - 25(x - 4)$

$= (x^2 - 25)(x - 4)$

$= (x - 5)(x + 5)(x - 4)$

$x = \pm 5, 4$

**45.** (a)

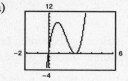

(b) $x = 0, \frac{5}{2}$

(c) $y = 4x^3 - 20x^2 + 25x$

$0 = 4x^3 - 20x^2 + 25x$

$0 = x(2x - 5)^2$

$x = 0$ or $x = \frac{5}{2}$   (multiplicity 2)

**47.**

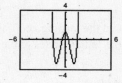

Zeros: $x \approx \pm 0.421, \pm 1.680$

Relative maximum: $(0, 1)$

Relative minimums: $(1.225, -3.5), (-1.225, -3.5)$

**49.**

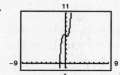

Zeros: $x \approx -1.178$

Relative maximum: $(-0.324, 6.218)$

Relative minimum: $(0.324, 5.782)$

**51.** $f(x) = (x - 0)(x - 4) = x^2 - 4x$

Note: $f(x) = a(x - 0)(x - 4) = ax(x - 4)$ has zeros 0 and 4 for all nonzero real numbers $a$.

**53.** $f(x) = (x - 0)(x + 2)(x + 3) = x^3 + 5x^2 + 6x$

Note: $f(x) = ax(x + 2)(x + 3)$ has zeros 0, $-2$, and $-3$ for all nonzero real numbers $a$.

**55.** $f(x) = (x - 4)(x + 3)(x - 3)(x - 0)$

$= (x - 4)(x^2 - 9)x$

$= x^4 - 4x^3 - 9x^2 + 36x$

Note: $f(x) = a(x^4 - 4x^3 - 9x^2 + 36x)$ has zeros 4, $-3$, 3, and 0 for all nonzero real numbers a.

**57.** $f(x) = \left[x - \left(1 + \sqrt{3}\right)\right]\left[x - \left(1 - \sqrt{3}\right)\right]$

$= \left[(x - 1) - \sqrt{3}\right]\left[(x - 1) + \sqrt{3}\right]$

$= (x - 1)^2 - \left(\sqrt{3}\right)^2$

$= x^2 - 2x + 1 - 3$

$= x^2 - 2x - 2$

Note: $f(x) = a(x^2 - 2x - 2)$ has zeros $1 + \sqrt{3}$ and $1 - \sqrt{3}$ for all nonzero real numbers a.

**59.** $f(x) = (x - 2)[x - (4 + \sqrt{5})][x - (4 - \sqrt{5})]$

$\quad = (x - 2)[(x - 4) - \sqrt{5}][(x - 4) + \sqrt{5}]$

$\quad = (x - 2)[(x - 4)^2 - 5]$

$\quad = x^3 - 10x^2 + 27x - 22$

Note: $f(x) = a(x - 2)[(x - 4)^2 - 5]$ has zeros $2, 4 + \sqrt{5}$, and $4 - \sqrt{5}$ for all nonzero real numbers $a$.

**61.** (a) The degree of $f$ is odd and the leading coefficient is 1. The graph falls to the left and rises to the right.

(b) $f(x) = x^3 - 9x = x(x^2 - 9) = x(x - 3)(x + 3)$

zeros: $0, 3, -3$

(c), (d)

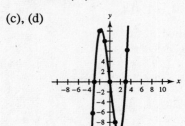

**63.** (a) The degree of $f$ is even and the leading coefficient is $\frac{1}{4}$. The graph rises to the left and to the right.

(b) $f(t) = \frac{1}{4}(t^2 - 2t + 15)$ has no real zeros.

(c), (d)

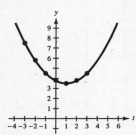

**65.** (a) The degree of $f$ is odd and the leading coefficient is 1. The graph falls to the left and rises to the right.

(b) $f(x) = x^3 - 3x^2 = x^2(x - 3)$; zeros: $0, 3$

(c), (d)

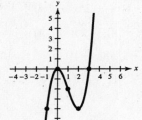

**67.** (a) The degree of $f$ is odd and the leading coefficient is $-1$. The graph rises to the left and falls to the right.

(b) $f(x) = -x^3 - 5x^2 = x^2(-x - 5)$; zeros: $0, -5$

(c), (d)

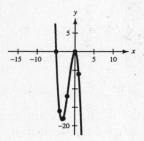

**69.** (a) The degree of $f$ is odd and the leading coefficient is 1. The graph falls to the left and rises to the right.

(b) $f(x) = x^2(x - 4)$; zeros: $0, 4$

(c), (d)

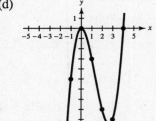

**71.** (a) The degree of $g$ is even (4) and the leading coefficient is $-\frac{1}{4}$. The graph falls to the left and to the right.

(b) $g(t) = -\frac{1}{4}(t - 2)^2(t + 2)^2$; zeros: $2, -2$

(c), (d)

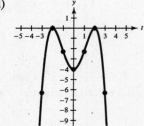

**73.** $f(x) = x^3 - 3x^2 + 3$

(a)

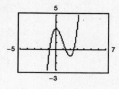

The function has three zeros. They are in the intervals $(-1, 0)$, $(1, 2)$ and $(2, 3)$.

(b) $-0.879, 1.347, 2.532$

(c)

| $x$ | $y1$ | $x$ | $y1$ | $x$ | $y1$ |
|---|---|---|---|---|---|
| $-0.9$ | $-0.159$ | 1.3 | 0.127 | 2.5 | $-0.125$ |
| $-0.89$ | $-0.0813$ | 1.31 | 0.09979 | 2.51 | $-0.087$ |
| $-0.88$ | $-0.0047$ | 1.32 | 0.07277 | 2.52 | $-0.0482$ |
| $-0.87$ | 0.0708 | 1.33 | 0.04594 | 2.53 | $-0.0084$ |
| $-0.86$ | 0.14514 | 1.34 | 0.0193 | 2.54 | 0.03226 |
| $-0.85$ | 0.21838 | 1.35 | $-0.0071$ | 2.55 | 0.07388 |
| $-0.84$ | 0.2905 | 1.36 | $-0.0333$ | 2.56 | 0.11642 |

**75.** $g(x) = 3x^4 + 4x^3 - 3$

(a)

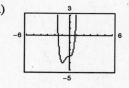

The function has two zeros. They are in the intervals $(-2, -1)$ and $(0, 1)$.

(b) $-1.585, 0.779$

(c)

| $x$ | $y1$ | $x$ | $y1$ |
|---|---|---|---|
| $-1.6$ | 0.2768 | 0.75 | $-0.3633$ |
| $-1.59$ | 0.09515 | 0.76 | $-0.2432$ |
| $-1.58$ | $-0.0812$ | 0.77 | $-0.1193$ |
| $-1.57$ | $-0.2524$ | 0.78 | 0.00866 |
| $-1.56$ | $-0.4184$ | 0.79 | 0.14066 |
| $-1.55$ | $-0.5795$ | 0.80 | 0.2768 |
| $-1.54$ | $-0.7356$ | 0.81 | 0.41717 |

**77.**

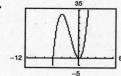

$f(x) = x^2(x + 6)$

No symmetry

Two $x$-intercepts

**79.**

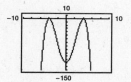

$g(t) = -\frac{1}{2}(t - 4)^2(t + 4)^2$

Symmetric about the $y$-axis

Two $x$-intercepts

**81.**

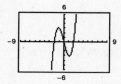

$f(x) = x^3 - 4x = x(x + 2)(x - 2)$

Symmetric to origin

Three $x$-intercepts

**83.**

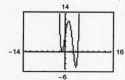

$g(x) = \frac{1}{5}(x + 1)^2(x - 3)(2x - 9)$

Three $x$-intercepts

No symmetry

**85.** (a) Volume = length × width × height

Because the box is made from a square, length = width.

Thus:

Volume = (length)$^2$ × height

$\quad = (36 - 2x)^2 x$

(c)

| Height, $x$ | Length and Width | Volume, $V$ |
|---|---|---|
| 1 | $36 - 2(1)$ | $1[36 - 2(1)]^2 = 1156$ |
| 2 | $36 - 2(2)$ | $2[36 - 2(2)]^2 = 2048$ |
| 3 | $36 - 2(3)$ | $3[36 - 2(3)]^2 = 2700$ |
| 4 | $36 - 2(4)$ | $4[36 - 2(4)]^2 = 3136$ |
| 5 | $36 - 2(5)$ | $5[36 - 2(5)]^2 = 3380$ |
| 6 | $36 - 2(6)$ | $6[36 - 2(6)]^2 = 3456$ |
| 7 | $36 - 2(7)$ | $7[36 - 2(7)]^2 = 3388$ |

Maximum volume 3456 for $x = 6$

(b) Domain: $0 < 36 - 2x < 36$

$\qquad -36 < -2x < 0$

$\qquad 18 > x > 0$

(d)

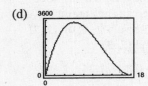

$x = 6$ when $V(x)$ is maximum.

**87.** The point of diminishing returns (where the graph changes from curving upward to curving downward) occurs when $x = 200$. The point is (200, 160) which corresponds to spending \$2,000,000 on advertising to obtain a revenue of \$160 million.

**89.** $y_1 = 0.1250t^2 - 1.446t^2 + 9.07t + 155.5$

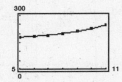

The model is a good fit.

**91.** For 2007, $t = 17$ and

$y_1 = 1250(17)^3 - 1.446(17)^2 + 9.07(17) + 155.5$

$\quad \approx 505.92, \quad$ or $\quad \$505,920$

$y_2 = -0.2(17)^3 + 5.155(17)^2 - 37.23(17) + 206.8$

$\quad \approx 81.085, \quad$ or $\quad \$81,085$

The answer for $y_2$ does not seem reasonable.

**93.** False. A sixth-degree polynomial can have at most five turning points.

**95.** $f(x) = -x^3$ matches graph (b)

No, a polynomial of odd degree must cross the $x$-axis.

**97.** $f(x) = x^3$ matches graph (a)

No, a polynomial of odd degree must cross the $x$-axis.

**99.** $(f + g)(-4) = f(-4) + g(-4)$

$\qquad\qquad\quad = -59 + 128 = 69$

**101.** $(fg)\left(-\dfrac{4}{7}\right) = f\left(-\dfrac{4}{7}\right)g\left(-\dfrac{4}{7}\right)$

$\qquad\qquad = (-11)\left(\dfrac{8 \cdot 16}{49}\right)$

$\qquad\qquad = -\dfrac{1408}{49} \approx -28.7347$

**103.** $(fg)(-1) = f(g(-1)) = f(8) = 109$

**105.** $3(x - 5) < 4x - 7$

$3x - 15 < 4x - 7$

$-8 < x$

**107.**

$$\frac{5x - 2}{x - 7} \leq 4$$

$$\frac{5x - 2}{x - 7} - 4 \leq 0$$

$$\frac{5x - 2 - 4(x - 7)}{x - 7} \leq 0$$

$$\frac{x + 26}{x - 7} \leq 0$$

$[x + 26 \geq 0 \text{ and } x - 7 < 0] \text{ or } [x + 26 \leq 0 \text{ and } x - 7 > 0]$

$[x \geq -26 \text{ and } x < 7] \text{ or } [x \leq -26 \text{ and } x > 7]$

impossible

$-26 \leq x < 7$

## Section 3.3 Real Zeros of Polynomial Functions

> You should know the following basic techniques and principles of polynomial division.
> - The Division Algorithm (Long Division of Polynomials)
> - Synthetic Division
> - $f(k)$ is equal to the remainder of $f(x)$ divided by $(x - k)$.
> - $f(k) = 0$ if and only if $(x - k)$ is a factor of $f(x)$.
> - The Rational Zero Test
> - The Upper and Lower Bound Rule

**Solutions to Odd-Numbered Exercises**

**1.**
$$\begin{array}{r} 2x + 4 \\ x + 3 \overline{)\, 2x^2 + 10x + 12} \\ -(2x^2 + 6x) \\ \hline 4x + 12 \\ -(4x + 12) \\ \hline 0 \end{array}$$

$$\frac{2x^2 + 10x + 12}{x + 3} = 2x + 4, \, x \neq -3$$

**3.**
$$\begin{array}{r} x^2 - 3x + 1 \\ 4x + 5 \overline{)\, 4x^3 - 7x^2 - 11x + 5} \\ -(4x^3 + 5x^2) \\ \hline -12x^2 - 11x \\ -(-12x^2 - 15x) \\ \hline 4x + 5 \\ -(4x + 5) \\ \hline 0 \end{array}$$

$$\frac{4x^3 - 7x^2 - 11x + 5}{4x + 5} = x^2 - 3x + 1, \quad x \neq -\frac{5}{4}$$

**5.**
$$x + 2 \overline{)\ 7x + 3\ }^{\ \ 7}$$
$$\underline{-\ (7x + 14)}$$
$$-\ 11$$

$$\frac{7x + 3}{x + 2} = 7 - \frac{11}{x + 2}$$

**7.**
$$2x^2 + 0x + 1 \overline{)\ 6x^3 + 10x^2 + x + 8\ }^{\ \ 3x + 5}$$
$$\underline{-\ (6x^3 + 0x^2 + 3x)}$$
$$10x^2 - 2x + 8$$
$$\underline{-\ (10x^2 + 0x + 5)}$$
$$-2x + 3$$

$$\frac{6x^3 + 10x^2 + x + 8}{2x^2 + 1} = 3x + 5 - \frac{2x - 3}{2x^2 + 1}$$

**9.**
$$x^2 + 1 \overline{)\ x^3 + 0x^2 + 0x - 9\ }^{\ \ x}$$
$$\underline{-\ (x^3 \qquad + x)}$$
$$-\ x - 9$$

$$\frac{x^3 - 9}{x^2 + 1} = x - \frac{x + 9}{x^2 + 1}$$

**11.**
$$x^2 - 2x + 1 \overline{)\ 2x^3 - 4x^2 - 15x + 5\ }^{\ \ 2x}$$
$$\underline{-(2x^3 - 4x^2 + 2x)}$$
$$-\ 17x + 5$$

$$\frac{2x^3 - 4x^2 - 15x + 5}{(x - 1)^2} = 2x - \frac{17x - 5}{(x - 1)^2}$$

**13.**

| 5 | 3 | $-17$ | 15 | $-25$ |
|---|---|---|---|---|
|   |   | 15 | $-10$ | 25 |
|   | 3 | $-2$ | 5 | 0 |

$$\frac{3x^3 - 17x^2 + 15x - 25}{x - 5} = 3x^2 - 2x + 5, \quad x \neq 5$$

**15.**

| 3 | 6 | 7 | $-1$ | 26 |
|---|---|---|---|---|
|   |   | 18 | 75 | 222 |
|   | 6 | 25 | 74 | 248 |

$$\frac{6x^3 + 7x^2 - x + 26}{x - 3} = 6x^2 + 25x + 74 + \frac{248}{x - 3}$$

**17.**

| 2 | 9 | $-18$ | $-16$ | 32 |
|---|---|---|---|---|
|   |   | 18 | 0 | $-32$ |
|   | 9 | 0 | $-16$ | 0 |

$$\frac{9x^3 - 18x^2 - 16x + 32}{x - 2} = 9x^2 - 16, \quad x \neq 2$$

**19.**

| $-8$ | 1 | 0 | 0 | 512 |
|---|---|---|---|---|
|   |   | $-8$ | 64 | $-512$ |
|   | 1 | $-8$ | 64 | 0 |

$$\frac{x^3 + 512}{x + 8} = x^2 - 8x + 64, \quad x \neq -8$$

**21.**

| $-\frac{1}{2}$ | 4 | 16 | $-23$ | $-15$ |
|---|---|---|---|---|
|   |   | $-2$ | $-7$ | 15 |
|   | 4 | 14 | $-30$ | 0 |

$$\frac{4x^3 + 16x^2 - 23x - 15}{x + \frac{1}{2}} = 4x^2 + 14x - 30,$$
$$x \neq -\frac{1}{2}$$

**23.**
$$y_2 = x - 2 + \frac{4}{x + 2}$$
$$= \frac{(x - 2)(x + 2) + 4}{x + 2}$$
$$= \frac{x^2 - 4 + 4}{x + 2}$$
$$= \frac{x^2}{x + 2}$$
$$= y_1$$

**25.** $f(x) = x^3 - x^2 - 14x + 11, \quad k = 4$

$$
\begin{array}{r|rrrr}
4 & 1 & -1 & -14 & 11 \\
  &   & 4 & 12 & -8 \\
\hline
  & 1 & 3 & -2 & 3
\end{array}
$$

$f(x) = (x - 4)(x^2 + 3x - 2) + 3$

$f(4) = (0)(26) + 3 = 3$

**27.**

$$
\begin{array}{r|rrrr}
\sqrt{2} & 1 & 3 & -2 & -14 \\
  &   & \sqrt{2} & 2 + 3\sqrt{2} & 6 \\
\hline
  & 1 & 3 + \sqrt{2} & 3\sqrt{2} & -8
\end{array}
$$

$f(x) = (x - \sqrt{2})(x^2 + (3 + \sqrt{2})x + 3\sqrt{2}) - 8$

$f(\sqrt{2}) = 0(4 + 6\sqrt{2}) - 8 = -8$

**29.**

$$
\begin{array}{r|rrrr}
1 - \sqrt{3} & 4 & -6 & -12 & -4 \\
  &   & 4 - 4\sqrt{3} & 10 - 2\sqrt{3} & 4 \\
\hline
  & 4 & -2 - 4\sqrt{3} & -2 - 2\sqrt{3} & 0
\end{array}
$$

$f(x) = (x - 1 + \sqrt{3})[4x^2 - (2 + 4\sqrt{3})x - (2 + 2\sqrt{3})]$

$f(1 - \sqrt{3}) = 0$

**31.** $f(x) = 4x^3 - 13x + 10$

(a)
$$
\begin{array}{r|rrrr}
1 & 4 & 0 & -13 & 10 \\
  &   & 4 & 4 & -9 \\
\hline
  & 4 & 4 & -9 & \underline{1} = f(1)
\end{array}
$$

(b)
$$
\begin{array}{r|rrrr}
-2 & 4 & 0 & -13 & 10 \\
   &   & -8 & 16 & -6 \\
\hline
   & 4 & -8 & 3 & \underline{4} = f(-2)
\end{array}
$$

(c)
$$
\begin{array}{r|rrrr}
\frac{1}{2} & 4 & 0 & -13 & 10 \\
  &   & 2 & 1 & -6 \\
\hline
  & 4 & 2 & -12 & \underline{4} = f\left(\frac{1}{2}\right)
\end{array}
$$

(d)
$$
\begin{array}{r|rrrr}
8 & 4 & 0 & -13 & 10 \\
  &   & 32 & 256 & 1944 \\
\hline
  & 4 & 32 & 243 & \underline{1954} = f(8)
\end{array}
$$

**33.** $h(x) = 3x^3 + 5x^2 - 10x + 1$

(a)
$$
\begin{array}{r|rrrr}
3 & 3 & 5 & -10 & 1 \\
  &   & 9 & 42 & 96 \\
\hline
  & 3 & 14 & 32 & \underline{97} = h(3)
\end{array}
$$

(b)
$$
\begin{array}{r|rrrr}
\frac{1}{3} & 3 & 5 & -10 & 1 \\
  &   & 1 & 2 & -\frac{8}{3} \\
\hline
  & 3 & 6 & -8 & \underline{-\frac{5}{3}} = h\left(\frac{1}{3}\right)
\end{array}
$$

(c)
$$
\begin{array}{r|rrrr}
-2 & 3 & 5 & -10 & 1 \\
   &   & -6 & 2 & 16 \\
\hline
   & 3 & -1 & -8 & \underline{17} = h(-2)
\end{array}
$$

(d)
$$
\begin{array}{r|rrrr}
-5 & 3 & 5 & -10 & 1 \\
   &   & -15 & 50 & -200 \\
\hline
   & 3 & -10 & 40 & \underline{-199} = h(-5)
\end{array}
$$

**35.**

$$
\begin{array}{r|rrrr}
2 & 1 & 0 & -7 & 6 \\
  &   & 2 & 4 & -6 \\
\hline
  & 1 & 2 & -3 & 0
\end{array}
$$

$x^3 - 7x + 6 = (x - 2)(x^2 + 2x - 3)$

$\qquad\qquad\quad = (x - 2)(x + 3)(x - 1)$

Zeros: $2, -3, 1$

**37.**

$$
\begin{array}{r|rrrr}
\frac{1}{2} & 2 & -15 & 27 & -10 \\
  &   & 1 & -7 & 10 \\
\hline
  & 2 & -14 & 20 & 0
\end{array}
$$

$2x^3 - 15x^2 + 27x - 10$

$\quad = \left(x - \frac{1}{2}\right)(2x^2 - 14x + 20)$

$\quad = (2x - 1)(x - 2)(x - 5)$

Zeros: $\frac{1}{2}, 2, 5$

**39.** (a)
$$
\begin{array}{r|rrrr}
-2 & 2 & 1 & -5 & 2 \\
   &   & -4 & 6 & -2 \\
\hline
   & 2 & -3 & 1 & 0
\end{array}
$$

$$
\begin{array}{r|rrr}
1 & 2 & -3 & 1 \\
  &   & 2 & -1 \\
\hline
  & 2 & -1 & 0
\end{array}
$$

(b) Remaining factor: $(2x - 1)$

(c) $f(x) = (x + 2)(x - 1)(2x - 1)$

(d) Real zeros: $-2, 1, \frac{1}{2}$

(e)

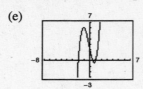

**41.** (a)
$$
\begin{array}{r|rrrrr}
5 & 1 & -4 & -15 & 58 & -40 \\
  &   & 5 & 5 & -50 & 40 \\
\hline
  & 1 & 1 & -10 & 8 & 0
\end{array}
$$

$$
\begin{array}{r|rrrr}
-4 & 1 & 1 & -10 & 8 \\
   &   & -4 & 12 & -8 \\
\hline
   & 1 & -3 & 2 & 0
\end{array}
$$

(b) $x^2 - 3x + 2 = (x - 2)(x - 1)$,

Remaining factors: $(x - 2), (x - 1)$

(c) $f(x) = (x - 5)(x + 4)(x - 2)(x - 1)$

(d) Real zeros: $5, -4, 2, 1$

(e)

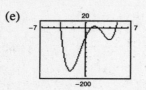

**43.** (a)
$$
\begin{array}{r|rrrr}
-\frac{1}{2} & 6 & 41 & -9 & -14 \\
   &   & -3 & -19 & 14 \\
\hline
   & 6 & 38 & -28 & 0
\end{array}
$$

$$
\begin{array}{r|rrr}
\frac{2}{3} & 6 & 38 & -28 \\
   &   & 4 & 28 \\
\hline
   & 6 & 42 & 0
\end{array}
$$

(b) $6x + 42$ Remaining factor $\left(\text{or } 6(x + 7)\right)$

(c) $f(x) = (2x + 1)(3x - 2)(x + 7)$

Note: Use $\frac{1}{6}(6x + 42) = x + 7$

(d) Real zeros: $-\frac{1}{2}, \frac{2}{3}, -7$

(e)

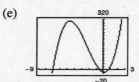

**45.** $f(x) = x^3 + 3x^2 - x - 3$

Possible rational zeros: $\pm 1, \pm 3$

Zeros shown on graph: $-3, -1, 1$

**47.** $f(x) = 2x^4 - 17x^3 + 35x^2 + 9x - 45$

Possible rational zeros: $\pm 1, \pm 3, \pm 5, \pm 9, \pm 15, \pm 45,$
$$\pm\frac{1}{2}, \pm\frac{3}{2}, \pm\frac{5}{2}, \pm\frac{9}{2}, \pm\frac{15}{2}, \pm\frac{45}{2}$$

Zeros shown of graph: $-1, \frac{3}{2}, 3, 5$

**49.** $z^4 - z^3 - 2z - 4 = 0$

Possible rational zeros: $\pm 1, \pm 2, \pm 4$

$$
\begin{array}{r|rrrrr}
-1 & 1 & -1 & 0 & -2 & -4 \\
   &   & -1 & 2 & -2 & 4 \\
\hline
   & 1 & -2 & 2 & -4 & 0
\end{array}
$$

$$
\begin{array}{r|rrrr}
2 & 1 & -2 & 2 & -4 \\
  &   & 2 & 0 & 4 \\
\hline
  & 1 & 0 & 2 & 0
\end{array}
$$

$z^4 - z^3 - 2z - 4 = (z + 1)(z - 2)(z^2 + 2) = 0$

The only real zeros are $-1$ and $2$. You can verify this by graphing the function $f(z) = z^4 - z^3 - 2z - 4$.

**51.** $2y^4 + 7y^3 - 26y^2 + 23y - 6 = 0$

Using a graphing utility and synthetic division, $1/2, 1,$ and $-6$ are rational zeros. Hence,

$$(y + 6)(y - 1)^2\left(y - \tfrac{1}{2}\right) = 0 \implies y = -6, 1, \tfrac{1}{2}.$$

**53.** $h(t) = t^3 - 2t^2 - 7t + 2$

(a) zeros: $-2, 3.732, 0.268$

(b)

$$-2 \begin{array}{|rrrr} 1 & -2 & -7 & 2 \\ & -2 & 8 & -2 \\ \hline 1 & -4 & 1 & 0 \end{array} \quad t = -2 \text{ is a zero}$$

(c) $h(t) = (t + 2)(t^2 - 4t + 1)$

$= (t + 2)\left[t - \left(\sqrt{3} + 2\right)\right]\left[t + \left(\sqrt{3} - 2\right)\right]$

**55.** $h(x) = x^5 - 7x^4 + 10x^3 + 14x^2 - 24x$

(a) $h(x) = x(x^4 - 7x^3 + 10x^2 + 14x - 24)$

From the calculator we have $x = 0, 3, 4$ and $x \approx \pm 1.414$.

(b)

$$3 \begin{array}{|rrrrr} 1 & -7 & 10 & 14 & -24 \\ & 3 & -12 & -6 & 24 \\ \hline 1 & -4 & -2 & 8 & 0 \end{array}$$

$$4 \begin{array}{|rrrr} 1 & -4 & -2 & 8 \\ & 4 & 0 & -8 \\ \hline 1 & 0 & -2 & 0 \end{array}$$

(c) $h(x) = x(x - 3)(x - 4)(x^2 - 2)$

$= x(x - 3)(x - 4)\left(x - \sqrt{2}\right)\left(x + \sqrt{2}\right)$

The exact roots are $x = 0, \ 3, \ 4, \ \pm\sqrt{2}$.

**57.** $f(x) = 2x^4 - x^3 + 6x^2 - x + 5$

4 variations in sign $\implies$ 4, 2 or 0 positive real zeros

$f(-x) = 2x^4 + x^3 + 6x^2 + x + 5$

0 variations in sign $\implies$ 0 negative real zeros

**59.** $g(x) = 4x^3 - 5x + 8$

2 variations in sign $\implies$ 2 or 0 positive real zeros

$g(-x) = -4x^3 + 5x + 8$

1 variation in sign $\implies$ 1 negative real zero

**61.** $f(x) = x^3 + x^2 - 4x - 4$

(a) $f(x)$ has 1 variation in sign $\implies$ 1 positive real zero

$f(-x) = -x^3 + x^2 + 4x - 4$ has 2 variations in sign $\implies$ 2 or 0 negative real zeros

(b) Possible rational zeros: $\pm 1, \pm 2, \pm 4$

(c)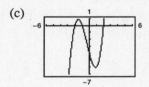

(d) Real zeros: $-2, -1, 2$

**63.** $f(x) = -2x^4 + 13x^3 - 21x^2 + 2x + 8$

(a) $f(x)$ has 3 variations in sign $\implies$ 3 or 1 positive real zeros

$f(-x) = -2x^4 - 13x^3 - 21x^2 - 2x + 8$ has 1 variation in sign $\implies$ 1 negative real zero

(b) Possible rational zeros: $\pm\frac{1}{2}, \pm 1, \pm 2, \pm 4, \pm 8$

(c)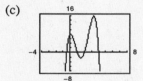

(d) Real zeros: $-\frac{1}{2}, 1, 2, 4$

**65.** $f(x) = 32x^3 - 52x^2 + 17x + 3$

(a) $f(x)$ has 2 variations in sign $\implies$ 2 or 0 positive real zeros

$f(-x) = -32x^3 - 52x^2 - 17x + 3$ has 1 variation in sign $\implies$ 1 negative real zero

(b) Possible rational zeros: $\pm\frac{1}{32}, \pm\frac{1}{16}, \pm\frac{1}{8}, \pm\frac{1}{4}, \pm\frac{1}{2},$

$\pm 1, \pm\frac{3}{32}, \pm\frac{3}{16}, \pm\frac{3}{8}, \pm\frac{3}{4}, \pm\frac{3}{2}, \pm 3$

(c)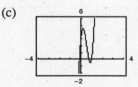

(d) Real zeros: $1, \frac{3}{4}, -\frac{1}{8}$

**67.** $f(x) = x^4 - 4x^3 + 15$

(a)
$$4 \begin{array}{|rrrrr} \phantom{-}1 & -4 & 0 & 0 & 15 \\ & 4 & 0 & 0 & 0 \\ \hline 1 & 0 & 0 & 0 & 15 \end{array}$$

4 is an upper bound.

(b)
$$-1 \begin{array}{|rrrrr} 1 & -4 & 0 & 0 & 15 \\ & -1 & 5 & -5 & 5 \\ \hline 1 & -5 & 5 & -5 & 20 \end{array}$$

$-1$ is a lower bound.

**69.** $f(x) = x^4 - 4x^3 + 16x - 16$

(a)
$$5 \begin{array}{|rrrrr} 1 & -4 & 0 & 16 & -16 \\ & 5 & 5 & 25 & 205 \\ \hline 1 & 1 & 5 & 41 & 189 \end{array}$$

5 is an upper bound.

(b)
$$-3 \begin{array}{|rrrrr} 1 & -4 & 0 & 16 & -16 \\ & -3 & 21 & -63 & 141 \\ \hline 1 & -7 & 21 & -47 & 125 \end{array}$$

$-3$ is a lower bound

**71.** $P(x) = x^4 - \frac{25}{4}x^2 + 9$

$\quad = \frac{1}{4}(4x^4 - 25x^2 + 36)$

$\quad = \frac{1}{4}(4x^2 - 9)(x^2 - 4)$

$\quad = \frac{1}{4}(2x + 3)(2x - 3)(x + 2)(x - 2)$

The rational zeros are $\pm\frac{3}{2}$ and $\pm 2$.

**73.** $f(x) = x^3 - \frac{1}{4}x^2 - x + \frac{1}{4}$

$\quad = \frac{1}{4}(4x^3 - x^2 - 4x + 1)$

$\quad = \frac{1}{4}[x^2(4x - 1) - 1(4x - 1)]$

$\quad = \frac{1}{4}(4x - 1)(x^2 - 1)$

$\quad = \frac{1}{4}(4x - 1)(x + 1)(x - 1)$

The rational zeros are $\frac{1}{4}$ and $\pm 1$.

**75.** $f(x) = x^3 - 1 = (x - 1)(x^2 + x + 1)$

Rational zeros: $1 \ (x = 1)$

Irrational zeros: $0$

Matches (d).

**77.** $f(x) = x^3 - x = x(x + 1)(x - 1)$

Rational zeros: $3 \ (x = 0, \pm 1)$

Irrational zeros: $0$

Matches (b).

**79.** $R = 0.03889t^3 - 0.9064t^2 + 8.327 - 0.92$

(a)

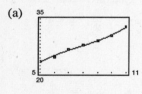

The model is a good approximation.

(b)

| Year | 1995 | 1996 | 1997 | 1998 | 1999 | 2000 | 2001 |
|------|------|------|------|------|------|------|------|
| $R$ | 23.07 | 24.41 | 26.48 | 27.81 | 28.92 | 30.37 | 32.87 |
| Model | 22.92 | 24.81 | 26.29 | 27.60 | 28.96 | 30.60 | 32.77 |

(c) For 2008, $t = 18$ and

$$18 \begin{array}{|rrrr} 0.03889 & -0.9064 & 8.327 & -0.92 \\ & 0.70002 & -3.7148 & 83.0189 \\ \hline .03889 & -0.20638 & 4.6122 & 82.099 \end{array}$$

$R(18) \approx 82.099$

No, the model has a cubic term which will grow rapidly.

**81.** (a) Combined length and width:

$$4x + y = 120 \implies y = 120 - 4x$$

$$\text{Volume} = l \cdot w \cdot h = x^2 y$$

$$= x^2(120 - 4x)$$

$$= 4x^2(30 - x)$$

(b)

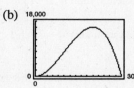

18,000

Dimensions with maximum volume: $20 \times 20 \times 40$

(c) $13{,}500 = 4x^2(30 - x)$

$$4x^3 - 120x^2 + 13{,}500 = 0$$

$$x^3 - 30x^2 + 3375 = 0$$

$$
\begin{array}{r|rrrr}
15 & 1 & -30 & 0 & 3375 \\
   &   & 15 & -225 & -3375 \\
\hline
   & 1 & -15 & -225 & 0
\end{array}
$$

$$(x - 15)(x^2 - 15x - 225) = 0$$

Using the Quadratic equation, $x = 15, \dfrac{15 \pm 15\sqrt{5}}{2}$

The value of $\dfrac{15 - 15\sqrt{5}}{2}$ is not possible because it is negative.

**83.** False, $-\frac{4}{7}$ is a zero of $f$.

**85.**

$$
\begin{array}{r}
x^{2n} + 6x^n + 9 \\
x^n + 3 \overline{\smash{\big)}\ x^{3n} + 9x^{2n} + 27x^n + 27} \\
\underline{x^{3n} + 3x^{2n}} \\
6x^{2n} + 27x^n \\
\underline{6x^{2n} + 18x^n} \\
9x^n + 27 \\
\underline{9x^n + 27} \\
0
\end{array}
$$

$$\frac{x^{3n} + 9x^{2n} + 27x^n + 27}{x^n + 3} = x^{2n} + 6x^n + 9,$$

$$x^n \neq -3$$

[Note: let $y = x^n$ and calculate $(y^3 + 9y^2 + 27y + 27) \div (y + 3)$]

**87.** (a) $\dfrac{x^2 - 1}{x - 1} = x + 1, \quad x \neq 1$

(b) $\dfrac{x^3 - 1}{x - 1} = x^2 + x + 1, \quad x \neq 1$

(c) $\dfrac{x^4 - 1}{x - 1} = x^3 + x^2 + x + 1, \quad x \neq 1$

In general,

$$\frac{x^n - 1}{x - 1} = x^{n-1} + x^{n-2} + \cdots + x + 1, \quad x \neq 1$$

**89.** $9x^2 - 25 = 0$

$$(3x + 5)(3x - 5) = 0$$

$$x = -\frac{5}{3}, \frac{5}{3}$$

**91.** $2x^2 + 6x + 3 = 0$

$$x = \frac{-6 \pm \sqrt{6^2 - 4(2)(3)}}{2(2)}$$

$$= \frac{-6 \pm \sqrt{12}}{4}$$

$$= \frac{-3 \pm \sqrt{3}}{2}$$

$$x = -\frac{3}{2} + \frac{\sqrt{3}}{2}, \quad -\frac{3}{2} - \frac{\sqrt{3}}{2}$$

**93.** $(x - 0)(x + 12) = x^2 + 12x$

**95.** $(x - 0)(x + 1)(x - 2)(x - 5)$

$$= (x^2 + x)(x^2 - 7x + 10)$$

$$= x^4 - 6x^3 + 3x^2 + 10x$$

## Section 3.4   The Fundamental Theorem of Algebra

---

- ■ You should know that if $f$ is a polynomial of degree $n > 0$, then $f$ has at least one zero in the complex number system. (Fundamental Theorem of Algebra)
- ■ You should know that if $a + bi$ is a complex zero of a polynomial $f$, with real coefficients, then $a - bi$ is also a complex zero of $f$.
- ■ You should know the difference between a factor that is irreducible over the rationals (such as $x^2 - 7$) and a factor that is irreducible over the reals (such as $x^2 + 9$).

---

**Solutions to Odd-Numbered Exercises**

**1.** $f(x) = x^2(x + 3)$

The three zeros are $x = 0$, $x = 0$ and $x = -3$

**3.** $f(x) = (x + 9)(x + 2i)(x - 2i)$

The three zeros are $x = -9$, $x = -2i$ and $x = 2i$

**5.** $f(x) = x^3 - 4x^2 + x - 4 = x^2(x - 4) + 1(x - 4)$

$\quad = (x - 4)(x^2 + 1)$   zeros: $4, \pm i$

The only real zero of $f(x)$ is $x = 4$. This corresponds to the $x$-intercept of $(4, 0)$ on the graph.

**7.** $f(x) = x^4 + 4x^2 + 4 = (x^2 + 2)^2$

zeros: $\pm \sqrt{2}i, \pm \sqrt{2}i$

$f(x)$ has no real zeros and the graph of $f(x)$ has no $x$-intercepts.

**9.** $h(x) = x^2 - 4x + 1$

$h$ has no rational zeros. By the Quadratic Formula, the zeros are $x = \dfrac{4 \pm \sqrt{16 - 4}}{2} = 2 \pm \sqrt{3}$.

$h(x) = \left[x - \left(2 + \sqrt{3}\right)\right]\left[x - \left(2 - \sqrt{3}\right)\right] = \left(x - 2 - \sqrt{3}\right)\left(x - 2 + \sqrt{3}\right)$

**11.** $f(x) = x^2 - 12x + 26$

$f$ has no rational zeros. By the Quadratic Formula, the zeros are

$x = \dfrac{12 \pm \sqrt{(-12)^2 - 4(26)}}{2} = 6 \pm \sqrt{10}$.

$f(x) = \left[x - \left(6 + \sqrt{10}\right)\right]\left[x - \left(6 - \sqrt{10}\right)\right]$

$\quad = \left(x - 6 - \sqrt{10}\right)\left(x - 6 + \sqrt{10}\right)$

**13.** $f(x) = x^2 + 25$

$\quad = (x + 5i)(x - 5i)$

The zeros of $f(x)$ are $x = \pm 5i$.

**15.** $f(x) = x^4 - 81$

$\quad = (x^2 - 9)(x^2 + 9)$

$\quad = (x + 3)(x - 3)(x + 3i)(x - 3i)$

The zeros of $f(x)$ are $x = \pm 3$ and $x = \pm 3i$.

**17.** $f(z) = z^2 - z + 56$

$z = \dfrac{1 \pm \sqrt{1 - 4(56)}}{2}$

$\quad = \dfrac{1 \pm \sqrt{-223}}{2}$

$\quad = \dfrac{1}{2} \pm \dfrac{\sqrt{223}}{2}i$

$f(z) = \left(z - \dfrac{1}{2} + \dfrac{\sqrt{223}i}{2}\right)\left(z - \dfrac{1}{2} - \dfrac{\sqrt{223}i}{2}\right)$

**19.** $f(t) = t^3 - 3t^2 - 15t + 125$

Possible rational zeros:  $\pm 1, \pm 5, \pm 25, \pm 125$

$$
\begin{array}{r|rrrr}
-5 & 1 & -3 & -15 & 125 \\
   &   & -5 & 40  & -125 \\
\hline
   & 1 & -8 & 25  & 0
\end{array}
$$

By the Quadratic Formula, the zeros of

$t^2 - 8t + 25$ are $t = \dfrac{8 \pm \sqrt{64 - 100}}{2} = 4 \pm 3i.$

The zeros of $f(t)$ are $t = -5$ and $t = 4 \pm 3i.$

$f(t) = [t - (-5)][t - (4 + 3i)][t - (4 - 3i)]$

$\quad = (t + 5)(t - 4 - 3i)(t - 4 + 3i)$

**21.** $f(x) = 5x^3 - 9x^2 + 28x + 6$

Possible rational zeros:  $\pm 6, \pm\frac{6}{5}, \pm 3, \pm\frac{3}{5}, \pm 2, \pm\frac{2}{5},$
$\pm 1, \pm\frac{1}{5}$

$$
\begin{array}{r|rrrr}
-\frac{1}{5} & 5 & -9 & 28 & 6 \\
             &   & -1 & 2  & -6 \\
\hline
             & 5 & -10 & 30 & 0
\end{array}
$$

By the Quadratic Formula, the zeros of
$5x^2 - 10x + 30$ are those of $x^2 - 2x + 6$:

$x = \dfrac{2 \pm \sqrt{4 - 4(6)}}{2} = 1 \pm \sqrt{5}\,i$

zeros: $-\frac{1}{5}, 1 \pm \sqrt{5}\,i$

$f(x) = 5\left(x + \frac{1}{5}\right)\left(x - \left(1 + \sqrt{5}\,i\right)\right)\left(x - \left(1 - \sqrt{5}\,i\right)\right)$

$\quad = (5x + 1)\left(x - 1 - \sqrt{5}\,i\right)\left(x - 1 + \sqrt{5}\,i\right)$

**23.** $f(x) = x^4 + 10x^2 + 9$

$\quad = (x^2 + 1)(x^2 + 9)$

$\quad = (x + i)(x - i)(x + 3i)(x - 3i)$

The zeros of $f(x)$ are $x = \pm i$ and $x = \pm 3i.$

**25.** $g(x) = x^4 - 4x^3 + 8x^2 - 16x + 16$

Possible rational zeros:  $\pm 1, \pm 2, \pm 4, \pm 8, \pm 16$

$$
\begin{array}{r|rrrrr}
2 & 1 & -4 & 8  & -16 & 16 \\
  &   & 2  & -4 & 8   & -16 \\
\hline
2 & 1 & -2 & 4  & -8  & 0 \\
  &   & 2  & 0  & 8   &   \\
\hline
  & 1 & 0  & 4  & 0   &
\end{array}
$$

$g(x) = (x - 2)(x - 2)(x^2 + 4)$

$\quad = (x - 2)^2(x + 2i)(x - 2i)$

The zeros of $g$ are $2, 2,$ and $\pm 2i.$

**27.** (a) $f(x) = x^2 - 14x + 46.$ By the Quadratic
Formula

$x = \dfrac{14 \pm \sqrt{(-14)^2 - 4(46)}}{2} = 7 \pm \sqrt{3}.$

The zeros are $7 + \sqrt{3}$ and $7 - \sqrt{3}.$

(b) $f(x) = \left[x - \left(7 + \sqrt{3}\right)\right]\left[x - \left(7 - \sqrt{3}\right)\right]$

$\quad = \left(x - 7 - \sqrt{3}\right)\left(x - 7 + \sqrt{3}\right)$

(c) $x$-intercepts: $\left(7 + \sqrt{3}, 0\right)$ and $\left(7 - \sqrt{3}, 0\right)$

(d)
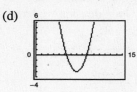

**29.** (a) $f(x) = x^2 + 14x + 44.$ By the Quadratic
Formula,

$x = \dfrac{-14 \pm \sqrt{14^2 - 4(44)}}{2} = -7 \pm \sqrt{5}$

The zeros are $-7 + \sqrt{5}$ and $-7 - \sqrt{5}.$

(b) $f(x) = \left[x - \left(-7 + \sqrt{5}\right)\right]\left[x - \left(-7 - \sqrt{5}\right)\right]$

$\quad = \left(x + 7 - \sqrt{5}\right)\left(x + 7 + \sqrt{5}\right)$

(c) $x$-intercepts: $\left(-7 + \sqrt{5}, 0\right), \left(-7 - \sqrt{5}, 0\right)$

(d)
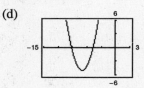

**31.** (a) $f(x) = x^3 - 11x + 150$

$\qquad = (x + 6)(x^2 - 6x + 25).$

Use the Quadratic Formula to find the zeros of $x^2 - 6x + 25$:

$$x = \frac{6 \pm \sqrt{(-6)^2 - 4(25)}}{2} = 3 \pm 4i.$$

The zeros are $-6, 3 + 4i,$ and $3 - 4i.$

(b) $f(x) = (x + 6)(x - 3 + 4i)(x - 3 - 4i)$

(c) $x$-intercept: $(-6, 0)$

(d)

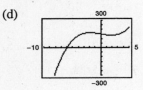

**33.** (a) $f(x) = x^4 + 25x^2 + 144$

$\qquad = (x^2 + 9)(x^2 + 16)$

The zeros are $\pm 3i, \pm 4i.$

(b) $f(x) = (x^2 + 9)(x^2 + 16)$

$\qquad = (x + 3i)(x - 3i)(x + 4i)(x - 4i)$

(c) No $x$-intercepts

(d)

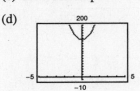

**35.** $f(x) = (x - 3)(x - i)(x + i)$

$\qquad = (x - 3)(x^2 + 1)$

$\qquad = x^3 - 3x^2 + x - 3$

Note: $f(x) = a(x^3 - 3x^2 + x - 3)$, where $a$ is any nonzero real number, has zeros $3, \pm i$

**37.** $f(x) = (x - 2)(x + 4 - i)(x + 4 + i)$

$\qquad = (x - 2)[(x + 4)^2 + 1]$

$\qquad = (x - 2)(x^2 + 8x + 17)$

$\qquad = x^3 + 6x^2 + x - 34$

**39.** If $1 + \sqrt{3}i$ is a zero, so is its conjugate $1 - \sqrt{3}i.$

$f(x) = (x + 5)^2(x - 1 + \sqrt{3}i)(x - 1 - \sqrt{3}i)$

$\qquad = (x^2 + 10x + 25)(x^2 - 2x + 4)$

$\qquad = x^4 + 8x^3 + 9x^2 - 10x + 100$

Note: $f(x) = a(x^4 + 8x^3 + 9x^2 - 10x + 100)$, where $a$ is any nonzero real number, has these zeros.

**41.** $f(x) = x^4 - 6x^2 - 7$

(a) $f(x) = (x^2 - 7)(x^2 + 1)$

(b) $f(x) = (x - \sqrt{7})(x + \sqrt{7})(x^2 + 1)$

(c) $f(x) = (x - \sqrt{7})(x + \sqrt{7})(x + i)(x - i)$

**43.** $f(x) = x^4 - 2x^3 - 3x^2 + 12x - 18$

(a) $f(x) = (x^2 - 6)(x^2 - 2x + 3)$

(b) $f(x) = (x + \sqrt{6})(x - \sqrt{6})(x^2 - 2x + 3)$

(c) $f(x) = (x + \sqrt{6})(x - \sqrt{6})(x - 1 - \sqrt{2}i)(x - 1 + \sqrt{2}i)$

**45.** $f(x) = 2x^3 + 3x^2 + 50x + 75$

Since $5i$ is a zero, so is $-5i.$

$$
\begin{array}{r|rrrr}
5i & 2 & 3 & 50 & 75 \\
   &   & 10i & -50 + 15i & -75 \\
\hline
   & 2 & 3 + 10i & 15i & 0 \\
-5i & 2 & 3 + 10i & 15i & \\
    &   & -10i & -15i & \\
\hline
   & 2 & 3 & 0 &
\end{array}
$$

The zero of $2x + 3$ is $x = -\frac{3}{2}.$ The zeros of $f$ are $x = -\frac{3}{2}$ and $x = \pm 5i.$

*Alternate Solution*

Since $x = \pm 5i$ are zeros of $f(x)$, $(x + 5i)(x - 5i) = x^2 + 25$ is a factor of $f(x)$. By long division we have:

$$
\begin{array}{r}
2x + 3 \\
x^2 + 0x + 25 \overline{\smash{)}\ 2x^3 + 3x^2 + 50x + 75} \\
\underline{2x^3 + 0x^2 + 50x} \\
3x^2 + 0x + 75 \\
\underline{3x^2 + 0x + 75} \\
0
\end{array}
$$

Thus, $f(x) = (x^2 + 25)(2x + 3)$ and the zeros of $f$ are $x = \pm 5i$ and $x = -\frac{3}{2}.$

**47.** $g(x) = x^3 - 7x^2 - x + 87$. Since $5 + 2i$ is a zero, so is $5 - 2i$.

| $5 + 2i$ | 1 | $-7$ | $-1$ | 87 |
|---|---|---|---|---|
| | | $5 + 2i$ | $-14 + 6i$ | $-87$ |
| | 1 | $-2 + 2i$ | $-15 + 6i$ | 0 |

| $5 - 2i$ | 1 | $-2 + 2i$ | $-15 + 6i$ |
|---|---|---|---|
| | | $5 - 2i$ | $15 - 6i$ |
| | 1 | 3 | 0 |

The zero of $x + 3$ is $x = -3$.

The zeros of $f$ are $-3, 5 \pm 2i$.

**49.** $h(x) = 3x^3 - 4x^2 + 8x + 8$. Since $1 - \sqrt{3}i$ is a zero, so is $1 + \sqrt{3}i$.

| $1 - \sqrt{3}i$ | 3 | $-4$ | 8 | 8 |
|---|---|---|---|---|
| | | $3 - 3\sqrt{3}i$ | $-10 - 2\sqrt{3}i$ | $-8$ |
| | 3 | $-1 - 3\sqrt{3}i$ | $-2 - 2\sqrt{3}i$ | 0 |

| $1 + \sqrt{3}i$ | 3 | $-1 - 3\sqrt{3}i$ | $-2 - 2\sqrt{3}i$ |
|---|---|---|---|
| | | $3 + 3\sqrt{3}i$ | $2 + 2\sqrt{3}i$ |
| | 3 | 2 | 0 |

The zero of $3x + 2$ is $x = -\frac{2}{3}$.

The zeros of $h$ are $x = -\frac{2}{3}, 1 \pm \sqrt{3}i$.

**51.** $h(x) = 8x^3 - 14x^2 + 18x - 9$. Since $\frac{1}{2}(1 - \sqrt{5}i)$ is a zero, so is $\frac{1}{2}(1 + \sqrt{5}i)$.

| $\frac{1}{2}(1 - \sqrt{5}i)$ | 8 | $-14$ | 18 | $-9$ |
|---|---|---|---|---|
| | | $4 - 4\sqrt{5}i$ | $-15 + 3\sqrt{5}i$ | 9 |
| | 8 | $-10 - 4\sqrt{5}i$ | $3 + 3\sqrt{5}i$ | 0 |

| $\frac{1}{2}(1 + \sqrt{5}i)$ | 8 | $-10 - 4\sqrt{5}i$ | $3 + 3\sqrt{5}i$ |
|---|---|---|---|
| | | $4 + 4\sqrt{5}i$ | $-3 - 3\sqrt{5}i$ |
| | 8 | $-6$ | 0 |

The zero of $8x - 6$ is $x = \frac{3}{4}$.

The zeros of $h$ are $x = \frac{3}{4}, \frac{1}{2}(1 \pm \sqrt{5}i)$.

**53.** $f(x) = x^4 + 3x^3 - 5x^2 - 21x + 22$

(a) The root feature yields the real roots 1 and 2, and the complex roots $-3 \pm 1.414i$.

(b) By synthetic division,

| 1 | 1 | 3 | $-5$ | $-21$ | 22 |
|---|---|---|---|---|---|
| | | 1 | 4 | $-1$ | $-22$ |
| | 1 | 4 | $-1$ | $-22$ | 0 |

| 2 | 1 | 4 | $-1$ | $-22$ |
|---|---|---|---|---|
| | | 2 | 12 | 22 |
| | 1 | 6 | 11 | 0 |

The complex roots of $x^2 + 6x + 11$ are

$$x = \frac{-6 \pm \sqrt{6^2 - 4(11)}}{2} = -3 \pm \sqrt{2}i.$$

**55.** $h(x) = 8x^3 - 14x^2 + 18x - 9$

(a) The root feature yields the real root 0.75, and the complex roots $0.5 \pm 1.118i$.

(b) By synthetic division,

| $\frac{3}{4}$ | 8 | $-14$ | 18 | $-9$ |
|---|---|---|---|---|
| | | 6 | $-6$ | 9 |
| | 8 | $-8$ | 12 | 0 |

The complex roots of $8x^2 - 8x + 12$ are

$$x = \frac{8 \pm \sqrt{64 - 4(8)(12)}}{2(8)} = \frac{1}{2} \pm \frac{\sqrt{5}}{2}i.$$

**57.** $\qquad -16t^2 + 48t = 64, \quad 0 \le t \le 3$

$$-16t^2 + 48t - 64 = 0$$

$$t = \frac{-48 \pm \sqrt{1792}i}{-32}$$

Since the roots are imaginary, the ball never will reach a height of 64 feet. You can verify this graphically by observing that $y_1 = -16t^2 + 48t$ and $y_2 = 64$ do not intersect.

**59.** False, a third degree polynomial must have at least one real zero.

**61.** $f(x) = x^4 - 4x^2 + k$

(a) $f$ has two real zeros each of multiplicity 2 for $k = 4$: $f(x) = x^4 - 4x^2 + 4 = (x^2 - 2)^2$.

(b) $f$ has two real zeros and two complex zeros if $k < 0$.

**63.** $f(x) = x^2 - 7x - 8 = \left(x^2 - 7x + \frac{49}{4}\right) - 8 - \frac{49}{4}$

$$= \left(x - \frac{7}{2}\right)^2 - \frac{81}{4}$$

Vertex: $\left(\frac{7}{2}, -\frac{81}{4}\right)$

$f(x) = (x - 8)(x + 1)$

Intercepts: $(8, 0), (-1, 0), (0, -8)$

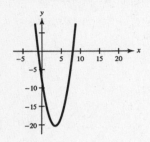

**65.** $f(x) = 6x^2 + 5x - 6 = (3x - 2)(2x + 3)$

Intercepts: $\left(\frac{2}{3}, 0\right), \left(-\frac{3}{2}, 0\right), (0, -6)$

$f(x) = 6x^2 + 5x - 6$

$$= 6\left(x^2 + \frac{5}{6}x + \frac{25}{144}\right) - 6 - \frac{25}{24}$$

$$= 6\left(x + \frac{5}{12}\right)^2 - \frac{169}{24}$$

Vertex: $\left(-\frac{5}{12}, -\frac{169}{24}\right)$

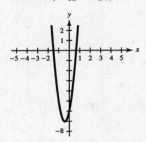

# Section 3.5    Rational Functions and Asymptotes

■  You should know the following basic facts about rational functions.

(a) A function of the form $f(x) = P(x)/Q(x)$, $Q(x) \neq 0$, where $P(x)$ and $Q(x)$ are polynomials, is called a rational function.

(b) The domain of a rational function is the set of all real numbers except those which make the denominator zero.

(c) If $f(x) = P(x)/Q(x)$ is in reduced form, and $a$ is a value such that $Q(a) = 0$, then the line $x = a$ is a vertical asymptote of the graph of $f$. $f(x) \to \infty$ or $f(x) \to -\infty$ as $x \to a$.

(d) The line $y = b$ is a horizontal asymptote of the graph of $f$ if $f(x) \to b$ as $x \to \infty$ or $x \to -\infty$.

(e) Let $f(x) = \dfrac{P(x)}{Q(x)} = \dfrac{a_n x^n + a_{n-1} x^{n-1} + \cdots + a_1 x + a_0}{b_m x^m + b_{m-1} x^{m-1} + \cdots + b_1 x + b_0}$ where $P(x)$ and $Q(x)$ have no common factors.

1. If $n < m$, then the $x$-axis ($y = 0$) is a horizontal asymptote.

2. If $n = m$, then $y = \dfrac{a_n}{b_m}$ is a horizontal asymptote.

3. If $n > m$, then there are no horizontal asymptotes.

**Solutions to Odd-Numbered Exercises**

**1.** $f(x) = \dfrac{1}{x-1}$

(a)

| $x$ | $f(x)$ |
|------|--------|
| 0.5 | $-2$ |
| 0.9 | $-10$ |
| 0.99 | $-100$ |
| 0.999 | $-1000$ |

| $x$ | $f(x)$ |
|------|--------|
| 1.5 | 2 |
| 1.1 | 10 |
| 1.01 | 100 |
| 1.001 | 1000 |

| $x$ | $f(x)$ |
|------|--------|
| 5 | 0.25 |
| 10 | $0.\overline{1}$ |
| 100 | $0.\overline{01}$ |
| 1000 | $0.\overline{001}$ |

| $x$ | $f(x)$ |
|------|--------|
| $-5$ | $-0.167$ |
| $-10$ | $-0.0909$ |
| $-100$ | $-0.0099$ |
| $-1000$ | $-0.001$ |

(b) The zero of the denominator is $x = 1$, so $x = 1$ is a vertical asymptote. The degree of the numerator is less than the degree of the denominator so the $x$-axis, or $y = 0$ is a horizontal asymptote.

(c) The domain is all real numbers except $x = 1$.

**3.** $f(x) = \dfrac{3x}{|x-1|}$

(a)

| $x$ | $f(x)$ |
|------|--------|
| 0.5 | 3 |
| 0.9 | 27 |
| 0.99 | 297 |
| 0.999 | 2997 |

| $x$ | $f(x)$ |
|------|--------|
| 1.5 | 9 |
| 1.1 | 33 |
| 1.01 | 303 |
| 1.001 | 3003 |

| $x$ | $f(x)$ |
|------|--------|
| 5 | 3.75 |
| 10 | $3.\overline{33}$ |
| 100 | $3.\overline{03}$ |
| 1000 | $3.\overline{003}$ |

| $x$ | $f(x)$ |
|------|--------|
| $-5$ | $-2.5$ |
| $-10$ | $-2.727$ |
| $-100$ | $-2.970$ |
| $-1000$ | $-2.997$ |

(b) The zero of the denominator is $x = 1$, so $x = 1$ is a vertical asymptote. Since $f(x) \rightarrow 3$ as $x \rightarrow \infty$ and $f(x) \rightarrow -3$ as $x \rightarrow -\infty$, both $y = 3$ and $y = -3$ are horizontal asymptotes.

(c) The domain is all real numbers except $x = 1$.

**5.** $f(x) = \dfrac{3x^2}{x^2-1}$

(a)

| $x$ | $f(x)$ |
|------|--------|
| 0.5 | $-1$ |
| 0.9 | $-12.79$ |
| 0.99 | $-148.79$ |
| 0.999 | $-1498$ |

| $x$ | $f(x)$ |
|------|--------|
| 1.5 | 5.4 |
| 1.1 | 17.29 |
| 1.01 | 152.3 |
| 1.001 | 1502.3 |

| $x$ | $f(x)$ |
|------|--------|
| 5 | 3.125 |
| 10 | $3.\overline{03}$ |
| 100 | $3.\overline{0003}$ |
| 1000 | 3 |

| $x$ | $f(x)$ |
|------|--------|
| $-5$ | 3.125 |
| $-10$ | $3.\overline{03}$ |
| $-100$ | $3.\overline{0003}$ |
| $-1000$ | 3 |

(b) The zeros of the denominator are $x = \pm 1$ so both $x = 1$ and $x = -1$ are vertical asymptotes.

Since the degree of the numerator equals the degree of the denominator, $y = \frac{3}{1} = 3$ is a horizontal asymptote.

(c) The domain is all real numbers except $x = \pm 1$.

**7.** $f(x) = \dfrac{2}{x+2}$

Vertical asymptote: $x = -2$

Horizontal asymptote: $y = 0$

Matches graph (a)

**9.** $f(x) = \dfrac{4x+1}{x}$

Vertical asymptote: $x = 0$

Horizontal asymptote: $y = 4$

Matches graph (c)

**11.** $f(x) = \dfrac{x - 2}{x - 4}$

Vertical asymptote: $x = 4$

Horizontal asymptote: $y = 1$

Matches graph (b)

**13.** $f(x) = \dfrac{1}{x^2}$

(a) Domain: all real numbers except $x = 0$

(b) Vertical asymptote: $x = 0$

  Horizontal asymptote: $y = 0$

  [Degree of $p(x) <$ degree of $q(x)$]

(c)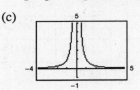

**15.** $f(x) = \dfrac{2 + x}{2 - x}$

(a) Domain: all real numbers except $x = 2$

(b) Vertical asymptote: $x = 2$

  Horizontal asymptote: $y = -1$

(c)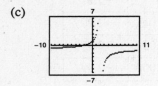

**17.** $f(x) = \dfrac{x^2 + 2x}{2x^2 - x} = \dfrac{x + 2}{2x - 1}, \quad x \neq 0$

(a) Domain: all real numbers except $x = 0, \ \frac{1}{2}$

(b) Vertical asymptote: $x = \frac{1}{2}$

  Horizontal asymptote: $y = \frac{1}{2}$

  [$x = 0$ is not a vertical asymptote]

(c)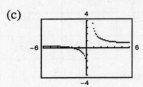

**19.** $f(x) = \dfrac{3x^2 + x - 5}{x^2 + 1}$

(a) Domain: all real numbers

(b) Vertical asymptote: none

  Horizontal asymptote: $y = 3$

(c)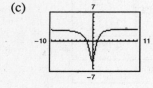

**21.** $f(x) = \dfrac{x - 3}{|x|}$

(a) Domain: all real numbers except $x = 0$

(b) Vertical asymptote: $x = 0$

  Horizontal asymptote:

  $y = 1$ to the right

  $y = -1$ (to the left)

(c)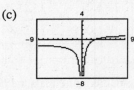

**23.** $f(x) = \dfrac{x^2 - 4}{x + 2}$, $g(x) = x - 2$

    (a) Domain of $f$:  all real numbers except $-2$         Domain of $g$:  all real numbers

    (b) Since $x + 2$ is a common factor of both the numerator and the denominator of $f(x)$, $x = -2$ is not a vertical asymptote of $f$. $f$ has no vertical asymptotes.

    (c)

| $x$ | $-4$ | $-3$ | $-2.5$ | $-2$ | $-1.5$ | $-1$ | $0$ |
|-----|------|------|--------|------|--------|------|-----|
| $f(x)$ | $-6$ | $-5$ | $-4.5$ | undef. | $-3.5$ | $-3$ | $-2$ |
| $g(x)$ | $-6$ | $-5$ | $-4.5$ | $-4$ | $-3.5$ | $-3$ | $-2$ |

    (d) $f$ and $g$ differ only where $f$ is undefined.

**25.** $f(x) = \dfrac{x - 3}{x^2 - 3x}$, $g(x) = \dfrac{1}{x}$

    (a) Domain of $f$:  all real number except $0$ and $3$         Domain of $g$:  all real numbers except $0$

    (b) Since $x - 3$ is a common factor of both the numerator and the denominator of $f$, $x = 3$ is not a vertical asymptote of $f$. The only vertical asymptote is $x = 0$.

    (c)

| $x$ | $-1$ | $-0.5$ | $0$ | $0.5$ | $2$ | $3$ | $4$ |
|-----|------|--------|-----|-------|-----|-----|-----|
| $f(x)$ | $-1$ | $-2$ | undef. | $2$ | $\frac{1}{2}$ | undef. | $\frac{1}{4}$ |
| $g(x)$ | $-1$ | $-2$ | undef. | $2$ | $\frac{1}{2}$ | $\frac{1}{3}$ | $\frac{1}{4}$ |

    (d) They differ only at $x = 3$, where $f$ is undefined and $g$ is defined.

**27.** $f(x) = 4 - \dfrac{1}{x}$

    (a) As $x \to \pm\infty, f(x) \to 4$

    (b) As $x \to \infty, f(x) \to 4$ but is less than $4$

    (c) As $x \to -\infty, f(x) \to 4$ but is greater than $4$

**29.** $f(x) = \dfrac{2x - 1}{x - 3}$

    (a) As $x \to \pm\infty, f(x) \to 2$

    (b) As $x \to \infty, f(x) \to 2$ but is greater than $2$

    (c) As $x \to -\infty, f(x) \to 2$ but is less than $2$

**31.** $g(x) = \dfrac{x^2 - 4}{x + 3} = \dfrac{(x - 2)(x + 2)}{x + 3}$

    The zeros of $g$ are the zeros of the numerator: $x = \pm 2$

**33.** $f(x) = 1 - \dfrac{2}{x - 5} = \dfrac{x - 7}{x - 5}$

    The zero of $f$ corresponds to the zero of the numerator and is $x = 7$.

**35.** $C = \dfrac{255p}{100 - p}$, $0 \le p < 100$

    (a) $C(10) = \dfrac{255(10)}{100 - 10} \approx 28.33$ million dollars

    (b) $C(40) = \dfrac{255(40)}{100 - 40} = 170$ million dollars

    (c) $C(75) = \dfrac{255(75)}{100 - 75} = 765$ million dollars

    (d)

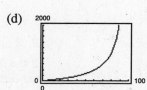

    (e) $C \to \infty$ as $x \to 100$. No, it would not be possible to remove 100% of the pollutants.

**37.** (a) Use data $\left(16, \frac{1}{3}\right), \left(32, \frac{1}{4.7}\right), \left(44, \frac{1}{9.8}\right),$

$\left(50, \frac{1}{19.7}\right), \left(60, \frac{1}{39.4}\right)$

$\frac{1}{y} = -0.00742x \times 0.445$

$y = \dfrac{1}{0.445 - 0.007x}$

(b)

| $x$ | 16 | 32 | 44 | 50 | 60 |
|---|---|---|---|---|---|
| $y$ | 3.0 | 4.5 | 7.3 | 10.5 | 40 |

(Answers will vary.)

(c) No, the function is negative for $x = 70$

**39.** $N = \dfrac{20(5 + 3t)}{1 + 0.04t}, \quad 0 \le t$

(a)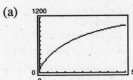

(b) $N(5) \approx 333$ deer

$N(10) = 500$ deer

$N(25) = 800$ deer

(c) The herd is limited by the horizontal asymptote:

$N = \dfrac{60}{0.04} = 1500$ deer

**41.** False. A rational function can have at most $n$ vertical asymptotes, where $n$ is the degree of the denominator.

**43.** One possible answer:

$f(x) = \dfrac{1}{(x + 2)(x - 1)}$

$= \dfrac{1}{x^2 + x - 2}$

**45.** One possible answer:

$f(x) = \dfrac{2x^2}{x^2 + 1}$

**47.** $y - 2 = \dfrac{-1 - 2}{0 - 3}(x - 3) = 1(x - 3)$

$y = x - 1$

$y - x + 1 = 0$

**49.** $y - 7 = \dfrac{10 - 7}{3 - 2}(x - 2) = 3(x - 2)$

$y = 3x + 1$

$3x - y + 1 = 0$

**51.**

$$
\begin{array}{r}
x + 9 \\
x - 4 \overline{\smash{)}\, x^2 + 5x + 6} \\
\underline{x^2 - 4x} \\
9x + 6 \\
\underline{9x - 36} \\
42
\end{array}
$$

$\dfrac{x^2 + 5x + 6}{x - 4} = x + 9 + \dfrac{42}{x - 4}$

**53.**

$$
\begin{array}{r}
2x - 9 \\
x + 5 \overline{\smash{)}\, 2x^2 + x - 11} \\
\underline{2x^2 + 10x} \\
-9x - 11 \\
\underline{-9x - 45} \\
34
\end{array}
$$

$\dfrac{2x^2 + x - 11}{x + 5} = 2x - 9 + \dfrac{34}{x + 5}$

## Section 3.6   Graphs of Rational Functions

■  You should be able to graph $f(x) = \dfrac{p(x)}{q(x)}$.

(a) Find the $x$- and $y$-intercepts.

(b) Find any vertical or horizontal asymptotes.

(c) Plot additional points.

(d) If the degree of the numerator is one more than the degree of the denominator, use long division to find the slant asymptote.

**Solutions to Odd-Numbered Exercises**

**1.** $g(x) = \dfrac{2}{x} + 1$

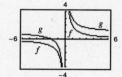

Vertical shift one unit upward

**3.** $g(x) = -\dfrac{2}{x}$

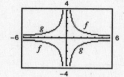

Reflection in the $x$-axis

**5.** $g(x) = \dfrac{2}{x^2} - 2$

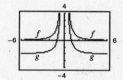

Vertical shift two units downward

**7.** $g(x) = \dfrac{2}{(x - 2)^2}$

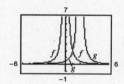

Horizontal shift two units to the right

**9.** $f(x) = \dfrac{1}{x + 2}$

$y$-intercept: $\left(0, \dfrac{1}{2}\right)$

Vertical asymptote: $x = -2$

Horizontal asymptote: $y = 0$

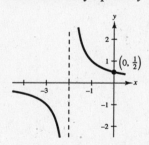

| $x$ | $-4$ | $-3$ | $-1$ | $0$ | $1$ |
|---|---|---|---|---|---|
| $y$ | $-\frac{1}{2}$ | $-1$ | $1$ | $\frac{1}{2}$ | $\frac{1}{3}$ |

**11.** $C(x) = \dfrac{5 + 2x}{1 + x} = \dfrac{2x + 5}{x + 1}$

$x$-intercept: $\left(-\dfrac{5}{2}, 0\right)$

$y$-intercept: $(0, 5)$

Vertical asymptote: $x = -1$

Horizontal asymptote: $y = 2$

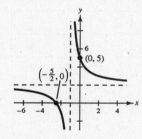

| $x$ | $-4$ | $-3$ | $-2$ | $0$ | $1$ | $2$ |
|---|---|---|---|---|---|---|
| $C(x)$ | $1$ | $\frac{1}{2}$ | $-1$ | $5$ | $\frac{7}{2}$ | $3$ |

**13.** $f(t) = \dfrac{1 - 2t}{t} = -\dfrac{2t - 1}{t}$

$t$-intercept: $\left(\dfrac{1}{2}, 0\right)$

Vertical asymptote: $t = 0$

Horizontal asymptote: $y = -2$

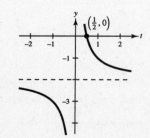

| $x$ | $-2$ | $-1$ | $\frac{1}{2}$ | $1$ | $2$ |
|---|---|---|---|---|---|
| $y$ | $-\frac{5}{2}$ | $-3$ | $0$ | $-1$ | $-\frac{3}{2}$ |

**15.** $f(x) = \dfrac{x^2}{x^2 - 4}$

Intercept: $(0, 0)$

Vertical asymptotes: $x = 2$, $x = -2$

Horizontal asymptote: $y = 1$

$y$-axis symmetry

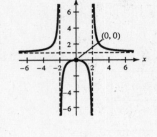

| $x$ | $\pm 4$ | $\pm 3$ | $\pm 1$ | 0 |
|---|---|---|---|---|
| $y$ | $\frac{4}{3}$ | $\frac{9}{5}$ | $-\frac{1}{3}$ | 0 |

**17.** $f(x) = \dfrac{x}{x^2 - 1} = \dfrac{x}{(x + 1)(x - 1)}$

Intercept: $(0, 0)$

Vertical asymptotes: $x = 1$ and $x = -1$

Horizontal asymptote: $y = 0$

Origin symmetry

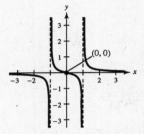

| $x$ | $-3$ | $-2$ | $-\frac{1}{2}$ | 0 | $\frac{1}{2}$ | 2 | 3 | 4 |
|---|---|---|---|---|---|---|---|---|
| $y$ | $-\frac{3}{8}$ | $-\frac{2}{3}$ | $\frac{2}{3}$ | 0 | $-\frac{2}{3}$ | $\frac{2}{3}$ | $\frac{3}{8}$ | $\frac{4}{15}$ |

**19.** $g(x) = \dfrac{4(x + 1)}{x(x - 4)}$

Intercept: $(-1, 0)$

Vertical asymptotes: $x = 0$ and $x = 4$

Horizontal asymptote: $y = 0$

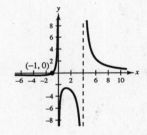

| $x$ | $-2$ | $-1$ | 1 | 2 | 3 | 5 | 6 |
|---|---|---|---|---|---|---|---|
| $y$ | $-\frac{1}{3}$ | 0 | $-\frac{8}{3}$ | $-3$ | $-\frac{16}{3}$ | $\frac{24}{5}$ | $\frac{7}{3}$ |

**21.** $f(x) = \dfrac{3x}{x^2 - x - 2} = \dfrac{3x}{(x + 1)(x - 2)}$

Intercept: $(0, 0)$

Vertical asymptotes: $x = -1, 2$

Horizontal asymptote: $y = 0$

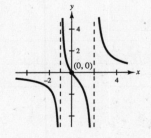

| $x$ | $-3$ | 0 | 1 | 3 | 4 |
|---|---|---|---|---|---|
| $y$ | $-\frac{9}{10}$ | 0 | $-\frac{3}{2}$ | $\frac{9}{4}$ | $\frac{6}{5}$ |

**23.** $f(x) = \dfrac{x^2 + 3x}{x^2 + x - 6} = \dfrac{x(x + 3)}{(x - 2)(x + 3)} = \dfrac{x}{x - 2},$

$x \neq -3$

Intercept: $(0, 0)$

Vertical asymptote: $x = 2$

   (There is a hole at $x = -3$)

Horizontal asymptote: $y = 1$

| $x$ | $-2$ | $-1$ | 0 | 1 | 2 | 3 |
|-----|------|------|---|---|---|---|
| $y$ | $\frac{1}{2}$ | $\frac{1}{3}$ | 0 | $-1$ | undef. | 3 |

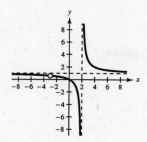

**25.** $f(x) = \dfrac{x^2 - 1}{x + 1} = \dfrac{(x + 1)(x - 1)}{x + 1} = x - 1,$

$x \neq -1$

The graph is a line, with a hole at $x = -1$.

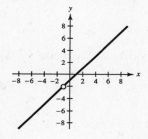

**27.** $f(x) = \dfrac{2 + x}{1 - x} = -\dfrac{x + 2}{x - 1}$

Vertical asymptote: $x = 1$

Horizontal asymptote: $y = -1$

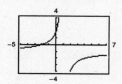

Domain: $x \neq 1$ or $(-\infty, 1) \cup (1, \infty)$

**29.** $f(t) = \dfrac{3t + 1}{t}$

Vertical asymptote: $t = 0$

Horizontal asymptote: $y = 3$

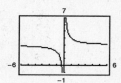

Domain: $t \neq 0$ or $(-\infty, 0) \cup (0, \infty)$

**31.** $h(t) = \dfrac{4}{t^2 + 1}$

Domain: all real numbers OR $(-\infty, \infty)$

Horizontal asymptote: $y = 0$

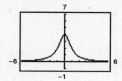

**33.** $f(x) = \dfrac{x + 1}{x^2 - x - 6} = \dfrac{x + 1}{(x - 3)(x + 2)}$

Domain: all real numbers except $x = 3, -2$

Vertical asymptotes: $x = 3,\ x = -2$

Horizontal asymptote: $y = 0$

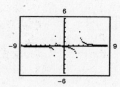

**35.** $f(x) = \dfrac{20x}{x^2 + 1} - \dfrac{1}{x} = \dfrac{19x^2 - 1}{x(x^2 + 1)}$

Domain:  all real numbers except 0,
OR $(-\infty, 0) \cup (0, \infty)$

Vertical asymptote:  $x = 0$

Horizontal asymptote:  $y = 0$

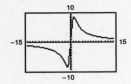

**37.** $h(x) = \dfrac{6x}{\sqrt{x^2 + 1}}$

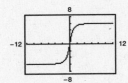

There are two horizontal asymptotes:  $y = \pm 6$.

**39.** $g(x) = \dfrac{4|x - 2|}{x + 1}$

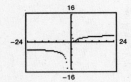

There are two horizontal asymptotes:  $y = \pm 4$.

One vertical asymptote:  $x = -1$.

**41.** $f(x) = \dfrac{4(x - 1)^2}{x^2 - 4x + 5}$

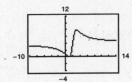

The graph crosses its horizontal asymptote:  $y = 4$.

**43.** $f(x) = \dfrac{2x^2 + 1}{x} = 2x + \dfrac{1}{x}$

Vertical asymptote:  $x = 0$

Slant asymptote:  $y = 2x$

Origin symmetry

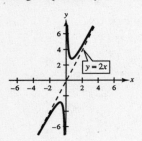

**45.** $h(x) = \dfrac{x^2}{x - 1} = x + 1 + \dfrac{1}{x - 1}$

Intercept:  $(0, 0)$

Vertical asymptote:  $x = 1$

Slant asymptote:  $y = x + 1$

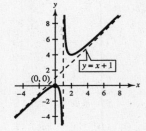

**47.** $g(x) = \dfrac{x^3}{2x^2 - 8} = \dfrac{1}{2}x + \dfrac{4x}{2x^2 - 8}$

Intercept:  $(0, 0)$

Vertical asymptotes:  $x = \pm 2$

Slant asymptote:  $y = \dfrac{1}{2}x$

Origin symmetry

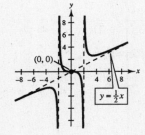

**49.** $f(x) = \dfrac{x^3 + 2x^2 + 4}{2x^2 + 1} = \dfrac{x}{2} + 1 + \dfrac{3 - \dfrac{x}{2}}{2x^2 + 1}$

Intercepts: $(-2.594, 0), (0, 4)$

Slant asymptote: $y = \dfrac{x}{2} + 1$

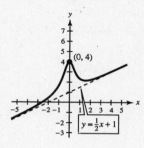

**51.** $y = \dfrac{x + 1}{x - 3}$

(a) $x$-intercept: $(-1, 0)$

(b)  $0 = \dfrac{x + 1}{x - 3}$

$0 = x + 1$

$-1 = x$

**53.** $y = \dfrac{1}{x} - x$

(a) $x$-intercepts: $(\pm 1, 0)$

(b)  $0 = \dfrac{1}{x} - x$

$x = \dfrac{1}{x}$

$x^2 = 1$

$x = \pm 1$

**55.** $y = \dfrac{2x^2 + x}{x + 1} = 2x - 1 + \dfrac{1}{x + 1}$

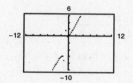

Domain: all real numbers except $x = -1$

Vertical asymptote: $x = -1$

Slant asymptote: $y = 2x - 1$

**57.** $y = \dfrac{1 + 3x^2 - x^3}{x^2} = \dfrac{1}{x^2} + 3 - x = -x + 3 + \dfrac{1}{x^2}$

Domain: all real numbers except 0

OR $(-\infty, 0) \cup (0, \infty)$

Vertical asymptote: $x = 0$

Slant asymptote: $y = -x + 3$

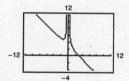

**59.** $y = \dfrac{1}{x + 5} + \dfrac{4}{x}$

(a)
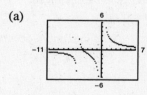

$x$-intercept: $(-4, 0)$

(b)   $0 = \dfrac{1}{x + 5} + \dfrac{4}{x}$

$-\dfrac{4}{x} = \dfrac{1}{x + 5}$

$-4(x + 5) = x$

$-4x - 20 = x$

$-5x = 20$

$x = -4$

**61.** $y = x - \dfrac{6}{x-1}$

(a)

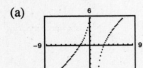

     *x*-intercept: $(-2, 0), (3, 0)$

(b) $\quad 0 = x - \dfrac{6}{x-1}$

$$\dfrac{6}{x-1} = x$$

$$6 = x(x-1)$$

$$0 = x^2 - x - 6$$

$$0 = (x+2)(x-3)$$

$$x = -2, \quad x = 3$$

**63.** (a) $.25(50) + .75(x) = C(50 + x)$

$$\dfrac{12.5 + .75x}{50 + x} = C$$

$$\dfrac{50 + 3x}{200 + 4x} = C$$

$$C = \dfrac{3x + 50}{4(x + 50)}$$

  (b) Domain: $x \geq 0$ and $x \leq 1000 - 50 = 950$

      Thus, $0 \leq x \leq 950$

(c)

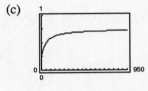

As the tank fills, the rate that the concentration is increasing slows down. It approaches the horizontal asymptote $C = \frac{3}{4} = 0.75$. When the tank is full $(x = 950)$, the concentration is $C = 0.725$.

**65.** (a) $A = xy$ and

$$(x-2)(y-4) = 30$$

$$y - 4 = \dfrac{30}{x-2}$$

$$y = 4 + \dfrac{30}{x-2} = \dfrac{4x + 22}{x-2}$$

    Thus, $A = xy = x\left(\dfrac{4x + 22}{x-2}\right) = \dfrac{2x(2x + 11)}{x-2}$.

(b) Domain: Since the margins on the left and right are each 1 inch, $x > 2$, OR $(2, \infty)$.

(c)

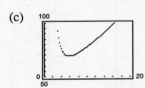

The area is minimum when $x \approx 5.87$ in. and $y \approx 11.75$ in.

**67.** $C = 100\left(\dfrac{200}{x^2} + \dfrac{x}{x + 30}\right), \ 1 \leq x$

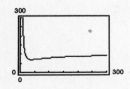

The minimum occurs when $x \approx 40.4 \approx 40$.

**69.** $C = \dfrac{3t^2 + t}{t^3 + 50}, \ 0 \leq t$

(a) The horizontal asymptote is the *t*-axis, or $C = 0$. This indicates that the chemical eventually dissipates.

(b)

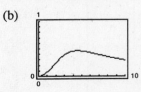

The maximum occurs when $t \approx 4.5$.

(c) Graph $C$ together with $y = 0.345$. The graphs intersect at $t \approx 2.65$ and $t \approx 8.32$. $C < 0.345$ when $0 \leq t < 2.65$ hours and when $t > 8.32$ hours

**71.** (a) $A = 0.44t + 1.8$

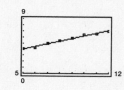

(b) $\dfrac{1}{A} = -0.016t + 0.32$

$A = \dfrac{1}{0.32 - 0.016t}$

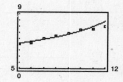

(c)

| $t$ | 5 | 6 | 7 | 8 | 9 | 10 | 11 | 12 |
|---|---|---|---|---|---|---|---|---|
| Linear | 4 | 4.4 | 4.9 | 5.3 | 5.8 | 6.2 | 6.6 | 7.1 |
| Rational | 4.2 | 4.5 | 4.8 | 5.2 | 5.7 | 6.3 | 6.9 | 7.8 |

The linear model is closer to the actual data.

**73.** False, you will have to lift your pencil to cross the vertical asymptote.

**75.** $h(x) = \dfrac{6 - 2x}{3 - x} = \dfrac{2(3 - x)}{3 - x} = 2, \quad x \neq 3$

Since $h(x)$ is not reduced and $(3 - x)$ is a factor of both the numerator and the denominator, $x = 3$ is not a horizontal asymptote.

There is a hole in the graph at $x = 3$.

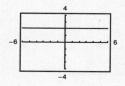

**77.** $y = x + 1 + \dfrac{a}{x - 2}$ has slant asymptote $y = x + 1$ and vertical asymptote $x = 2$.

$0 = -2 + 1 + \dfrac{a}{-2 - 2}$

$1 = \dfrac{a}{-4}$

$a = -4$

Hence, $y = x + 1 + \dfrac{-4}{x - 2} = \dfrac{x^2 - x - 6}{x - 2}$

**79.** $\left(\dfrac{x}{8}\right)^{-3} = \left(\dfrac{8}{x}\right)^3 = \dfrac{512}{x^3}$

**81.** $\dfrac{3x^3y^2}{15xy^4} = \dfrac{x^2}{5y^2}, x \neq 0$

**83.** $\dfrac{3^{7/6}}{3^{1/6}} = 3^{6/6} = 3$

**85.**

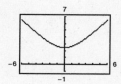

Domain: all $x$

Range: $y \geq \sqrt{6}$

**87.**

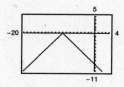

Domain: all $x$

Range: $y \leq 0$

# Section 3.7   Exploring Data: Quadratic Models

**Solutions to Odd-Numbered Exercises**

**1.** A quadratic model is better

**3.** A linear model is better

**5.** Neither linear nor quadratic

**7.** (a)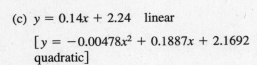

(b) Linear model is better

(c) $y = 0.14x + 2.24$ linear

$[y = -0.00478x^2 + 0.1887x + 2.1692$ quadratic]

(d)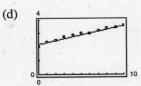

(e)

| $x$ | 0 | 1 | 2 | 3 | 4 | 5 | 6 | 7 | 8 | 9 | 10 |
|---|---|---|---|---|---|---|---|---|---|---|---|
| $y$ | 2.1 | 2.4 | 2.5 | 2.8 | 2.9 | 3.0 | 3.0 | 3.2 | 3.4 | 3.5 | 3.6 |
| Model | 2.2 | 2.4 | 2.5 | 2.7 | 2.8 | 2.9 | 3.1 | 3.2 | 3.4 | 3.5 | 3.6 |

**9.** (a)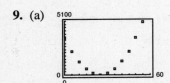

(b) Quadratic model is better

(c) $y = 5.55x^2 - 277.5x + 3478$

(d)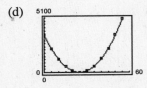

(e)

| $x$ | 0 | 5 | 10 | 15 | 20 | 25 | 30 | 35 | 40 | 45 | 50 | 55 |
|---|---|---|---|---|---|---|---|---|---|---|---|---|
| $y$ | 3480 | 2235 | 1250 | 565 | 150 | 12 | 145 | 575 | 1275 | 2225 | 3500 | 5010 |
| Model | 3478 | 2229 | 1258 | 564 | 148 | 9 | 140 | 564 | 1258 | 2229 | 3478 | 5004 |

**11.** (a)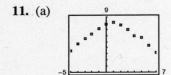

(b) Quadratic model is better

(c) $y = -0.1203x^3 + 0.208x + 7.53$

(d)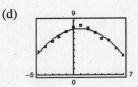

(e)

| $x$ | $-5$ | $-4$ | $-3$ | $-2$ | $-1$ | 0 | 1 | 2 | 3 | 4 | 5 | 6 | 7 |
|---|---|---|---|---|---|---|---|---|---|---|---|---|---|
| $y$ | 3.8 | 4.7 | 5.5 | 6.2 | 7.1 | 7.9 | 8.1 | 7.7 | 6.9 | 6 | 5.6 | 4.4 | 3.2 |
| Model | 3.5 | 4.8 | 5.8 | 6.6 | 7.2 | 7.5 | 7.6 | 7.5 | 7.1 | 6.4 | 5.6 | 4.4 | 3.1 |

**13.** (a)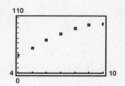

(b) $P = -1.298t^2 + 28.99t - 61.3$

(c)

The model is a good fit.

(d) $P = 100$ when $t \approx 10.48$, or during 2000.

(e) No, the model turns downward as time increases.

**15.** (a)

(b) $H = -1.682t^2 + 40.63t + 6903.2$

(c)

The model is a good fit.

(d) $H$ is a maximum of 7148.6 when $t \approx 12.08$, or 1972.

(e) No, the model decreases as time increases.

**17.** (a) $y = 2.477x + 1.13$          Linear

$y = 0.071x^2 + 1.69x + 2.7$    Quadratic

(b) 0.989 for linear model

0.9952 for quadratic model

(c) Quadratic fits better

**19.** (a) $y = -0.892x + 0.24$          Linear

$y = 0.0012x^2 - 0.895x + 0.203$    Quadratic

(b) 0.999824 for linear model

0.999873 for quadratic model

(c) Quadratic fits slightly better

(d) $S = 6.66t^2 + 35.1t - 261$

(e)

(f) Both models fit well. For the linear model, 2007 corresponds to $t = 17$ and $S \approx 1861$.

**21.** (a)

(b) $S = 155.014t - 774.16$

(c)

**23.** True.

**25.** (a) $(f \circ g)(x) = f(x^2 + 3) = 2(x^2 + 3) - 1 = 2x^2 + 5$

(b) $(g \circ f)(x) = g(2x - 1) = (2x - 1)^2 + 3 = 4x^2 - 4x + 4$

**27.** (a) $(f \circ g)(x) = f\left(\sqrt[3]{x + 1}\right) = x + 1 - 1 = x$

(b) $(g \circ f)(x) = g(x^3 - 1) = \sqrt[3]{x^3 - 1 + 1} = x$

**29.** $f$ is one-to-one

$$y = 2x + 5$$

$$x = 2y + 5$$

$$2y = x - 5$$

$$y = (x - 5)/2 \implies f^{-1}x = \frac{x - 5}{2}$$

**31.** $f$ is one-to-one on $[0, \infty)$

$$y = x^2 + 5 \qquad x \geq 0$$

$$x = y^2 + 5 \qquad y \geq 0$$

$$y^2 = x - 5$$

$$y = \sqrt{x - 5} \implies f^{-1}(x) = \sqrt{x - 5}$$

**33.**

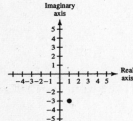

**35.**

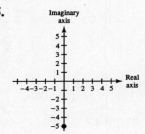

# Review Exercises for Chapter 3

**Solutions to Odd-Numbered Exercises**

**1.**

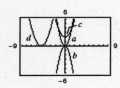

(a) $y = 2x^2$ is a vertical stretch

(b) $y = -2x^2$ is a vertical stretch and reflection in the $x$-axis

(c) $y = x^2 + 2$ is a vertical shift 2 units upward

(d) $y = (x + 5)^2$ is a horizontal shift 5 units to the left

**3.** $f(x) = \left(x + \frac{3}{2}\right)^2 + 1$

Vertex: $\left(-\frac{3}{2}, 1\right)$

$y$-intercept: $\left(0, \frac{13}{4}\right)$

No $x$-intercepts

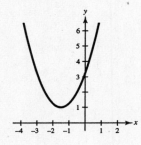

**5.** $f(x) = \frac{1}{3}(x^2 + 5x - 4)$

$$= \frac{1}{3}\left(x^2 + 5x + \frac{25}{4} - \frac{25}{4} - 4\right)$$

$$= \frac{1}{3}\left[\left(x + \frac{5}{2}\right)^2 - \frac{41}{4}\right]$$

$$= \frac{1}{3}\left(x + \frac{5}{2}\right)^2 - \frac{41}{12}$$

Vertex: $\left(-\frac{5}{2}, -\frac{41}{12}\right)$

$y$-intercepts: $\left(0, -\frac{4}{3}\right)$

$x$-intercepts: $0 = \frac{1}{3}(x^2 + 5x - 4)$

$$0 = x^2 + 5x - 4$$

$$x = \frac{-5 \pm \sqrt{41}}{2} \qquad \text{Use the Quadratic Formula.}$$

$$\left(\frac{-5 \pm \sqrt{41}}{2}, 0\right)$$

**7.** Vertex: $(1, -4) \Rightarrow f(x) = a(x - 1)^2 - 4$

Point: $(2, -3) \Rightarrow -3 = a(2 - 1)^2 - 4$

$$1 = a$$

Thus, $f(x) = (x - 1)^2 - 4$.

**9.** Vertex: $(-2, -2) \Rightarrow f(x) = a(x + 2)^2 - 2$

Point: $(-1, 0) \Rightarrow 0 = a(-1 + 2)^2 - 2$

$$a = 2$$

Thus, $f(x) = 2(x + 2)^2 - 2$.

**11.** (a) $A = xy = x\left(\dfrac{8 - x}{2}\right)$, since $x + 2y - 8 = 0 \Rightarrow y = \dfrac{8 - x}{2}$.

Since the figure is in the first quadrant and $x$ and $y$ must be positive, the domain of

$A = x\left(\dfrac{8 - x}{2}\right)$ is $0 < x < 8$.

(b)

| $x$ | $y$ | Area |
|---|---|---|
| 1 | $4 - \frac{1}{2}(1)$ | $(1)\left[4 - \frac{1}{2}(1)\right] = \frac{7}{2}$ |
| 2 | $4 - \frac{1}{2}(2)$ | $(2)\left[4 - \frac{1}{2}(2)\right] = 6$ |
| 3 | $4 - \frac{1}{2}(3)$ | $(3)\left[4 - \frac{1}{2}(3)\right] = \frac{15}{2}$ |
| 4 | $4 - \frac{1}{2}(4)$ | $(4)\left[4 - \frac{1}{2}(4)\right] = 8$ |
| 5 | $4 - \frac{1}{2}(5)$ | $(5)\left[4 - \frac{1}{2}(5)\right] = \frac{15}{2}$ |
| 6 | $4 - \frac{1}{2}(6)$ | $(6)\left[4 - \frac{1}{2}(6)\right] = 6$ |

The dimensions that will produce a maximum area seem to be $x = 4$ and $y = 2$.

(c)

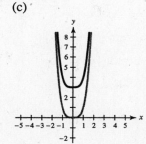

The maximum area of 8 occurs at the vertex when $x = 4$ and $y = \dfrac{8 - 4}{2} = 2$.

(d) $A = x\left(\dfrac{8 - x}{2}\right)$

$$= \frac{1}{2}(8x - x^2)$$

$$= -\frac{1}{2}(x^2 - 8x)$$

$$= -\frac{1}{2}(x^2 - 8x + 16 - 16)$$

$$= -\frac{1}{2}\left[(x - 4)^2 - 16\right]$$

$$= -\frac{1}{2}(x - 4)^2 + 8 \qquad \text{The maximum area of 8 occurs when } x = 4 \text{ and } y = \dfrac{8 - 4}{2} = 2.$$

(e) The answers are the same.

**13.** $y = x^4$

(a)

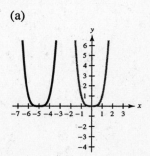

(b)

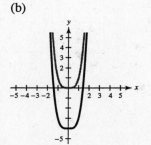

(c)

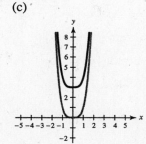

(d)

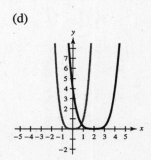

**15.** $y = x^6$

(a)

(b)

(c)

(d)

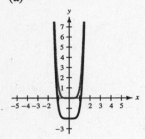

**17.** $f(x) = \frac{1}{2}x^3 - 2x + 1$; $g(x) = \frac{1}{2}x^3$

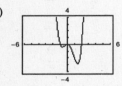

**19.** $f(x) = -x^2 + 6x + 9$

The degree is even and the leading coefficient is negative. The graph falls to the left and right.

**21.** $f(x) = \frac{3}{4}(x^4 + 3x^2 + 2)$

The degree is even and the leading coefficient is positive. The graph rises to the left and right.

**23.** $g(x) = x^4 - x^3 - 2x^2$

(a)

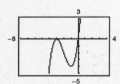

(b) zeros: $0, 2, -1$

(c) $0 = x^4 - x^3 - 2x^2$

$0 = x^2(x^2 - x - 2)$

$0 = x^2(x - 2)(x + 1)$

zeros: $0, 0, 2, -1$

**25.** $f(t) = t^3 - 3t$

(a)

(b) zeros: $\pm 1.732, 0$

(c) $0 = t^3 - 3t$

$0 = t(t^2 - 3)$

zeros: $0, \pm\sqrt{3}$

**27.** $f(x) = x(x + 3)^2$

(a)

(b) zeros: $0, -3$

(c) $0 = (x + 3)^2$

zeros: $0, -3, -3$

**29.** $f(x) = x^3 + 2x^2 - x - 1$

(a) $f(-3) < 0$, $f(-2) > 0 \Rightarrow$ zero in $[-3, -2]$

$f(-1) > 0$, $f(0) < 0 \Rightarrow$ zero in $[-1, 0]$

$f(0) < 0$, $f(1) > 0 \Rightarrow$ zero in $[0, 1]$

(b) zeros: $-2.247, -0.555, 0.802$

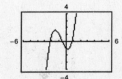

**31.** $f(x) = x^4 - 6x^2 - 4$

(a) $f(-3) > 0$, $f(-2) < 0 \implies$ zero in $[-3, -2]$

$f(2) < 0$, $f(3) > 0 \implies$ zero in $[2, 3]$

(b) zeros: $\pm 2.570$

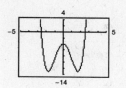

**33.** $y_1 = \dfrac{x^2}{x - 2}$

$y_2 = x + 2 + \dfrac{4}{x - 2}$

$= \dfrac{(x + 2)(x - 2)}{x - 2} + \dfrac{4}{x - 2}$

$= \dfrac{x^2 - 4}{x - 2} + \dfrac{4}{x - 2}$

$= \dfrac{x^2}{x - 2}$

$= y_1$

**35.**
$$
\begin{array}{r}
8x + 5 \\
3x - 2 \overline{) \, 24x^2 - \phantom{0}x - 8} \\
\underline{24x^2 - 16x} \\
15x - \phantom{0}8 \\
\underline{15x - 10} \\
2
\end{array}
$$

Thus, $\dfrac{24x^2 - x - 8}{3x - 2} = 8x + 5 + \dfrac{2}{3x - 2}$.

**37.**
$$
\begin{array}{r}
x^2 - 2 \\
x^2 - 1 \overline{) \, x^4 - 3x^2 + 2} \\
\underline{x^4 - \phantom{0}x^2} \\
-2x^2 + 2 \\
\underline{-2x^2 + 2} \\
0
\end{array}
$$

Thus, $\dfrac{x^4 - 3x^2 + 2}{x^2 - 1} = x^2 - 2 \quad (x \neq \pm 1)$

**39.**
$$
\begin{array}{r}
5x + 2 \\
x^2 - 3x + 1 \overline{) \, 5x^3 - 13x^2 - \phantom{0}x + 2} \\
\underline{5x^3 - 15x^2 + 5x} \\
2x^2 - 6x + 2 \\
\underline{2x^2 - 6x + 2} \\
0
\end{array}
$$

Thus, $\dfrac{5x^3 - 13x^2 - x + 2}{x^2 - 3x + 1} = 5x + 2$,

$x \neq \dfrac{1}{2}\left(3 \pm \sqrt{5}\right)$

**41.**
$$
\begin{array}{r}
3x^2 + 5x + 8 \\
2x^2 + 0x - 1 \overline{) \, 6x^4 + 10x^3 + 13x^2 - 5x + 2} \\
\underline{6x^4 + \phantom{0}0x^3 - \phantom{0}3x^2} \\
10x^3 + 16x^2 - 5x \\
\underline{10x^3 + \phantom{0}0x^2 - 5x} \\
16x^2 - \phantom{0}0 + 2 \\
\underline{16x^2 + \phantom{0}0 - 8} \\
10
\end{array}
$$

$\dfrac{6x^4 + 10x^3 + 13x^2 - 5x + 2}{2x^2 - 1} = 3x^2 + 5x + 8 + \dfrac{10}{2x^2 - 1}$

**43.**
$$
\begin{array}{r|rrrrr}
-2 & 0.25 & -4 & 0 & 0 & 0 \\
   &      & -\frac{1}{2} & 9 & -18 & 36 \\
\hline
   & \frac{1}{4} & -\frac{9}{2} & 9 & -18 & 36 \\
\end{array}
$$

Hence,

$$
\frac{0.25x^4 - 4x^3}{x + 2} = \frac{1}{4}x^3 - \frac{9}{2}x^2 + 9x - 18 + \frac{36}{x + 2}
$$

**45.**
$$
\begin{array}{r|rrrrr}
\frac{2}{3} & 6 & -4 & -27 & 18 & 0 \\
            &   & 4 & 0 & -18 & 0 \\
\hline
            & 6 & 0 & -27 & 0 & 0 \\
\end{array}
$$

Thus,

$$
\frac{6x^4 - 4x^3 - 27x^2 + 18x}{x - \left(\frac{2}{3}\right)} = 6x^3 - 27x, \ x \neq \frac{2}{3}
$$

**47.**
$$
\begin{array}{r|rrrr}
4 & 3 & -10 & 12 & -22 \\
  &   & 12 & 8 & 80 \\
\hline
  & 3 & 2 & 20 & 58 \\
\end{array}
$$

Thus, $\dfrac{3x^3 - 10x^2 + 12x - 22}{x - 4} = 3x^2 + 2x + 20 + \dfrac{58}{x - 4}$

**49.** (a)
$$
\begin{array}{r|rrrrr}
-3 & 1 & 10 & -24 & 20 & 44 \\
   &   & -3 & -21 & 135 & -465 \\
\hline
   & 1 & 7 & -45 & 155 & \underline{-421} \\
\end{array}
$$

$f(-3) = -421$

(b)
$$
\begin{array}{r|rrrrr}
-1 & 1 & 10 & -24 & 20 & 44 \\
   &   & -1 & -9 & 33 & -53 \\
\hline
   & 1 & 9 & -33 & 53 & \underline{-9} \\
\end{array}
$$

$f(-1) = -9$

**51.** $f(x) = x^3 + 4x^2 - 25x - 28$

(a)
$$
\begin{array}{r|rrrr}
4 & 1 & 4 & -25 & -28 \\
  &   & 4 & 32 & 28 \\
\hline
  & 1 & 8 & 7 & 0 \\
\end{array}
$$

$(x - 4)$ is a factor

(b) $x^2 + 8x + 7 = (x + 1)(x + 7)$

Remaining factors: $(x + 1)$, $(x + 7)$

(c) $f(x) = (x - 4)(x + 1)(x + 7)$

(d) Zeros: $4. -1, -7$

(e)

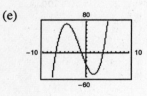

**53.** $f(x) = x^4 - 4x^3 - 7x^2 + 22x + 24$

(a)
$$
\begin{array}{r|rrrrr}
-2 & 1 & -4 & -7 & 22 & 24 \\
   &   & -2 & 12 & -10 & -24 \\
\hline
   & 1 & -6 & 5 & 12 & 0 \\
\end{array}
$$

$(x + 2)$ is a factor

$$
\begin{array}{r|rrrr}
3 & 1 & -6 & 5 & 12 \\
  &   & 3 & -9 & -12 \\
\hline
  & 1 & -3 & -4 & 0 \\
\end{array}
$$

$(x - 3)$ is a factor

(b) $x^2 - 3x - 4 = (x - 4)(x + 1)$

Remaining factors: $(x - 4)$, $(x + 1)$

(c) $f(x) = (x + 2)(x - 3)(x - 4)(x + 1)$

(d) zeros: $-2, 3, 4, -1$

(e)

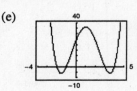

**55.** Possible rational zeros: $\pm 3, \pm\frac{3}{2}, \pm\frac{3}{4}, \pm 1, \pm\frac{1}{2}, \pm\frac{1}{4}$

zeros: $1, 1, \frac{3}{4}$

**57.** $f(x) = 6x^3 - 5x^2 + 24x - 20$

Graphing $f(x)$ with a graphing utility suggests that $x = \frac{5}{6}$ is a zero.

$$
\begin{array}{r|rrrr}
\frac{5}{6} & 6 & -5 & 24 & -20 \\
 & & 5 & 0 & 20 \\
\hline
 & 6 & 0 & 24 & 0
\end{array}
$$

The quadratic $6x^2 + 24 = 0$ has complex zeros $x = \pm 2i$. Thus, the zeros are $\frac{5}{6}, 2i, -2i$.

**59.** $f(x) = 6x^4 - 25x^3 + 14x^2 + 27x - 18$

Possible Rational Zeros: $\pm 1, \pm 2, \pm 3, \pm 6, \pm 9, \pm 18, \pm\frac{1}{2}, \pm\frac{3}{2}, \pm\frac{9}{2}, \pm\frac{1}{3}, \pm\frac{2}{3}, \pm\frac{1}{6}$. Use a graphing utility to see that $x = -1$ and $x = 3$ are probably zeros.

$$
\begin{array}{r|rrrrr}
-1 & 6 & -25 & 14 & 27 & -18 \\
 & & -6 & 31 & -45 & 18 \\
\hline
 & 6 & -31 & 45 & -18 & 0
\end{array}
$$

$$
\begin{array}{r|rrrr}
3 & 6 & -31 & 45 & -18 \\
 & & 18 & -39 & 18 \\
\hline
 & 6 & -13 & 6 & 0
\end{array}
$$

$$
\begin{aligned}
6x^4 - 25x^3 + 14x^2 + 27x - 18 &= (x + 1)(x - 3)(6x^2 - 13x + 6) \\
&= (x + 1)(x - 3)(3x - 2)(2x - 3)
\end{aligned}
$$

Thus, the zeros of $f$ are $x = -1$, $x = 3$, $x = \frac{2}{3}$, and $x = \frac{3}{2}$.

**61.** $g(x) = 5x^3 + 3x^2 - 6x + 9$ has two variations in sign $\implies$ 0 or 2 positive real zeros.

$g(-x) = -5x^3 + 3x^2 + 6x + 9$ has one variation in sign $\implies$ 1 negative real zero.

**63.**
$$
\begin{array}{r|rrrr}
1 & 4 & -3 & 4 & -3 \\
 & & 4 & 1 & 5 \\
\hline
 & 4 & 1 & 5 & 2
\end{array}
$$

All entries positive. $x = 1$ is upper bound.

$$
\begin{array}{r|rrrr}
-\frac{1}{4} & 4 & -3 & 4 & -3 \\
 & & -1 & 1 & -\frac{5}{4} \\
\hline
 & 4 & -4 & 5 & -\frac{17}{4}
\end{array}
$$

Alternating signs. $x = -\frac{1}{4}$ is lower bound.

**65.** $f(x) = 3x(x - 2)^2$

zeros: $0, 2, 2$

**67.** $f(x) = (x + 4)(x - 6)(x - 2i)(x + 2i)$

zeros: $-4, 6, 2i, -2i$

**69.** $f(x) = 2x^4 - 5x^3 + 10x - 12$

$$
\begin{array}{r|rrrrr}
2 & 2 & -5 & 0 & 10 & -12 \\
 & & 4 & -2 & -4 & 12 \\
\hline
 & 2 & -1 & -2 & 6 & 0
\end{array}
$$

$x = 2$ is a zero

$$
\begin{array}{r|rrrr}
-\frac{3}{2} & 2 & -1 & -2 & 6 \\
 & & -3 & 6 & -6 \\
\hline
 & 2 & -4 & 4 & 0
\end{array}
$$

$x = -\frac{3}{2}$ is a zero

**—CONTINUED—**

**69. —CONTINUED—**

$f(x) = (x - 2)(x + \frac{3}{2})(2x^2 - 4x + 4)$

$\qquad = (x - 2)(2x + 3)(x^2 - 2x + 2)$

By the Quadratic Formula, applied to $x^2 - 2x + 2$,

$$x = \frac{2 \pm \sqrt{4 - 4(2)}}{2} = 1 \pm i$$

zeros: $2, -\frac{3}{2}, 1 \pm i$

$f(x) = (x - 2)(2x + 3)(x - 1 + i)(x - 1 - i)$

**71.** $h(x) = x^3 - 7x^2 + 18x - 24$

$$
\begin{array}{r|rrrr}
4 & 1 & -7 & 18 & -24 \\
  &   & 4 & -12 & 24 \\
\hline
  & 1 & -3 & 6 & 0
\end{array}
$$

$x = 4$ is a zero. Applying the Quadratic Formula on $x^2 - 3x + 6$,

$$x = \frac{3 \pm \sqrt{9 - 4(6)}}{2} = \frac{3}{2} \pm \frac{\sqrt{15}}{2}i$$

zeros: $4, \frac{3}{2} + \frac{\sqrt{15}}{2}i, \frac{3}{2} - \frac{\sqrt{15}}{2}i$

$$h(x) = (x - 4)\left(x - \frac{3 + \sqrt{15}i}{2}\right)\left(x - \frac{3 - \sqrt{15}i}{2}\right)$$

**73.** $f(x) = x^3 - 4x^2 + 6x - 4$

(a) $x^3 - 4x^2 + 6x - 4 = (x - 2)(x^2 - 2x + 2)$

By the Quadratic Formula, for $x^2 - 2x + 2$,

$$x = \frac{2 \pm \sqrt{(-2)^2 - 4(2)}}{2} = 1 \pm i$$

zeros: $2, 1 + i, 1 - i$

(b) $f(x) = (x - 2)(x - 1 - i)(x - 1 + i)$

(c) $x$-intercept: $(2, 0)$

(d)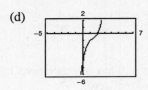

**75.** $f(x) = x^3 + 6x^2 + 11x + 12$

(a) $x^3 + 6x^2 + 11x + 12 = (x + 4)(x^2 + 2x + 3)$

By the Quadratic Formula for $x^2 + 2x + 3$,

$$x = \frac{-2 \pm \sqrt{4 - 4(3)}}{2} = -1 \pm \sqrt{2}i$$

zeros: $-4, -1 \pm \sqrt{2}i$

(b) $f(x) = (x + 4)(x + 1 - \sqrt{2}i)(x + 1 + \sqrt{2}i)$

(c) $x$-intercept: $(-4, 0)$

(d)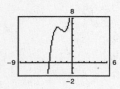

**77.** $f(x) = x^4 + 34x^2 + 225$

(a) $x^4 + 34x^2 + 225 = (x^2 + 9)(x^2 + 25)$

zeros: $\pm 3i, \pm 5i$

(b) $(x + 3i)(x - 3i)(x + 5i)(x - 5i)$

(c) No $x$-intercepts

(d)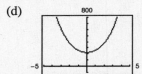

**79.** $f(x) = (x + 2)(x + 2)(x + 5i)(x - 5i)$

$\qquad = (x^2 + 4x + 4)(x^2 + 25)$

$\qquad = x^4 + 4x^3 + 29x^2 + 100x + 100$

**81.** $f(x) = (x - 1)(x + 4)(x + 3 - 5i)(x + 3 + 5i)$

$\qquad = (x^2 + 3x - 4)((x + 3)^2 + 25)$

$\qquad = (x^2 + 3x - 4)(x^2 + 6x + 34)$

$\qquad = x^4 + 9x^3 + 48x^2 + 78x - 136$

**83.** $f(x) = x^4 + 2x^2 - 8$

    (a) $f(x) = (x^2 + 4)(x^2 - 2)$

    (b) $f(x) = (x^2 + 4)(x + \sqrt{2})(x - \sqrt{2})$

    (c) $f(x) = (x + 2i)(x - 2i)(x + \sqrt{2})(x - \sqrt{2})$

**85.** $f(x) = x^4 - 2x^3 + 8x^2 - 18x - 9$

    (a) $f(x) = (x^2 + 9)(x^2 - 2x - 1)$

    For the quadratic $x^2 - 2x - 1$, $x = \dfrac{2 \pm \sqrt{(-2)^2 - 4(-1)}}{2} = 1 \pm \sqrt{2}$

    (b) $f(x) = (x^2 + 9)(x - 1 + \sqrt{2})(x - 1 - \sqrt{2})$

    (c) $f(x) = (x + 3i)(x - 3i)(x - 1 + \sqrt{2})(x - 1 - \sqrt{2})$

**87.** $f(x) = \dfrac{x - 8}{1 - x}$

    (a) Domain: all $x \neq 1$

    (b) Horizontal asymptote: $y = -1$

        Vertical asymptote: $x = 1$

**89.** $f(x) = \dfrac{2}{x^2 - 3x - 18} = \dfrac{2}{(x - 6)(x + 3)}$

    (a) Domain: all $x \neq 6, \, -3$

    (b) Horizontal asymptote: $y = 0$

        Vertical asymptotes: $x = 6$, $x = -3$

**91.** $f(x) = \dfrac{7 + x}{7 - x}$

    (a) Domain: all $x \neq 7$

    (b) Horizontal asymptote: $y = -1$

        Vertical asymptote: $x = 7$

**93.** $f(x) = \dfrac{4x^2}{2x^2 - 3}$

    (a) Domain: all $x \neq \pm\sqrt{\frac{3}{2}}$

    (b) Horizontal asymptote: $y = 2$

        Vertical asymptote: $x = \pm\sqrt{\frac{3}{2}}$

**95.** $f(x) = \dfrac{2x - 10}{x^2 - 2x - 15} = \dfrac{2(x - 5)}{(x - 5)(x + 3)} = \dfrac{2}{x + 3}$,

    $x \neq 5$

    (a) Domain: all $x \neq 5, -3$

    (b) Vertical asymptote: $x = -3$

        (There is a hole at $x = 5$)

        Horizontal asymptote: $y = 0$

**97.** $f(x) = \dfrac{x - 2}{|x| + 2}$

    (a) Domain: all real numbers

    (b) No vertical asymptotes

        Horizontal asymptotes: $y = 1, y = -1$

**99.** $C = \dfrac{528p}{100 - p}$,   $0 \leq p < 100$

    (a) When $p = 25$, $C = \dfrac{528(25)}{100 - 25} = 176$ million

        When $p = 50$, $C = \dfrac{528(50)}{100 - 50} = 528$ million

        When $p = 75$, $C = \dfrac{528(75)}{100 - 75} = 1584$ million

    (b)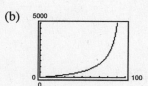

    (c) No. As $p \to 100$, $C$ tends to infinity.

**101.** $f(x) = \dfrac{2x - 1}{x - 5}$

Intercepts: $\left(0, \frac{1}{5}\right), \left(\frac{1}{2}, 0\right)$

Vertical asymptote: $x = 5$

Horizontal asymptote: $y = 2$

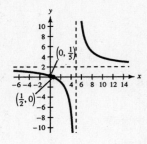

**103.** $f(x) = \dfrac{2x}{x^2 + 4}$

Intercept: $(0, 0)$

Origin symmetry

Horizontal asymptote: $y = 0$

| $x$ | $-2$ | $-1$ | $0$ | $1$ | $2$ |
|---|---|---|---|---|---|
| $y$ | $-\frac{1}{2}$ | $-\frac{2}{5}$ | $0$ | $\frac{2}{5}$ | $\frac{1}{2}$ |

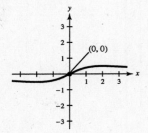

**105.** $f(x) = \dfrac{x^2}{x^2 + 1}$

Intercept: $(0, 0)$

$y$-axis symmetry

Horizontal asymptote: $y = 1$

| $x$ | $\pm 3$ | $\pm 2$ | $\pm 1$ | $0$ |
|---|---|---|---|---|
| $y$ | $\frac{9}{10}$ | $\frac{4}{5}$ | $\frac{1}{2}$ | $0$ |

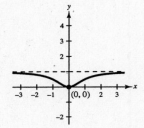

**107.** $f(x) = \dfrac{2}{(x + 1)^2}$

Intercept: $(0, 2)$

Horizontal asymptote: $y = 0$

Vertical asymptote: $x = -1$

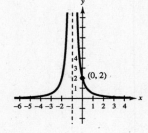

**109.** $f(x) = \dfrac{2(x^2 - 16)}{x^2 + 2x - 8} = \dfrac{2(x + 4)(x - 4)}{(x + 4)(x - 2)} = \dfrac{2(x - 4)}{x - 2}$,

$x \neq -4$

Intercepts: $(0, 4), (4, 0)$

Horizontal asymptote: $y = 2$

Vertical asymptote: $x = 2$

Hole at $x = -4$

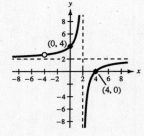

**111.** $f(x) = \dfrac{2x^3}{x^2 + 1} = 2x - \dfrac{2x}{x^2 + 1}$

Intercept: $(0, 0)$

Origin symmetry

Slant asymptote: $y = 2x$

| $x$ | $-2$ | $-1$ | $0$ | $1$ | $2$ |
|---|---|---|---|---|---|
| $y$ | $-\frac{16}{5}$ | $-1$ | $0$ | $1$ | $\frac{16}{5}$ |

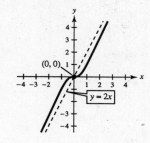

**113.** $f(x) = \dfrac{x^2 - x + 1}{x - 3}$

$\qquad = x + 2 + \dfrac{7}{x - 3}$

Intercept: $\left(0, -\frac{1}{3}\right)$

Vertical asymptote: $x = 3$

Slant asymptote: $y = x + 2$

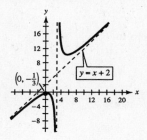

**115.** $N = \dfrac{20(4 + 3t)}{1 + 0.05t}, \quad t \geq 0$

(a)
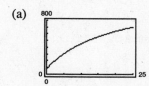

(b) $N(5)\ = 304,000$ fish

$\quad N(10) \approx 453,333$ fish

$\quad N(25) \approx 702,222$ fish

(c) The limit is

$\quad \dfrac{60}{0.05} = 1,200,000$ fish,

the horizontal asymptote.

**117.** Quadratic model

**119.** Linear model

**121.** (a)
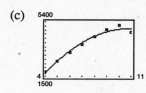

(b) $R = -68.71t^2 + 1485.3t - 3127$

(c)

The model is a fair fit to the date

(d) The model gives a maximum at $t = 10.8$, or near 2001

(e) No, the model begins to decrease after $t = 10.8$.

**123.** False. The degree of the numerator is two more than the degree of the denominator.

**125.** It means that the divisor is a factor of the dividend.

# Chapter 3   Practice Test

1. Sketch the graph of $f(x) = x^2 - 6x + 5$ by hand and identify the vertex and the intercepts.

2. Find the number of units $x$ that produce a minimum cost $C$ if $C = 0.01x^2 - 90x + 15,000$.

3. Find the quadratic function that has a maximum at $(1, 7)$ and passes through the point $(2, 5)$.

4. Find two quadratic functions that have $x$-intercepts $(2, 0)$ and $\left(\frac{4}{3}, 0\right)$.

5. Use the leading Coefficient Test to determine the right-hand and left-hand behavior of the graph of the polynomial function $f(x) = -3x^5 + 2x^3 - 17$.

6. Find all the real zeros of $f(x) = x^5 - 5x^3 + 4x$. Verify your answer with a graphing utility.

7. Find the polynomial function with 0, 3, and $-2$ as zeros.

8. Sketch $f(x) = x^3 - 12x$ by hand.

9. Divide $3x^4 - 7x^2 + 2x - 10$ by $x - 3$ using long division.

10. Divide $x^3 - 11$ by $x^2 + 2x - 1$.

11. Use synthetic division to divide $3x^5 + 13x^4 + 12x - 1$ by $x + 5$.

12. Use synthetic division to find $f(-6)$ when $f(x) = 7x^3 + 40x^2 - 12x + 15$.

13. Find the real zeros of $f(x) = x^3 - 19x - 30$.

14. Find the real zeros of $f(x) = x^4 + x^3 - 8x^2 - 9x - 9$.

15. List all possible rational zeros of the function $f(x) = 6x^3 - 5x^2 + 4x - 15$.

16. Find the rational zeros of the polynomial $f(x) = x^3 - \frac{20}{3}x^2 + 9x - \frac{10}{3}$.

17. Write $f(x) = x^4 + x^3 + 3x^2 + 5x - 10$ as a product of linear factors.

18. Find a polynomial with real coefficients that has 2, $3 + i$, and $3 - 2i$ as zeros.

19. Use synthetic division to show that $3i$ is a zero of $f(x) = x^3 + 4x^2 + 9x + 36$.

20. Find a mathematical model for the statement, "$z$ varies directly as the square of $x$ and inversely as the square root of $y$".

21. Sketch the graph of $f(x) = \dfrac{x - 1}{2x}$ and label all intercepts and asymptotes.

22. Sketch the graph of $f(x) = \dfrac{3x^2 - 4}{x}$ and label all intercepts and asymptotes.

23. Find all the asymptotes of $f(x) = \dfrac{8x^2 - 9}{x^2 + 1}$.

24. Find all the asymptotes of $f(x) = \dfrac{4x^2 - 2x + 7}{x - 1}$.

25. Sketch the graph of $f(x) = \dfrac{x - 5}{(x - 5)^2}$.

# CHAPTER 4
# Exponential and Logarithmic Functions

# CHAPTER 4
# Exponential and Logarithmic Functions

## Section 4.1    Exponential Functions and Their Graphs

- You should know that a function of the form $y = a^x$, where $a > 0$, $a \neq 1$, is called an exponential function with base $a$.
- You should be able to graph exponential functions.
- You should be familiar with the number $e$ and the natural exponential function $f(x) = e^x$.
- You should know formulas for compound interest.

  (a) For $n$ compoundings per year: $A = P\left(1 + \dfrac{r}{n}\right)^{nt}$.

  (b) For continuous compoundings: $A = Pe^{rt}$.

### Solutions to Odd-Numbered Exercises

**1.** $(3.4)^{6.8} \approx 4112.033$

**3.** $5^{-\pi} \approx 0.006$

**5.** $17^{2(\sqrt{3})} \approx 18,297.851$

**7.** $g(x) = 5^x$

Asymptote: $y = 0$

Intercept: $(0, 1)$

Increasing

| $x$ | $-2$ | $-1$ | 0 | 1 | 2 |
|---|---|---|---|---|---|
| $y$ | $\frac{1}{25}$ | $\frac{1}{5}$ | 1 | 5 | 25 |

**9.** $f(x) = \left(\frac{1}{5}\right)^x = 5^{-x}$

Asymptote: $y = 0$

Intercepts: $(0, 1)$

Decreasing

| $x$ | $-2$ | $-1$ | 0 | 1 | 2 |
|---|---|---|---|---|---|
| $y$ | 25 | 5 | 1 | $\frac{1}{5}$ | $\frac{1}{25}$ |

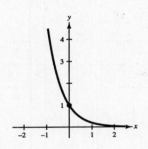

**11.** $h(x) = 5^{x-2}$

Asymptote: $y = 0$

Intercepts: $\left(0, \frac{1}{25}\right)$

Increasing

| $x$ | $-1$ | 0 | 1 | 2 | 3 |
|---|---|---|---|---|---|
| $y$ | $\frac{1}{125}$ | $\frac{1}{25}$ | $\frac{1}{5}$ | 1 | 5 |

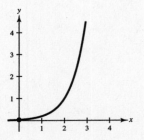

**13.** $g(x) = 5^{-x} - 3$

Asymptote: $y = -3$

Intercepts: $(0, -2), (-0.683, 0)$

Decreasing

| $x$ | $-1$ | 0 | 1 | 2 |
|---|---|---|---|---|
| $y$ | 2 | $-2$ | $-2\frac{4}{5}$ | $-2\frac{24}{25}$ |

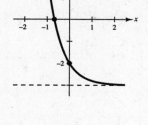

**15.** $f(x) = 2^{x-2}$ rises to the right.

Asymptote: $y = 0$

Intercept: $\left(0, \frac{1}{4}\right)$

Matches graph (d).

**17.** $f(x) = 2^x - 4$ rises to the right.

Asymptote: $y = -4$

Intercept: $(0, -3)$

Matches graph (c).

**19.** $f(x) = 3^x$

$g(x) = 3^{x-5} = f(x - 5)$

Horizontal shift five units to the right

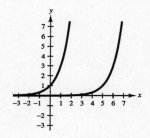

**21.** $f(x) = \left(\frac{3}{5}\right)^x$

$g(x) = -\left(\frac{3}{5}\right)^{x+4} = -f(x + 4)$

Horizontal shift 4 units to the left, followed by reflection in $x$-axis.

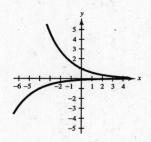

**23.** $e^{9.2} \approx 9897.129$

**25.** $50e^{4(0.02)} \approx 54.164$

**27.** $2.5e^{\left(-\frac{1}{2}\right)} \approx 1.516$

**29.** $f(x) = \left(\frac{5}{2}\right)^x$

| $x$ | $-1$ | 0 | 1 | 2 | 3 |
|---|---|---|---|---|---|
| $f(x)$ | 0.4 | 1 | 2.5 | 6.25 | 15.625 |

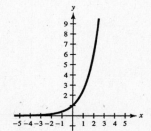

**31.** $f(x) = 6^x$

| $x$ | $-1$ | 0 | 1 | 2 |
|---|---|---|---|---|
| $f(x)$ | 0.167 | 1 | 6 | 36 |

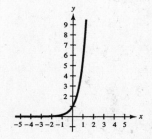

**33.** $f(x) = 3^{x+2} = 9 \cdot 3^x$

| $x$    | $-1$ | $-2$ | $0$ | $1$  |
|--------|------|------|-----|------|
| $f(x)$ | 3    | 1    | 9   | 27   |

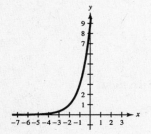

**35.** $f(x) = 3e^{x+4}$

| $x$    | $-7$  | $-6$  | $-5$  | $-4$ | $-3$  |
|--------|-------|-------|-------|------|-------|
| $f(x)$ | 0.149 | 0.406 | 1.104 | 3    | 8.155 |

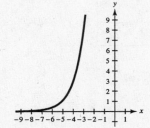

**37.** $f(x) = 2 + e^{x-5}$

| $x$    | 2    | 3     | 4     | 5 | 6     | 7     |
|--------|------|-------|-------|---|-------|-------|
| $f(x)$ | 2.05 | 2.135 | 2.368 | 3 | 4.718 | 9.389 |

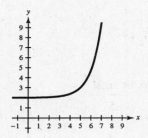

**39.** $y = 2^{-x^2}$

Asymptote: $y = 0$

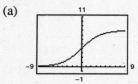

**41.** $f(x) = 3^{x-2} + 1$

Asymptote: $y = 1$

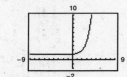

**43.** $g(x) = 2 - e^{-x}$

Asymptote: $y = 2$

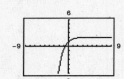

**45.** $s(t) = 2e^{0.12t}$

Asymptote: $y = 0$

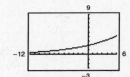

**47.** $f(x) = \dfrac{8}{1 + e^{-0.5x}}$

(a)

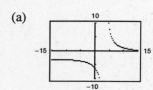

(b)

| $x$    | $-30$     | $-20$     | $-10$ | $0$ | $10$ | $20$      | $30$      |
|--------|-----------|-----------|-------|-----|------|-----------|-----------|
| $f(x)$ | $\approx 0$ | $\approx 0$ | 0.05  | 4   | 7.95 | $\approx 8$ | $\approx 8$ |

Horizontal asymptotes: $y = 0, y = 8$

**49.** $f(x) = \dfrac{-6}{2 - e^{0.2x}}$

(a)

(b)

| $x$    | $-20$ | $-10$ | $0$ | $3$  | $3.4$ | $3.46$ |
|--------|-------|-------|-----|------|-------|--------|
| $f(x)$ | $-3.03$ | $-3.22$ | $-6$ | $-34$ | $-230$ | $-2617$ |

| $x$    | $3.47$ | $4$  | $5$ | $10$ | $20$ |
|--------|--------|------|-----|------|------|
| $f(x)$ | 3516   | 26.6 | 8.4 | 1.11 | 0.11 |

Horizontal asymptotes: $y = -3, y = 0$

Vertical asymptote: $x \approx 3.46$

**51.** $f(x) = x^2 e^{-x}$

(a)

(b) Decreasing: $(-\infty, 0)$, $(2, \infty)$
Increasing: $(0, 2)$

(c) Relative maximum: $(2, 4e^{-2}) \approx (2, 0.541)$
Relative minimum: $(0, 0)$

**53.** $f(x) = x(2^{3-x})$

(a)

(b) Decreasing: $(1.44, \infty)$
Increasing: $(-\infty, 1.44)$

(c) Relative maximum: $(1.44, 4.25)$

**55.** $P = 2500, r = 2.5\% = 0.025, t = 10$

Compounded $n$ times per year: $A = P\left(1 + \dfrac{r}{n}\right)^{nt} = 2500\left(1 + \dfrac{0.025}{n}\right)^{10n}$

Compounded continuously: $A = Pe^{rt} = 2500e^{(0.025)(10)}$

| $n$ | 1 | 2 | 4 | 12 | 365 | Continuos |
|---|---|---|---|---|---|---|
| $A$ | 3200.21 | 3205.09 | 3207.57 | 3209.23 | 3210.04 | 3210.06 |

**57.** $P = 2500, r = 4\% = 0.04, t = 20$

Compounded $n$ times per year: $A = P\left(1 + \dfrac{r}{n}\right)^{nt} = 2500\left(1 + \dfrac{.04}{n}\right)^{20n}$

Compounded continuously: $A = Pe^{rt} = 2500e^{(.04)(20)}$

| $n$ | 1 | 2 | 4 | 12 | 365 | Continuos |
|---|---|---|---|---|---|---|
| $A$ | 5477.81 | 5520.10 | 5541.79 | 5556.46 | 5563.61 | 5563.85 |

**59.** $P = 12,000, r = 4\% = 0.04$

$A = Pe^{rt} = 12000e^{0.04t}$

| $t$ | 1 | 10 | 20 | 30 | 40 | 50 |
|---|---|---|---|---|---|---|
| $A$ | 12489.73 | 17901.90 | 26706.49 | 39841.40 | 59436.39 | 88668.67 |

**61.** $P = 12,000, r = 3.5\% = 0.035$

$A = Pe^{rt} = 12000e^{0.035t}$

| $t$ | 1 | 10 | 20 | 30 | 40 | 50 |
|---|---|---|---|---|---|---|
| $A$ | 12427.44 | 17028.81 | 24165.03 | 34291.81 | 48662.40 | 69055.23 |

**63.** $p = 5000\left(1 - \dfrac{4}{4 + e^{-0.002x}}\right)$

(a)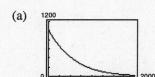

(b) If $x = 500, p \approx \$421.12.$

(c) For $x = 600, p \approx \$350.13.$

(d)

| $x$ | 100 | 200 | 300 | 400 | 500 | 600 | 700 |
|---|---|---|---|---|---|---|---|
| $p$ | 849.53 | 717.64 | 603.25 | 504.94 | 421.12 | 350.13 | 290.35 |

**65.** $Q = 25\left(\frac{1}{2}\right)^{t/1620}$

(a) When $t = 0$, $Q = 25\left(\frac{1}{2}\right)^{0/1620} = 25(1) = 25$ grams.

(b) When $t = 1000$, $Q = 25\left(\frac{1}{2}\right)^{1000/1620} \approx 16.30$ grams.

(c)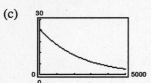

(d) Never, $Q \to 0$ as $t \to \infty$, but $Q$ never reaches 0.

**67.** $P(t) = 100e^{0.2197t}$

(a)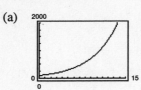

(b) $P(0) = 100$

$P(5) \approx 300$

$P(10) \approx 900$

(c) $P(0) = 100e^{0.2197(0)} = 100$

$P(5) = 100e^{0.2197(5)} = 299.966 \approx 300$

$P(10) = 100e^{0.2197(10)} = 899.798 \approx 900$

**69.** $C(t) = P(1.04)^t$

(a)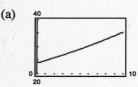

(b) $C(10) \approx 35.45$

(c) $C(10) = 23.95(1.04)^{10} \approx 35.45$

**71.** True. $f(x) = 1^x$ is not an exponential function.

**73.** $y_1 = e^x$

$y_2 = x^2$

$y_3 = x^3$

$y_4 = \sqrt{x}$

$y_5 = |x|$

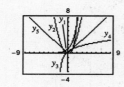

(a) $y_1 = e^x$ increases at the fastest rate.

(b) For any positive integer $n$, $e^x > x^n$ for $x$ sufficiently large. That is, $e^x$ grows faster than $x^n$.

(c) A quantity is growing exponentially if its growth rate is of the form $y = ce^{rx}$. This is a faster rate than any polynomial growth rate.

**75.** $f(x) = \left(1 + \dfrac{0.5}{x}\right)^x$ and

$g(x) = e^{0.5} \approx 1.6487$

(Horizontal line)

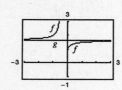

As $x \to \infty$, $f(x) \to g(x)$.

**77.** $f$ has an inverse because $f$ is one-to-one.

$y = 5x - 7$

$x = 5y - 7$

$x + 7 = 5y$

$f^{-1}(x) = \frac{1}{5}(x + 7)$

**79.** $f$ has an inverse because $f$ is one-to-one.

$y = \sqrt[3]{x + 8}$

$x = \sqrt[3]{y + 8}$

$x^3 = y + 8$

$x^3 - 8 = y$

$f^{-1}(x) = x^3 - 8$

**81.** $f(x) = \dfrac{2x}{x-7}$

Vertical asymptote: $x = 7$

Horizontal asymptote: $y = 2$

Intercept: $(0, 0)$

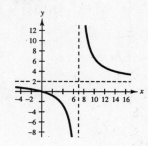

## Section 4.2  Logarithmic Functions and Their Graphs

- ■ You should know that a function of the form $y = \log_a x$, where $a \geq 0$, $a \neq 1$, and $x > 0$, is called a logarithm of $x$ to base $a$.
- ■ You should be able to convert from logarithmic form to exponential form and vice versa.

  $$y = \log_a x \iff a^y = x$$
- ■ You should know the following properties of logarithms.
  - (a) $\log_a 1 = 0$ since $a^0 = 1$.
  - (b) $\log_a a = 1$ since $a^1 = a$.
  - (c) $\log_a a^x = x$ since $a^x = a^x$.
  - (d) If $\log_a x = \log_a y$, then $x = y$.
- ■ You should know the definition of the natural logarithmic function.

  $$\log_e x = \ln x, \, x > 0$$
- ■ You should know the properties of the natural logarithmic function.
  - (a) $\ln 1 = 0$ since $e^0 = 1$.
  - (b) $\ln e = 1$ since $e^1 = e$.
  - (c) $\ln e^x = x$ since $e^x = e^x$.
  - (d) If $\ln x = \ln y$, then $x = y$.
- ■ You should be able to graph logarithmic functions.

**Solutions to Odd-Numbered Exercises**

**1.** $\log_4 64 = 3 \implies 4^3 = 64$

**3.** $\log_7 \frac{1}{49} = -2 \implies 7^{-2} = \frac{1}{49}$

**5.** $\log_{32} 4 = \frac{2}{5} \implies 32^{2/5} = 4$

**7.** $\ln 1 = 0 \implies e^0 = 1$

**9.** $5^3 = 125 \implies \log_5 125 = 3$

**11.** $81^{1/4} = 3 \implies \log_{81} 3 = \frac{1}{4}$

**13.** $6^{-2} = \frac{1}{36} \implies \log_6 \frac{1}{36} = -2$

**15.** $e^3 = 20.0855 \ldots \implies \ln 20.0855 \ldots = 3$

**17.** $\log_2 16 = \log_2 2^4 = 4$

**19.** $\log_{10} 0.01 = \log_{10} 10^{-2} = -2$

**21.** $\log_{10} 345 \approx 2.538$

**23.** $6 \log_{10} 14.8 \approx 7.022$

**25.** $\log_7 x = \log_7 9$

  $x = 9$

**27.** $\log_6 6^2 = x$

  $2\log_6 6 = x$

  $2 = x$

**29.** $\log_8 x = \log_8 10^{-1}$

  $x = 10^{-1} = \frac{1}{10}$

**31.** $f(x) = 3^x$, $g(x) = \log_3 x$

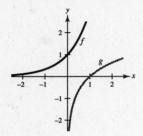

f and g are inverses. Their graphs are reflected about the line $y = x$.

**33.** $f(x) = e^x$, $g(x) = \ln x$

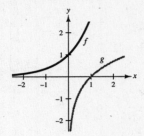

f and g are inverses. Their graphs are reflected about the line $y = x$.

**35.** $f(x) = \log_4 x$

Domain: $x > 0 \Rightarrow$ The domain is $(0, \infty)$.

Vertical asymptote: $x = 0$

$x$-intercept: $(1, 0)$

$y = \log_4 x \Rightarrow 4^y = x$

| $x$ | $\frac{1}{4}$ | 1 | 4 | 2 |
|---|---|---|---|---|
| $y$ | $-1$ | 0 | 1 | $\frac{1}{2}$ |

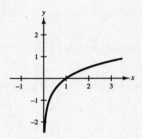

**37.** $f(x) = \log_{10}\left(\dfrac{x}{5}\right)$

Domain: $\dfrac{x}{5} > 0 \Rightarrow x > 0$

The domain is $(0, \infty)$.

Vertical asymptote: $\dfrac{x}{5} = 0 \Rightarrow x = 0$

The vertical asymptote is the $y$-axis.

$x$-intercept: $\log_{10}\left(\dfrac{x}{5}\right) = 0$

$$\frac{x}{5} = 10^0$$

$$\frac{x}{5} = 1 \Rightarrow x = 5$$

The $x$-intercept is $(5, 0)$.

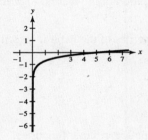

**39.** $h(x) = \log_4(x - 3)$

Domain: $x - 3 > 0$ or $(3, \infty)$

Vertical asymptote: $x = 3$

Intercept: $(4, 0)$

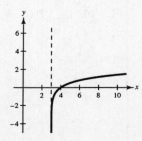

**41.** $y = -\log_{10} x + 2$

Domain: $x > 0$   or   $(0, \infty)$

Vertical asymptote: $x = 0$

$x$-intercept: $(100, 0)$

| $x$ | $\frac{1}{10}$ | 1 | 10 | 100 |
|---|---|---|---|---|
| $y$ | 3 | 2 | 1 | 0 |

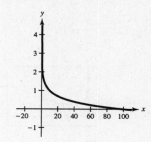

**43.** $f(x) = 6 + \log_6 (x - 3)$

Domain: $(3, \infty)$

Vertical asymptote: $x = 3$

$x$-intercept: $\log_6(x - 3) = -6$

$6^{-6} = x - 3$

$x = 3 + 6^{-6} \approx 3$

| $x$ | 4 | 9 | $3\frac{1}{6}$ |
|---|---|---|---|
| $y$ | 6 | 7 | 5 |

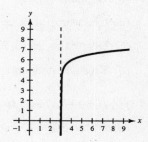

**45.** $f(x) = \log_3 x + 2$

Asymptote: $x = 0$

Point on graph: $(1, 2)$

Matches graph (b).

**47.** $f(x) = -\log_3(x + 2)$

Asymptote: $x = -2$

Point on graph: $(-1, 0)$

Matches graph (d).

**49.** $f(x) = \log_{10} x$

$g(x) = -\log_{10} x$ is a reflection in the $x$-axis of the graph of $f$.

**51.** $f(x) = \log_2 x$

$g(x) = 4 - \log_2 x$ is obtained from $f$ by a reflection in the $x$-axis followed by a vertical shift 4 units upward.

**53.** $\ln\sqrt{42} \approx 1.869$

**55.** $-\ln\left(\frac{1}{2}\right) \approx 0.693$

**57.** $\ln e^2 = 2$

(inverse property)

**59.** $e^{\ln 1.8} = 1.8$

(inverse property)

**61.** $f(x) = \ln(x - 1)$

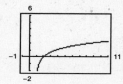

Domain: $x > 1$

Vertical asymptote: $x = 1$

$x$-intercept: $(2, 0)$

**63.** $g(x) = \ln(-x)$

Domain: $-x > 0 \implies x < 0$

The domain is $(-\infty, 0)$.

Vertical asymptote: $-x = 0 \implies x = 0$

$x$-intercept:   $0 = \ln(-x)$

$e^0 = -x$

$-1 = x$

The $x$-intercept is $(-1, 0)$.

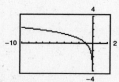

**65.** $f(x) = \dfrac{x}{2} - \ln\dfrac{x}{4}$

(a)

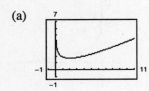

(b) Domain: $(0, \infty)$

(c) Increasing on $(2, \infty)$

Decreasing on $(0, 2)$

(d) Relative minimum: $(2, 1.693)$

**67.** $h(x) = 4x \ln x$

(a)

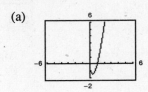

(b) Domain: $(0, \infty)$

(c) Increasing on $(0.368, \infty)$

Decreasing on $(0, 0.368)$

(d) Relative minimum: $(0.368, -1.472)$

**69.** $f(t) = 80 - 17 \log_{10}(t + 1)$, $0 \le t \le 12$

(a) $f(0) = 80 - 17 \log_{10}(0 + 1) = 80$

(b) $f(4) = 80 - 17 \log_{10}(4 + 1) \approx 68.1$

(c) $f(10) = 80 - 17 \log_{10}(10 + 1) \approx 62.3$

(d)

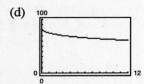

**71.** $t = \dfrac{\ln K}{0.055}$

(a)

| $K$ | 1 | 2 | 4 | 6 | 8 | 10 | 12 |
|---|---|---|---|---|---|---|---|
| $t$ | 0 | 12.6 | 25.2 | 32.6 | 37.8 | 41.9 | 45.2 |

As the amount increases, the time increases, but at a lesser rate.

(b)

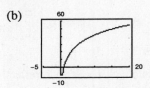

**73.** $\beta = 10 \log_{10}\left(\dfrac{I}{10^{-12}}\right)$

(a) $I = 1$: $\beta = 10 \log_{10}\left(\dfrac{1}{10^{-12}}\right) = 10 \cdot \log_{10}(10^{12}) = 10(12) = 120$ decibels

(b) $I = 10^{-2}$: $\beta = 10 \log_{10}\left(\dfrac{10^{-2}}{10^{-12}}\right) = 10 \log_{10}(10^{10}) = 10(10) = 100$ decibels

(c) No, this is a logarithmic scale.

**75.** $y = 80.4 - 11 \ln x$

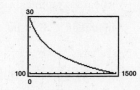

$y(300) = 80.4 - 11 \ln 300 \approx 17.66$ ft$^3$/min

**77.** False. You would reflect $y = 6^x$ in the line $y = x$.

**79.** $f(x) = \log_a x$ is the inverse of $g(x) = a^x$, where $a > 0$, $a \ne 1$

**81.** (a) False, $y$ is not an exponential function of $x$. ($y$ can never be 0.)

(b) True, $y$ could be $\log_2 x$.

(c) True, $x$ could be $2^y$.

(d) False, $y$ is not linear. (The points are not collinear.)

**83.** $f(x) = \dfrac{\ln x}{x}$

(a)

| $x$ | 1 | 5 | 10 | $10^2$ | $10^4$ | $10^6$ |
|-----|---|-----|-------|-------|---------|-----------|
| $f(x)$ | 0 | 0.322 | 0.230 | 0.046 | 0.00092 | 0.0000138 |

(b) As $x$ increases without bound, $f(x)$ approaches 0.

(c)

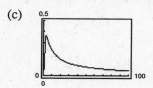

**85.** $x^2 + 2x - 3 = (x + 3)(x - 1)$

**87.** $12x^2 + 5x - 3 = (4x + 3)(3x - 1)$

**89.** $16x^2 - 25 = (4x + 5)(4x - 5)$

**91.** $2x^3 + x^2 - 45x = x(2x^2 + x - 45) = x(2x - 9)(x + 5)$

**93.** $(f + g)(2) = f(2) + g(2) = [3(2) + 2] + [2^3 - 1] = 8 + 7 = 15$

**95.** $(fg)(6) = f(6)g(6) = [3(6) + 2][6^3 - 1] = [20][215] = 4300$

**97.** $5x - 7 = x + 4$

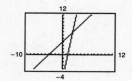

The graphs of $y = 5x - 7$ and $y_2 = x + 4$
intersect when $x = 2.75$ or $\frac{11}{4}$.

**99.** $\sqrt{3x - 2} = 9$

The graphs of $y_1 = \sqrt{3x - 2}$ and $y_2 = 9$ intersect
when $x \approx 27.667$ or $\frac{83}{3}$.

**101.** $f(x) = \dfrac{4}{-8 - x} = \dfrac{-4}{x + 8}$

Vertical asymptote: $x = -8$
Horizontal asymptote: $y = 0$

**103.** $f(x) = \dfrac{x + 5}{(2x^2 + x - 15)} = \dfrac{x + 5}{(2x - 5)(x + 3)}$

Vertical asymptotes: $x = \dfrac{5}{2}, -3$

Horizontal asymptote: $y = 0$

**105.** $g(x) = \dfrac{x^2 - 4}{x^2 - 4x - 12}$

$= \dfrac{(x + 2)(x - 2)}{(x - 6)(x + 2)}$

$= \dfrac{x - 2}{x - 6}, \quad x \neq -2$

Vertical asymptote: $x = 6$
Horizontal asymptote: $y = 1$

## Section 4.3    Properties of Logarithms

---

■   You should know the following properties of logarithms.

(a) $\log_a x = \dfrac{\log_b x}{\log_b a}$

(b) $\log_a (uv) = \log_a u + \log_a v$      $\ln (uv) = \ln u + \ln v$

(c) $\log_a (u/v) = \log_a u - \log_a v$      $\ln (u/v) = \ln u - \ln v$

(d) $\log_a u^n = n \log_a u$           $\ln u^n = n \ln u$

■   You should be able to rewrite logarithmic expressions using these properties.

---

**Solutions to Odd-Numbered Exercises**

**1.** (a) $\log_5 x = \dfrac{\log_{10} x}{\log_{10} 5}$

    (b) $\log_5 x = \dfrac{\ln x}{\ln 5}$

**3.** (a) $\log_{1/5} x = \dfrac{\log_{10} x}{\log_{10} \frac{1}{5}} = \dfrac{\log_{10} x}{-\log_{10} 5}$

    (b) $\log_{1/5} x = \dfrac{\ln x}{\ln \frac{1}{5}} = \dfrac{\ln x}{-\ln 5}$

**5.** (a) $\log_a \left(\dfrac{3}{10}\right) = \dfrac{\log_{10}\left(\frac{3}{10}\right)}{\log_{10} a}$

    (b) $\log_a \left(\dfrac{3}{10}\right) = \dfrac{\ln\left(\frac{3}{10}\right)}{\ln a}$

**7.** (a) $\log_{2.6} x = \dfrac{\log_{10} x}{\log_{10} 2.6}$

    (b) $\log_{2.6} x = \dfrac{\ln x}{\ln 2.6}$

**9.** $\log_3 7 = \dfrac{\ln 7}{\ln 3} \approx 1.771$

**11.** $\log_{1/2} 4 = \dfrac{\ln 4}{\ln \frac{1}{2}} = -2$

**13.** $\log_9 (0.8) = \dfrac{\ln(0.8)}{\ln 9} \approx -0.102$

**15.** $\log_{15} 1460 = \dfrac{\ln 1460}{\ln 15} \approx 2.691$

**17.** $\log_4 8 = \log_4 2^3 = 3 \log_4 2$

           $= 3 \log_4 4^{1/2} = 3\left(\frac{1}{2}\right)\log_4 4$

           $= \frac{3}{2}$

**19.** $\ln(5e^6) = \ln 5 + \ln e^6 = \ln 5 + 6 = 6 + \ln 5$

**21.** $\log_5 \frac{1}{250} = \log_5 1 - \log_5 250 = 0 - \log_5 (125 \cdot 2)$

        $= -\log_5 (5^3 \cdot 2) = -[\log_5 5^3 + \log_5 2]$

        $= -[3 \log_5 5 + \log_5 2] = -3 - \log_5 2$

**23.** $\log_{10} 5x = \log_{10} 5 + \log_{10} x$

**25.** $\log_{10} \dfrac{5}{x} = \log_{10} 5 - \log_{10} x$

**27.** $\log_8 x^4 = 4 \log_8 x$

**29.** $\ln \sqrt{z} = \ln z^{1/2} = \frac{1}{2} \ln z$

**31.** $\ln xyz = \ln x + \ln y + \ln z$

**33.** $\ln\left(a^2\sqrt{a-1}\right) = \ln a^2 + \ln(a-1)^2$

$$= 2\ln a + \tfrac{1}{2}\ln(a-1),\ a > 1$$

**35.** $\ln\sqrt[3]{\dfrac{x}{y}} = \dfrac{1}{3}\ln\dfrac{x}{y}$

$$= \dfrac{1}{3}\left[\ln x - \ln y\right]$$

$$= \dfrac{1}{3}\ln x - \dfrac{1}{3}\ln y$$

**37.** $\ln\left(\dfrac{x^2-1}{x^3}\right) = \ln(x^2-1) - \ln x^3$

$$= \ln[(x-1)(x+1)] - 3\ln x$$

$$= \ln(x-1) + \ln(x+1) - 3\ln x,$$

$$x > 1$$

**39.** $\ln\left(\dfrac{x^4\sqrt{y}}{z^5}\right) = \ln x^4\sqrt{y} - \ln z^5$

$$= \ln x^4 + \ln\sqrt{y} - \ln z^5$$

$$= 4\ln x + \dfrac{1}{2}\ln y - 5\ln z$$

**41.** $\log_b\left(\dfrac{x^2}{y^2 z^3}\right) = \log_b x^2 - \log_b y^2 z^3$

$$= \log_b x^2 - \left[\log_b y^2 + \log_b z^3\right]$$

$$= 2\log_b x - 2\log_b y - 3\log_b z$$

**43.** $y_1 = \ln[x^3(x+4)]$

$y_2 = 3\ln x + \ln(x+4)$

(a)

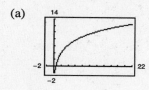

(b)

| $x$ | 0.5 | 1 | 1.5 | 2 | 3 | 10 |
|---|---|---|---|---|---|---|
| $y_1$ | $-0.5754$ | 1.6094 | 2.9211 | 3.8712 | 5.2417 | 9.5468 |
| $y_2$ | $-0.5754$ | 1.6094 | 2.9211 | 3.8712 | 5.2417 | 9.5468 |

(c) The graphs and table suggest that
$y_1 = y_2$ for $x > 0$. In fact,

$$y_1 = \ln[x^3(x+4)] = \ln x^3 + \ln(x+4)$$

$$= 3\ln x + \ln(x+4) = y_2$$

**45.** $\ln x + \ln 4 = \ln 4x$

**47.** $\log_4 z - \log_4 y = \log_4 \dfrac{z}{y}$

**49.** $2\log_2(x+3) = \log_2(x+3)^2$

**51.** $\tfrac{1}{3}\log_3 7x = \log_3(7x)^{1/3} = \log_3 \sqrt[3]{7x}$

**53.** $\ln x - 3\ln(x+1) = \ln x - \ln(x+1)^3$

$$= \ln\dfrac{x}{(x+1)^3}$$

**55.** $\ln(x-2) - \ln(x+2) = \ln\left(\dfrac{x-2}{x+2}\right)$

**57.** $\ln x - 2[\ln(x+2) + \ln(x-2)] = \ln x - 2\ln[(x+2)(x-2)]$

$$= \ln x - 2\ln(x^2-4)$$

$$= \ln x - \ln(x^2-4)^2$$

$$= \ln\dfrac{x}{(x^2-4)^2}$$

**59.** $\frac{1}{3}[2\ln(x+3)+\ln x-\ln(x^2-1)]=\frac{1}{3}[\ln(x+3)^2+\ln x-\ln(x^2-1)]$

$$=\frac{1}{3}[\ln[x(x+3)^2]-\ln(x^2-1)]$$

$$=\frac{1}{3}\ln\frac{x(x+3)^2}{x^2-1}$$

$$=\ln\sqrt[3]{\frac{x(x+3)^2}{x^2-1}}$$

**61.** $\frac{1}{3}[\ln y+2\ln(y+4)]-\ln(y-1)=\frac{1}{3}[\ln y+\ln(y+4)^2]-\ln(y-1)$

$$=\frac{1}{3}\ln[y(y+4)^2]-\ln(y-1)$$

$$=\ln\sqrt[3]{y(y+4)^2}-\ln(y-1)$$

$$=\ln\frac{\sqrt[3]{y(y+4)^2}}{y-1}$$

**63.** $y_1=2[\ln 8-\ln(x^2+1)]$

$y_2=\ln\left[\dfrac{64}{(x^2+1)^2}\right]$

(a)

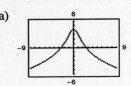

(c) The graphs and table suggest that $y_1=y_2$. In fact,

$$y_1=2[\ln 8-\ln(x^2+1)]$$

$$=2\ln\frac{8}{x^2+1}=\ln\frac{64}{(x^2+1)^2}=y_2$$

(b)

| $x$ | $-8$ | $-4$ | $-2$ | $0$ | $2$ | $4$ | $8$ |
|---|---|---|---|---|---|---|---|
| $y_1$ | $-4.1899$ | $-1.5075$ | $0.9400$ | $4.1589$ | $0.9400$ | $-1.5075$ | $-4.1899$ |
| $y_2$ | $-4.1899$ | $-1.5075$ | $0.9400$ | $4.1589$ | $0.9400$ | $-1.5075$ | $-4.1899$ |

**65.** $y_1=\ln x^2$

$y_2=2\ln x$

(a)

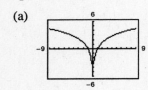

(The domain of $y_2$ is $x>0$)

(b)

| $x$ | $-8$ | $-4$ | $1$ | $2$ | $4$ |
|---|---|---|---|---|---|
| $y_1$ | $4.1589$ | $2.7726$ | $0$ | $1.3863$ | $2.7726$ |
| $y_2$ | undefined | undefined | $0$ | $1.3863$ | $2.7726$ |

(c) The graphs and table suggest that $y_1=y_2$ for $x>0$. The functions are not equivalent because the domains are different.

**67.** $\log_3 9=2\log_3 3=2$

**69.** $\log_4 16^{3.4}=3.4\log_4(4^2)=6.8\log_4 4=6.8$

**71.** $\log_2(-4)$ is undefined. $-4$ is not in the domain of $f(x)=\log_2 x$

**73.** $\log_5 75-\log_5 3=\log_5\frac{75}{3}=\log_5 25=\log_5 5^2=2$

**75.** $\ln e^3 - \ln e^7 = 3 - 7 = -4$

**77.** $2 \ln e^4 = 2(4)\ln e = 8$

**79.** $\ln\left(\dfrac{1}{\sqrt{e}}\right) = \ln(1) - \ln e^{1/2} = 0 - \dfrac{1}{2}\ln e = -\dfrac{1}{2}$

**81.** (a) $\beta = 10 \cdot \log_{10}\left(\dfrac{I}{10^{-12}}\right) = 10[\log_{10} I - \log_{10} 10^{-12}]$

      $= 10[\log_{10} I - (-12)\log_{10} 10]$

      $= 10\,[\log_{10} I + 12] = 120 + 10 \cdot \log_{10} I$

(b)

| $I$ | $10^{-4}$ | $10^{-6}$ | $10^{-8}$ | $10^{-10}$ | $10^{-12}$ | $10^{-14}$ |
|---|---|---|---|---|---|---|
| $\beta$ | 80 | 60 | 40 | 20 | 0 | $-20$ |

(c) $\beta(10^{-4}) = 120 + 10 \cdot \log_{10} 10^{-4} = 120 - 40 = 80$

    $\beta(10^{-6}) = 120 + 10 \cdot \log_{10} 10^{-6} = 120 - 60 = 60$

    $\beta(10^{-8}) = 120 + 10 \cdot \log_{10} 10^{-8} = 120 - 80 = 40$

    $\beta(10^{-10}) = 120 + 10 \cdot \log_{10} 10^{-10} = 120 - 100 = 20$

    $\beta(10^{-12}) = 120 + 10 \cdot \log_{10} 10^{-12} = 120 - 120 = 0$

    $\beta(10^{-14}) = 120 + 10 \cdot \log_{10} 10^{-14} = 120 - 140 = -20$

**83.** (a)

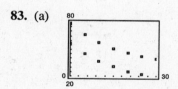

(b) $T - 21 = 54.4(0.964)^t$

    $T = 21 + 54.4(0.964)^t$

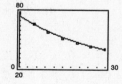

The data $(t,\, T - 21)$ fits the model
$T - 21 = 54.4(0.964)^t$ The model
$T = 21 + 54.4(0.964)^t$ fits the original data.

(c) $\ln(T - 21) = -0.0372t + 3.9971$    linear model

    $T - 21 = e^{-0.0372t + 3.9971}$

        $T = 21 + 54.4e^{-0.0372t}$

          $= 21 + 54.4(.964)^t$

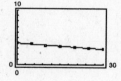

(d)

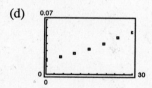

$\dfrac{1}{T - 21} = 0.00121t + 0.01615$    linear model

$T - 21 = \dfrac{1}{0.00121t + 0.01615}$

$T = 21 + \dfrac{1}{0.00121t + 0.01615}$

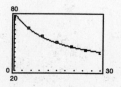

**85.** $f(x) = \ln x$

False, $f(0) \neq 0$ since 0 is not in the domain of $f(x)$. $f(1) = \ln 1 = 0$

**87.** True.

**89.** False. $\sqrt{\ln x} \neq \frac{1}{2} \ln x$

In fact, $\ln x^{1/2} = \frac{1}{2} \ln x$

**91.** True.

**93.** Let $x = \log_b u$.

Then $b^x = u$ and $u^n = b^{nx}$.

Hence,

$\log_b u^n = \log_b b^{nx} = nx = n \log_b u$.

**95.** $f(x) = \log_2 x = \dfrac{\ln x}{\ln 2}$

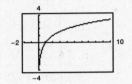

**97.** $f(x) = \log_3 \sqrt{x} = \dfrac{1}{2} \dfrac{\ln x}{\ln 3}$

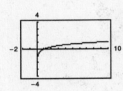

**99.** $f(x) = \log_5\left(\dfrac{x}{3}\right) = \dfrac{\ln\left(\dfrac{x}{3}\right)}{\ln 5}$

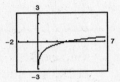

**101.** $f(x) = \ln \dfrac{x}{2}$

$g(x) = \dfrac{\ln x}{\ln 2}$

$h(x) = \ln x - \ln 2$

$f(x) = h(x)$ by Property 2.

**103.** $\dfrac{24xy^{-2}}{16x^{-3}y} = \dfrac{24xx^3}{16yy^2} = \dfrac{3x^4}{2y^3}$

**105.** $(18x^3y^4)^{-3}(18x^3y^4)^3 = \dfrac{(18x^3y^4)^3}{(18x^3y^4)^3} = 1$ if $x \neq 0, y \neq 0$.

**107.** $x^2 - 6x + 2 = 0$

$x = \dfrac{6 \pm \sqrt{36 - 4(2)}}{2} = 3 \pm \sqrt{7}$

**109.** $x^4 - 19x^2 + 48 = 0$

$(x^2 - 16)(x^2 - 3) = 0$

$(x - 4)(x + 4)(x - \sqrt{3})(x + \sqrt{3}) = 0$

$x = \pm 4, \pm \sqrt{3}$

**111.** $x^3 - 6x^2 - 4x + 24 = 0$

$x^2(x - 6) - 4(x - 6) = 0$

$(x^2 - 4)(x - 6) = 0$

$(x - 2)(x + 2)(x - 6) = 0$

$x = 2, -2, 6$

# Section 4.4    Solving Exponential and Logarithmic Equations

- To solve an exponential equation, isolate the exponential expression, then take the logarithm of both sides. Then solve for the variable.
    1. $\log_a a^x = x$
    2. $\ln e^x = x$
- To solve a logarithmic equation, rewrite it in exponential form. Then solve for the variable.
    1. $a^{\log_a x} = x$
    2. $e^{\ln x} = x$
- If $a > 0$ and $a \neq 1$ we have the following:
    1. $\log_a x = \log_a y \implies x = y$
    2. $a^x = a^y \implies x = y$
- Use your graphing utility to approximate solutions.

## Solutions to Odd-Numbered Exercises

**1.** $4^{2x-7} = 64$

(a) $x = 5$

$$4^{2(5)-7} = 4^3 = 64$$

Yes, $x = 5$ is a solution.

(b) $x = 2$

$$4^{2(2)-7} = 4^{-3} = \tfrac{1}{64} \neq 64$$

No, $x = 2$ is not a solution.

**3.** $3e^{x+2} = 75$

(a) $x = -2 + e^{25}$

$$3e^{(-2+e^{25})+2} = 3e^{e^{25}} \neq 75$$

No, $x = -2 + e^{25}$ is not a solution.

(b) $x = -2 + \ln 25$

$$3e^{(-2+\ln 25)+2} = 3e^{\ln 25} = 3(25) = 75$$

Yes, $x = -2 + \ln 25$ is a solution.

(c) $x \approx 1.2189$

$$3e^{1.2189+2} = 3e^{3.2189} \approx 75$$

Yes, $x \approx 1.2189$ is a solution.

**5.** $\log_4(3x) = 3$

$$4^3 = 3x$$

$$x = \tfrac{64}{3} \approx 21.333$$

(a) $x = 21.3560$ is an approximate solution

(b) No, $x = -4$ is not a solution

(c) Yes, $x = \tfrac{64}{3}$ is a solution

**7.** $\ln(x - 1) = 3.8$

(a) $x = 1 + e^{3.8}$

$$\ln(1 + e^{3.8} - 1) = \ln e^{3.8} = 3.8$$

Yes, $x = 1 + e^{3.8}$ is a solution.

(b) $x \approx 45.7012$

$$\ln(45.7012 - 1) = \ln(44.7012) \approx 3.8$$

Yes, $x \approx 45.7012$ is a solution.

(c) $x = 1 + \ln 3.8$

$$\ln(1 + \ln 3.8 - 1) = \ln(\ln 3.8) \approx 0.289$$

No, $x = 1 + \ln 3.8$ is not a solution.

**9.**

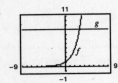

Intersection Point: $(3, 8)$

Algebraically, $2^x = 8$

$$2^x = 2^3$$

$$x = 3 \implies y = 8 \implies (3, 8)$$

**11.**

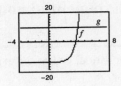

Intersection: $(4, 10)$

Algebraically, $5^{x-2} - 15 = 10$

$$5^{x-2} = 25 = 5^2$$

$$x - 2 = 2$$

$$x = 4$$

$(4, 10)$

**13.**

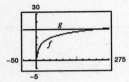

Intersection: $(243, 20)$

Algebraically, $4\log_3 x = 20$

$$\log_3 x = 5$$

$$x = 3^5 = 243$$

$(243, 20)$

**15.**

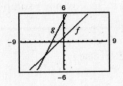

Intersection: $(-4, -3)$

Algebraically, $\ln e^{x+1} = 2x + 5$

$$x + 1 = 2x + 5$$

$$-4 = x$$

$(-4, -3)$

**17.** $4^x = 16$

$$4^x = 4^2$$

$$x = 2$$

**19.** $5^x = \dfrac{1}{625}$

$$5^x = \dfrac{1}{5^4} = 5^{-4}$$

$$x = -4$$

**21.** $\left(\dfrac{1}{8}\right)^x = 64$

$$8^{-x} = 8^2$$

$$-x = 2$$

$$x = -2$$

**23.** $\left(\dfrac{2}{3}\right)^x = \dfrac{81}{16}$

$$\left(\dfrac{3}{2}\right)^{-x} = \left(\dfrac{3}{2}\right)^4$$

$$-x = 4$$

$$x = -4$$

**25.** $e^x = 4$

$$x = \ln 4 \approx 1.386$$

**27.** $\ln x - \ln 5 = 0$

$$\ln x = \ln 5$$

$$x = 5$$

**29.** $\ln x = -7$

$$x = e^{-7}$$

**31.** $\log_x 625 = 4$

$$x^4 = 625$$

$$x^4 = 5^4$$

$$x = 5$$

**33.** $\log_{10} x = -1$

$$x = 10^{-1}$$

$$x = \dfrac{1}{10}$$

**35.** $\ln(2x - 1) = 5$

$$2x - 1 = e^5$$

$$x = \dfrac{1 + e^5}{2} \approx 74.707$$

**37.** $\ln e^{x^2} = x^2 \ln e = x^2$

**39.** $e^{\ln(5x+2)} = 5x + 2$

**41.** $e^{\ln x^2} = x^2$

**43.**
$$8^{3x} = 360$$
$$\ln 8^{3x} = \ln 360$$
$$3x \ln 8 = \ln 360$$
$$3x = \frac{\ln 360}{\ln 8}$$
$$x = \frac{1}{3} \frac{\ln 360}{\ln 8}$$
$$x \approx 0.944$$

**45.** $2e^{5x} = 18$
$$e^{5x} = 9$$
$$5x = \ln 9$$
$$x = \frac{1}{5} \ln 9$$
$$x \approx 0.439$$

**47.** $500e^{-x} = 300$
$$e^{-x} = \frac{3}{5}$$
$$-x = \ln \frac{3}{5}$$
$$x = -\ln \frac{3}{5} = \ln \frac{5}{3} \approx 0.511$$

**49.** $7 - 2e^x = 5$
$$-2e^x = -2$$
$$e^x = 1$$
$$x = \ln 1 = 0$$

**51.** $5^{-t/2} = 0.20 = \frac{1}{5}$
$$-\frac{t}{2} \ln 5 = \ln\left(\frac{1}{5}\right)$$
$$-\frac{t}{2} \ln 5 = -\ln 5$$
$$\frac{t}{2} = 1$$
$$t = 2$$

**53.** $2^{3-x} = 565$
$$(3 - x) \ln 2 = \ln 565$$
$$-x \ln 2 = \ln 565 - 3 \ln 2$$
$$x = (3 \ln 2 - \ln 565)/\ln 2 \approx -6.142$$

**55.**
$$e^{2x} - 4e^x - 5 = 0$$
$$(e^x - 5)(e^x + 1) = 0$$
$$e^x = 5 \text{ or } e^x = -1$$
$$x = \ln 5 \approx 1.609$$
$$(e^x = -1 \text{ is impossible.})$$

**57.** $\dfrac{400}{1 + e^{-x}} = 350$
$$1 + e^{-x} = \frac{400}{350} = \frac{8}{7}$$
$$e^{-x} = \frac{1}{7}$$
$$-x = \ln\left(\frac{1}{7}\right) = -\ln 7$$
$$x = \ln 7 \approx 1.946$$

**59.** $\left(1 + \dfrac{0.10}{12}\right)^{12t} = 2$
$$\left(\frac{12.1}{12}\right)^{12t} = 2$$
$$(12t)\ln\left(\frac{12.1}{12}\right) = \ln 2$$
$$t = \frac{1}{12} \frac{\ln 2}{\ln\left(\dfrac{12.1}{12}\right)}$$
$$t \approx 6.960$$

**61.** $e^{3x} = 12$

| $x$ | 0.6 | 0.7 | 0.8 | 0.9 | 1.0 |
|-----|-----|-----|-----|-----|-----|
| $f(x)$ | 6.05 | 8.17 | 11.02 | 14.88 | 20.09 |

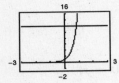

$x \approx 0.828$

**63.** $20(100 - e^{x/12}) = 500$

| $x$ | 5 | 6 | 7 | 8 | 9 |
|-----|-----|-----|-----|-----|-----|
| $f(x)$ | 1756 | 1598 | 1338 | 908 | 200 |

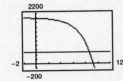

$x \approx 8.635$

**65.** $\left(1 + \dfrac{0.065}{365}\right)^{365t} = 4 \implies t = 21.330$

**67.** $\dfrac{3000}{2 + e^{2x}} = 2$

$1500 = 2 + e^{2x}$

$1498 = e^{2x}$

$\ln 1498 = \ln e^{2x}$

$\ln 1498 = 2x$

$\dfrac{\ln 1498}{2} = x \approx 3.656$

**69.** $g(x) = 6e^{1-x} - 25$

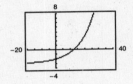

zero at $x = -0.427$

**71.** $g(t) = e^{0.09t} - 3$

zero at $t = 12.207$

**73.** $\ln x = -3$

$x = e^{-3} \approx 0.050$

**75.** $\ln 4x = 2.1$

$4x = e^{2.1}$

$x = \tfrac{1}{4}e^{2.1}$

$\approx 2.042$

**77.** $-2 + 2 \ln 3x = 17$

$2 \ln 3x = 19$

$\ln 3x = \tfrac{19}{2}$

$3x = e^{19/2}$

$x = \tfrac{1}{3}e^{19/2}$

$x \approx 4453.242$

**79.** $\log_{10}(z - 3) = 2$

$z - 3 = 10^2$

$z = 10^2 + 3 = 103$

**81.** $7 \log_4 (0.6x) = 12$

$\log_4(0.6x) = \dfrac{12}{7}$

$4^{12/7} = 0.6x = \dfrac{3}{5}x$

$x = \dfrac{5}{3} 4^{12/7} \approx 17.945$

**83.** $\ln\sqrt{x + 2} = 1$

$\sqrt{x + 2} = e^1$

$x + 2 = e^2$

$x = e^2 - 2 \approx 5.389$

**85.** $\ln(x + 1)^2 = 2$

$$e^{\ln(x+1)^2} = e^2$$

$$(x + 1)^2 = e^2$$

$$x + 1 = e \text{ or } x + 1 = -e$$

$$x = e - 1 \approx 1.718$$

or

$$x = -e - 1 \approx -3.718$$

**87.** $\log_4 x - \log_4(x - 1) = \dfrac{1}{2}$

$$\log_4\left(\dfrac{x}{x - 1}\right) = \dfrac{1}{2}$$

$$4^{\log_4(x/x-1)} = 4^{1/2}$$

$$\dfrac{x}{x - 1} = 2$$

$$x = 2(x - 1)$$

$$x = 2x - 2$$

$$2 = x$$

**89.**   $\ln(x + 5) = \ln(x - 1) - \ln(x + 1)$

$$\ln(x + 5) = \ln\left(\dfrac{x - 1}{x + 1}\right)$$

$$x + 5 = \dfrac{x - 1}{x + 1}$$

$$(x + 5)(x + 1) = x - 1$$

$$x^2 + 6x + 5 = x - 1$$

$$x^2 + 5x + 6 = 0$$

$$(x + 2)(x + 3) = 0$$

$$x = -2 \text{ or } x = -3$$

Both of these solutions are extraneous, so the equation has no solution.

**91.** $\log_{10} 8x - \log_{10}\left(1 + \sqrt{x}\right) = 2$

$$\log_{10} \dfrac{8x}{1 + \sqrt{x}} = 2$$

$$\dfrac{8x}{1 + \sqrt{x}} = 10^2$$

$$8x = 100 + 100\sqrt{x}$$

$$8x - 100\sqrt{x} - 100 = 0$$

$$2x - 25\sqrt{x} - 25 = 0$$

$$\sqrt{x} = \dfrac{25 \pm \sqrt{25^2 - 4(2)(-25)}}{4}$$

$$= \dfrac{25 \pm 5\sqrt{33}}{4}$$

Choosing the positive value, we have $\sqrt{x} \approx 13.431$ and $x \approx 180.384$.

**93.** $\ln 2x = 2.4$

| $x$ | 2 | 3 | 4 | 5 | 6 |
|-----|------|------|------|------|------|
| $f(x)$ | 1.39 | 1.79 | 2.08 | 2.30 | 2.48 |

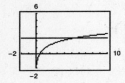

$x \approx 5.512$

**95.** $6 \log_3(0.5x) = 11$

| $x$ | 12 | 13 | 14 | 15 | 16 |
|-----|------|-------|-------|-------|-------|
| $f(x)$ | 9.79 | 10.22 | 10.63 | 11.00 | 11.36 |

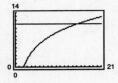

$x \approx 14.988$

**97.** $\log_{10} x = x^3 - 3$

Graphing $y = \log_{10} x - x^3 + 3$, you obtain 2 zeros

$x \approx 1.469$ and $x \approx 0.001$

**99.** $\log_3 x + \log_3(x - 3) = 1$

Graphing $y = \dfrac{\log x}{\log 3} + \dfrac{\log(x - 3)}{\log 3} - 1$,

you obtain $x \approx 3.791$

**101.** $\ln(x - 3) + \ln(x + 3) = 1$

Graphing $y = \ln(x - 3) + \ln(x + 3) - 1$, you obtain $x \approx 3.423$

**103.** $y_1 = 7$

$y_2 = 2^{x-1} - 5$

Intersection: $(4.585, 7)$

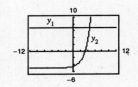

**105.** $y_1 = 80$

$y_2 = 4e^{-0.2x}$

Intersection: $(-14.979, 80)$

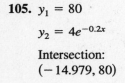

**107.** $y_1 = 3.25$

$y_2 = \frac{1}{2}\ln(x + 2)$

Intersection: $(663.142, 3.25)$

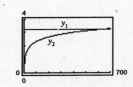

**109.** (a) $\quad A = Pe^{rt}$

$2000 = 1000e^{0.085t}$

$2 = e^{0.085t}$

$\ln 2 = 0.085t$

$\dfrac{\ln 2}{0.085} = t$

$t \approx 8.2$ years

(b) $\quad 3000 = 1000e^{0.085t}$

$3 = e^{0.085t}$

$\ln 3 = 0.085t$

$\dfrac{\ln 3}{0.085} = t$

$t \approx 12.9$ years

**111.** $p = 500 - 0.5(e^{0.004x})$

(a) $\quad p = 350$

$350 = 500 - 0.5(e^{0.004x})$

$300 = e^{0.004x}$

$0.004x = \ln 300$

$x \approx 1426$ units

(b) $\quad p = 300$

$300 = 500 - 0.5(e^{0.004x})$

$400 = e^{0.004x}$

$0.004x = \ln 400$

$x \approx 1498$ units

**113.** $N = 68(10^{-0.04x})$

When $N = 21$:

$$21 = 68(10^{-0.04x})$$

$$\frac{21}{68} = 10^{-0.04x}$$

$$\log_{10} \frac{21}{68} = -0.04x$$

$$x = -\frac{\log(21/68)}{0.04} \approx 12.76 \text{ inches}$$

**115.** (a)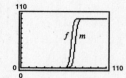

(b) From the graph we see horizontal asymptotes at $y = 0$ and $y = 100$. These represent the lower and upper percent bounds.

(c) Males:

$$50 = \frac{100}{1 + e^{-0.6114(x-69.71)}}$$

$$1 + e^{-0.6114(x-69.71)} = 2$$

$$e^{-0.6114(x-69.71)} = 1$$

$$-0.6114(x - 69.71) = \ln 1$$

$$-0.6114(x - 69.71) = 0$$

$$x = 69.71 \text{ inches}$$

Females:

$$50 = \frac{100}{1 + e^{-0.66607(x-64.51)}}$$

$$1 + e^{-0.66607(x-64.51)} = 2$$

$$e^{-0.66607(x-64.51)} = 1$$

$$-0.66607(x - 64.51) = \ln 1$$

$$-0.66607(x - 64.51) = 0$$

$$x = 64.51 \text{ inches}$$

**117.** $T = 20[1 + 7(2^{-h})]$

(a)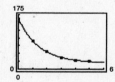

(b) We see a horizontal asymptote at $y = 20$. This represents the room temperature.

(c)
$$100 = 20[1 + 7(2^{-h})]$$

$$5 = 1 + 7(2^{-h})$$

$$4 = 7(2^{-h})$$

$$\frac{4}{7} = 2^{-h}$$

$$\ln\left(\frac{4}{7}\right) = \ln 2^{-h}$$

$$\ln\left(\frac{4}{7}\right) = -h \ln 2$$

$$\frac{\ln(4/7)}{-\ln 2} = h$$

$$h \approx 0.81 \text{ hour}$$

**119.** True

**121.** Answers will vary

**123.** Yes. The doubling time is given by

$$2P = Pe^{rt}$$

$$2 = e^{rt}$$

$$\ln 2 = rt$$

$$t = \frac{\ln 2}{r}$$

The time to quadruple is given by

$$4P = Pe^{rt}$$

$$4 = e^{rt}$$

$$\ln 4 = rt$$

$$t = \frac{\ln 4}{r} = \frac{\ln 2^2}{r} = \frac{2 \ln 2}{r} = 2\left[\frac{\ln 2}{r}\right]$$

which is twice as long.

**125.** $f(x) = 3x^3 - 4$

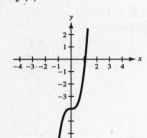

**127.** $f(x) = |x| + 9$

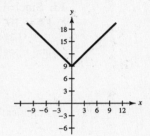

**129.** $f(x) = \begin{cases} 2x & x < 0 \\ -x^2 + 4 & x \geq 0 \end{cases}$

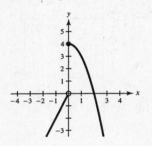

## Section 4.5     Exponential and Logarithmic Models

- You should be able to solve compound interest problems.

  1. $A = P\left(1 + \dfrac{r}{n}\right)^{nt}$

  2. $A = Pe^{rt}$

- You should be able to solve growth and decay problems.

  (a) Exponential growth if $b > 0$ and $y = ae^{bx}$.

  (b) Exponential decay if $b > 0$ and $y = ae^{-bx}$.

- You should be able to use the Gaussian model

  $y = ae^{-(x-b)^2/c}$.

- You should be able to use the logistics growth model

  $y = \dfrac{a}{1 + be^{-(x-c)/d}}$.

- You should be able to use the logarithmic models

  $y = \ln(ax + b)$ and $y = \log_{10}(ax + b)$.

### Solutions to Odd-Numbered Exercises

**1.** $y = 2e^{x/4}$

This is an exponential growth model.

Matches graph (c).

**3.** $y = 6 + \log_{10}(x + 2)$

This is a logarithmic model, and contains $(-1, 6)$.

Matches graph (b).

**5.** $y = \ln(x + 1)$

This is a logarithmic model.

Matches graph (d).

**7.** Since $A = 1000e^{0.035t}$, the time to double is given by

$$2000 = 1000e^{0.035t}$$

$$2 = e^{0.035t}$$

$$\ln 2 = 0.035t$$

$$t = \frac{1}{0.035} \ln 2 \approx 19.8 \text{ years}$$

Amount after 10 years:

$$A = 1000e^{0.035(10)} \approx \$1419.07$$

**9.** Since $A = 750e^{rt}$ and $A = 1500$ when $t = 7.75$, we have the following.

$$1500 = 750e^{7.75r}$$

$$r = \frac{\ln 2}{7.75} \approx 0.0894 = 8.94\%$$

Amount after 10 years:

$$A = 750e^{0.0894(10)} \approx \$1833.67$$

**11.** Since $A = 500e^{rt}$ and $A = 1292.85$ when $t = 10$, we have the following:

$$1292.85 = 500e^{10r}$$

$$r = \frac{\ln(1292.85/500)}{10} \approx 0.0950 = 9.5\%$$

The time to double is given by

$$1000 = 500e^{0.095t}$$

$$2 = e^{0.095t}$$

$$\ln 2 = 0.095t$$

$$t = \frac{\ln 2}{0.095} \approx 7.30 \text{ years}$$

**13.** Since $A = Pe^{0.045t}$ and $A = 10,000.00$ when $t = 10$, we have the following:

$$10,000.00 = Pe^{0.045(10)}$$

$$\frac{10,000.00}{e^{0.045(10)}} = P \approx 6376.28$$

The time to double is given by

$$12,752.56 = 6376.28 e^{0.045t}$$

$$t = \frac{\ln 2}{0.045} \approx 15.40 \text{ years.}$$

**15.** $\quad 3P = Pe^{rt}$

$$3 = e^{rt}$$

$$\ln 3 = rt$$

$$\frac{\ln 3}{r} = t$$

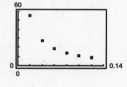

| $r$ | 2% | 4% | 6% | 8% | 10% | 12% |
|---|---|---|---|---|---|---|
| $t = \dfrac{\ln 3}{r}$ | 54.93 | 27.47 | 18.31 | 13.73 | 10.99 | 9.16 |

**17.**

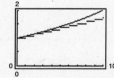

Continuous compounding results in faster growth.

$$A = 1 + 0.075[\![t]\!]$$

and $A = e^{0.07t}$

**19.** $\dfrac{1}{2} C = Ce^{k(1600)}$

$$k = \frac{\ln 0.5}{1600}$$

Given $C = 10$ and $t = 1000$,

$$y = Ce^{kt}$$

$$= 10e^{\left[\frac{\ln 0.5}{1600} 1000\right]}$$

$$\approx 6.484 \text{ grams}$$

**21.** $\dfrac{1}{2} C = Ce^{k(5730)}$

$$k = \frac{\ln 0.5}{5730}$$

Given $C = 3$ grams, after 1000 years, we have

$$y = 3e^{[(\ln 0.5)/5730](1000)}$$

$$y \approx 2.66 \text{ grams.}$$

**23.**   $y = ae^{bx}$

$1 = ae^{b(0)} \implies 1 = a$

$10 = e^{b(3)}$

$\ln 10 = 3b$

$\dfrac{\ln 10}{3} = b \quad \implies \quad b \approx 0.7675$

Thus, $y = e^{0.7675x}$.

**25.**   $(0, 4) \implies a = 4$

$(5, 1) \implies 1 = 4e^{b(5)} \implies b = \frac{1}{5}\ln\left(\frac{1}{4}\right)$

$= -\frac{1}{5}\ln 4 \approx -0.2773$

$y = 4e^{-0.2773x}$

**27.** (a)

Australia:  $(0, 19.2)\ (10, 20.9)$

$a = 19.2$ and $20.9 = 19.2e^{b(10)} \implies b = 0.008484$

$y = 19.2e^{0.008484t}$      For 2030, $y \approx 24.8$ million

Canada:  $(0, 31.3)\ (10, 34.3)$

$a = 31.3$ and $34.3 = 31.3e^{b(10)} \implies b = 0.009153$

$y = 31.3e^{0.009153t}$      For 2030, $y \approx 41.2$ million

Philippines:  $(0, 81.2)\ (10, 97.9)$

$a = 81.2$ and $97.9 = 81.2e^{b(10)} \implies b = 0.0187$

$y = 81.2e^{0.0187t}$      For 2030, $y \approx 142.3$ million

South Africa:  $(0, 43.4)\ (10, 41.1)$

$a = 43.4$ and $41.1 = 43.4e^{b(10)} \implies b = -0.005445$

$y = 43.4e^{-0.005445t}$      For 2030, $y \approx 36.9$ million

Turkey:  $(0, 65.7)\ (10, 73.3)$

$a = 65.7$ and $73.3 = 65.7e^{b(10)} \implies b = 0.01095$

$y = 65.7e^{0.01095t}$      For 2030, $y \approx 91.2$ million

(b) The constant $b$ gives the growth rates.

(c) The constant $b$ is negative for South Africa.

**29.**   $N = 100e^{kt}$

$300 = 100e^{5k}$

$k = \dfrac{\ln 3}{5} \approx 0.2197$

$N = 100e^{0.2197t}$

$200 = 100e^{0.2197t}$

$t = \dfrac{\ln 2}{0.2197} \approx 3.15$ hours

**31.**   $y = Ce^{kt}$

$\dfrac{1}{2}C = Ce^{(1620)k}$

$\ln\dfrac{1}{2} = 1620k$

$k = \dfrac{\ln(1/2)}{1620}$

When $t = 100$, we have

$y = Ce^{[\ln(1/2)/1620](100)} \approx 0.958C = 95.8\%C$.

After 100 years, approximately 95.8% of the radioactive radium will remain.

**33.** (a) $V = mt + b$,   $V(0) = 32{,}000 \implies b = 32{,}000$.

$V(2) = 18{,}000 \implies 18{,}000 = 2m + 32{,}000 \implies m = -7000$

$V(t) = -7000t + 32{,}000$

(b) $V = ae^{kt}$,   $V(0) = 32{,}000 \implies a = 32{,}000$

$V(2) = 18{,}000 \implies 18{,}000 = 32{,}000e^{2k} \implies \frac{18}{32} = e^{2k} \implies k = \frac{1}{2}\ln\left(\frac{18}{32}\right) \approx -0.2877$

$V(t) = 32{,}000e^{-0.2877t}$

(c)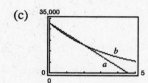

The exponential model depreciates faster in the first year.

(e) The negative slope means the car depreciates $7000 per year.

(d) Straight line:  $V(1) = \$25{,}000$

$V(3) = \$11{,}000$

Exponential:  $V(1) \approx \$23{,}999.57$

$V(3) \approx \$13{,}499.27$

**35.** $S(t) = 100(1 - e^{kt})$

(a)   $15 = 100(1 - e^{k(1)})$

$-85 = -100e^{k}$

$k = \ln 0.85$

$k \approx -0.1625$

$S(t) = 100(1 - e^{-0.1625t})$

(b)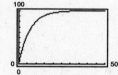

(c) $S(5) = 100(1 - e^{-0.1625(5)})$

$\approx 55.625 = 55{,}625$ units

**37.** $y = 0.0266e^{-(x-100)^2/450}$,   $70 \le x \le 115$

(a)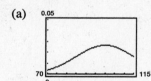

(b) Maximum point is $x = 100$, the average IQ score.

**39.** $p(t) = \dfrac{1000}{1 + 9e^{-0.1656t}}$

(a)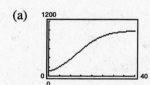

The horizontal asymptotes are $y = 0$ and $y = 1000$. The asymptote with the larger $p$-value, $y = 1000$, indicates that the population size will approach 1000 as time increases.

(b) $p(5) = \dfrac{1000}{1 + 9e^{-0.1656(5)}} \approx 203$ animals

(c)    $500 = \dfrac{1000}{1 + 9e^{-0.1656t}}$

$1 + 9e^{-0.1656t} = 2$

$9e^{-0.1656t} = 1$

$e^{-0.1656t} = \dfrac{1}{9}$

$t = -\dfrac{\ln(1/9)}{0.1656} \approx 13$ months

**41.** $R = \log_{10}\left(\dfrac{I}{I_0}\right) = \log_{10}(I)$

$I = 10^R$

(a) $I = 10^{6.5} \approx 3{,}162{,}278$

(b) $I = 10^{7.9} \approx 79{,}432{,}823$

(c) $I = 10^{5.2} \approx 158{,}489$

**43.** $\beta(I) = 10\log_{10}(I/I_0)$, where $I_0 = 10^{-12}$ watt per square meter.

(a) $\beta(10^{-10}) = 10 \cdot \log_{10}\left(\dfrac{10^{-10}}{10^{-12}}\right) = 10\log_{10}10^2 = 20$ decibels

(b) $\beta(10^{-5}) = 10 \cdot \log_{10}\left(\dfrac{10^{-5}}{10^{-12}}\right) = 10\log_{10}10^7 = 70$ decibels

(c) $\beta(10^0) = 10 \cdot \log_{10}\left(\dfrac{10^0}{10^{-12}}\right) = 10\log_{10}10^{12} = 120$ decibels

**45.** $\beta = 10\log_{10}\left(\dfrac{I}{I_0}\right)$

$10^{\beta/10} = \dfrac{I}{I_0}$

$I = I_0\,10^{\beta/10}$

$\%\text{ decrease} = \dfrac{I_0\,10^{8.8} - I_0\,10^{7.2}}{I_0\,10^{8.8}} \times 100$

$= 97.5\%$

**47.** $\text{pH} = -\log_{10}[H^+] = -\log_{10}[2.3 \times 10^{-5}] \approx 4.64$

**49.** $pH = -\log_{10}[H^+]$

$-pH = \log_{10}[H^+]$

$10^{-ph} = [H^+]$

$\dfrac{\text{Hydrogen ion concentration of grape}}{\text{Hydrogen ion concentration of milk of magnesia}} = \dfrac{10^{-3.5}}{10^{-10.5}} = 10^7$

**51.** (a) $P = 120{,}000, t = 30, r = 0.075, M = 839.06$

$u = M - \left(M - \dfrac{Pr}{12}\right)\left(1 + \dfrac{r}{12}\right)^{12t}$

$= 839.06 - (839.06 - 750)(1 + 0.00625)^{12t}$

$v = (839.06 - 750)(1.00625)^{12t}$

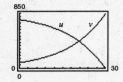

(b) In the early years, the majority of the monthly payment goes toward interest. The interest and principle are equal when $t \approx 20.729 \approx 21$ years.

(c) $P = 120{,}000, t = 20, r = 0.075, M = 966.71$

$u = 966.71 - (966.71 - 750)(1.00625)^{12t}$

$v = (966.71 - 750)(1.00625)^{12t}$

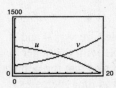

$u = v$ when $t \approx 10.73$ years

**53.** $t = -10 \ln\left(\dfrac{T - 70}{98.6 - 70}\right)$

At 9:00 A.M. we have

$t = -10 \ln (85.7 - 70/98.6 - 70) \approx 6$ hours

Thus, we can conclude that the person died 6 hours before 9 A.M., or 3:00 A.M.

**55.** False. The domain could be all real numbers.

**57.** True. For the Gaussian model, $y > 0$.

**59.** $4x - 3y - 9 = 0 \implies y = \frac{1}{3}(4x - 9)$

Slope: $\frac{4}{3}$

Matches (a).

Intercepts: $(0, -3), \left(\frac{9}{4}, 0\right)$

**61.** $y = 25 - 2.25x$

Slope: $-2.25$

Matches (d).

Intercepts: $(0, 25), \left(\frac{100}{9}, 0\right)$

**63.** $f(x) = 2x^3 - 3x^2 + x - 1$

The graph falls to the left and rises to the right.

**65.** $g(x) = -1.6x^5 + 4x^2 - 2$

The graph rises to the left and falls to the right.

**67.**

$$
\begin{array}{r|rrrr}
4 & 2 & -8 & 3 & -9 \\
  &   & 8 & 0 & 12 \\
\hline
  & 2 & 0 & 3 & 3
\end{array}
$$

$$\frac{2x^3 - 8x^2 + 3x - 9}{x - 4} = 2x^2 + 3 + \frac{3}{x - 4}$$

# Section 4.6    Exploring Data:  Nonlinear Models

**Solutions to Odd-Numbered Exercises**

**1.** Logarithmic model

**3.** Quadratic model

**5.** Exponential model

**7.** Quadratic model

**9.**

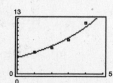

Logarithmic model

**11.**

Exponential model

**13.**

Linear model

**15.** $y = 3.807(1.3057)^x$

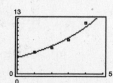

**17.** $y = 8.463(0.7775)^x$

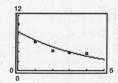

**19.** $y = 2.083 + 1.257 \ln x$

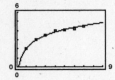

**21.** $y = 9.826 - 4.097 \ln x$

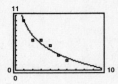

**23.** $y = 1.985x^{0.760}$

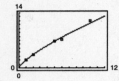

**25.** $y = 16.103x^{-3.174}$

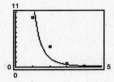

**27.** (a) Quadratic model: $R = 0.0125x^2 + 1.635x + 94.86$

Exponential model: $R = 95.324(1.0165)^x$

Power model: $R = 81.230x^{0.1682}$

(b)

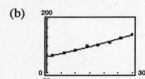 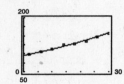 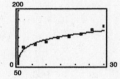

(c) The quadratic model fits best

(d) For 2004, $t = 34$ and $R \approx 164.9$ or 164.9 million

**29.** (a) $y = 3.127x + 250.87$

(b) $y = 251.453(1.0116)^x$

(c) No, the linear model is better. Its $r$-value is closer to one.

(d) Linear model: 307 million

Exponential model: 309 million

[Answer will vary depending on number of digits.]

**31.** (a) $T = -1.239t + 73.02$

No, the data does not appear linear

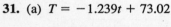

(c) Subtracting 21 from the $T$ values, the exponential model is

$y = 54.438(0.9635)^x$

Adding back 21,

$T = 54.438(0.9635)^t + 21$

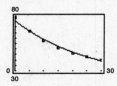

(b) $T = 0.034t^2 - 2.26t + 77.3$

Yes, the data appears quadratic.

But, for $t = 60$, the graph is increasing, which is incorrect.

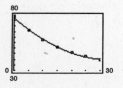

(d) Answers will vary.

**33.** (a) $S = 925.734(1.155)^x$

(b)

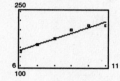

(c) The model is a good fit

(d) For 2007, $x = 17$ and $S = 10,725$ million

[Answers will vary.]

**35.** (a) Linear model: $y = 15.79x + 47.9$

Logarithmic model: $y = -97.5 + 131.92 \ln x$

Quadratic model: $y = -1.968x^2 + 49.24x - 88.6$

Exponential model: $y = 83.94(1.09)^x$

Power model: $y = 36.51x^{0.7525}$

(b) Linear model:

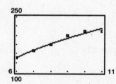

Logarithmic model:

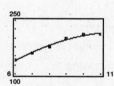

Quadratic model:

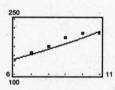

Exponential model:

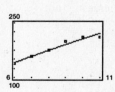

Power model:

The quadratic model appears to fit best.

(c) Linear model: 217.2

Logarithmic model: 120.91

Quadratic model: 72.7

Exponential model: 532.2

Power model: 189.0

The quadratic model fits best because 72.7 is smallest.

(d) Linear model: 0.9526

Logarithmic model: 0.9736

Quadratic model: 0.9841

Exponential model: 0.9402

Power model: 0.9697

The quadratic model has the largest $r^2$ value.

(e) The quadratic model is best.

**37.** True

**39.** $2x + 5y = 10$

$$5y = -2x + 10$$

$$y = -\tfrac{2}{5}x + 2$$

slope: $-\tfrac{2}{5}$

$y$-intercept: $(0, 2)$

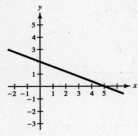

**41.** $1.2x + 3.5y = 10.5$

$$35y = -12x + 105$$

$$y = -\tfrac{12}{35}x + \tfrac{105}{35}$$

$$= -\tfrac{12}{35}x + 3$$

slope: $-\tfrac{12}{35}$

$y$-intercept: $(0, 3)$

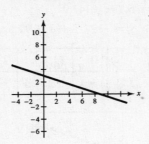

# Review Exercises for Chapter 4

### Solutions to Odd-Numbered Exercises

**1.** $(1.45)^{2\pi} \approx 10.3254$

**3.** $60^{2(-1.1)} = 60^{-2.2} \approx 0.0001225 \approx 0.0$

**5.** $f(x) = 4^x$

Intercept: $(0, 1)$

Horizontal asymptote: $x$-axis

Increasing on: $(-\infty, \infty)$

Matches graph (c).

**7.** $f(x) = -4^x$

Intercept: $(0, -1)$

Horizontal asymptote: $x$-axis

Decreasing on: $(-\infty, \infty)$

Matches graph (b).

**9.** $f(x) = 6^x$

Intercept: $(0, 1)$

Horizontal asymptote: $x$-axis

Increasing on: $(-\infty, \infty)$

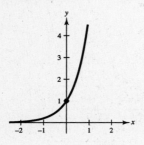

**11.** $g(x) = 1 + 6^{-x}$

Intercept: $(0, 2)$

Horizontal asymptote: $y = 1$

Decreasing on: $(-\infty, \infty)$

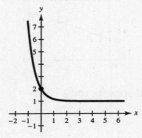

**13.** $e^8 \approx 2980.958$

**15.** $e^{-2.1} \approx 0.122$

**17.** $h(x) = e^{x-1}$

| $x$ | $-4$ | $-2$ | $0$ | $1$ | $2$ | $4$ |
|------|--------|--------|--------|---|--------|--------|
| $h(x)$ | 0.0067 | 0.0498 | 0.3679 | 1 | 2.7183 | 20.086 |

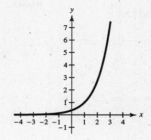

**19.** $h(x) = -e^x$

| $x$ | $-4$ | $-2$ | $0$ | $1$ | $2$ | $4$ |
|------|---------|---------|-----|--------|--------|--------|
| $h(x)$ | $-0.0183$ | $-0.1353$ | $-1$ | $-2.718$ | $-7.389$ | $-54.6$ |

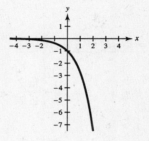

**21.** $f(x) = 4e^{-0.5x}$

| $x$ | $-4$ | $-2$ | $0$ | $1$ | $2$ |
|------|--------|--------|---|--------|--------|
| $f(x)$ | 29.556 | 10.873 | 4 | 2.4261 | 1.4715 |

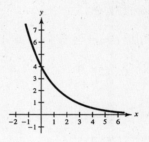

**23.** $g(t) = 8 - 0.5e^{-t/4}$

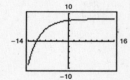

Horizontal asymptote: $y = 8$

**25.** $g(x) = 200e^{4/x}$

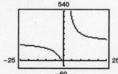

Vertical asymptote: $x = 0$
Horizontal asymptote: $y = 200$

**27.** $f(x) = \dfrac{10}{1 + 2^{-0.05x}}$

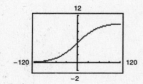

Horizontal asymptotes: $y = 0$, $y = 10$

**29.** $A = Pe^{rt} = 10,000e^{0.08t}$

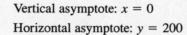

| $t$ | 1 | 10 | 20 |
|-----|-----------|-----------|-----------|
| $A$ | 10,832.87 | 22,255.41 | 49,530.32 |

| $t$ | 30 | 40 | 50 |
|-----|------------|------------|------------|
| $A$ | 110,231.76 | 245,325.30 | 545,981.50 |

**31.** $V(t) = 26,000\left(\frac{3}{4}\right)^t$

(a)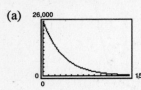

(b) For $t = 2$, $V(2) = \$14,625$

(c) The car depreciates most rapidly at the beginning, which is realistic.

**33.**  $4^3 = 64$

$\log_4 64 = 3$

**35.** $25^{3/2} = 125$

$\log_{25} 125 = \frac{3}{2}$

**37.** $\log_6 216 = \log_6 6^3 = 3\log_6 6 = 3$

**39.** $\log_4\left(\frac{1}{4}\right) = \log_4(4^{-1}) = -\log_4 4 = -1$

**41.** $g(x) = -\log_2 x + 5 = 5 - \dfrac{\ln x}{\ln 2}$

Domain: $x > 0$

Vertical asymptote: $x = 0$

$x$-intercept: $(32, 0)$

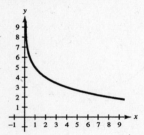

**43.** $f(x) = \log_2(x - 1) + 6 = 6 + \dfrac{\ln(x - 1)}{\ln (2)}$

Domain: $x > 1$

Vertical asymptote: $x = 1$

$x$-intercept: $(1.016, 0)$

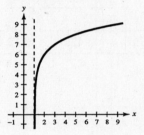

**45.** $\ln(21.5) \approx 3.068$

**47.** $\ln(e^7) = 7$

**49.** $\ln\sqrt{6} \approx 0.896$

**51.** $f(x) = \ln x + 3$

Domain: $(0, \infty)$

Vertical asymptote: $x = 0$

$x$-intercept: $(0.05, 0)$

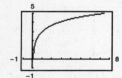

**53.** $h(x) = \frac{1}{2}\ln x$

Domain: $x > 0$

Vertical asymptote: $x = 0$

$x$-intercept: $(1, 0)$

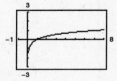

**55.** $t = 50\log_{10}\dfrac{18,000}{18,000 - h}$

(a) $0 \le h < 18,000$

(c) The plane climbs at a faster rate as it approaches its absolute ceiling.

(d) If $h = 4000$, $t = 50\log_{10}\dfrac{18,000}{18,000 - 4000} \approx 5.46$ minutes.

(b)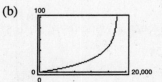

Vertical asymptote: $h = 18,000$

**57.** $\log_4 9 = \dfrac{\log_{10} 9}{\log_{10} 4} \approx 1.585$

$\qquad \log_4 9 = \dfrac{\ln 9}{\ln 4} \approx 1.585$

**59.** $\log_{12} 200 = \dfrac{\log_{10} 200}{\log_{10} 12} \approx 2.132$

$\qquad \log_{12} 200 = \dfrac{\ln 200}{\ln 12} \approx 2.132$

**61.** $\ln 20 = \ln[4 \cdot 5] = \ln 4 + \ln 5$

**63.** $\log_5\left(\tfrac{1}{15}\right) = \log_5(15)^{-1}$

$\qquad\qquad = -\log_5(3 \cdot 5)$

$\qquad\qquad = -\log_5 3 - \log_5 5$

$\qquad\qquad = -\log_5 3 - 1$

**65.** $\log_5 5x^2 = \log_5 5 + \log_5 x^2$

$\qquad\qquad = 1 + 2\log_5 x$

**67.** $\log_{10} \dfrac{5\sqrt{y}}{x^2} = \log_{10} 5\sqrt{y} - \log_{10} x^2$

$\qquad\qquad = \log_{10} 5 + \log_{10}\sqrt{y} - \log_{10} x^2$

$\qquad\qquad = \log_{10} 5 + \dfrac{1}{2}\log_{10} y - 2\log_{10} x$

**69.** $\ln\left(\dfrac{x+3}{xy}\right) = \ln(x+3) - \ln(xy)$

$\qquad\qquad = \ln(x+3) - \ln x - \ln y$

**71.** $\log_2 5 + \log_2 x = \log_2 5x$

**73.** $\dfrac{1}{2}\ln(2x-1) - 2\ln(x+1) = \ln\sqrt{2x-1} - \ln(x+1)^2$

$\qquad\qquad\qquad\qquad\qquad\quad = \ln\dfrac{\sqrt{2x-1}}{(x+1)^2}$

**75.** $\ln 3 + \dfrac{1}{3}\ln(4-x^2) - \ln x = \ln\left[\dfrac{3(4-x^2)^{1/3}}{x}\right] = \ln\left[\dfrac{3\sqrt[3]{4-x^2}}{x}\right]$

**77.** $s = 25 - \dfrac{13\ln\left(\dfrac{h}{12}\right)}{\ln 3}$

(a)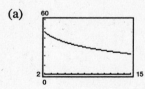

(b)

| $h$ | 4 | 6 | 8 | 10 | 12 | 14 |
|---|---|---|---|---|---|---|
| $s$ | 38 | 33.2 | 29.8 | 27.2 | 25 | 23.2 |

(c) As the depth increases, the number of miles of roads cleared decreases.

**79.** $8^x = 512 = 8^3 \implies x = 3$

**81.** $6^x = \dfrac{1}{216} = \dfrac{1}{6^3} = 6^{-3} \implies x = -3$

**83.** $\log_7 x = 4 \implies x = 7^4 = 2401$

**85.** $\ln x = 4$

$\qquad x = e^4 \approx 54.598$

**87.** $e^x = 12$

$x = \ln 12 \approx 2.485$

**89.** $3e^{-5x} = 132$

$e^{-5x} = 44$

$-5x = \ln 44$

$x = -\dfrac{\ln 44}{5} \approx -0.757$

**91.** $2^x + 13 = 35$

$2^x = 22$

$x \ln 2 = \ln 22$

$x = \dfrac{\ln 22}{\ln 2} \approx 4.459$

**93.** $-4(5^x) = -68$

$5^x = 17$

$x \ln 5 = \ln 17$

$x = \dfrac{\ln 17}{\ln 5} \approx 1.760$

**95.** $e^{2x} - 7e^x + 10 = 0$

$(e^x - 5)(e^x - 2) = 0$

$e^x = 5 \implies x = \ln 5 \approx 1.609$

$e^x = 2 \implies x = \ln 2 \approx 0.693$

**97.** $\ln 3x = 8.2$

$3x = e^{8.2}$

$x = \dfrac{e^{8.2}}{3} \approx 1213.650$

**99.** $2 \ln 4x = 15$

$\ln 4x = \dfrac{15}{2}$

$4x = e^{15/2}$

$x = \dfrac{1}{4} e^{15/2} \approx 452.011$

**101.** $\ln x - \ln 3 = 2$

$\ln \dfrac{x}{3} = 2$

$\dfrac{x}{3} = e^2$

$x = 3e^2 \approx 22.167$

**103.** $\ln \sqrt{x + 1} = 2$

$\frac{1}{2} \ln(x + 1) = 2$

$\ln(x + 1) = 4$

$x + 1 = e^4$

$x = e^4 - 1 \approx 53.598$

**105.** $\log_{10}(x - 1) = \log_{10}(x - 2) - \log_{10}(x + 2)$

$\log_{10}(x - 1) = \log_{10}\left(\dfrac{x - 2}{x + 2}\right)$

$x - 1 = \dfrac{x - 2}{x + 2}$

$(x - 1)(x + 2) = x - 2$

$x^2 + x - 2 = x - 2$

$x^2 = 0$

$x = 0$

Since $x = 0$ is not in the domain of $\ln(x - 1)$ or of $\ln(x - 2)$, it is an extraneous solution. The equation has no solution. You can verify this by graphing each side of the equation and observing that the two curves do not intersect.

**107.** $\log_{10}(1 - x) = -1$

$10^{-1} = 1 - x$

$x = 1 - 10^{-1} = 0.9$

**109.** $3(7550) = 7550e^{0.0725t}$

$3 = e^{0.0725t}$

$\ln 3 = 0.0725t$

$t = \dfrac{\ln 3}{0.0725} \approx 15.2 \text{ years}$

**111.** $y = 3e^{-2x/3}$.

Decreasing exponential.

Matches graph (e).

**113.** $y = \ln(x + 3)$

Logarithmic function shifted to left.

Matches graph (f).

**115.** $y = 2e^{-(x + 4)^2/3}$

Gaussian model.

Matches graph (a).

**117.**
$$y = ae^{bx}$$
$$2 = ae^{b(0)} \implies a = 2$$
$$3 = 2e^{b(4)}$$
$$1.5 = e^{4b}$$
$$\ln 1.5 = 4b \implies b \approx 0.1014$$
Thus, $y = 2e^{0.1014x}$.

**119.**
$$y = ae^{bx}$$
$$\tfrac{1}{2} = ae^{b(0)} \implies a = \tfrac{1}{2}$$
$$5 = \tfrac{1}{2}e^{b(5)}$$
$$10 = e^{5b}$$
$$\ln 10 = 5b \implies b \approx 0.4605$$
Thus, $y = \tfrac{1}{2}e^{0.4605x}$.

**121.** $P = 361e^{kt}$

$t = 0$ corresponds to 2000

$(-20, 215)$:

$$215 = 361e^{k(-20)}$$

$$\frac{215}{361} = e^{-20k}$$

$$-20k = \ln\left(\frac{215}{361}\right)$$

$$k = -\tfrac{1}{20}\ln\left(\frac{215}{361}\right) = \tfrac{1}{20}\ln\left(\frac{361}{215}\right) \approx 0.02591$$

$P = 361e^{0.02591t}$

For 2020, $P(20) = 361e^{0.02591(20)} \approx 606.1$

606,100 population in 2020

**123.** (a) $20,000 = 10,000e^{r(12)}$
$$2 = e^{12r}$$
$$\ln 2 = 12r$$
$$r = \frac{\ln 2}{12} \approx 0.0578 \text{ or } 5.78\%$$

(b) $10,000e^{0.0578(1)} \approx \$10,595.03$

**125.** $N = \dfrac{157}{1 + 5.4e^{-0.12t}}$

(a) When $N = 50$:

$$50 = \frac{157}{1 + 5.4e^{-0.12t}}$$

$$1 + 5.4e^{-0.12t} = \frac{157}{50}$$

$$5.4e^{-0.12t} = \frac{107}{50}$$

$$e^{-0.12t} = \frac{107}{270}$$

$$-0.12t = \ln\frac{107}{270}$$

$$t = \frac{\ln(107/270)}{-0.12} \approx 7.7 \text{ weeks}$$

(b) When $N = 75$:

$$75 = \frac{157}{1 + 5.4e^{-0.12t}}$$

$$1 + 5.4e^{-0.12t} = \frac{157}{75}$$

$$5.4e^{-0.12t} = \frac{82}{75}$$

$$e^{-0.12t} = \frac{82}{405}$$

$$-0.12t = \ln\frac{82}{405}$$

$$t = \frac{\ln(82/405)}{-0.12} \approx 13.3 \text{ weeks}$$

**127.** Logistics model

**129.** Logarithmic model

**131.** (a) Quadratic model: $y = 0.025x^2 + 0.12x + 10.1$

Exponential model: $y = 8.73(1.05)^x$

Power model: $y = 3.466x^{0.647}$

(b) Quadratic model:

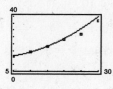

Exponential model:

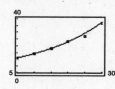

Power model:

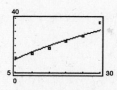

(c) The exponential model appears to fit best

(d) For 2007, $x = 37$ and $y \approx 53.1$ or 53,100 screens.

**133.** (a) $P = \dfrac{9999.887}{1 + 19.0e^{-0.2x}}$

(b)

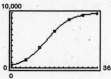

(c) The model is a good fit

(d) The limiting size is $\dfrac{9999.887}{1 + 0} \approx 10,000$ fish

**135.** True; by the inverse properties, $\log_b b^{2x} = 2x$.

**137.** False; $\ln x + \ln y = \ln(xy) \neq \ln(x + y)$

**139.** False. The domain of $f(x) = \ln(x)$ is $x > 0$

**141.** (a)

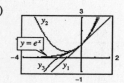

(b) Pattern $\displaystyle\sum_{i=0}^{n} \frac{x^i}{i!}$

$$y_4 = 1 + x + \frac{x^2}{2!} + \frac{x^3}{3!} + \frac{x^4}{4!}$$

The graph of $y_4$ closely approximates $y = e^x$ near $(0, 1)$.

## Chapter 4 Practice Test

1. Solve for $x$: $x^{3/5} = 8$

2. Solve for $x$: $3^{x-1} = \frac{1}{81}$

3. Graph $f(x) = 2^{-x}$ by hand.

4. Graph $g(x) = e^x + 1$ by hand.

5. If \$5000 is invested at 9% interest, find the amount after three years if the interest is compounded

   (a) monthly     (b) quarterly     (c) continuously.

6. Write the equation in logarithmic form: $7^{-2} = \frac{1}{49}$

7. Solve for $x$: $x - 4 = \log_2 \frac{1}{64}$

8. Given $\log_b 2 = 0.3562$ and $\log_b 5 = 0.8271$, evaluate $\log_b \sqrt[4]{8/25}$.

9. Write $5 \ln x - \frac{1}{2} \ln y + 6 \ln z$ as a single logarithm.

10. Using your calculator and the change of base formula, evaluate $\log_9 28$.

11. Use your calculator to solve for $N$: $\log_{10} N = 0.6646$

12. Graph $y = \log_4 x$ by hand.

13. Determine the domain of $f(x) = \log_3(x^2 - 9)$.

14. Graph $y = \ln(x - 2)$ by hand.

15. True or false: $\dfrac{\ln x}{\ln y} = \ln(x - y)$

16. Solve for $x$: $5^x = 41$

17. Solve for $x$: $x - x^2 = \log_5 \frac{1}{25}$

18. Solve for $x$: $\log_2 x + \log_2(x - 3) = 2$

19. Solve for $x$: $\dfrac{e^x + e^{-x}}{3} = 4$

20. Six thousand dollars is deposited into a fund at an annual percentage rate of 13%. Find the time required for the investment to double if the interest is compounded continuously.

21. Use a graphing utility to find the points of intersection of the graphs of $y = \ln(3x)$ and $y = e^x - 4$.

22. Use a graphing utility to find the power model $y = ax^b$ for the data $(1, 1)$, $(2, 5)$, $(3, 8)$, and $(4, 17)$.

# C H A P T E R   5
# Linear Systems and Matrices

# C H A P T E R   5
## Linear Systems and Matrices

### Section 5.1    Solving Systems of Equations

---

- ■ You should be able to solve systems of equations by the method of substitution.
    1. Solve one of the equations for one of the variables.
    2. Substitute this expression into the other equation and solve.
    3. Back-substitute into the first equation to find the value of the other variable.
    4. Check your answer in each of the original equations.
- ■ You should be able to find solutions graphically. (See Example 5 in textbook.)

---

**Solutions to Odd-Numbered Exercises**

**1.** (a) $4(0) - (-3) \overset{?}{=} 1$

　　　$6(0) + (-3) \overset{?}{=} -6$

　　　$3 \neq 1$

　　　$-3 \neq -6$

　　　No, $(0, -3)$ is not a solution.

(b) $4(-1) - (-5) \overset{?}{=} 1$

　　$6(-1) + (-5) \overset{?}{=} -6$

　　　$1 = 1$

　　　$-11 \neq -6$

　　　No, $(-1, -5)$ is not a solution.

(c) $4\left(-\frac{3}{2}\right) - (3) \overset{?}{=} 1$

　　$6\left(-\frac{3}{2}\right) + (3) \overset{?}{=} -6$

　　　$-9 \neq 1$

　　　$-6 = -6$

　　　No, $\left(-\frac{3}{2}, 3\right)$ is not a solution.

(d) $4\left(-\frac{1}{2}\right) - (-3) \overset{?}{=} 1$

　　$6\left(-\frac{1}{2}\right) + (-3) \overset{?}{=} -6$

　　　$1 = 1$

　　　$-6 = -6$

　　　Yes, $\left(-\frac{1}{2}, -3\right)$ is a solution.

**3.** (a) 　　　$0 \overset{?}{=} -2e^{-2}$

　　　$3(-2) - 0 \overset{?}{=} 2$

　　　$0 \neq -2e^{-2}$

　　　$-6 \neq 2$

　　　No, $(-2, 0)$ is not a solution.

(b) 　　　$-2 \overset{?}{=} -2e^{0}$

　　　$3(0) - (-2) \overset{?}{=} 2$

　　　$-2 = -2$

　　　$2 = 2$

　　　Yes, $(0, -2)$ is a solution.

(c) 　　　$-3 \overset{?}{=} -2e^{0}$

　　　$3(0) - (-3) \overset{?}{=} 2$

　　　$-3 \neq -2$

　　　$3 \neq 2$

　　　No, $(0, -3)$ is not a solution.

(d) 　　　$-5 \overset{?}{=} -2e^{-1}$

　　　$3(-1) - (-5) \overset{?}{=} 2$

　　　$-5 \neq -2e^{-1}$

　　　$2 = 2$

　　　No, $(-1, -5)$ is not a solution.

**5.** $\begin{cases} 2x + y = 6 & \text{Equation 1} \\ -x + y = 0 & \text{Equation 2} \end{cases}$

Solve for $y$ in Equation 1:  $y = 6 - 2x$

Substitute for $y$ in Equation 2:  $-x + (6 - 2x) = 0$

Solve for $x$:  $-3x + 6 = 0 \implies x = 2$

Back-substitute $x = 2$:  $y = 6 - 2(2) = 2$

*Answer:* $(2, 2)$

**7.** $\begin{cases} x - y = -4 & \text{Equation 1} \\ x^2 - y = -2 & \text{Equation 2} \end{cases}$

Solve for $y$ in Equation 1:  $y = x + 4$

Substitute for $y$ in Equation 2:  $x^2 - (x + 4) = -2$

Solve for $x$:  $x^2 - x - 2 = 0 \implies (x + 1)(x - 2) = 0 \implies x = -1, 2$

Back-substitute $x = -1$:  $y = -1 + 4 = 3$

Back-substitute $x = 2$:  $y = 2 + 4 = 6$

*Answers:*  $(-1, 3), (2, 6)$

**9.** $\begin{cases} 3x + y = 2 & \text{Equation 1} \\ x^3 - 2 + y = 0 & \text{Equation 2} \end{cases}$

Solve for $y$ in Equation 1:  $y = 2 - 3x$

Substitute for $y$ in Equation 2:  $x^3 - 2 + (2 - 3x) = 0$

Solve for $x$:  $x^3 - 3x = 0 \implies x(x^2 - 3) = 0 \implies x = 0, \pm\sqrt{3}$

Back-substitute:  $x = 0$:  $y = 2$

$$x = \sqrt{3}: \ y = 2 - 3\sqrt{3}$$
$$x = -\sqrt{3}: \ y = 2 + 3\sqrt{3}$$

Solutions: $(0, 2), \left(\sqrt{3}, 2 - 3\sqrt{3}\right), \left(-\sqrt{3}, 2 + 3\sqrt{3}\right)$

**11.** $\begin{cases} x^2 + y = 0 & \text{Equation 1} \\ x^2 - 4x - y = 0 & \text{Equation 2} \end{cases}$

Solve for $y$ in Equation 1:  $y = -x^2$

Substitute for $y$ in Equation 2:  $x^2 - 4x - (-x^2) = 0$

Solve for $x$:  $2x^2 - 4x = 0 \implies 2x(x - 2) = 0 \implies x = 0, 2$

Back-substitute $x = 0$:  $y = -0^2 = 0$

Back-substitute $x = 2$:  $y = -2^2 = -4$

*Answers:*  $(0, 0), (2, -4)$

**13.** $\begin{cases} -\frac{7}{2}x - y = -18 & \text{Equation 1} \\ 8x^2 - 2y^3 = 0 & \text{Equation 2} \end{cases}$

Solve for $x$ in Equation 1: $-\frac{7}{2}x = y - 18 \implies x = -\frac{2}{7}y + \frac{36}{7}$

Substitute for $x$ in Equation 2: $8\left(-\frac{2}{7}y + \frac{36}{7}\right)^2 - 2y^3 = 0$

Solve for $x$: $-2y^3 + 8\left(\frac{4}{49}y^2 - \frac{144}{49}y + \frac{36^2}{49}\right) = 0$

$$49y^3 - 16y^2 + 576y - 5184 = 0$$

$$(y - 4)(49y^2 + 180y + 1296) = 0$$

Hence, $y = 4$ and $x = -\frac{2}{7}(4) + \frac{36}{7} = 4$.

Solution: $(4, 4)$

**15.** $\begin{cases} x - y = 0 & \text{Equation 1} \\ 5x - 3y = 10 & \text{Equation 2} \end{cases}$

Solve for $y$ in Equation 1: $y = x$

Substitute for $y$ in Equation 2: $5x - 3x = 10$

Solve for $x$: $2x = 10 \implies x = 5$

Back-substitute in Equation 1: $y = x = 5$

*Answer:* $(5, 5)$

**17.** $\begin{cases} 2x - y + 2 = 0 & \text{Equation 1} \\ 4x + y - 5 = 0 & \text{Equation 2} \end{cases}$

Solve for $y$ in Equation 1: $y = 2x + 2$

Substitute for $y$ in Equation 2: $4x + (2x + 2) - 5 = 0$

Solve for $x$: $4x + (2x + 2) - 5 = 0 \implies 6x - 3 = 0 \implies x = \frac{1}{2}$

Back-substitute $x = \frac{1}{2}$: $y = 2x + 2 = 2\left(\frac{1}{2}\right) + 2 = 3$

*Answer:* $\left(\frac{1}{2}, 3\right)$

**19.** $\begin{cases} 1.5x + 0.8y = 2.3 \implies 15x + 8y = 23 \\ 0.3x - 0.2y = 0.1 \implies 3x - 2y = 1 \end{cases}$

Solve for $y$ in Equation 2: $-2y = 1 - 3x$

$$y = \frac{3x - 1}{2}$$

Substitute for $y$ in Equation 1: $15x + 8\left(\frac{3x - 1}{2}\right) = 23$

$$15x + 12x - 4 = 23$$

$$27x = 27$$

$$x = 1$$

Then, $y = \frac{3x - 1}{2} = \frac{3(1) - 1}{2} = 1$.

Solution: $(1, 1)$

**21.** $\begin{cases} \frac{1}{5}x + \frac{1}{2}y = 8 & \text{Equation 1} \\ x + y = 20 & \text{Equation 2} \end{cases}$

Solve for $x$ in Equation 2: $x = 20 - y$

Substitute for $x$ in Equation 1: $\frac{1}{5}(20 - y) + \frac{1}{2}y = 8$

Solve for $y$: $4 + \frac{3}{10}y = 8 \implies y = \frac{40}{3}$

Back-substitute $y = \frac{40}{3}$: $x = 20 - y$

$$= 20 - \frac{40}{3} = \frac{20}{3}$$

*Answer:* $\left(\frac{20}{3}, \frac{40}{3}\right)$

**23.** $\begin{cases} 6x + 5y = -3 & \text{Equation 1} \\ -x - \frac{5}{6}y = -7 & \text{Equation 2} \end{cases}$

Solve for $x$ in Equation 2: $x = 7 - \frac{5}{6}y$

Substitute for $x$ in Equation 1: $6\left(7 - \frac{5}{6}y\right) + 5y = -3$

Solve for $y$: $\quad\quad\quad\quad 42 - 5y + 5y = -3$

$$42 = -3 \quad \text{Inconsistent}$$

No solution.

**25.** $\begin{cases} -\frac{5}{3}x + y = 5 & \text{Equation 1} \\ -5x + 3y = 6 & \text{Equation 2} \end{cases}$

Solve for $y$ in Equation 1: $y = 5 + \frac{5}{3}x$

Substitute for $y$ in Equation 2: $-5x + 3\left(5 + \frac{5}{3}x\right) = 6$

Solve for $x$: $-5x + 15 + 5x = 6$

$$15 \neq 6 \quad \text{Inconsistent}$$

No solution

**27.** $\begin{cases} x^3 - y = 0 & \text{Equation 1} \\ x - y = 0 & \text{Equation 2} \end{cases}$

Solve for $y$ in Equation 2: $y = x$

Substitute for $y$ in Equation 1: $x^3 - x = 0$

Solve for $x$: $x(x - 1)(x + 1) = 0 \implies x = 0, 1, -1$

Back-substitute: $x = 0 \implies y = 0$

$$x = 1 \implies y = 1$$
$$x = -1 \implies y = -1$$

*Answer:* $(0, 0), (1, 1), (-1, -1)$

**29.** $\begin{cases} -x + 2y = 2 \\ 3x + y = 15 \end{cases}$

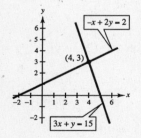

Point of intersection: $(4, 3)$

**31.** $\begin{cases} x - 3y = -2 \\ 5x + 3y = 17 \end{cases}$

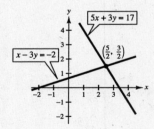

Point of intersection: $\left(\frac{5}{2}, \frac{3}{2}\right)$

**33.** $\begin{cases} x + y = 4 \\ x^2 + y^2 - 4x = 0 \end{cases}$

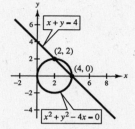

Points of intersection: $(2, 2), (4, 0)$

**35.** $\begin{cases} x - y + 3 = 0 \\ x^2 - 4x + 7 = y \end{cases}$

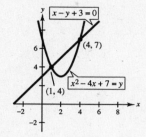

Points of intersection:  $(1, 4), (4, 7)$

**37.** $\begin{cases} 7x + 8y = 24 \implies y_1 = -\frac{7}{8}x + 3 \\ x - 8y = 8 \implies y_2 = \frac{1}{8}x - 1 \end{cases}$

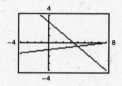

Point of intersection: $\left(4, -\frac{1}{2}\right)$

**39.** $\begin{cases} 2x - y + 3 = 0 \implies y_1 = 2x + 3 \\ x^2 + y^2 - 4x = 0 \implies y_2 = \sqrt{4x - x^2}, y_3 = -\sqrt{4x - x^2} \end{cases}$

No points of intersection

**41.** $\begin{cases} x^2 + y^2 = 8 \implies y_1 = \sqrt{8 - x^2} \text{ and } y_2 = -\sqrt{8 - x^2} \\ y = x^2 \implies y_3 = x^2 \end{cases}$

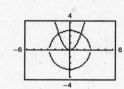

Points of intersection: $\left( \pm\sqrt{\dfrac{-1 + \sqrt{33}}{2}}, \dfrac{-1 + \sqrt{33}}{2} \right) \approx (\pm 1.54, 2.37)$

**43.** $\begin{cases} y = e^x \\ x - y + 1 = 0 \implies y = x + 1 \end{cases}$

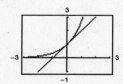

Point of intersection: $(0, 1)$

**45.** $\begin{cases} x + 2y = 8 \qquad \implies y_1 = 4 - x/2 \\ \qquad y = 2 + \ln x \implies y_2 = 2 + \ln x \end{cases}$

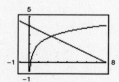

Point of intersection: Approximately $(2.318, 2.841)$

**47.** $\begin{cases} y = \sqrt{x} + 4 \\ y = 2x + 1 \end{cases}$

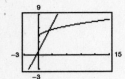

Points of intersection: $\left(\frac{9}{4}, \frac{11}{2}\right)$

**49.** $\begin{cases} x^2 + y^2 = 169 \implies y_1 = \sqrt{169 - x^2} \text{ and} \\ \qquad\qquad\qquad\quad y_2 = -\sqrt{169 - x^2} \\ x^2 - 8y = 104 \implies y_3 = \frac{1}{8}x^2 - 13 \end{cases}$

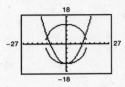

Points of intersection: $(0, -13), (\pm 12, 5)$

**51.** $\begin{cases} y = 2x & \text{Equation 1} \\ y = x^2 + 1 & \text{Equation 2} \end{cases}$

Substitute for $y$ in Equation 2: $2x = x^2 + 1$

Solve for $x$: $x^2 - 2x + 1 = (x - 1)^2 = 0 \implies x = 1$

Back-substitute $x = 1$ in Equation 1: $y = 2x = 2$

*Answer:* $(1, 2)$

**53.** $\begin{cases} 3x - 7y + 6 = 0 & \text{Equation 1} \\ \quad x^2 - y^2 = 4 & \text{Equation 2} \end{cases}$

Solve for $y$ in Equation 1: $y = \dfrac{3x + 6}{7}$

Substitute for $y$ in Equation 2: $x^2 - \left(\dfrac{3x + 6}{7}\right)^2 = 4$

Solve for $x$: $\quad x^2 - \left(\dfrac{9x^2 + 36x + 36}{49}\right) = 4$

$$49x^2 - (9x^2 + 36x + 36) = 196$$

$$40x^2 - 36x - 232 = 0$$

$10x^2 - 9x - 58 = 0 \implies x = \dfrac{9 \pm \sqrt{81 + 40(58)}}{20} \implies x = \dfrac{29}{10}, -2$

Back-substitute $x = \dfrac{29}{10}$: $y = \dfrac{3x + 6}{7} = \dfrac{3(29/10) + 6}{7} = \dfrac{21}{10}$

Back-substitute $x = -2$: $y = \dfrac{3x + 6}{7} = 0$

*Answers:* $\left(\dfrac{29}{10}, \dfrac{21}{10}\right), (-2, 0)$

**55.** $\begin{cases} y = 2x + 1 \\ y = \sqrt{x + 2} \end{cases}$

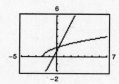

Point of intersection: $\left(\frac{1}{4}, \frac{3}{2}\right)$

**57.** $\begin{cases} y - e^{-x} = 1 \implies y = e^{-x} + 1 \\ y - \ln x = 3 \implies y = \ln x + 3 \end{cases}$

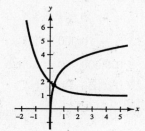

Point of intersection: Approximately (0.287, 1.751)

**59.** $\begin{cases} y = x^3 - 2x^2 + 1 & \text{Equation 1} \\ y = 1 - x^2 & \text{Equation 2} \end{cases}$

Substitute for $y$ in Equation 2: $x^3 - 2x^2 + 1 = 1 - x^2$

Solve for $x$:    $x^3 - x^2 = 0$

$\qquad x^2(x - 1) = 0 \implies x = 0, 1$

Back-substitute: $x = 0 \implies y = 1$

$\qquad\qquad x = 1 \implies y = 0$

*Answers:* (0, 1), (1, 0)

**61.** $\begin{cases} xy - 1 = 0 & \text{Equation 1} \\ 2x - 4y + 7 = 0 & \text{Equation 2} \end{cases}$

Solve for $y$ in Equation 1: $y = \dfrac{1}{x}$

Substitute for $y$ in Equation 2: $2x - 4\left(\dfrac{1}{x}\right) + 7 = 0$

Solve for $x$: $2x^2 - 4 + 7x = 0 \implies (2x - 1)(x + 4) = 0 \implies x = \dfrac{1}{2}, -4$

Back-substitute $x = \dfrac{1}{2}$: $y = \dfrac{1}{1/2} = 2$

Back-substitute $x = -4$: $y = \dfrac{1}{-4} = -\dfrac{1}{4}$

*Answers:* $\left(\dfrac{1}{2}, 2\right), \left(-4, -\dfrac{1}{4}\right)$

**63.**    $C = 8650x + 250,000, \quad R = 9950x$

$\qquad R = C$

$9950x = 8650x + 250,000$

$1300x = 250,000$

$\qquad x \approx 192 \text{ units}$

$R \approx \$1,910,400$

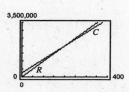

**65.** $C = 5.5\sqrt{x} + 10,000$, $R = 3.29x$

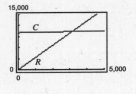

$$R = C$$

$$3.29x = 5.5\sqrt{x} + 10,000$$

$$3.29x - 10,000 = 5.5\sqrt{x}$$

$$10.8241x^2 - 65,800x + 100,000,000 = 30.25x$$

$$10.8241x^2 - 65,830.25x + 100,000,000 = 0$$

$$x \approx 3133 \text{ units}$$

In order for the revenue to break even with the cost, 3133 units must be sold, $R \approx \$10,308$.

**67.** (a) $C = 35.45x + 16,000$

$R = 55.95x$

(b)

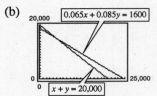

$C = R$ for $x \approx 780$ units

(c) $35.45x + 16,000 = 55.95x$

$$16,000 = 20.5x$$

$$x = \frac{16,000}{20.5} \approx 780 \text{ units}$$

**69.** $0.06x = 0.03x + 350$

$0.03x = 350$

$x \approx \$11,666.67$

To make the straight commission offer better, you would have to sell more than $\$11,666.67$ per week.

**71.** (a) $\begin{cases} x + y = 20,000 \\ 0.065x + 0.085y = 1600 \end{cases}$

(c) The curves intersect at $x = 5000$. Thus, $\$5000$ should be invested at 6.5%.

(b)

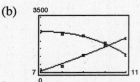

As $x$ increases, $y$ decreases and the amount of interest decreases.

**73.** (a) $F_{\text{VCR}} = -107.86x^2 + 1583.6x - 3235$

$F_{\text{DVD}} = 21.14x^2 + 143.8x - 1939$

(b)

3500

7          11
0

(c) The curves intersect at $x = 10.17$, so DVD sales exceed VCR sales in 2000.

(d) $21.14x^2 + 143.8x - 1939 > 107.86x^2 + 1583.6x - 3235$

$$129x^2 - 1439.8x + 1296 > 0$$

Solving this inequality for $x$, you obtain $x > 10.17$.

(e) The answers are the same.

**75.** $2l + 2w = 30 \Rightarrow l + w = 15$

$\quad l = w + 3 \Rightarrow (w + 3) + w = 15$

$\qquad\qquad\qquad\qquad 2w = 12$

$\qquad\qquad\qquad\qquad\ w = 6$

$l = w + 3 = 9$

Dimensions: 6 meters $\times$ 9 meters

**77.** $2l + 2w = 40 \Rightarrow l + w = 20 \Rightarrow w = 20 - l$

$\quad lw = 96 \Rightarrow l(20 - l) = 96$

$\qquad\qquad\qquad 20l - l^2 = 96$

$\qquad\qquad\qquad 0 = l^2 - 20l + 96$

$\qquad\qquad\qquad 0 = (l - 8)(l - 12)$

$\qquad\qquad\qquad l = 8 \text{ or } l = 12$

$l = 12, w = 8$

Since the length is supposed to be greater than the width, we have $l = 12$ miles and $w = 8$ miles.

**79.** False. You could solve for $x$ first.

**81.** The system has no solution if you arrive at a false statement, ie. $4 = 8$, or you have a quadratic equation with a negative discriminant, which would yield imaginary roots.

**83.** (a) The line $y = 2x$ intersects the parabola $y = x^2$ at two points, $(0, 0)$ and $(2, 4)$.

(b) The line $y = 0$ intersects $y = x^2$ at $(0, 0)$ only.

(c) The line $y = x - 2$ does not intersect $y = x^2$.

(Other answers possible.)

**85.** $(-2, 7), (5, 5)$

$m = \dfrac{5 - 7}{5 - (-2)} = -\dfrac{2}{7}$

$\qquad y - 7 = -\dfrac{2}{7}(x - (-2))$

$\quad 7y - 49 = -2x - 4$

$2x + 7y - 45 = 0$

**87.** $(6, 3), (10, 3)$

$m = \dfrac{3 - 3}{10 - 6} = 0 \Rightarrow$ The line is horizontal.

$y = 3 \Rightarrow y - 3 = 0$

**89.** $\left(\dfrac{3}{5}, 0\right), (4, 6)$

$m = \dfrac{6 - 0}{4 - \frac{3}{5}} = \dfrac{6}{\frac{17}{5}} = \dfrac{30}{17}$

$\qquad y - 6 = \dfrac{30}{17}(x - 4)$

$\quad 17y - 102 = 30x - 120$

$\qquad\qquad\ 0 = 30x - 17y - 18$

**91.** Domain: all $x \neq 6$

Vertical asymptotes: $x = 6$

Horizontal asymptote: $y = 0$

**93.** Domain: all $x \neq \pm 4$

Vertical asymptotes: $x = \pm 4$

Horizontal asymptote: $y = 1$

# Section 5.2  Systems of Linear Equations in Two Variables

■  You should be able to solve a linear system by the method of elimination.

  1. Obtain coefficients for either $x$ or $y$ that differ only in sign. This is done by multiplying all the terms of one or both equations by appropriate constants.

  2. Add the equations to eliminate one of the variables and then solve for the remaining variable.

  3. Use back-substitution into either original equation and solve for the other variable.

  4. Check your answer.

■  You should know that for a system of two linear equations, one of the following is true.

  (a) There are infinitely many solutions; the lines are identical. The system is consistent.

  (b) There is no solution; the lines are parallel. The system is inconsistent.

  (c) There is one solution; the lines intersect at one point. The system is consistent.

## Solutions to Odd-Numbered Exercises

**1.** $\begin{cases} 2x + y = 5 & \text{Equation 1} \\ x - y = 1 & \text{Equation 2} \end{cases}$

Add to eliminate $y$: $3x = 6 \implies x = 2$

Substitute $x = 2$ in Equation 2:
$2 - y = 1 \implies y = 1$

*Answer:* $(2, 1)$

**3.** $\begin{cases} x + y = 0 & \text{Equation 1} \\ 3x + 2y = 1 & \text{Equation 2} \end{cases}$

Multiply Equation 1 by $-2$: $-2x - 2y = 0$

Add this to Equation 2 to eliminate $y$: $x = 1$

Substitute $x = 1$ in Equation 1:
$1 + y = 0 \implies y = -1$

*Answer:* $(1, -1)$

**5.** $\begin{cases} x - y = 2 & \text{Equation 1} \\ -2x + 2y = 5 & \text{Equation 2} \end{cases}$

Multiply Equation 1 by 2: $2x - 2y = 4$

Add this to Equation 2: $0 = 9$

There are no solutions.

**7.** $\begin{cases} x + 2y = 4 & \text{Equation 1} \\ x - 2y = 1 & \text{Equation 2} \end{cases}$

Add to eliminate $y$:

$$2x = 5$$

$$x = \tfrac{5}{2}$$

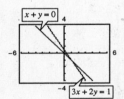

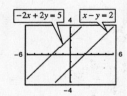

Substitute $x = \frac{5}{2}$ in Equation 1:

$$\tfrac{5}{2} + 2y = 4 \implies y = \tfrac{3}{4}$$

*Answer:* $\left(\tfrac{5}{2}, \tfrac{3}{4}\right)$

**9.** $\begin{cases} 2x + 3y = 18 & \text{Equation 1} \\ 5x - y = 11 & \text{Equation 2} \end{cases}$

Multiply Equation 2 by 3: $15x - 3y = 33$

Add this to Equation 1 to eliminate $y$:

$17x = 51 \implies x = 3$

Substitute $x = 3$ in Equation 1:

$6 + 3y = 18 \implies y = 4$

*Answer:* $(3, 4)$

**11.** $\begin{cases} 3r + 2s = 10 & \text{Equation 1} \\ 2r + 5s = 3 & \text{Equation 2} \end{cases}$

Multiply Equation 1 by 2 and Equation 2 by $-3$:

$6r + 4s = 20$

$-6r - 15s = -9$

Add to eliminate $r$:

$-11s = 11 \implies s = -1$

Substitute $s = -1$ in Equation 1:

$3r + 2(-1) = 10 \implies r = 4$

*Answer:* $(4, -1)$

**13.** $\begin{cases} 5u + 6v = 24 & \text{Equation 1} \\ 3u + 5v = 18 & \text{Equation 2} \end{cases}$

Multiply Equation 1 by 3 and Equation 2 by $(-5)$: $\quad 15u + 18v = 72$

$\qquad\qquad -15u - 25v = -90$

Add to eliminate $u$: $-7v = -18 \implies v = \frac{18}{7}$

Substitute $v = \frac{18}{7}$ in Equation 2: $3u + 5\left(\frac{18}{7}\right) = 18 \implies u = \frac{12}{7}$

*Answer:* $\left(\frac{12}{7}, \frac{18}{7}\right)$

**15.** $\begin{cases} 1.8x + 1.2y = 4 & \text{Equation 1} \\ 9x + 6y = 3 & \text{Equation 2} \end{cases}$

Multiply Equation 1 by $(-5)$: $-9x - 6y = -20$

Add this to Equation 2: $0 = -17$

*Inconsistent; no solution*

**17.** $\begin{cases} 2x - 5y = 0 \implies y = \frac{2}{5}x & \text{passes through } (0, 0) \\ x - y = 3 \implies y = x - 3 & y\text{-intercept } (0, -3) \end{cases}$

Matches (b)

**19.** $\begin{cases} 2x - 5y = 0 \implies y = \frac{2}{5}x & \text{passes through } (0, 0) \\ 2x - 3y = -4 \implies y = \frac{2}{3}x + \frac{4}{3} & y\text{-intercept } \left(0, \frac{4}{3}\right) \end{cases}$

Matches (c)

**21.** $\begin{cases} 4x + 3y = 3 & \text{Equation 1} \\ 3x + 11y = 13 & \text{Equation 2} \end{cases}$

Multiply Equation 1 by 3 and Equation 2 by $-4$:

$\begin{cases} 12x + 9y = 9 \\ -12x - 44y = -52 \end{cases}$

Add to eliminate $x$: $-35y = -43 \implies y = \frac{43}{35}$

Substitute $y = \frac{43}{35}$ into Equation 1:

$4x + 3\left(\frac{43}{35}\right) = 3 \implies x = -\frac{6}{35}$

*Answer:* $\left(-\frac{6}{35}, \frac{43}{35}\right)$

**23.** $\begin{cases} \frac{2}{5}x - \frac{3}{2}y = 4 & \text{Equation 1} \\ \frac{1}{5}x - \frac{3}{4}y = -2 & \text{Equation 2} \end{cases}$

Multiply Equation 2 by $-2$ and add to Equation 1:

$0 = 8$

Inconsistent. no solution

**25.** $\begin{cases} \frac{3}{4}x + y = \frac{1}{8} & \text{Equation 1} \\ \frac{9}{4}x + 3y = \frac{3}{8} & \text{Equation 2} \end{cases}$

Multiply Equation 1 by $-3$: $-\frac{9}{4}x - 3y = -\frac{3}{8}$

Add this to Equation 2:          $0 = 0$

There are an infinite number of solutions.

The solutions consist of all $(x, y)$ satisfying

$\frac{3}{4}x + y = \frac{1}{8}$, or $6x + 8y = 1$.

**27.** $\begin{cases} \dfrac{x+3}{4} + \dfrac{y-1}{3} = 1 & \text{Equation 1} \\ \quad\quad 2x - y = 12 & \text{Equation 2} \end{cases}$

Multiply Equation 1 by 12 and Equation 2 by 4

$\begin{cases} 3x + 4y = 7 \\ 8x - 4y = 48 \end{cases}$

Add to eliminate $y$: $11x = 55 \implies x = 5$

Substitute $x = 5$ into Equation 2:

$2(5) - y = 12 \implies y = -2$

*Answer:* $(5, -2)$

**29.** $\begin{cases} 2.5x - 3y = 1.5 & \text{Equation 1} \\ 10x - 12y = 6 & \text{Equation 2 multiplied by 5} \end{cases}$

Multiply Equation 1 by $(-4)$:

$-10x + 12y = -6$

Add this to Equation 2 to eliminate $x$:

$0 = 0$

The solution set consists of all points lying on the line

$10x - 12y = 6$.

All points on the line

$5x - 6y = 3$

Let $x = a$, then $y = \frac{5}{6}a - \frac{1}{2}$.

*Answer:* $\left(a, \frac{5}{6}a - \frac{1}{2}\right)$, where $a$ is any real number.

**31.** $\begin{cases} 0.2x - 0.5y = -27.8 & \text{Equation 1} \\ 0.3x + 0.4y = 68.7 & \text{Equation 2} \end{cases}$

Multiply Equation 1 by 40 and Equation 2 by 50:

$\begin{cases} 8x - 20y = -1112 \\ 15x + 20y = 3435 \end{cases}$

Adding the equation eliminates $y$:

$23x = 2323 \implies x = 101$

Substitute $x = 101$ into Equation 1:

$8(101) - 20y = -1112 \implies y = 96$

Solution: $(101, 96)$

**33.** $\begin{cases} 2x - 5y = 0 \implies y = \frac{2}{5}x \\ x - y = 3 \implies y = x - 3 \end{cases}$

The system is consistent. There is one solution, $(5, 2)$.

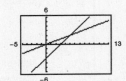

**35.** $\begin{cases} \frac{3}{5}x - y = 3 \implies y = \frac{3}{5}x - 3 \\ -3x + 5y = 9 \implies y = \frac{1}{5}(3x + 9) = \frac{3}{5}x + \frac{9}{5} \end{cases}$

The lines are parallel. The system is inconsistent.

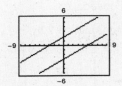

**37.** $\begin{cases} 8x - 14y = 5 \implies y = (8x - 5)/14 = \frac{4}{7}x - \frac{5}{14} \\ 2x - 3.5y = 1.25 \implies y = (2x - 1.25)/3.5 = \frac{4}{7}x - \frac{5}{14} \end{cases}$

The system is consistent. The solution set consists of all points on the line $y = \frac{4}{7}x - \frac{5}{14}$, or $8x - 14y = 5$.

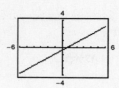

**39.** $\begin{cases} 6y = 42 \implies y = 7 \\ 6x - y = 16 \implies y = 6x - 16 \end{cases}$

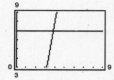

Solution: $\left(\frac{23}{6}, 7\right) \approx (3.833, 7)$

**41.** $\begin{cases} \frac{3}{2}x - \frac{1}{5}y = 8 \\ -2x + 3y = 3 \end{cases}$

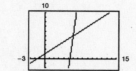

Solution: $(6, 5)$

**43.** $\begin{cases} 0.5x + 2.2y = 9 \\ 6x + 0.4y = -22 \end{cases}$

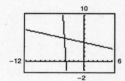

Solution: $(-4, 5)$

**45.** $\begin{cases} 3x - 5y = 7 & \text{Equation 1} \\ 2x + y = 9 & \text{Equation 2} \end{cases}$

Multiply Equation 2 by 5:

$10x + 5y = 45$

Add this to Equation 1:

$13x = 52 \implies x = 4$

Back-substitute $x = 4$ into Equation 2:

$2(4) + y = 9 \implies y = 1$

Solution: $(4, 1)$

**47.** $\begin{cases} y = 4x + 3 \\ y = -5x - 12 \end{cases}$

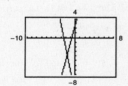

The lines intersect at

$\left(-\frac{5}{3}, -\frac{11}{3}\right) \approx (-1.667, -3.667)$

**49.** $\begin{cases} x - 5y = 21 \\ 6x + 5y = 21 \end{cases}$

Adding the equations, $7x = 42 \implies x = 6$.

Back-substituting, $x - 5y = 6 - 5y = 21 \implies$

$-5y = 15 \implies y = -3$

Solution: $(6, -3)$

**51.** $\begin{cases} -2x + 8y = 19 & \text{Equation 1} \\ y = x - 3 & \text{Equation 2} \end{cases}$

Substituting into Equation 1,

$-2x + 8(x - 3) = 19 \implies 6x = 43$

$\implies x = \frac{43}{6}$

Back-substituting, $y = x - 3 = \frac{43}{6} - 3 = \frac{25}{6}$

Solution: $\left(\frac{43}{6}, \frac{25}{6}\right)$

**53.** There are infinitely many systems that have the solution $(0, 8)$. One possible system is

$$\begin{cases} x + y = 8 \\ -x + y = 8 \end{cases}$$

**55.** There are infinitely many systems that have the solution $\left(3, \frac{5}{2}\right)$. One possible system is:

$$2(3) + 2\left(\tfrac{5}{2}\right) = 11 \implies 2x + 2y = 11$$

$$3 - 4\left(\tfrac{5}{2}\right) = -7 \implies x - 4y = -7$$

**57.** Demand = Supply

$$50 - 0.5x = 0.125x$$

$$50 = 0.625x$$

$$x = 80 \text{ units}$$

$$p = \$10$$

*Answer:* $(80, 10)$

**59.** Demand = Supply

$$140 - 0.00002x = 80 + 0.00001x$$

$$60 = 0.00003x$$

$$x = 2,000,000 \text{ units}$$

$$p = \$100.00$$

*Answer:* $(2,000,000, 100)$

**61.** Let $x$ = the ground speed and $y$ = the wind speed.

$$\begin{cases} 3.6(x - y) = 1800 \quad \text{Equation 1} \\ 3(x + y) = 1800 \quad \text{Equation 2} \end{cases}$$

$$\begin{aligned} x - y &= 500 \\ x + y &= 600 \\ \hline 2x \phantom{+ y} &= 110 \\ x \phantom{+ y} &= 550 \end{aligned}$$

Substituting $x = 550$ in Equation 2: $550 + y = 600$

$$y = 50$$

*Answer:* $x = 550$ mph, $y = 50$ mph

**63.** (a) Let $x$ = the number of liters of 40%

$y$ = the number of liters of 65%

$$\begin{cases} x + y = 20 \\ 0.4x + 0.65y = 0.5(20) = 10 \end{cases}$$

(b) $y_1 = 20 - x$

$$y_2 = \frac{10 - 0.4x}{0.65}$$

As $x$ increases, $y$ decreases

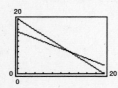

(c) The lines intersect at $(12, 8)$.

12 liters of 40% solution, 8 liters of 65% solution

**65.** Let $x$ = amount invested at 7.5%

Let $y$ = amount invested at 6%

$$\begin{cases} x + y = 15,000 \\ 0.075x + 0.06y = 990 \end{cases}$$

From Equation 1, $y = 15,000 - x$. Then,

$$0.075x + 0.06(15,000 - x) = 990$$

$$0.015x = 90 \implies x = 6000, y = 9000$$

$9000 in the 6% fund.

**67.** Let $x$ = number of adult tickets sold, $y$ = number of child tickets sold.

$$\begin{cases} x + y = 500 \\ 7.5x + 4y = \$3312.50 \end{cases}$$

Equation 1

Equation 2

$$\begin{aligned} -4x - 4y &= -2000.00 \\ 7.5x + 4y &= \underline{\phantom{-}3312.50} \\ 3.5x &= 1312.50 \\ x &= 375 \\ 375 + y &= 500 \\ y &= 125 \end{aligned}$$

*Answer:* $x = 375$ adult tickets, $y = 125$ child tickets

**69.** $\begin{cases} 5b + 10a = 20.2 \implies -10b - 20a = -40.4 \\ 10b + 30a = 50.1 \implies \underline{\phantom{-}10b + 30a = \phantom{-}50.1} \end{cases}$

$$\begin{aligned} 10a &= 9.7 \\ a &= 0.97 \\ b &= 2.10 \end{aligned}$$

Least squares regression line: $y = 0.97x + 2.10$

**71.** (a) $\begin{cases} 4b + 7a = 174 \implies 28b + 49a = 1218 \\ 7b + 13.5a = 322 \implies -28b - 54a = -1288 \end{cases}$

Adding, $-5a = -70 \implies a = 14$, $b = 19$.

Thus, $y = 14x + 19$

(c)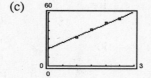

(b) Using a graphing utility, you obtain $y = 14x + 19$.

(d) If $x = 1.6$, (160 pounds/acre),
$y = 14(1.6) + 19 = 41.4$ bushels per acre.

**73.** True. A consistent linear system has either one solution or an infinite number of solutions.

**75.** $\begin{cases} 100y - x = 200 \\ 99y - x = -198 \end{cases}$

Equation 1

Equation 2

Subtract Equation 2 from Equation 1 to eliminate

$x$: $y = 398$

Substitute $y = 398$ into Equation 1:

$100(398) - x = 200 \implies x = 39,600$

Solution: $(39,600, 398)$

The lines are not parallel. The scale on the axes must be changed to see the point of intersection.

**77.** No, it is not possible for a consistent system of linear equations to have exactly two solutions. Either the lines will intersect once or they will coincide and then the system would have infinite solutions.

**79.** $\begin{cases} 4x - 8y = -3 & \text{Equation 1} \\ 2x + ky = \phantom{-}16 & \text{Equation 2} \end{cases}$

Multiply Equation 2 by $-2$: $-4x - 2ky = -32$

Add this to Equation 1: $-8y - 2ky = -35$

The system is inconsistent if $-8y - 2ky = 0$.

This occurs when $k = -4$. Note that for $k = -4$, the two original equations represent parallel lines.

**81.** Subtracting the two equations:

$$vxe^x - v(x + 1)e^x = -e^x \ln x$$

$$vxe^x - vxe^x - ve^x = -e^x \ln x$$

$$ve^x = e^x \ln x$$

$$v = \ln x$$

Finally, $ue^x + vxe^x = ue^x + \ln x \cdot x \cdot e^x = 0$

$$\implies ue^x = -x \ln x \cdot e^x$$

$$\implies u \phantom{e^x} = -x \ln x$$

**83.** $-11 - 6x \geq 33$

$\phantom{-11} -6x \geq 44$

$\phantom{-11-6} x \leq -\frac{44}{6} = -\frac{22}{3}$

**85.** $|x - 8| < 10$

$\phantom{|x-8|} -10 < x - 8 < 10$

$\phantom{|x-8|-1} -2 < x < 18$

**87.** $2x^2 + 3x - 35 < 0$

$(2x - 7)(x + 5) < 0$

Critical numbers: $\frac{7}{2}, -5$.
Testing the three

intervals, $-5 < x < \frac{7}{2}$.

**89.** $\ln x + \ln 6 = \ln 6x$

**91.** $\log_9 12 - \log_9 x = \log_9 \dfrac{12}{x}$

# Section 5.3    Multivariable Linear Systems

---

- ■   You should know the operations that lead to equivalent systems of linear equations:
  - (a) Interchange any two equations.
  - (b) Multiply all terms of an equation by a nonzero constant.
  - (c) Replace an equation by the sum of itself and a constant multiple of any other equation in the system.
- ■   You should be able to use the method of elimination.

---

**Solutions to Odd-Numbered Exercises**

**1.** (a) $3(2) - 5 + 0 \overset{?}{=} 1$    Yes

$\phantom{(a)} 2(2) - 3(0) \overset{?}{=} -14$    No

$\phantom{(a)} 5(5) + 2(0) \overset{?}{=} 8$    No

No, $(2, 5, 0)$ is not a solution.

 (b) $3(-2) - 0 + 4 \overset{?}{=} 1$    No

$\phantom{(b)} 2(-2) - 3(4) \overset{?}{=} -14$    No

$\phantom{(b)} 5(0) + 2(4) \overset{?}{=} 8$    Yes

No, $(-2, 0, 4)$ is not a solution.

**—CONTINUED—**

**1. —CONTINUED—**

(c) $3(0) - (-1) + 3 \stackrel{?}{=} 1$     No

$2(0) - 3(3) \stackrel{?}{=} -14$  No

$5(-1) + 2(3) \stackrel{?}{=} 8$     No

No, $(0, -1, 3)$ is not a solution.

(d) $3(-1) - 0 + 4 \stackrel{?}{=} 1$     Yes

$2(-1) - 3(4) \stackrel{?}{=} -14$  Yes

$5(0) + 2(4) \stackrel{?}{=} 8$     Yes

Yes, $(-1, 0, 4)$ is a solution.

**3.** (a)  $4(0) + 1 - 1 \stackrel{?}{=} 0$     Yes

$-8(0) - 6(1) + 1 \stackrel{?}{=} -\frac{7}{4}$  No

$3(0) - 1 \stackrel{?}{=} -\frac{9}{4}$  No

No, $(0, 1, 1)$ is not a solution.

(b)  $4\left(-\frac{3}{2}\right) + \frac{5}{4} - \left(-\frac{5}{4}\right) \stackrel{?}{=} 0$   No

$-8\left(-\frac{3}{2}\right) - 6\left(\frac{5}{4}\right) + \left(-\frac{5}{4}\right) \stackrel{?}{=} -\frac{7}{4}$  No

$3\left(-\frac{3}{2}\right) - \left(\frac{5}{4}\right) \stackrel{?}{=} -\frac{9}{4}$  No

No, $\left(-\frac{3}{2}, \frac{5}{4}, -\frac{5}{4}\right)$ is not a solution.

(c)  $4\left(-\frac{1}{2}\right) + \frac{3}{4} - \left(-\frac{5}{4}\right) \stackrel{?}{=} 0$   Yes

$-8\left(-\frac{1}{2}\right) - 6\left(\frac{3}{4}\right) - \frac{5}{4} \stackrel{?}{=} -\frac{7}{4}$  Yes

$3\left(-\frac{1}{2}\right) - \frac{3}{4} \stackrel{?}{=} -\frac{9}{4}$  Yes

Yes, $\left(-\frac{1}{2}, \frac{3}{4}, -\frac{5}{4}\right)$ is a solution.

(d)  $4\left(-\frac{1}{2}\right) + 2 - 0 \stackrel{?}{=} 0$   Yes

$-8\left(-\frac{1}{2}\right) - 6(2) + 0 \stackrel{?}{=} -\frac{7}{2}$  No

$3\left(-\frac{1}{2}\right) - 2 \stackrel{?}{=} -\frac{9}{4}$  No

No, $\left(-\frac{1}{2}, 2, 0\right)$ is not a solution.

**5.**  $\begin{cases} 2x - y + 5z = 24 & \text{Equation 1} \\ y + 2z = 4 & \text{Equation 2} \\ z = 6 & \text{Equation 3} \end{cases}$

Back-substitute $z = 6$ into Equation 2

$y + 2(6) = 4$

$y = -8$

Back-substitute $y = -8$ and $z = 6$ into Equation 1

$2x - (-8) + 5(6) = 24$

$2x = -14$

$x = -7$

*Answer:* $(-7, -8, 6)$

**7.**  $\begin{cases} 2x + y - 3z = 10 & \text{Equation 1} \\ y + z = 12 & \text{Equation 2} \\ z = 2 & \text{Equation 3} \end{cases}$

Back-substitute $z = 2$ into Equation 2

$y + 2 = 12$

$y = 10$

Back-substitute $y = 10$ and $z = 2$ into Equation 1

$2x + 10 - 3(2) = 10$

$2x = 6$

$x = 3$

*Answer:* $(3, 10, 2)$

**9.**  $\begin{cases} 4x - 2y + z = 8 & \text{Equation 1} \\ -y + z = 4 & \text{Equation 2} \\ z = 2 & \text{Equation 3} \end{cases}$

Back-substitute $z = 2$ into Equation 2

$-y + 2 = 4$

$y = -2$

Back-substitute $y = -2$ and $z = 2$ into Equation 1

$4x - 2(-2) + 2 = 8$

$4x = 2$

$x = \frac{1}{2}$

*Answer:* $\left(\frac{1}{2}, -2, 2\right)$

**11.**  $\begin{cases} x - 2y + 3z = 5 & \text{Equation 1} \\ -x + 3y - 5z = 4 & \text{Equation 2} \\ 2x - 3z = 0 & \text{Equation 3} \end{cases}$

Add Equation 1 to Equation 2.

$y - 2z = 9$   New Equation 2

This is the first step in putting the system in row-echelon form.

$\begin{cases} x - 2y + 3z = 5 \\ y - 2z = 9 \\ 2x - 3z = 0 \end{cases}$

**13.** $\begin{cases} x + y + z = 6 \\ 2x - y + z = 3 \\ 3x \quad\;\; - z = 0 \end{cases}$    Equation 1

                Equation 2

                Equation 3

$\begin{cases} x + y + z = 6 \\ \quad -3y - z = -9 \\ \quad -3y - 4z = -18 \end{cases}$    $(-2)$ Eq. 1 + Eq. 2

                      $(-3)$ Eq. 1 + Eq. 3

$\begin{cases} x + y + z = 6 \\ \quad -3y - z = -9 \\ \quad\quad\;\; -3z = -9 \end{cases}$    $(-1)$ Eq. 2 + Eq. 3

$-3z = -9 \implies z = 3$

$-3y - 3 = -9 \implies y = 2$

$x + 2 + 3 = 6 \implies x = 1$

*Answer:* $(1, 2, 3)$

**15.** $\begin{cases} 2x \quad\quad + 2z = 2 \\ 5x + 3y \quad\quad = 4 \\ \quad\;\; 3y - 4z = 4 \end{cases}$    Equation 1

                Equation 2

                Equation 3

$\begin{cases} x \quad\quad + z = 1 \\ 5x + 3y \quad\quad = 4 \\ \quad\;\; 3y - 4z = 4 \end{cases}$    $\left(\frac{1}{2}\right)$ Eq. 1

$\begin{cases} x \quad\quad + z = 1 \\ \quad 3y - 5z = -1 \\ \quad 3y - 4z = 4 \end{cases}$    $(-5)$ Eq. 1 + Eq. 2

$\begin{cases} x \quad\quad + z = 1 \\ \quad 3y - 5z = -1 \\ \quad\quad\quad z = 5 \end{cases}$    $(-1)$ Eq. 2 + Eq. 3

$3y - 5(5) = -1 \implies y = 8$

$x + 5 = 1 \implies x = -4$

*Answer:* $(-4, 8, 5)$

**17.** $\begin{cases} \quad\quad 6y + 4z = -18 \\ 3x + 3y \quad\quad = 9 \\ 2x \quad\quad - 3z = 12 \end{cases}$    Equation 1

                Equation 2

                Equation 3

$\begin{cases} 3x + 3y \quad\quad = 9 \\ \quad\quad 6y + 4z = -18 \\ 2x \quad\quad - 3z = 12 \end{cases}$    Interchange equations 1 and 2

$\begin{cases} x + y \quad\quad = 3 \\ \quad\quad 6y + 4z = -18 \\ 2x \quad\quad - 3z = 12 \end{cases}$    $\frac{1}{3}$ (New Eq. 1)

$\begin{cases} x + y \quad\quad = 3 \\ \quad\quad 6y + 4z = -18 \\ \quad -2y - 3z = 6 \end{cases}$    $(-2)$ Eq. 1 + Eq. 3

$\begin{cases} x + y \quad\quad = 3 \\ \quad -2y - 3z = 6 \\ \quad\quad 6y + 4z = -18 \end{cases}$    Interchange the equations

$\begin{cases} x + y \quad\quad = 3 \\ \quad -2y - 3z = 6 \\ \quad\quad\quad -5z = 0 \end{cases}$    3 Eq. 2 + Eq. 3

$-5z = 0 \implies z = 0$

$-2y - 3(0) = 6 \implies y = -3$

$x + (-3) = 3 \implies x = 6$

*Answer:* $(6, -3, 0)$

**19.** $\begin{cases} x + y - 2z = 3 \\ 3x - 2y + 4z = 1 \\ 2x - 3y + 6z = 8 \end{cases}$    Interchange the equations.

$\begin{cases} x + y - 2z = 3 \\ \quad -5y + 10z = -8 \\ \quad -5y + 10z = 10 \end{cases}$    $-3$ Eq. 1 + Eq. 2

                      $-2$ Eq. 1 + Eq. 3

$\begin{cases} x + y - 2z = 3 \\ \quad -5y + 10z = -8 \\ \quad\quad\quad\quad 0 = 18 \end{cases}$    $-$Eq. 2 + Eq. 3

No solution, inconsistent

**21.** $\begin{cases} 3x + 3y + 5z = 1 \\ 3x + 5y + 9z = 0 \\ 5x + 9y + 17z = 0 \end{cases}$

$\begin{cases} 6x + 6y + 10z = 2 \\ 3x + 5y + 9z = 0 \\ 5x + 9y + 17z = 0 \end{cases}$    2 Eq. 1

$\begin{cases} x - 3y - 7z = 2 \\ 3x + 5y + 9z = 0 \\ 5x + 9y + 17z = 0 \end{cases}$    $-$Eq. 3 + Eq. 1

$\begin{cases} x - 3y - 7z = 2 \\ 14y + 30z = -6 \\ 24y + 52z = -10 \end{cases}$    $-3$ Eq. 1 + Eq. 2
$-5$ Eq. 1 + Eq. 3

$\begin{cases} x - 3y - 7z = 2 \\ 84y + 180z = -36 \\ 84y + 182z = -35 \end{cases}$    6 Eq. 2
3.5 Eq. 3

$\begin{cases} x - 3y - 7z = 2 \\ 84y + 180z = -36 \\ 2z = 1 \end{cases}$    $-$Eq. 2 + Eq. 3

$2z = 1 \implies z = \frac{1}{2}$

$84y + 180\left(\frac{1}{2}\right) = -36 \implies y = -\frac{3}{2}$

$x - 3\left(-\frac{3}{2}\right) - 7\left(\frac{1}{2}\right) = 2 \implies x = 1$

*Answer:* $\left(1, -\frac{3}{2}, \frac{1}{2}\right)$

**23.** $\begin{cases} x + 2y - 7z = -4 \\ 2x + y + z = 13 \\ 3x + 9y - 36z = -33 \end{cases}$

$\begin{cases} x + 2y - 7z = -4 \\ -3y + 15z = 21 \\ 3y - 15z = -21 \end{cases}$    $-2$ Eq. 1 + Eq. 2
$-3$ Eq. 1 + Eq. 3

$\begin{cases} x + 2y - 7z = -4 \\ -3y + 15z = 21 \\ 0 = 0 \end{cases}$    Eq. 2 + Eq. 3

$\begin{cases} x + 2y - 7z = -4 \\ y - 5z = -7 \end{cases}$    $-\frac{1}{3}$ Eq. 2

$\begin{cases} x + 3z = 10 \\ y - 5z = -7 \end{cases}$    $-2$ Eq. 2 + Eq. 1

Let $z = a$, then:

$y = 5a - 7$

$x = -3a + 10$

*Answer:* $(-3a + 10, 5a - 7, a)$

**25.** $\begin{cases} 3x - 3y + 6z = 6 \\ x + 2y - z = 5 \\ 5x - 8y + 13z = 7 \end{cases}$

$\begin{cases} x - y + 2z = 2 \\ x + 2y - z = 5 \\ 5x - 8y + 13z = 7 \end{cases}$

$\begin{cases} x - y + 2z = 2 \\ 3y - 3z = 3 \\ -3y + 3z = -3 \end{cases}$

$\begin{cases} x - y + 2z = 2 \\ y - z = 1 \\ 0 = 0 \end{cases}$

$\begin{cases} x + z = 3 \\ y - z = 1 \end{cases}$

Let $z = a$, then:

$y = a + 1$

$x = -a + 3$

*Answer:* $(-a + 3, a + 1, a)$

**27.** $\begin{cases} x - 2y + 3z = 4 \\ 3x - y + 2z = 0 \\ x + 3y - 4z = -2 \end{cases}$    Equation 1
Equation 2
Equation 3

$\begin{cases} x - 2y + 3z = 4 \\ 5y - 7z = -12 \\ 5y - 7z = -6 \end{cases}$    $-3$ Eq. 1 + Eq. 2
$-1$ Eq. 1 + Eq. 3

$\begin{cases} x - 2y + 3z = 4 \\ 5y - 7z = -12 \\ 0 = 6 \end{cases}$    $-$Eq. 2 + Eq. 3

No solution, inconsistent.

**29.** $\begin{cases} x + 4z = 1 \\ x + y + 10z = 10 \\ 2x - y + 2z = -5 \end{cases}$

$\begin{cases} x + 4z = 1 \\ y + 6z = 9 \\ -y - 6z = -7 \end{cases}$    $-$Eq. 1 + Eq. 2
$-2$ Eq. + Eq. 3

$\begin{cases} x + 4z = 1 \\ y + 6z = 9 \\ 0 = 2 \end{cases}$    Eq. 2 + Eq. 3

No solution, inconsistent.

**31.** $\begin{cases} x - 2y + 5z = 2 \\ 4x \quad - z = 0 \end{cases}$

$\begin{cases} x - 2y + 5z = 2 \\ \quad 8y - 21z = -8 \end{cases}$    $-4$ Eq. 1 + Eq. 2

$\begin{cases} x - 2y + 5z = 2 \\ \quad y - \frac{21}{8}z = -1 \end{cases}$    $\frac{1}{8}$ Eq. 3

$\begin{cases} x \quad - \frac{1}{4}z = 0 \\ \quad y - \frac{21}{8}z = -1 \end{cases}$    $2$ Eq. 2 + Eq. 1

Let $z = a$. Then $y = \frac{21}{8}a - 1$ and $x = \frac{1}{4}a$

*Solution:* $\left(\frac{1}{4}a, \frac{21}{8}a - 1, a\right)$

**33.** $\begin{cases} 2x - 3y + z = -2 \\ -4x + 9y \quad = 7 \end{cases}$

$\begin{cases} 2x - 3y + z = -2 \\ \quad 3y + 2z = 3 \end{cases}$    $2$ Eq. 1 + Eq. 2

$\begin{cases} 2x \quad + 3z = 1 \\ \quad 3y + 2z = 3 \end{cases}$    Eq. 2 + Eq. 1

Let $z = a$, then:

$y = -\frac{2}{3}a + 1$

$x = -\frac{3}{2}a + \frac{1}{2}$

*Answer:* $\left(-\frac{3}{2}a + \frac{1}{2}, -\frac{2}{3}a + 1, a\right)$

**35.** $\begin{cases} x - 3y + 2z = 18 \\ 5x - 13y + 12z = 80 \end{cases}$    Equation 1 / Equation 2

$\begin{cases} x - 3y + 2z = 18 \\ \quad 2y + 2z = -10 \end{cases}$    $-5$ Eq. 1 + Eq. 2

$\begin{cases} x - 3y + 2z = 18 \\ \quad y + z = -5 \end{cases}$    $\frac{1}{2}$ Eq. 2

$\begin{cases} x \quad + 5z = 3 \\ \quad y + z = -5 \end{cases}$    $3$ Eq. 2 + Eq. 1

Let $z = a$, then $y = -a - 5$, and $x = -5a + 3$.

*Answer:* $(-5a + 3, -a - 5, a)$

**37.** $\begin{cases} x \quad\quad + 3w = 4 \\ \quad 2y - z - w = 0 \\ \quad 3y \quad - 2w = 1 \\ 2x - y + 4z \quad = 5 \end{cases}$

$\begin{cases} x \quad\quad + 3w = 4 \\ \quad 2y - z - w = 0 \\ \quad 3y \quad - 2w = 1 \\ \quad - y + 4z - 6w = -3 \end{cases}$    $-2$ Eq. 1 + Eq. 4

$\begin{cases} x \quad\quad + 3w = 4 \\ \quad y - 4z + 6w = 3 \\ \quad 2y - z - w = 0 \\ \quad 3y \quad - 2w = 1 \end{cases}$    $-$Eq. 4 and interchange the equations

$\begin{cases} x \quad\quad + 3w = 4 \\ \quad y - 4z + 6w = 3 \\ \quad 7z - 13w = -6 \\ \quad 12z - 20w = -8 \end{cases}$    $-2$ Eq. 2 + Eq. 3 / $-3$ Eq. 2 + Eq. 4

$\begin{cases} x \quad\quad + 3w = 4 \\ \quad y - 4z + 6w = 3 \\ \quad z - 3w = -2 \\ \quad 12z - 20w = -8 \end{cases}$    $-\frac{1}{2}$ Eq. 4 + Eq. 3

$\begin{cases} x \quad\quad + 3w = 4 \\ \quad y - 4z + 6w = 3 \\ \quad z - 3w = -2 \\ \quad 16w = 16 \end{cases}$    $-12$ Eq. 3 + Eq. 4

$16w = 16 \implies w = 1$

$z - 3(1) = -2 \implies z = 1$

$y - 4(1) + 6(1) = 3 \implies y = 1$

$x + 3(1) = 4 \implies x = 1$

*Answer:* $(1, 1, 1, 1)$

**39.** There are an infinite number of linear systems that has $(4, -1, 2)$ as their solution. One such system is as follows:

$$3(4) + (-1) - (2) = 9 \implies 3x + y - z = 9$$
$$(4) + 2(-1) - (2) = 0 \implies x + 2y - z = 0$$
$$-(4) + (-1) + 3(2) = 1 \implies -x + y + 3z = 1$$

**41.** There are infinite numbers of linear systems that have $\left(3, -\frac{1}{2}, \frac{7}{4}\right)$ as their solution. One such system is as follows:

$$1(3) + 2\left(-\tfrac{1}{2}\right) + 4\left(\tfrac{7}{4}\right) = 9 \implies x + 2y + 4z = 9$$
$$4\left(-\tfrac{1}{2}\right) + 8\left(\tfrac{7}{4}\right) = 12 \implies 4y + 8z = 12$$
$$4\left(\tfrac{7}{4}\right) = 7 \implies 4z = 7$$

**43.** Plane: $2x + 3y + 4z = 12$

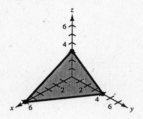

Four points are:

$$(6, 0, 0), \ (0, 4, 0), \ (0, 0, 3), \ (4, 0, 1)$$

**45.** Plane: $2x + y + z = 4$

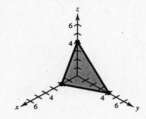

Four points are:

$$(2, 0, 0), \ (0, 4, 0), \ (0, 0, 4), \ (0, 2, 2)$$

**47.** $\dfrac{7}{x^2 - 14x} = \dfrac{7}{x(x - 14)} = \dfrac{A}{x} + \dfrac{B}{x - 14}$

**49.** $\dfrac{12}{x^3 - 10x^2} = \dfrac{12}{x^2(x - 10)} = \dfrac{A}{x} + \dfrac{B}{x^2} + \dfrac{C}{x - 10}$

**51.** $\dfrac{4x^2 + 3}{(x - 5)^3} = \dfrac{A}{(x - 5)} + \dfrac{B}{(x - 5)^2} + \dfrac{C}{(x - 5)^3}$

**53.** (a) $\dfrac{1}{x^2 - 1} = \dfrac{A}{x + 1} + \dfrac{B}{x - 1}$

$$1 = A(x - 1) + B(x + 1) = (A + B)x + (B - A)$$

$$\begin{cases} A + B = 0 \\ -A + B = 1 \end{cases}$$

$$2B = \implies B = \tfrac{1}{2} \implies A = -\tfrac{1}{2}$$

$$\dfrac{1}{x^2 - 1} = \dfrac{-\frac{1}{2}}{x + 1} + \dfrac{\frac{1}{2}}{x - 1} = \dfrac{1}{2}\left[\dfrac{1}{x - 1} - \dfrac{1}{x + 1}\right]$$

(c)

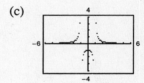

(b) Checking, $\dfrac{1}{2}\left[\dfrac{1}{x - 1} - \dfrac{1}{x + 1}\right] = \dfrac{1}{2}\left[\dfrac{x + 1 - (x - 1)}{x^2 - 1}\right] = \dfrac{1}{x^2 - 1}$

**55.** (a) $\dfrac{1}{x^2 + x} = \dfrac{1}{x(x + 1)} = \dfrac{A}{x} + \dfrac{B}{x + 1}$

$$1 = A(x + 1) + Bx = (A + B)x + A$$

$$\begin{cases} A + B = 0 \\ A = 1 \implies B = -1 \end{cases}$$

$$\dfrac{1}{x^2 + x} = \dfrac{1}{x} + \dfrac{-1}{x + 1} = \dfrac{1}{x} - \dfrac{1}{x + 1}$$

(c)

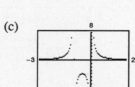

(b) Checking, $\dfrac{1}{x} - \dfrac{1}{x + 1} = \dfrac{(x + 1) - x}{x(x + 1)} = \dfrac{1}{x^2 + x}$

**57.** (a) $\dfrac{1}{2x^2 + x} = \dfrac{1}{x(2x + 1)} = \dfrac{A}{2x + 1} + \dfrac{B}{x}$

$1 = Ax + B(2x + 1) = (A + 2B)x + B$

$\begin{cases} A + 2B = 0 \\ \quad\;\; B = 1 \Rightarrow A = -2 \end{cases}$

$\dfrac{1}{2x^2 + x} = \dfrac{-2}{2x + 1} + \dfrac{1}{x} = \dfrac{1}{x} - \dfrac{2}{2x + 1}$

(b) Checking, $\dfrac{1}{x} - \dfrac{2}{2x + 1} = \dfrac{2x + 1 - 2x}{x(2x + 1)} = \dfrac{1}{2x^2 + x}$

(c)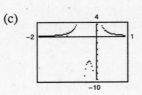

**59.** (a) $\dfrac{5 - x}{2x^2 + x - 1} = \dfrac{5 - x}{(2x - 1)(x + 1)} = \dfrac{A}{2x - 1} + \dfrac{B}{x + 1}$

$5 - x = A(x + 1) + B(2x - 1) = (A + 2B)x + (A - B)$

$\begin{cases} A + 2B = -1 \Rightarrow A = -1 - 2B \\ A - \;\; B = \;\; 5 \end{cases}$

$(-1 - 2B) - B = 5 \Rightarrow B = -2 \quad\text{and}\quad A = 3$

$\dfrac{5 - x}{2x^2 + x - 1} = \dfrac{3}{2x - 1} + \dfrac{-2}{x + 1}$

(b) Checking, $\dfrac{3}{2x - 1} + \dfrac{-2}{x + 1} = \dfrac{3(x + 1) - 2(2x - 1)}{(2x - 1)(x + 1)} = \dfrac{-x + 5}{2x^2 + x - 1}$

(c)

**61.** (a) $\dfrac{x^2 + 12x + 12}{x^3 - 4x} = \dfrac{x^2 + 12x + 12}{x(x - 2)(x + 2)} = \dfrac{A}{x} + \dfrac{B}{x + 2} + \dfrac{C}{x - 2}$

$x^2 + 12x + 12 = A(x + 2)(x - 2) + Bx(x - 2) + Cx(x + 2)$

$\qquad\qquad\qquad\quad = (A + B + C)x^2 + (-2B + 2C)x + (-4A)$

$\begin{cases} A + \;\; B + \;\; C = \;\; 1 \\ \quad\;\; -2B + 2C = 12 \\ -4A \qquad\qquad\;\; = 12 \Rightarrow A = -3 \end{cases}$

$\begin{cases} \;\; B + C = 4 \\ -B + C = 6 \end{cases}$

$2C = 10 \Rightarrow C = 5 \Rightarrow B = -1$

$\dfrac{x^2 + 12x + 12}{x^3 - 4x} = \dfrac{-3}{x} + \dfrac{-1}{x + 2} + \dfrac{5}{x - 2}$

(c)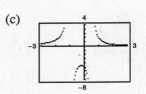

(b) Checking, $\dfrac{-3}{x} + \dfrac{-1}{x + 2} + \dfrac{5}{x - 2} = \dfrac{-3(x^2 - 4) - 1(x)(x - 2) + 5x(x + 2)}{x(x^2 - 4)}$

$\qquad\qquad\qquad\qquad\qquad\qquad = \dfrac{x^2 + 12x + 12}{x^3 - 4x}$

**63.** (a) $\dfrac{4x^2 + 2x - 1}{x^2(x + 1)} = \dfrac{A}{x} + \dfrac{B}{x^2} + \dfrac{C}{x + 1}$

$4x^2 + 2x - 1 = Ax(x + 1) + B(x + 1) + Cx^2$

$\qquad\qquad\quad = (A + C)x^2 + (A + B)x + B$

$\begin{cases} A \quad\ + C = \quad 4 \\ A + B \quad\ = \quad 2 \\ \quad B \quad\ = -1 \end{cases}$

$B = -1 \implies A = 3 \implies C = 1$

$\dfrac{4x^2 + 2x - 1}{x^2(x + 1)} = \dfrac{3}{x} + \dfrac{-1}{x^2} + \dfrac{1}{x + 1}$

(b) Checking, $\dfrac{3}{x} + \dfrac{-1}{x^2} + \dfrac{1}{x + 1} = \dfrac{3x(x + 1) - (x + 1) + x^2}{x^2(x + 1)} = \dfrac{4x^2 + 2x - 1}{x^2(x + 1)}$

(c)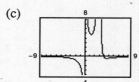

**65.** (a) $\dfrac{27 - 7x}{x(x - 3)^2} = \dfrac{A}{x} + \dfrac{B}{x - 3} + \dfrac{C}{(x - 3)^2}$

$27 - 7x = A(x - 3)^2 + Bx(x - 3) + Cx$

$\qquad\quad = (A + B)x^2 + (-6A - 3B + C)x + 9a$

$\begin{cases} A + B \quad\quad\ = \quad 0 \\ -6A - 3B + C = -7 \\ 9A \quad\quad\quad = \quad 27 \end{cases}$

$A = 3 \implies B = -3 \implies C = -7 + 18 - 9 = 2$

$\dfrac{27 - 7x}{x(x - 3)^2} = \dfrac{3}{x} + \dfrac{-3}{x - 3} + \dfrac{2}{(x - 3)^2}$

(b) Checking, $\dfrac{3}{x} + \dfrac{-3}{x - 3} + \dfrac{2}{(x - 3)^2} = \dfrac{3(x - 3)^2 - 3(x)(x - 3) + 2x}{x(x - 3)^2} = \dfrac{-7x + 27}{x(x - 3)^2}$

(c)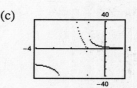

**67.** (a) $\dfrac{2x^3 - x^2 + x + 5}{x^2 + 3x + 2} = 2x - 7 + \dfrac{18x + 19}{(x + 1)(x + 2)}$

$\dfrac{18x + 19}{(x + 1)(x + 2)} = \dfrac{A}{x + 1} + \dfrac{B}{x + 2}$

$18x + 19 = A(x + 2) + B(x + 1) = (A + B)x + (2A + B)$

$\begin{cases} A + B = 18 \\ 2A + B = 19 \end{cases}$

$A = 1 \implies B = 17$

$\dfrac{2x^3 - x^2 + x + 5}{x^2 + 3x + 2} = 2x - 7 + \dfrac{1}{x + 1} + \dfrac{17}{x + 2}$

(b) Checking,

$2x - 7 + \dfrac{1}{x + 1} + \dfrac{17}{x + 2} = \dfrac{(2x - 7)(x^2 + 3x + 2) + (x + 2) + 17(x + 1)}{x^2 + 3x + 2} = \dfrac{2x^3 - x^2 + x + 5}{x^2 + 3x + 2}$

(c)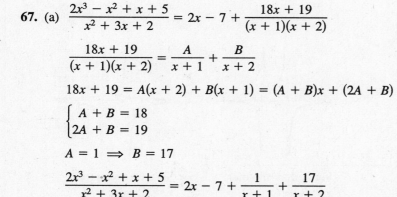

**69. (a)** $\dfrac{x^4}{(x-1)^3} = x + 3 + \dfrac{6x^2 - 8x + 3}{(x-1)^3}$

$\dfrac{6x^2 - 8x + 3}{(x-1)^3} = \dfrac{A}{x-1} + \dfrac{B}{(x-1)^2} + \dfrac{C}{(x-1)^3}$

$6x^2 - 8x + 3 = A(x-1)^2 + B(x-1) + C$

$\qquad\qquad = Ax^2 + (-2A + B)x + (A - B + C)$

$\begin{cases} A & = & 6 \\ -2A + B & = & -8 \\ A - B + C & = & 3 \end{cases}$

$A = 6 \implies B = -8 + 2(6) = 4 \implies C = 3 - 6 + 4 = 1$

$\dfrac{x^4}{(x-1)^3} = \dfrac{6}{x-1} + \dfrac{4}{(x-1)^2} + \dfrac{1}{(x-1)^3} + x + 3$

**(b)** Checking,

$\dfrac{6}{x-1} + \dfrac{4}{(x-1)^2} + \dfrac{1}{(x-1)^3} + x + 3 = \dfrac{6(x-1)^2 + 4(x-1) + 1 + (x-1)^3(x+3)}{(x-1)^3}$

$\qquad\qquad\qquad = \dfrac{6x^2 - 12x + 6 + 4x - 3 + (x^4 - 6x^2 + 8x - 3)}{(x-1)^3}$

$\qquad\qquad\qquad = \dfrac{x^4}{(x-1)^3}$

**(c)**

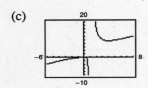

**71.** $\dfrac{x-12}{x(x-4)} = \dfrac{A}{x} + \dfrac{B}{x-4}$

$x - 12 = A(x-4) + Bx$

$\begin{cases} A + B = & 1 \\ -4A & = -12 \end{cases} \implies A = 3, B = -2$

$\dfrac{x-12}{x(x-4)} = \dfrac{3}{x} - \dfrac{2}{x-4}$

$y = \dfrac{x-12}{x(x-4)} \qquad\qquad y = \dfrac{3}{x}, y = -\dfrac{2}{x-4}$

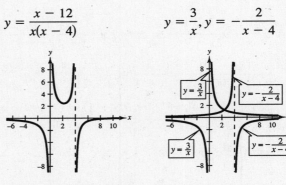

Vertical asymptotes: $\qquad$ Vertical asymptotes:

$x = 0$ and $x = 4$ $\qquad\qquad x = 0$ and $x = 4$

The combination of the vertical asymptotes of the terms of the decompositions are the same as the vertical asymptotes of the rational function.

**73.** $s = \frac{1}{2}at^2 + v_0t + s_0$

(1, 128), (2, 80), (3, 0)

$$\begin{cases} 128 = \frac{1}{2}a + v_0 + s_0 \implies a + 2v_0 + 2s_0 = 256 \\ 80 = 2a + 2v_0 + s_0 \implies 2a + 2v_0 + s_0 = 80 \\ 0 = \frac{9}{2}a + 3v_0 + s_0 \implies 9a + 6v_0 + 2s_0 = 0 \end{cases}$$

Solving this system yields $a = -32$, $v_0 = 0$, $s_0 = 144$.

Thus, $s = \frac{1}{2}(-32)t^2 + (0)t + 144$

$\qquad = -16t^2 + 144.$

**75.** $s = \frac{1}{2}at^2 + v_0t + a_0$

(1, 452), (2, 372), (3, 260)

$$\begin{cases} 452 = \frac{1}{2}a + v_0 + s_0 \implies a + 2v_0 + 2s_0 = 904 \\ 372 = 2a + 2v_0 + s_0 \implies 2a + 2v_0 + s_0 = 372 \\ 260 = \frac{9}{2}a + 3v_0 + s_0 \implies 9a + 6v_0 + 2s_0 = 520 \end{cases}$$

Solving this system yields $a = -32$, $v_0 = -32$, $s_0 = 500$

Thus, $s = \frac{1}{2}(-32)t^2 - 32t + 500$

$\qquad = -16t^2 - 32t + 500$

**77.** $y = ax^2 + bx + c$ passing through $(0, 0)$, $(2, -2)$, $(4, 0)$

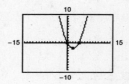

$(0, \ 0)$: $\quad 0 = \qquad\qquad c$

$(2, -2)$: $-2 = 4a + 2b + c \implies -1 = 2a + b$

$(4, \ 0)$: $\quad 0 = 16a + 4b + c \implies 0 = 4a + b$

*Answer:* $a = \frac{1}{2}$, $b = -2$, $c = 0$

The equation of the parabola is $y = \frac{1}{2}x^2 - 2x$.

**79.** $y = ax^2 + bx + c$ passing through $(2, 0)$, $(3, -1)$, $(4, 0)$

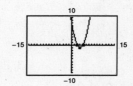

$(2, \ 0)$: $\quad 0 = 4a + 2b + c$

$(3, -1)$: $-1 = 9a + 3b + c \implies -1 = 5a + b$

$(4, \ 0)$: $\quad 0 = 16a + 4b + c \implies 0 = 12a + 2b$

*Answer:* $a = 1$, $b = -6$, $c = 8$

The equation of the parabola is $y = x^2 - 6x + 8$.

**81.** $x^2 + y^2 + Dx + Ey + F = 0$ passing through $(0, 0)$, $(2, 2)$, $(4, 0)$

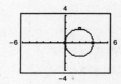

$(0, 0)$: $\qquad\qquad F = 0$

$(2, 2)$: $\ 8 + 2D + 2E + F = 0 \implies D + E = -4$

$(4, 0)$: $16 + 4D \qquad + F = 0 \implies D = -4$ and $E = 0$

The equation of the circle is $x^2 + y^2 - 4x = 0$.

To graph, let $y_1 = \sqrt{4x - x^2}$ and $y_2 = -\sqrt{4x - x^2}$.

**83.** $x^2 + y^2 + Dx + Ey + F = 0$ passing through $(-3, -1), (2, 4), (-6, 8)$

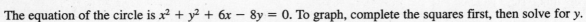

$(-3, -1)$: $10 - 3D - E + F = 0 \implies 10 = 3D + E - F$

$(2, 4)$: $20 + 2D + 4E + F = 0 \implies 20 = -2D - 4E - F$

$(-6, 8)$: $100 - 6D + 8E + F = 0 \implies 100 = 6D - 8E - F$

*Answer:* $D = 6, E = -8, F = 0$

The equation of the circle is $x^2 + y^2 + 6x - 8y = 0$. To graph, complete the squares first, then solve for $y$.

$$(x^2 + 6x + 9) + (y^2 - 8y + 16) = 0 + 9 + 16$$

$$(x + 3)^2 + (y - 4)^2 = 25$$

$$(y - 4)^2 = 25 - (x + 3)^2$$

$$y - 4 = \pm\sqrt{25 - (x + 3)^2}$$

$$y = 4 \pm \sqrt{25 - (x + 3)^2}$$

Let $y_1 = 4 + \sqrt{25 - (x + 3)^2}$ and $y_2 = 4 - \sqrt{25 - (x + 3)^2}$.

**85.** Let $x =$ amount at 8%.

Let $y =$ amount at 9%.

Let $z =$ amount at 10%.

$$\begin{cases} x + y + z = 775{,}000 \\ 0.08x + 0.09y + 0.10z = 67{,}000 \\ x = 4z \end{cases}$$

$$\begin{cases} x + y + z = 775{,}000 \\ 8x + 9y + 10z = 6{,}700{,}000 \\ x - 4z = 0 \end{cases}$$

Solving this system, you obtain

$x \approx 366{,}666.67, y \approx 316{,}666.67, z \approx 91{,}666.67$

*Answer:* $x = \$366{,}666.67$ at 8%

$y = \$316{,}666.67$ at 9%

$z = \$91{,}666.67$ at 10%

**87.** Let $C =$ amount in certificates of deposit

Let $M =$ amount in municipal bonds

Let $B =$ amount in blue chip stocks

Let $G =$ amount in growth or speculative stocks

$$\begin{cases} C + M + B + G = 500{,}000 \\ 0.08C + 0.09M + 0.12B + 0.15G = 0.10(500{,}000) \\ M = \frac{1}{4}(500{,}000) \end{cases}$$

Solving this system, you obtain $M = 125{,}000$ and

$C = 156{,}250 + 0.75s$

$M = 125{,}000$

$B = 218750 - 1.75s$

$G = s$

**89.** Let $x =$ gallons of spray X.

Let $y =$ gallons of spray Y.

Let $z =$ gallons of spray Z.

Chemical A: $\frac{1}{5}x + \frac{1}{2}z = 12$
Chemical B: $\frac{2}{5}x + \frac{1}{2}z = 16$ $\left.\right\} \implies x = 20, z = 16$
Chemical C: $\frac{2}{5}x + y = 26$ $\implies y = 18$

*Answer:* 20 liters of spray X

18 liters of spray Y

16 liters of spray Z

**91.**

|  | Product | |
|---|---|---|
| Truck | A | B |
| Large | 6 | 3 |
| Medium | 4 | 4 |
| Small | 0 | 3 |

Possible solutions:

(1)  4 medium trucks

(2)  2 large trucks, 1 medium truck, 2 small trucks

(3)  3 large trucks, 1 medium truck, 1 small truck

(4)  3 large trucks, 3 small trucks

**93.** $\begin{cases} I_1 - I_2 + I_3 = 0 & \text{Equation 1} \\ 3I_1 + 2I_2 \phantom{+ I_3} = 7 & \text{Equation 2} \\ \phantom{I_1 +} 2I_2 + 4I_3 = 8 & \text{Equation 3} \end{cases}$

$\begin{cases} I_1 - I_2 + I_3 = 0 \\ \phantom{I_1} 5I_2 - 3I_3 = 7 \\ \phantom{I_1} 2I_2 + 4I_3 = 8 \end{cases}$ $\quad -3$ Eq. 1 + Eq. 2

$\begin{cases} I_1 - I_2 + I_3 = 0 \\ \phantom{I_1} 10I_2 - 6I_3 = 14 \\ \phantom{I_1} 10I_2 + 20I_3 = 40 \end{cases}$ $\quad \begin{matrix} 2 \text{ Eq. 2} \\ 5 \text{ Eq. 3} \end{matrix}$

$\begin{cases} I_1 - I_2 + I_3 = 0 \\ \phantom{I_1} 10I_2 - 6I_3 = 14 \\ \phantom{I_1 10I_2} 26I_3 = 26 \end{cases}$ $\quad -$Eq. 2 + Eq. 3

$26I_3 = 26 \implies I_3 = 1$

$10I_2 - 6(1) = 14 \implies I_2 = 2$

$I_1 - 2 + 1 = 0 \implies I_1 = 1$

*Answer:* $I_1 = 1$ ampere, $I_2 = 2$ amperes,

$\phantom{Answer:} I_3 = 1$ ampere

**95.** Least squares regression parabola through
$(-4, 5), (-2, 6), (2, 6), (4, 2)$

$\begin{cases} 4c \phantom{+ 40b} + 40a = 19 \\ \phantom{4c +} 40b \phantom{+ 544a} = -12 \\ 40c \phantom{+ 40b} + 544a = 160 \end{cases}$

Solving this system yields

$a = -\frac{5}{24}, b = -\frac{3}{10}$, and $c = \frac{41}{6}$.

Thus, $y = -\frac{5}{24}x^2 - \frac{3}{10}x + \frac{41}{6}$.

**97.** Least squares regression parabola through
$(0, 0), (2, 2), (3, 6), (4, 12)$

$\begin{cases} 4c + 9b + 29a = 20 \\ 9c + 29b + 99a = 70 \\ 29c + 99b + 353a = 254 \end{cases}$

Solving this system yields

$a = 1, b = -1$, and $c = 0$. Thus, $y = x^2 - x$.

**99.** (a) $\begin{cases} a(30)^2 + b(30) + c = 55 \\ a(40)^2 + b(40) + c = 105 \\ a(50)^2 + b(50) + c = 188 \end{cases}$

Solving the system, $a = 0.165, b = -6.55$ and
$c = 103$

$y = 0.165x^2 - 6.55x + 103$

(b)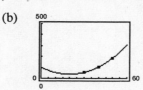

(c) For $x = 70, y = 453$ feet.

**101.** (a) $\dfrac{2000(4 - 3x)}{(11 - 7x)(7 - 4x)} = \dfrac{A}{11 - 7x} + \dfrac{B}{7 - 4x}$, $0 \le x \le 1$

$\qquad 2000(4 - 3x) = A(7 - 4x) + B(11 - 7x)$

$\qquad \begin{cases} -6000 = -4A - 7B \\ 8000 = 7A + 11B \end{cases} \Longrightarrow \begin{matrix} A = -2000 \\ B = 2000 \end{matrix}$

$\qquad \dfrac{2000(4 - 3x)}{(11 - 7x)(7 - 4x)} = \dfrac{-2000}{11 - 7x} + \dfrac{2000}{7 - 4x}$

$\qquad\qquad\qquad\qquad = \dfrac{2000}{7 - 4x} - \dfrac{2000}{11 - 7x}$

(b) $y_1 = \dfrac{2000}{7 - 4x}$

$\quad y_2 = \dfrac{2000}{11 - 7x}$

**103.** False. The coefficient of $y$ in the second equation is not 1.

**105.** False. The correct form is

$$\dfrac{A}{x + 10} + \dfrac{B}{x - 10} + \dfrac{C}{(x - 10)^2}.$$

**107.** $\dfrac{1}{a^2 - x^2} = \dfrac{1}{(a + x)(a - x)} = \dfrac{A}{a + x} + \dfrac{B}{a - x}$

$\qquad 1 = A(a - x) + B(a + x) = (-A + B)x + (Aa + Ba)$

$\qquad \begin{cases} -A + B = 0 \\ Aa + Ba = 1 \end{cases} \Longrightarrow A = \dfrac{1}{2a}, B = \dfrac{1}{2a}$

$\qquad \dfrac{1}{a^2 - x^2} = \dfrac{\frac{1}{2a}}{a + x} + \dfrac{\frac{1}{2a}}{a - x} = \dfrac{1}{2a}\left[\dfrac{1}{a + x} + \dfrac{1}{a - x}\right]$

**109.** $\dfrac{1}{y(a - y)} = \dfrac{A}{y} + \dfrac{B}{a - y}$

$\qquad 1 = A(a - y) + By = (-A + B)y + aA$

$\qquad A = \dfrac{1}{a}, B = \dfrac{1}{a}$

$\qquad \dfrac{1}{y(a - y)} = \dfrac{1}{a}\left(\dfrac{1}{y} + \dfrac{1}{a - y}\right)$

**111.** No, they are not equivalent. In the second system, the constant in the second equation should be $-11$ and the coefficient of $z$ in the third equation should be 2.

**113.** $\begin{cases} y + \quad\ \lambda = \\ \quad x + \ \lambda = \\ x + y - 10 = \end{cases} \Longrightarrow \begin{matrix} x = y = -\lambda \\ \ \\ 2x - 10 = 0 \end{matrix}$

$\qquad\qquad\qquad\qquad x = 5$

$\qquad\qquad\qquad\qquad y = 5$

$\qquad\qquad\qquad\qquad \lambda = -5$

**115.** $\begin{cases} 2x - 2x\lambda = 0 \Longrightarrow x = x\lambda \\ -2y + \lambda = 0 \Longrightarrow 2y = \lambda \\ y - x^2 = 0 \Longrightarrow y = x^2 \end{cases}$

From the first equation, $x = 0$ or $\lambda = 1$.

If $x = 0$, then $y = 0^2 = 0$ and $\lambda = 0$.

If $x \ne 0$, then $\lambda = 1 \Longrightarrow y = \frac{1}{2}$ and $x = \pm\sqrt{\frac{1}{2}}$.

Thus, the solutions are:

(1) $x = y = \lambda = 0$

(2) $x = \dfrac{\sqrt{2}}{2}, y = \dfrac{1}{2}, \lambda = 1$

(3) $x = -\dfrac{\sqrt{2}}{2}, y = \dfrac{1}{2}, \lambda = 1$

**117.** $y = -3x + 7$

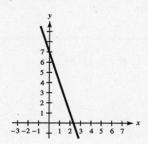

**119.** $y = -2x^2$

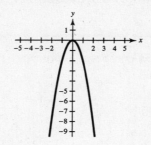

**121.** $y = -x^2(x - 3)$

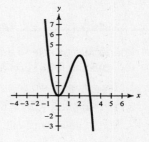

**123.** $f(x) = x^3 + x^2 - 12x = x(x^2 + x - 12) = x(x + 4)(x - 3) \Longrightarrow x = 0, -4, 3$

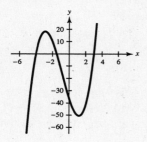

**125.** $f(x) = 2x^3 + 5x^2 - 21x - 36 = (2x + 3)(x + 4)(x - 3) \Longrightarrow x = -\frac{3}{2}, -4, 3$

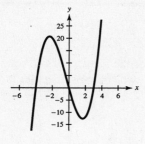

**127.** $y = 4^{-x-4} - 5$

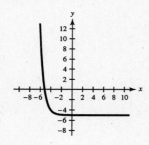

| $x$ | $-7$ | $-6$ | $-5$ | $-4$ | $-3$ | $-1$ | $0$ |
|---|---|---|---|---|---|---|---|
| $y$ | 59 | 11 | $-1$ | $-4$ | $-4.75$ | $-4.98$ | $-4.996$ |

**129.** $y = 2.9^{0.8x} - 3$

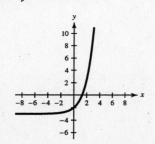

| $x$ | $-3$ | $-2$ | $-1$ | $0$ | $1$ | $2$ | $3$ |
|---|---|---|---|---|---|---|---|
| $y$ | $-2.9$ | $-2.8$ | $-2.6$ | $-2$ | $-.66$ | $2.5$ | $9.9$ |

## Section 5.4    Matrices and Systems of Equations

- ■ You should be able to use elementary row operations to produce a row-echelon form (or reduced row-echelon form) of a matrix.
    1. Interchange two rows
    2. Multiply a row by a nonzero constant
    3. Add a multiple of one row to another row
- ■ You should be able to use either Gaussian elimination with back-substitution or Gauss-Jordan elimination to solve a system of linear equations.

**Solutions to Odd-Numbered Exercises**

**1.** Since the matrix has one row and two columns, its order is $1 \times 2$.

**3.** Since the matrix has three rows and one column, its order is $3 \times 1$.

**5.** Since the matrix has two rows and two columns, its order is $2 \times 2$.

**7.** $\begin{cases} 4x - 5y = 33 \\ -x + 5y = -27 \end{cases}$

$$\begin{bmatrix} 4 & -5 & \vdots & 33 \\ -1 & 5 & \vdots & -27 \end{bmatrix}$$

**9.** $\begin{cases} x + 10y - 2z = 2 \\ 5x - 3y + 4z = 0 \\ 2x + y = 6 \end{cases}$

$$\begin{bmatrix} 1 & 10 & -2 & \vdots & 2 \\ 5 & -3 & 4 & \vdots & 0 \\ 2 & 1 & 0 & \vdots & 6 \end{bmatrix}$$

**11.** $\begin{bmatrix} 1 & 2 & \vdots & 7 \\ 2 & -3 & \vdots & 4 \end{bmatrix}$

$\begin{cases} x + 2y = 7 \\ 2x - 3y = 4 \end{cases}$

**13.** $\begin{bmatrix} 9 & 12 & 3 & \vdots & 0 \\ -2 & 18 & 5 & \vdots & -10 \\ 1 & 7 & -8 & \vdots & -4 \end{bmatrix}$

$\begin{cases} 9x + 12y + 3z = 0 \\ -2x + 18y + 5z = 10 \\ x + 7y - 8z = -4 \end{cases}$

**15.** $\begin{bmatrix} 1 & 4 & 3 \\ 2 & 10 & 5 \end{bmatrix}$

$-2R_1 + R_2 \to \begin{bmatrix} 1 & 4 & 3 \\ 0 & \boxed{2} & -1 \end{bmatrix}$

**17.** $\begin{bmatrix} 1 & 1 & 4 & -1 \\ 3 & 8 & 10 & 3 \\ -2 & 1 & 12 & 6 \end{bmatrix} \begin{matrix} \\ -3R_1 + R_2 \to \\ 2R_1 + R_3 \to \end{matrix} \begin{bmatrix} 1 & 1 & 4 & -1 \\ 0 & 5 & \boxed{-2} & \boxed{6} \\ 0 & 3 & \boxed{20} & \boxed{4} \end{bmatrix} \frac{1}{5}R_2 \to \begin{bmatrix} 1 & 1 & 4 & -1 \\ 0 & 1 & -\frac{2}{5} & \frac{6}{5} \\ 0 & 3 & 20 & 4 \end{bmatrix}$

**19.** Add 5 times Row 2 to Row 1.

**21.** Interchange Rows 1 and 2.

**23.** $\begin{bmatrix} 1 & 0 & 0 & 0 \\ 0 & 1 & 1 & 5 \\ 0 & 0 & 0 & 0 \end{bmatrix}$

This matrix is in reduced row-echelon form.

**25.** $\begin{bmatrix} 2 & 0 & 4 & 0 \\ 0 & -1 & 3 & 6 \\ 0 & 0 & 1 & 5 \end{bmatrix}$

The first nonzero entries in rows one and two are not one. The matrix is not in row-echelon form.

**27.** $\begin{bmatrix} 1 & 2 & 3 \\ 2 & -1 & -4 \\ 3 & 1 & -1 \end{bmatrix}$

**29.** (See Exercise 27.) (Answer is series of screens.)

(a) $\begin{bmatrix} 1 & 2 & 3 \\ 0 & -5 & -10 \\ 3 & 1 & -1 \end{bmatrix}$ (b) $\begin{bmatrix} 1 & 2 & 3 \\ 0 & -5 & -10 \\ 0 & -5 & -10 \end{bmatrix}$

(c) $\begin{bmatrix} 1 & 2 & 3 \\ 0 & -5 & -10 \\ 0 & 0 & 0 \end{bmatrix}$ (d) $\begin{bmatrix} 1 & 2 & 3 \\ 0 & 1 & 2 \\ 0 & 0 & 0 \end{bmatrix}$

(e) $\begin{bmatrix} 1 & 0 & -1 \\ 0 & 1 & 2 \\ 0 & 0 & 0 \end{bmatrix}$  This matrix is in reduced row-echelon form.

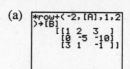

(a) 
```
*row+(-2,[A],1,2
)→[B]
[[1 2  3 ]
 [0 -5 -10]
 [3 1  -1 ]]
```
(b)
```
*row+(-3,[B],1,3
)→[C]
[[1 2  3 ]
 [0 -5 -10]
 [0 -5 -10]]
```

(c)
```
*row+(-1,[C],2,3
)→[D]
[[1 2  3 ]
 [0 -5 -10]
 [0 0  0 ]]
```
(d)
```
*row(-1/5,[D],2)
→[E]
[[1 2 3]
 [0 1 2]
 [0 0 0]]
```

(e)
```
*row+(-2,[E],2,1
)
[[1 0 -1]
 [0 1 2 ]
 [0 0 0 ]]
```

**31.** $\begin{bmatrix} 1 & 1 & 0 & 5 \\ -2 & -1 & 2 & -10 \\ 3 & 6 & 7 & 14 \end{bmatrix}$

$\begin{matrix} 2R_1 + R_2 \rightarrow \\ -3R_1 + R_3 \rightarrow \end{matrix}\begin{bmatrix} 1 & 1 & 0 & 5 \\ 0 & 1 & 2 & 0 \\ 0 & 3 & 7 & -1 \end{bmatrix}$

$\begin{matrix} \\ \\ -3R_2 + R_3 \rightarrow \end{matrix}\begin{bmatrix} 1 & 1 & 0 & 5 \\ 0 & 1 & 2 & 0 \\ 0 & 0 & 1 & -1 \end{bmatrix}$

**33.** $\begin{bmatrix} 1 & -1 & -1 & 1 \\ 5 & -4 & 1 & 8 \\ -6 & 8 & 18 & 0 \end{bmatrix}$

$\begin{matrix} -5R_1 + R_2 \rightarrow \\ 6R_1 + R_3 \rightarrow \end{matrix}\begin{bmatrix} 1 & -1 & -1 & 1 \\ 0 & 1 & 6 & 3 \\ 0 & 2 & 12 & 6 \end{bmatrix}$

$\begin{matrix} \\ -2R_2 + R_3 \rightarrow \end{matrix}\begin{bmatrix} 1 & -1 & -1 & 1 \\ 0 & 1 & 6 & 3 \\ 0 & 0 & 0 & 0 \end{bmatrix}$

**35.** $\begin{bmatrix} 3 & 3 & 3 \\ -1 & 0 & -4 \\ 2 & 4 & -2 \end{bmatrix}$

$\frac{1}{3}R_1 \rightarrow \begin{bmatrix} 1 & 1 & 1 \\ -1 & 0 & -4 \\ 2 & 4 & -2 \end{bmatrix}$

$\begin{matrix} R_1 + R_2 \rightarrow \\ -2R_1 + R_3 \rightarrow \end{matrix}\begin{bmatrix} 1 & 1 & 1 \\ 0 & 1 & -3 \\ 0 & 2 & -4 \end{bmatrix}$

$\begin{matrix} -R_2 + R_1 \rightarrow \\ \\ -2R_2 + R_3 \rightarrow \end{matrix}\begin{bmatrix} 1 & 0 & 4 \\ 0 & 1 & -3 \\ 0 & 0 & 2 \end{bmatrix}$

$\begin{matrix} \\ \\ \frac{1}{2}R_3 \rightarrow \end{matrix}\begin{bmatrix} 1 & 0 & 4 \\ 0 & 1 & -3 \\ 0 & 0 & 1 \end{bmatrix}$

$\begin{matrix} -4R_3 + R_1 \rightarrow \\ 3R_3 + R_2 \rightarrow \\ \\ \end{matrix}\begin{bmatrix} 1 & 0 & 0 \\ 0 & 1 & 0 \\ 0 & 0 & 1 \end{bmatrix}$

**37.** $\begin{bmatrix} -3 & 5 & 1 & 12 \\ 1 & -1 & 1 & 4 \end{bmatrix}$

$\begin{matrix} R_1 \rightarrow \\ R_2 \rightarrow \end{matrix}\begin{bmatrix} 1 & -1 & 1 & 4 \\ -3 & 5 & 1 & 12 \end{bmatrix}$

$3R_1 + R_2 \rightarrow \begin{bmatrix} 1 & -1 & 1 & 4 \\ 0 & 2 & 4 & 24 \end{bmatrix}$

$\frac{1}{2}R_2 \rightarrow \begin{bmatrix} 1 & -1 & 1 & 4 \\ 0 & 1 & 2 & 12 \end{bmatrix}$

$R_2 + R_1 \rightarrow \begin{bmatrix} 1 & 0 & 3 & 16 \\ 0 & 1 & 2 & 12 \end{bmatrix}$

**39.** $\begin{cases} x - 2y = 4 \\ \quad\quad y = -3 \end{cases}$

$x - 2(-3) = 4$

$\quad\quad\quad x = -2$

*Answer:* $(-2, -3)$

**41.** $\begin{cases} x - y + 2z = 4 \\ \quad y - z = 2 \\ \quad\quad\quad z = -2 \end{cases}$

$y - (-2) = 2$

$\quad\quad y = 0$

$x - 0 + 2(-2) = 4$

$\quad\quad\quad\quad x = 8$

*Answer:* $(8, 0, -2)$

**43.** $\begin{bmatrix} 1 & 0 & \vdots & 7 \\ 0 & 1 & \vdots & -5 \end{bmatrix}$

$x = 7$

$y = -5$

*Answer:* $(7, -5)$

**45.** $\begin{bmatrix} 1 & 0 & 0 & \vdots & -4 \\ 0 & 1 & 0 & \vdots & -8 \\ 0 & 0 & 1 & \vdots & 2 \end{bmatrix}$

$x = -4$

$y = -8$

$z = 2$

*Answer:* $(-4, -8, 2)$

**47.** $\begin{cases} x + 2y = 7 \\ 2x + y = 8 \end{cases}$

$\begin{bmatrix} 1 & 2 & \vdots & 7 \\ 2 & 1 & \vdots & 8 \end{bmatrix} \begin{matrix} \\ -2R_1 + R_2 \rightarrow \end{matrix} \begin{bmatrix} 1 & 2 & \vdots & 7 \\ 0 & -3 & \vdots & -6 \end{bmatrix} \begin{matrix} \\ -\frac{1}{3}R_2 \rightarrow \end{matrix} \begin{bmatrix} 1 & 2 & \vdots & 7 \\ 0 & 1 & \vdots & 2 \end{bmatrix}$

$y = 2$

$x + 2(2) = 7 \implies x = 3$

*Answer:* $(3, 2)$

**49.** $\begin{cases} -x + y = -22 \\ 3x + 4y = 4 \\ 4x - 8y = 32 \end{cases}$

$\begin{bmatrix} -1 & 1 & \vdots & -22 \\ 3 & 4 & \vdots & 4 \\ 4 & -8 & \vdots & 32 \end{bmatrix}$

$\begin{matrix} \\ 3R_1 + R_2 \\ 4R_1 + R_3 \end{matrix} \begin{bmatrix} -1 & 1 & \vdots & -22 \\ 0 & 7 & \vdots & -62 \\ 0 & -4 & \vdots & -56 \end{bmatrix}$

$\begin{matrix} \\ \curvearrowright R_2 \\ \curvearrowleft R_3 \end{matrix} \begin{bmatrix} -1 & 1 & & -22 \\ 0 & -4 & & -56 \\ 0 & 7 & & -62 \end{bmatrix}$

$\begin{matrix} \\ -\frac{1}{4}R_2 \\ -7R_2 + R_3 \end{matrix} \begin{bmatrix} -1 & 1 & & -22 \\ 0 & 1 & & 14 \\ 0 & 0 & & -160 \end{bmatrix}$

No solution, Inconsistent

**51.** $\begin{cases} 8x - 4y = 13 \\ 5x + 2y = 7 \end{cases}$

$\begin{bmatrix} 8 & -4 & 13 \\ 5 & 2 & 7 \end{bmatrix} \begin{matrix} 3R_1 \rightarrow \\ 5R_2 \rightarrow \end{matrix} \begin{bmatrix} 24 & -12 & 39 \\ 25 & 10 & 35 \end{bmatrix}$

$\begin{matrix} -R_2 + R_1 \rightarrow \\ \frac{1}{5}R_2 \rightarrow \end{matrix} \begin{bmatrix} -1 & -22 & 4 \\ 5 & 2 & 7 \end{bmatrix}$

$\begin{matrix} \\ 5R_1 + R_2 \rightarrow \end{matrix} \begin{bmatrix} -1 & -22 & 4 \\ 0 & -108 & 27 \end{bmatrix}$

$\begin{matrix} -1R_1 \rightarrow \\ -\frac{1}{108}R_2 \rightarrow \end{matrix} \begin{bmatrix} 1 & 22 & -4 \\ 0 & 1 & -\frac{1}{4} \end{bmatrix}$

$y = -\frac{1}{4}$

$x + 22\left(-\frac{1}{4}\right) = -4 \implies x = \frac{3}{2}$

*Answer:* $\left(\frac{3}{2}, -\frac{1}{4}\right)$

**53.** $\begin{cases} -x + 2y = 1.5 \\ 2x - 4y = 3.0 \end{cases}$

$$\begin{bmatrix} -1 & 2 & \vdots & 1.5 \\ 2 & -4 & \vdots & 3.0 \end{bmatrix}$$

$$2R_1 + R_2 \rightarrow \begin{bmatrix} -1 & 2 & \vdots & 1.5 \\ 0 & 0 & \vdots & 6.0 \end{bmatrix}$$

The system is inconsistent and there is no solution.

**57.** $\begin{cases} x + y - 5z = 3 \\ x \quad\;\; - 2z = 1 \\ 2x - y - z = 0 \end{cases}$

$$\begin{bmatrix} 1 & 1 & -5 & \vdots & 3 \\ 1 & 0 & -2 & \vdots & 1 \\ 2 & -1 & -1 & \vdots & 0 \end{bmatrix}$$

$$\begin{matrix} -R_1 + R_2 \rightarrow \\ -2R_1 + R_3 \rightarrow \end{matrix} \begin{bmatrix} 1 & 1 & -5 & \vdots & 3 \\ 0 & -1 & 3 & \vdots & -2 \\ 0 & -3 & 9 & \vdots & -6 \end{bmatrix}$$

$$-3R_2 + R_3 \rightarrow \begin{bmatrix} 1 & 1 & -5 & \vdots & 3 \\ 0 & -1 & 3 & \vdots & -2 \\ 0 & 0 & 0 & \vdots & 0 \end{bmatrix}$$

$$\begin{matrix} R_2 + R_1 \rightarrow \\ -R_2 \rightarrow \end{matrix} \begin{bmatrix} 1 & 0 & -2 & \vdots & 1 \\ 0 & 1 & -3 & \vdots & 2 \\ 0 & 0 & 0 & \vdots & 0 \end{bmatrix}$$

Let $z = a$, any real number

$y - 3a = 2 \implies y = 3a + 2$

$x - 2a = 1 \implies x = 2a + 1$

*Answer:* $(2a + 1, 3a + 2, a)$

**55.** $\begin{cases} x \quad\quad - 3z = -2 \\ 3x + y - 2z = 5 \\ 2x + 2y + z = 4 \end{cases}$

$$\begin{bmatrix} 1 & 0 & -3 & \vdots & -2 \\ 3 & 1 & -2 & \vdots & 5 \\ 2 & 2 & 1 & \vdots & 4 \end{bmatrix}$$

$$\begin{matrix} -3R_1 + R_2 \rightarrow \\ -2R_1 + R_3 \rightarrow \end{matrix} \begin{bmatrix} 1 & 0 & -3 & \vdots & -2 \\ 0 & 1 & 7 & \vdots & 11 \\ 0 & 2 & 7 & \vdots & 8 \end{bmatrix}$$

$$-2R_2 + R_3 \rightarrow \begin{bmatrix} 1 & 0 & -3 & \vdots & -2 \\ 0 & 1 & 7 & \vdots & 11 \\ 0 & 0 & -7 & \vdots & -14 \end{bmatrix}$$

$$-\tfrac{1}{7}R_3 \rightarrow \begin{bmatrix} 1 & 0 & -3 & \vdots & -2 \\ 0 & 1 & 7 & \vdots & 11 \\ 0 & 0 & 1 & \vdots & 2 \end{bmatrix}$$

$$z = 2$$

$y + 7(2) = 11 \implies y = -3$

$x - 3(2) = -2 \implies x = 4$

*Answer:* $(4, -3, 2)$

**59.** $\begin{cases} -x + y - z = -14 \\ 2x - y + z = 21 \\ 3x + 2y + z = 19 \end{cases}$

$$\begin{bmatrix} -1 & 1 & -1 & \vdots & -14 \\ 2 & -1 & 1 & \vdots & 21 \\ 3 & 2 & 1 & \vdots & 19 \end{bmatrix}$$

$$\begin{matrix} 2R_1 + R_2 \\ 3R_1 + R_3 \end{matrix} \begin{bmatrix} -1 & 1 & -1 & \vdots & -14 \\ 0 & 1 & -1 & \vdots & -7 \\ 0 & 5 & -2 & \vdots & -23 \end{bmatrix}$$

$$\begin{matrix} -R_1 \\ -5R_2 + R_3 \end{matrix} \begin{bmatrix} 1 & -1 & 1 & \vdots & 14 \\ 0 & 1 & -1 & \vdots & -7 \\ 0 & 0 & 3 & \vdots & 12 \end{bmatrix}$$

$$3z = 12 \implies z = 4$$

$$y - 4 = 7 \implies y = -3$$

$x - (-3) + 4 = 14 \implies x = 7$

*Answer:* $(7, -3, 4)$

**61.** $\begin{cases} 3x + 3y + 12z = 6 \\ x + y + 4z = 2 \\ 2x + 5y + 20z = 10 \\ -x + 2y + 8z = 4 \end{cases}$   $\begin{bmatrix} 3 & 3 & 12 & \vdots & 6 \\ 1 & 1 & 4 & \vdots & 2 \\ 2 & 5 & 20 & \vdots & 10 \\ -1 & 2 & 8 & \vdots & 4 \end{bmatrix} \Rightarrow \begin{bmatrix} 1 & 0 & 0 & \vdots & 0 \\ 0 & 0 & 0 & \vdots & 0 \\ 0 & 1 & 4 & \vdots & 2 \\ 0 & 0 & 0 & \vdots & 0 \end{bmatrix}$

Let $z = a$, any real number

$y = -4a + 2$

$x = 0$

*Answer:* $(0, -4a + 2, a)$

**63.** $\begin{bmatrix} 2 & 1 & -1 & 2 & \vdots & -6 \\ 3 & 4 & 0 & 1 & \vdots & 1 \\ 1 & 5 & 2 & 6 & \vdots & -3 \\ 5 & 2 & -1 & -1 & \vdots & 3 \end{bmatrix}$ row reduces to $\begin{bmatrix} 1 & 0 & 0 & 0 & \vdots & 1 \\ 0 & 1 & 0 & 0 & \vdots & 0 \\ 0 & 0 & 1 & 0 & \vdots & 4 \\ 0 & 0 & 0 & 1 & \vdots & -2 \end{bmatrix}$

*Answer:* $(1, 0, 4, -2)$

**65.** $\begin{cases} x + y + z = 0 \\ 2x + 3y + z = 0 \\ 3x + 5y + z = 0 \end{cases}$   $\begin{bmatrix} 1 & 1 & 1 & \vdots & 0 \\ 2 & 3 & 1 & \vdots & 0 \\ 3 & 5 & 1 & \vdots & 0 \end{bmatrix} \Rightarrow \begin{bmatrix} 1 & 0 & 2 & \vdots & 0 \\ 0 & 1 & -1 & \vdots & 0 \\ 0 & 0 & 0 & \vdots & 0 \end{bmatrix}$

Let $z = a$, any real number

$y = a$

$x = -2a$

*Answer:* $(-2a, a, a)$

**67.** Yes, the systems yield the same solutions.

(a) $z = -3$; $y = 5(-3) + 16 = 1$;
$x = 2(1) - (-3) - 6 = -1$

*Answer:* $(-1, 1, -3)$

(b) $z = -3$, $y = -3(-3) - 8 = 1$,
$x = -1 + 2(-3) + 6 = -1$

*Answer:* $(-1, 1, -3)$

**69.** No, solutions are different.

(a) $z = 8$, $y = 7(8) - 54 = 2$,
$x = 4(2) - 5(8) + 27 = -5$

*Answer:* $(-5, 2, 8)$

(b) $z = 8$, $y = -5(8) + 42 = 2$,
$x = 6(2) - 8 + 15 = 19$

*Answer:* $(19, 2, 8)$

**71.** $f(x) = ax^2 + bx + c$

$$\begin{cases} f(1) = a + b + c = 8 \\ f(2) = 4a + 2b + c = 13 \\ f(3) = 9a + 3b + c = 20 \end{cases}$$

$$\begin{bmatrix} 1 & 1 & 1 & \vdots & 8 \\ 4 & 2 & 1 & \vdots & 13 \\ 9 & 3 & 1 & \vdots & 20 \end{bmatrix}$$

$$\begin{matrix} \\ -4R_1 + R_2 \longrightarrow \\ -9R_1 + R_3 \longrightarrow \end{matrix} \begin{bmatrix} 1 & 1 & 1 & \vdots & 8 \\ 0 & -2 & -3 & \vdots & -19 \\ 0 & -6 & -8 & \vdots & -52 \end{bmatrix}$$

$$\begin{matrix} \\ -\frac{1}{2}R_2 \longrightarrow \\ -3R_2 + R_3 \longrightarrow \end{matrix} \begin{bmatrix} 1 & 1 & 1 & \vdots & 8 \\ 0 & 1 & \frac{3}{2} & \vdots & \frac{19}{2} \\ 0 & 0 & 1 & \vdots & 5 \end{bmatrix}$$

$$c = 5$$

$$b + \tfrac{3}{2}(5) = \tfrac{19}{2} \implies b = 2$$

$$a + 2 + 5 = 8 \implies a = 1$$

*Answer:* $y = x^2 + 2x + 5$

**73.** $x = $ amount at 7%

$y = $ amount at 8%

$z = $ amount at 10%

$$\begin{cases} x + y + z = 1{,}500{,}000 \\ 0.07x + 0.08y + 0.1z = 130{,}500 \\ 4x - z = 0 \end{cases}$$

$$\begin{bmatrix} 1 & 1 & 1 & \vdots & 1{,}500{,}000 \\ 0.07 & 0.08 & 0.1 & \vdots & 130{,}500 \\ 4 & 0 & -1 & \vdots & 0 \end{bmatrix}$$

$$\begin{matrix} -0.07R_1 + R_2 \\ -4R_1 + R_2 \end{matrix} \begin{bmatrix} 1 & 1 & 1 & \vdots & 1{,}500{,}000 \\ 0 & 0.01 & 0.03 & \vdots & 25{,}500 \\ 0 & -4 & -5 & \vdots & -6{,}000{,}000 \end{bmatrix}$$

$$\begin{matrix} 100R_2 \\ 4R_2 + R_3 \end{matrix} \begin{bmatrix} 1 & 1 & 1 & \vdots & 1{,}500{,}000 \\ 0 & 1 & 3 & \vdots & 2{,}550{,}000 \\ 0 & 0 & 7 & \vdots & 4{,}200{,}000 \end{bmatrix}$$

$7z = 4{,}200{,}000 \implies z = 600{,}00$

$y + 3(600{,}000) = 2{,}550{,}000 \implies y = 750{,}000$

$x + 750{,}000 + 600{,}000 = 1{,}500{,}000 \implies x = 150{,}000$

*Answers:*  $150,000 at 7%

$750,000 at 8%

$600,000 at 10%

**75.** $\begin{cases} I_1 - I_2 + I_3 = 0 \\ 2I_1 + 2I_2 = 7 \\ 2I_2 + 4I_3 = 8 \end{cases}$

$$\begin{bmatrix} 1 & -1 & 1 & \vdots & 0 \\ 2 & 2 & 0 & \vdots & 7 \\ 0 & 2 & 4 & \vdots & 8 \end{bmatrix}$$

$$-2R_1 + R_2 \longrightarrow \begin{bmatrix} 1 & -1 & 1 & \vdots & 0 \\ 0 & 4 & -2 & \vdots & 7 \\ 0 & 2 & 4 & \vdots & 8 \end{bmatrix}$$

$$\begin{matrix} R_3 \longrightarrow \\ R_2 \longrightarrow \end{matrix} \begin{bmatrix} 1 & -1 & 1 & \vdots & 0 \\ 0 & 2 & 4 & \vdots & 8 \\ 0 & 4 & -2 & \vdots & 7 \end{bmatrix}$$

$$\tfrac{1}{2}R_2 \longrightarrow \begin{bmatrix} 1 & -1 & 1 & \vdots & 0 \\ 0 & 1 & 2 & \vdots & 4 \\ 0 & 4 & -2 & \vdots & 7 \end{bmatrix}$$

$$-4R_2 + R_3 \longrightarrow \begin{bmatrix} 1 & -1 & 1 & \vdots & 0 \\ 0 & 1 & 2 & \vdots & 4 \\ 0 & 0 & -10 & \vdots & -9 \end{bmatrix}$$

$$-\tfrac{1}{10}R_3 \longrightarrow \begin{bmatrix} 1 & -1 & 1 & \vdots & 0 \\ 0 & 1 & 2 & \vdots & 4 \\ 0 & 0 & 1 & \vdots & \frac{9}{10} \end{bmatrix}$$

$I_3 = \frac{9}{10}$ amperes

$I_2 + 2\left(\frac{9}{10}\right) = 4 \implies I_2 = \frac{11}{5}$ amperes

$I_1 - \frac{11}{5} + \frac{9}{10} = 0 \implies I_1 = \frac{13}{10}$ amperes

**77.** (a)   (9, 43.90)    $\begin{cases} 81a + 9b + c = 43.9 \\ 100a + 10b + c = 42.1 \\ 121a + 11b + c = 34.10 \end{cases}$
      (10, 42.10)
      (11, 34.10)

$$\begin{bmatrix} 81 & 9 & 1 & \vdots & 43.9 \\ 100 & 10 & 1 & \vdots & 42.1 \\ 121 & 11 & 1 & \vdots & 34.1 \end{bmatrix} \Rightarrow \begin{bmatrix} 1 & 0 & 0 & \vdots & -3.1 \\ 0 & 1 & 0 & \vdots & 57.1 \\ 0 & 0 & 1 & \vdots & -218.9 \end{bmatrix}$$

$a = 3.1, b = 57.1, c = -218.9$

$y = -3.1t^2 + 57.1t - 218.9$

(b)

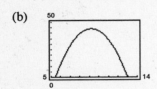

(c) For 2002, $t = 12$ and $y = -3.1(12)^2 + 57.1(12) - 218.9 = 19.9$

(d) For 2005, $t = 15$ and $y = -59.9$, which is unreasonable.

**79.** (a) $\begin{cases} x_1 + x_3 = 600 \\ x_1 = x_2 + x_4 \Rightarrow x_1 - x_2 - x_4 = 0 \\ x_2 + x_5 = 500 \\ x_3 + x_6 = 600 \\ x_4 + x_7 = x_6 \Rightarrow x_4 - x_6 + x_7 = 0 \\ x_5 + x_7 = 500 \end{cases}$

$$\begin{bmatrix} 1 & 0 & 1 & 0 & 0 & 0 & 0 & \vdots & 600 \\ 1 & -1 & 0 & -1 & 0 & 0 & 0 & \vdots & 0 \\ 0 & 1 & 0 & 0 & 1 & 0 & 0 & \vdots & 500 \\ 0 & 0 & 1 & 0 & 0 & 1 & 0 & \vdots & 600 \\ 0 & 0 & 0 & 1 & 0 & -1 & 1 & \vdots & 0 \\ 0 & 0 & 0 & 0 & 1 & 0 & 1 & \vdots & 500 \end{bmatrix}$$

$$\begin{array}{l} -R_1 + R_2 \rightarrow \\ R_2 + R_3 \rightarrow \\ R_3 + R_4 \rightarrow \\ R_4 + R_5 \rightarrow \\ -R_5 + R_6 \rightarrow \end{array} \begin{bmatrix} 1 & 0 & 1 & 0 & 0 & 0 & 0 & \vdots & 600 \\ 0 & -1 & -1 & -1 & 0 & 0 & 0 & \vdots & -600 \\ 0 & 0 & -1 & -1 & 1 & 0 & 0 & \vdots & -100 \\ 0 & 0 & 0 & -1 & 1 & 1 & 0 & \vdots & 500 \\ 0 & 0 & 0 & 0 & 1 & 0 & 1 & \vdots & 500 \\ 0 & 0 & 0 & 0 & 0 & 0 & 0 & \vdots & 0 \end{bmatrix}$$

$$\begin{array}{l} -R_3 + R_2 \rightarrow \\ -R_4 + R_3 \rightarrow \\ -R_4 \rightarrow \end{array} \begin{bmatrix} 1 & 0 & 1 & 0 & 0 & 0 & 0 & \vdots & 600 \\ 0 & -1 & 0 & 0 & -1 & 0 & 0 & \vdots & -500 \\ 0 & 0 & -1 & 0 & 0 & -1 & 0 & \vdots & -600 \\ 0 & 0 & 0 & 1 & -1 & -1 & 0 & \vdots & -500 \\ 0 & 0 & 0 & 0 & 1 & 0 & 1 & \vdots & 500 \\ 0 & 0 & 0 & 0 & 0 & 0 & 0 & \vdots & 0 \end{bmatrix}$$

Let $x_7 = t$ and $x_6 = s$, then:

$x_5 = 500 - t$

$x_4 = -500 + s + (500 - t) = s - t$

$x_3 = 600 - s$

$x_2 = 500 - (500 - t) = t$

$x_1 = 600 - (600 - s) = s$

(b) If $x_6 = x_7 = 0$, then $s = t = 0$, and

     $x_1 = 0$

     $x_2 = 0$

     $x_3 = 600$

     $x_4 = 0$

     $x_5 = 500$

     $x_6 = x_7 = 0$

(c) If $x_5 = 1000$ and $x_6 = 0$, then $s = 0$ and $t = -500$.

    Thus, $x_1 = 0$

     $x_2 = -500$

     $x_3 = 600$

     $x_4 = 500$

     $x_5 = 1000$

     $x_6 = 0$

     $x_7 = -500$

**81.** False. It is a $2 \times 4$ matrix.

**83.** $\begin{cases} x + 3z = -2 & \text{Equation 1} \\ y + 4z = \phantom{-}1 & \text{Equation 2} \end{cases}$

(Equation 1) + (Equation 2) $\rightarrow$ new Equation 1

(Equation 1) + 2(Equation 2) $\rightarrow$ new Equation 2

2(Equation 1) + (Equation 2) $\rightarrow$ new Equation 3

$\begin{cases} x + \phantom{2}y + \phantom{1}7z = -1 \\ x + 2y + 11z = \phantom{-}0 \\ 2x + \phantom{2}y + 10z = -3 \end{cases}$

**85.** The row operation $-2R_1 + R_2$ was not performed on the last column. Nor was $-R_2 + R_1$.

**87.** $f(x) = \dfrac{7}{-x - 1}$

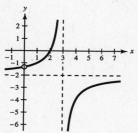

Asymptotes: $x = -1, y = 0$

**89.** $f(x) = \dfrac{x^2 - 2x - 3}{x - 4} = x + 2 + \dfrac{5}{x - 4}$

Asymptotes: $x = 4, y = x + 2$

**91.** $f(x) = \dfrac{2x^2 - 4x}{3x - x^2} = \dfrac{2x(x - 2)}{x(3 - x)} = \dfrac{2(x - 2)}{3 - x}, \quad x \neq 0$

Asymptotes: $x = 3, y = -2$

Hole at $x = 0$

## Section 5.5   Operations with Matrices

- $A = B$ if and only if they have the same order and $a_{ij} = b_{ij}$.
- You should be able to perform the operations of matrix addition, scalar multiplication, and matrix multiplication.
- Some properties of matrix addition and scalar multiplication are:
  - (a) $A + B = B + A$
  - (b) $A + (B + C) = (A + B) + C$
  - (c) $(cd)A = c(dA)$
  - (d) $1A = A$
  - (e) $c(A + B) = cA + cB$
  - (f) $(c + d)A = cA + dA$
- Some properties of matrix multiplication are:
  - (a) $A(BC) = (AB)C$
  - (b) $A(B + C) = AB + AC$
  - (c) $(A + B)C = AC + BC$
  - (d) $c(AB) = (cA)B = A(cB)$
- You should remember that $AB \neq BA$ in general.

### Solutions to Odd-Numbered Exercises

**1.** $x = -4, \; y = 22$

**3.** $2x + 7 = 5 \implies x = -1$

$$3y = 12 \implies y = 4$$

$$3z - 14 = 4 \implies z = 6$$

**5.** (a) $A + B = \begin{bmatrix} 1 & -1 \\ 2 & -1 \end{bmatrix} + \begin{bmatrix} 2 & -1 \\ -1 & 8 \end{bmatrix} = \begin{bmatrix} 1+2 & -1-1 \\ 2-1 & -1+8 \end{bmatrix} = \begin{bmatrix} 3 & -2 \\ 1 & 7 \end{bmatrix}$

(b) $A - B = \begin{bmatrix} 1 & -1 \\ 2 & -1 \end{bmatrix} - \begin{bmatrix} 2 & -1 \\ -1 & 8 \end{bmatrix} = \begin{bmatrix} 1-2 & -1+1 \\ 2+1 & -1-8 \end{bmatrix} = \begin{bmatrix} -1 & 0 \\ 3 & -9 \end{bmatrix}$

(c) $3A = 3\begin{bmatrix} 1 & -1 \\ 2 & -1 \end{bmatrix} = \begin{bmatrix} 3(1) & 3(-1) \\ 3(2) & 3(-1) \end{bmatrix} = \begin{bmatrix} 3 & -3 \\ 6 & -3 \end{bmatrix}$

(d) $3A - 2B = \begin{bmatrix} 3 & -3 \\ 6 & -3 \end{bmatrix} - 2\begin{bmatrix} 2 & -1 \\ -1 & 8 \end{bmatrix} = \begin{bmatrix} 3 & -3 \\ 6 & -3 \end{bmatrix} + \begin{bmatrix} -4 & 2 \\ 2 & -16 \end{bmatrix} = \begin{bmatrix} -1 & -1 \\ 8 & -19 \end{bmatrix}$

**7.** $A = \begin{bmatrix} 8 & -1 \\ 2 & 3 \\ -4 & 5 \end{bmatrix}, B = \begin{bmatrix} 1 & 6 \\ -1 & -5 \\ 1 & 10 \end{bmatrix}$

(a) $A + B = \begin{bmatrix} 9 & 5 \\ 1 & -2 \\ -3 & 15 \end{bmatrix}$
(b) $A - B = \begin{bmatrix} 7 & -7 \\ 3 & 8 \\ -5 & -5 \end{bmatrix}$
(c) $3A = \begin{bmatrix} 24 & -3 \\ 6 & 9 \\ -12 & 15 \end{bmatrix}$

(d) $3A - 2B = \begin{bmatrix} 24 & -3 \\ 6 & 9 \\ -12 & 15 \end{bmatrix} - \begin{bmatrix} 2 & 12 \\ -2 & -10 \\ 2 & 20 \end{bmatrix} = \begin{bmatrix} 22 & -15 \\ 8 & 19 \\ -14 & -5 \end{bmatrix}$

**9.** $A = \begin{bmatrix} 4 & 5 & -1 & 3 & 4 \\ 1 & 2 & -2 & -1 & 0 \end{bmatrix}$

$B = \begin{bmatrix} 1 & 0 & -1 & 1 & 0 \\ -6 & 8 & 2 & -3 & -7 \end{bmatrix}$

(a) $A + B = \begin{bmatrix} 5 & 5 & -2 & 4 & 4 \\ -5 & 10 & 0 & -4 & -7 \end{bmatrix}$

(b) $A - B = \begin{bmatrix} 3 & 5 & 0 & 2 & 4 \\ 7 & -6 & -4 & 2 & 7 \end{bmatrix}$

(c) $3A = \begin{bmatrix} 12 & 15 & -3 & 9 & 12 \\ 3 & 6 & -6 & -3 & 0 \end{bmatrix}$

(d) $3A - 2B = \begin{bmatrix} 12 & 15 & -3 & 9 & 12 \\ 3 & 6 & -6 & -3 & 0 \end{bmatrix} - \begin{bmatrix} 2 & 0 & -2 & 2 & 0 \\ -12 & 16 & 4 & -6 & -14 \end{bmatrix}$

$= \begin{bmatrix} 10 & 15 & -1 & 7 & 12 \\ 15 & -10 & -10 & 3 & 14 \end{bmatrix}$

**11.** $A = \begin{bmatrix} 6 & 0 & 3 \\ -1 & -4 & 0 \end{bmatrix}$, $B = \begin{bmatrix} 8 & -1 \\ 4 & -3 \end{bmatrix}$

(a) $A + B$ not possible  (b) $A - B$ not possible  (c) $3A = \begin{bmatrix} 18 & 0 & 9 \\ -3 & -12 & 0 \end{bmatrix}$  (d) $3A - 2B$ not possible

**13.** $\begin{bmatrix} -5 & 0 \\ 3 & -6 \end{bmatrix} + \begin{bmatrix} 7 & 1 \\ -2 & -1 \end{bmatrix} + \begin{bmatrix} -10 & -8 \\ 14 & 6 \end{bmatrix}$

$= \begin{bmatrix} -5 & 0 \\ 3 & -6 \end{bmatrix} + \begin{bmatrix} -3 & -7 \\ 12 & 5 \end{bmatrix} = \begin{bmatrix} -8 & -7 \\ 15 & -1 \end{bmatrix}$

**15.** $4\left( \begin{bmatrix} -4 & 0 & 1 \\ 0 & 2 & 3 \end{bmatrix} - \begin{bmatrix} 2 & 1 & -2 \\ 3 & -6 & 0 \end{bmatrix} \right) = 4\begin{bmatrix} -6 & -1 & 3 \\ -3 & 8 & 3 \end{bmatrix} = \begin{bmatrix} -24 & -4 & 12 \\ -12 & 32 & 12 \end{bmatrix}$

**17.** $\begin{bmatrix} 2 & 5 \\ -1 & -4 \end{bmatrix} + \begin{bmatrix} -3 & 0 \\ 2 & 2 \end{bmatrix} = \begin{bmatrix} -1 & 5 \\ 1 & -2 \end{bmatrix}$

**19.** $-\frac{1}{2}\begin{bmatrix} 3.211 & 6.829 \\ -1.004 & 4.914 \\ 0.055 & -3.889 \end{bmatrix} - 8\begin{bmatrix} 1.630 & -3.090 \\ 5.256 & 8.335 \\ -9.768 & 4.251 \end{bmatrix} = \begin{bmatrix} -14.645 & 21.305 \\ -41.546 & -69.137 \\ 78.117 & -32.064 \end{bmatrix}$

**21.** $-3\left( \begin{bmatrix} 0 & -3 \\ 7 & 2 \end{bmatrix} + \begin{bmatrix} -6 & 3 \\ 8 & 1 \end{bmatrix} \right) - \begin{bmatrix} 4 & -4 \\ 7 & -9 \end{bmatrix} = -3\begin{bmatrix} -6 & 0 \\ 15 & 3 \end{bmatrix} - \begin{bmatrix} 4 & -4 \\ 7 & -9 \end{bmatrix} = \begin{bmatrix} 18 & 0 \\ -45 & -9 \end{bmatrix} - \begin{bmatrix} 4 & -4 \\ 7 & -9 \end{bmatrix}$

$= \begin{bmatrix} 14 & 4 \\ -52 & 0 \end{bmatrix}$

**23.** $X = 3\begin{bmatrix} -2 & -1 \\ 1 & 0 \\ 3 & -4 \end{bmatrix} - 2\begin{bmatrix} 0 & 3 \\ 2 & 0 \\ -4 & -1 \end{bmatrix} = \begin{bmatrix} -6 & -3 \\ 3 & 0 \\ 9 & -12 \end{bmatrix} - \begin{bmatrix} 0 & 6 \\ 4 & 0 \\ -8 & -2 \end{bmatrix} = \begin{bmatrix} -6 & -9 \\ -1 & 0 \\ 17 & -10 \end{bmatrix}$

**25.** $X = -\frac{3}{2}A + \frac{1}{2}B = -\frac{3}{2}\begin{bmatrix} -2 & -1 \\ 1 & 0 \\ 3 & -4 \end{bmatrix} + \frac{1}{2}\begin{bmatrix} 0 & 3 \\ 2 & 0 \\ -4 & -1 \end{bmatrix} = \begin{bmatrix} 3 & 3 \\ -\frac{1}{2} & 0 \\ -\frac{13}{2} & \frac{11}{2} \end{bmatrix}$

**27.** $A$ is $3 \times 2$ and $B$ is $3 \times 3 \implies AB$ is not defined.    **29.** $AB = \begin{bmatrix} -1 & 6 \\ -4 & 5 \\ 0 & 3 \end{bmatrix}\begin{bmatrix} 2 & 3 \\ 0 & 9 \end{bmatrix} = \begin{bmatrix} -2 & 51 \\ -8 & 33 \\ 0 & 27 \end{bmatrix}$

**31.** $A$ is $3 \times 3$, $B$ is $3 \times 3 \implies AB$ is $3 \times 3$.

$AB = \begin{bmatrix} 5 & 0 & 0 \\ 0 & -8 & 0 \\ 0 & 0 & 7 \end{bmatrix}\begin{bmatrix} \frac{1}{5} & 0 & 0 \\ 0 & -\frac{1}{8} & 0 \\ 0 & 0 & \frac{1}{2} \end{bmatrix} = \begin{bmatrix} 1 & 0 & 0 \\ 0 & 1 & 0 \\ 0 & 0 & \frac{7}{2} \end{bmatrix}$

**33.** $AB = \begin{bmatrix} 5 \\ 6 \end{bmatrix}\begin{bmatrix} -3 & -1 & -5 & -9 \end{bmatrix} = \begin{bmatrix} -15 & -5 & -25 & -45 \\ -18 & -6 & -30 & -54 \end{bmatrix}$

**35. (a)** $AB = \begin{bmatrix} 1 & 2 \\ 5 & 2 \end{bmatrix}\begin{bmatrix} 2 & -1 \\ -1 & 8 \end{bmatrix} = \begin{bmatrix} 2-2 & -1+16 \\ 10-2 & -5+16 \end{bmatrix} = \begin{bmatrix} 0 & 15 \\ 8 & 11 \end{bmatrix}$

**(b)** $BA = \begin{bmatrix} 2 & -1 \\ -1 & 8 \end{bmatrix}\begin{bmatrix} 1 & 2 \\ 5 & 2 \end{bmatrix} = \begin{bmatrix} 2-5 & 4-2 \\ -1+40 & -2+16 \end{bmatrix} = \begin{bmatrix} -3 & 2 \\ 39 & 14 \end{bmatrix}$

**(c)** $A^2 = \begin{bmatrix} 1 & 2 \\ 5 & 2 \end{bmatrix}\begin{bmatrix} 1 & 2 \\ 5 & 2 \end{bmatrix} = \begin{bmatrix} 1+10 & 2+4 \\ 5+10 & 10+4 \end{bmatrix} = \begin{bmatrix} 11 & 6 \\ 15 & 14 \end{bmatrix}$

**37. (a)** $AB = \begin{bmatrix} 3 & -1 \\ 1 & 3 \end{bmatrix}\begin{bmatrix} 1 & -3 \\ 3 & 1 \end{bmatrix} = \begin{bmatrix} 3-3 & -9-1 \\ 1+9 & -3+3 \end{bmatrix} = \begin{bmatrix} 0 & -10 \\ 10 & 0 \end{bmatrix}$

**(b)** $BA = \begin{bmatrix} 1 & -3 \\ 3 & 1 \end{bmatrix}\begin{bmatrix} 3 & -1 \\ 1 & 3 \end{bmatrix} = \begin{bmatrix} 3-3 & -1-9 \\ 9+1 & -3+3 \end{bmatrix} = \begin{bmatrix} 0 & -10 \\ 10 & 0 \end{bmatrix}$

**(c)** $A^2 = \begin{bmatrix} 3 & -1 \\ 1 & 3 \end{bmatrix}\begin{bmatrix} 3 & -1 \\ 1 & 3 \end{bmatrix} = \begin{bmatrix} 9-1 & -3-3 \\ 3+3 & -1+9 \end{bmatrix} = \begin{bmatrix} 8 & -6 \\ 6 & 8 \end{bmatrix}$

**39. (a)** $AB = \begin{bmatrix} 7 \\ 8 \\ -1 \end{bmatrix}\begin{bmatrix} 1 & 1 & 2 \end{bmatrix} = \begin{bmatrix} 7 & 7 & 14 \\ 8 & 8 & 16 \\ -1 & -1 & -2 \end{bmatrix}$

**(b)** $BA = \begin{bmatrix} 1 & 1 & 2 \end{bmatrix}\begin{bmatrix} 7 \\ 8 \\ -1 \end{bmatrix} = [7+8-2] = [13]$    **(c)** $A^2$ is not defined.

**41.** $AB = \begin{bmatrix} 70 & -17 & 73 \\ 32 & 11 & 6 \\ 16 & -38 & 70 \end{bmatrix}$

**43.** $\begin{bmatrix} -3 & 8 & -6 & 8 \\ -12 & 15 & 9 & 6 \\ 5 & -1 & 1 & 5 \end{bmatrix}\begin{bmatrix} 3 & 1 & 6 \\ 24 & 15 & 14 \\ 16 & 10 & 21 \\ 8 & -4 & 10 \end{bmatrix} = \begin{bmatrix} 151 & 25 & 48 \\ 516 & 279 & 387 \\ 47 & -20 & 87 \end{bmatrix}$

**45.** $A$ is $2 \times 4$ and $B$ is $2 \times 4 \implies AB$ is not defined.

**47.** $\left( \begin{bmatrix} 3 & 1 \\ 0 & -2 \end{bmatrix} \begin{bmatrix} 1 & 0 \\ -2 & 2 \end{bmatrix} \right) \begin{bmatrix} 1 & 0 \\ 2 & 4 \end{bmatrix} = \begin{bmatrix} 1 & 2 \\ 4 & -4 \end{bmatrix} \begin{bmatrix} 1 & 0 \\ 2 & 4 \end{bmatrix} = \begin{bmatrix} 5 & 8 \\ -4 & -16 \end{bmatrix}$

**49.** $\begin{bmatrix} 0 & 2 & -2 \\ 4 & 1 & 2 \end{bmatrix} \left( \begin{bmatrix} 4 & 0 \\ 0 & -1 \\ -1 & 2 \end{bmatrix} + \begin{bmatrix} -2 & 3 \\ -3 & 5 \\ 0 & -3 \end{bmatrix} \right) = \begin{bmatrix} 0 & 2 & -2 \\ 4 & 1 & 2 \end{bmatrix} \begin{bmatrix} 2 & 3 \\ -3 & 4 \\ -1 & -1 \end{bmatrix} = \begin{bmatrix} -4 & 10 \\ 3 & 14 \end{bmatrix}$

**51.** $\begin{bmatrix} 1 & 2 & \vdots & 4 \\ 3 & 2 & \vdots & 0 \end{bmatrix}$

(a) $\begin{bmatrix} 1 & 2 \\ 3 & 2 \end{bmatrix} \begin{bmatrix} 2 \\ 1 \end{bmatrix} = \begin{bmatrix} 4 \\ 8 \end{bmatrix} \implies \begin{bmatrix} 2 \\ 1 \end{bmatrix}$ is not a solution.    (b) $\begin{bmatrix} 1 & 2 \\ 3 & 2 \end{bmatrix} \begin{bmatrix} -2 \\ 3 \end{bmatrix} = \begin{bmatrix} 4 \\ 0 \end{bmatrix} \implies \begin{bmatrix} -2 \\ 3 \end{bmatrix}$ is a solution.

(c) $\begin{bmatrix} 1 & 2 \\ 3 & 2 \end{bmatrix} \begin{bmatrix} -4 \\ 4 \end{bmatrix} = \begin{bmatrix} 4 \\ -4 \end{bmatrix} \implies \begin{bmatrix} -4 \\ 4 \end{bmatrix}$ is not a    (d) $\begin{bmatrix} 1 & 2 \\ 3 & 2 \end{bmatrix} \begin{bmatrix} 2 \\ -3 \end{bmatrix} = \begin{bmatrix} -4 \\ 0 \end{bmatrix} \implies \begin{bmatrix} 2 \\ -3 \end{bmatrix}$ is not a

solution.    solution.

**53.** $\begin{bmatrix} -2 & -3 & \vdots & -6 \\ 4 & 2 & \vdots & 20 \end{bmatrix}$

(a) $\begin{bmatrix} -2 & -3 \\ 4 & 2 \end{bmatrix} \begin{bmatrix} 3 \\ 0 \end{bmatrix} = \begin{bmatrix} -6 \\ 12 \end{bmatrix} \implies \begin{bmatrix} 3 \\ 0 \end{bmatrix}$ is not a    (b) $\begin{bmatrix} -2 & -3 \\ 4 & 2 \end{bmatrix} \begin{bmatrix} 6 \\ -2 \end{bmatrix} = \begin{bmatrix} -6 \\ 20 \end{bmatrix} \implies \begin{bmatrix} 6 \\ -2 \end{bmatrix}$ is a

solution.    solution.

(c) $\begin{bmatrix} -2 & -3 \\ 4 & 2 \end{bmatrix} \begin{bmatrix} -6 \\ 6 \end{bmatrix} = \begin{bmatrix} -6 \\ -12 \end{bmatrix} \implies \begin{bmatrix} -6 \\ 6 \end{bmatrix}$ is not    (d) $\begin{bmatrix} -2 & -3 \\ 4 & 2 \end{bmatrix} \begin{bmatrix} 4 \\ 2 \end{bmatrix} = \begin{bmatrix} -14 \\ 20 \end{bmatrix} \implies \begin{bmatrix} 4 \\ 2 \end{bmatrix}$ is not

a solution.    a solution.

**55.** (a) $A = \begin{bmatrix} -1 & 1 \\ -2 & 1 \end{bmatrix}$, $X = \begin{bmatrix} x_1 \\ x_2 \end{bmatrix}$, $B = \begin{bmatrix} 4 \\ 0 \end{bmatrix}$

(b) By Gauss-Jordan elimination on

$\begin{bmatrix} -1 & 1 & \vdots & 4 \\ -2 & 1 & \vdots & 0 \end{bmatrix} \begin{matrix} -R_1 \to \\ 2R_1 + R_2 \to \end{matrix} \begin{bmatrix} 1 & -1 & \vdots & -4 \\ 0 & -1 & \vdots & -8 \end{bmatrix} \begin{matrix} R_2 + R_1 \to \\ -R_2 \to \end{matrix} \begin{bmatrix} 1 & 0 & \vdots & 4 \\ 0 & 1 & \vdots & 8 \end{bmatrix}$,

we have $x_1 = 4$ and $x_2 = 8$. Thus, $X = \begin{bmatrix} 4 \\ 8 \end{bmatrix}$.

**57.** (a) $A = \begin{bmatrix} -2 & -3 \\ 6 & 1 \end{bmatrix}$, $X = \begin{bmatrix} x_1 \\ x_2 \end{bmatrix}$, $B = \begin{bmatrix} -4 \\ -36 \end{bmatrix}$

(b) $\begin{bmatrix} -2 & -3 & \vdots & -4 \\ 6 & 1 & \vdots & -36 \end{bmatrix}$

$3R_1 + R_2 \begin{bmatrix} -2 & -3 & \vdots & -4 \\ 0 & -8 & \vdots & -48 \end{bmatrix}$

$(-\frac{1}{8})R_2 \begin{bmatrix} -2 & -3 & \vdots & -4 \\ 0 & 1 & \vdots & 6 \end{bmatrix}$

$3R_2 + R_1 \begin{bmatrix} -2 & 0 & \vdots & 14 \\ 0 & 1 & \vdots & 6 \end{bmatrix}$

$-\frac{1}{2}R_1 \begin{bmatrix} 1 & 0 & \vdots & -7 \\ 0 & 1 & \vdots & 6 \end{bmatrix}$

$x_1 = -7, x_2 = 6$. Thus, $X = \begin{bmatrix} -7 \\ 6 \end{bmatrix}$.

**59.** (a) $A = \begin{bmatrix} 1 & -2 & 3 \\ -1 & 3 & -1 \\ 2 & -5 & 5 \end{bmatrix}$, $X = \begin{bmatrix} x_1 \\ x_2 \\ x_3 \end{bmatrix}$, $B = \begin{bmatrix} 9 \\ -6 \\ 17 \end{bmatrix}$

(b) $\begin{bmatrix} 1 & -2 & 3 & \vdots & 9 \\ -1 & 3 & -1 & \vdots & -6 \\ 2 & -5 & 5 & \vdots & 17 \end{bmatrix}$

$\begin{matrix} R_1 + R_2 \rightarrow \\ -2R_1 + R_3 \rightarrow \end{matrix} \begin{bmatrix} 1 & -2 & 3 & \vdots & 9 \\ 0 & 1 & 2 & \vdots & 3 \\ 0 & -1 & -1 & \vdots & -1 \end{bmatrix}$

$2R_2 + R_1 \rightarrow \begin{bmatrix} 1 & 0 & 7 & \vdots & 15 \\ 0 & 1 & 2 & \vdots & 3 \\ 0 & 0 & 1 & \vdots & 2 \end{bmatrix}$
$R_2 + R_3 \rightarrow$

$\begin{matrix} -7R_3 + R_1 \rightarrow \\ -2R_3 + R_2 \rightarrow \end{matrix} \begin{bmatrix} 1 & 0 & 0 & \vdots & 1 \\ 0 & 1 & 0 & \vdots & -1 \\ 0 & 0 & 1 & \vdots & 2 \end{bmatrix}$

$x_1 = 1, x_2 = -1, x_3 = 2$. Thus, $X = \begin{bmatrix} 1 \\ -1 \\ 2 \end{bmatrix}$.

**61.** (a) $A = \begin{bmatrix} 1 & -5 & 2 \\ -3 & 1 & -1 \\ 0 & -2 & 5 \end{bmatrix}$, $X = \begin{bmatrix} x_1 \\ x_2 \\ x_3 \end{bmatrix}$, $B = \begin{bmatrix} -20 \\ 8 \\ -16 \end{bmatrix}$

(b) $\begin{bmatrix} 1 & -5 & 2 & \vdots & -20 \\ -3 & 1 & -1 & \vdots & 8 \\ 0 & -2 & 5 & \vdots & -16 \end{bmatrix}$

$3R_1 + R_2 \begin{bmatrix} 1 & -5 & 2 & \vdots & -20 \\ 0 & -14 & 5 & \vdots & -52 \\ 0 & -2 & 5 & \vdots & -16 \end{bmatrix}$

$\begin{matrix} R_2 \\ R_3 \end{matrix} \begin{bmatrix} 1 & -5 & 2 & \vdots & -20 \\ 0 & -2 & 5 & \vdots & -16 \\ 0 & -14 & 5 & \vdots & -52 \end{bmatrix}$

$-7R_2 + R_3 \begin{bmatrix} 1 & -5 & 2 & \vdots & -20 \\ 0 & -2 & 5 & \vdots & -16 \\ 0 & 0 & -30 & \vdots & 60 \end{bmatrix}$

$-\frac{1}{30}R_3 \begin{bmatrix} 1 & -5 & 2 & \vdots & -20 \\ 0 & -2 & 5 & \vdots & -16 \\ 0 & 0 & 1 & \vdots & -2 \end{bmatrix}$

$\begin{matrix} -2R_3 + R_1 \\ -5R_3 + R_2 \end{matrix} \begin{bmatrix} 1 & -5 & 0 & \vdots & -16 \\ 0 & -2 & 0 & \vdots & -6 \\ 0 & 0 & 1 & \vdots & -2 \end{bmatrix}$

$(-\frac{1}{2})R_2 \begin{bmatrix} 1 & -5 & 0 & \vdots & -16 \\ 0 & 1 & 0 & \vdots & 3 \\ 0 & 0 & 1 & \vdots & -2 \end{bmatrix}$

$5R_2 + R_1 \begin{bmatrix} 1 & 0 & 0 & \vdots & -1 \\ 0 & 1 & 0 & \vdots & 3 \\ 0 & 0 & 1 & \vdots & -2 \end{bmatrix}$

$x_1 = -1, x_2 = 3, x_3 = -2$. Thus, $X = \begin{bmatrix} -1 \\ 3 \\ -2 \end{bmatrix}$.

**63.** $A = \begin{bmatrix} 2 & 0 \\ 4 & 5 \end{bmatrix}$

$$f(A) = A^2 - 5A + 2I = \begin{bmatrix} 2 & 0 \\ 4 & 5 \end{bmatrix}\begin{bmatrix} 2 & 0 \\ 4 & 5 \end{bmatrix} - 5\begin{bmatrix} 2 & 0 \\ 4 & 5 \end{bmatrix} + 2\begin{bmatrix} 1 & 0 \\ 0 & 1 \end{bmatrix} = \begin{bmatrix} -4 & 0 \\ 8 & 2 \end{bmatrix}$$

**65.** $1.20\begin{bmatrix} 70 & 50 & 25 \\ 35 & 100 & 70 \end{bmatrix} = \begin{bmatrix} 84 & 60 & 30 \\ 42 & 120 & 84 \end{bmatrix}$

**67.** $BA = \begin{bmatrix} 3.50 & 6.00 \end{bmatrix}\begin{bmatrix} 125 & 100 & 75 \\ 100 & 175 & 125 \end{bmatrix}$

$\qquad = \begin{bmatrix} 1037.50 & 1400 & 1012.50 \end{bmatrix}$

The entries in the last matrix $BA$ represent the profit for both crops at each of the three outlets.

**69.** $ST = \begin{bmatrix} 3 & 2 & 2 & 3 & 0 \\ 0 & 2 & 3 & 4 & 3 \\ 4 & 2 & 1 & 3 & 2 \end{bmatrix}\begin{bmatrix} 840 & 1100 \\ 1200 & 1350 \\ 1450 & 1650 \\ 2650 & 3000 \\ 3050 & 3200 \end{bmatrix} = \begin{bmatrix} \$15{,}770 & \$18{,}300 \\ \$26{,}500 & \$29{,}250 \\ \$21{,}260 & \$24{,}150 \end{bmatrix}$

The entries represent the wholesale and retail prices of the inventory at each outlet.

**71.** $P^2 = \begin{bmatrix} 0.6 & 0.1 & 0.1 \\ 0.2 & 0.7 & 0.1 \\ 0.2 & 0.2 & 0.8 \end{bmatrix}\begin{bmatrix} 0.6 & 0.1 & 0.1 \\ 0.2 & 0.7 & 0.1 \\ 0.2 & 0.2 & 0.8 \end{bmatrix} = \begin{bmatrix} 0.40 & 0.15 & 0.15 \\ 0.28 & 0.53 & 0.17 \\ 0.32 & 0.32 & 0.68 \end{bmatrix}$

This product represents the changes in party affiliation after *two* elections.

**73.** True

For 75–81, $A$ is of order $2 \times 3$, $B$ is of order $2 \times 3$, $C$ is of order $3 \times 2$ and $D$ is of order $2 \times 2$.

**75.** $A + 2C$ is not possible. $A$ and $C$ are not of the same order.

**77.** $AB$ is not possible. The number of columns of $A$ does not equal the number of rows of $B$.

**79.** $BC - D$ is possible. The resulting order is $2 \times 2$.

**81.** $D(A - 3B)$ is possible. The resulting order is $2 \times 3$.

**83.** $AC = \begin{bmatrix} 0 & 1 \\ 0 & 1 \end{bmatrix}\begin{bmatrix} 2 & 3 \\ 2 & 3 \end{bmatrix} = \begin{bmatrix} 2 & 3 \\ 2 & 3 \end{bmatrix}$

$\quad\;\; BC = \begin{bmatrix} 1 & 0 \\ 1 & 0 \end{bmatrix}\begin{bmatrix} 2 & 3 \\ 2 & 3 \end{bmatrix} = \begin{bmatrix} 2 & 3 \\ 2 & 3 \end{bmatrix}$

$\quad\;\; AC = BC$, but $A \neq B$.

**85.** (a) $A^2 = \begin{bmatrix} i & 0 \\ 0 & i \end{bmatrix} \begin{bmatrix} i & 0 \\ 0 & i \end{bmatrix} = \begin{bmatrix} -1 & 0 \\ 0 & -1 \end{bmatrix}$ and $i^2 = -1$

$A^3 = A^2A = \begin{bmatrix} -1 & 0 \\ 0 & -1 \end{bmatrix} \begin{bmatrix} i & 0 \\ 0 & i \end{bmatrix} = \begin{bmatrix} -i & 0 \\ 0 & -i \end{bmatrix}$ and $i^3 = -i$

$A^4 = A^3A = \begin{bmatrix} -i & 0 \\ 0 & -i \end{bmatrix} \begin{bmatrix} i & 0 \\ 0 & i \end{bmatrix} = \begin{bmatrix} 1 & 0 \\ 0 & 1 \end{bmatrix}$ and $i^4 = 1$

(b) $B^2 = \begin{bmatrix} 0 & -i \\ i & 0 \end{bmatrix} \begin{bmatrix} 0 & -i \\ i & 0 \end{bmatrix} = \begin{bmatrix} 1 & 0 \\ 0 & 1 \end{bmatrix}$,

The identity matrix

**87.** (a) $A = \begin{bmatrix} 0 & 2 \\ 0 & 0 \end{bmatrix}$, $B = \begin{bmatrix} 0 & 2 & 3 \\ 0 & 0 & 4 \\ 0 & 0 & 0 \end{bmatrix}$

(b) $A^2$ and $B^3$ are both zero matrices.

(c) If $A$ is $4 \times 4$, then $A^4$ will be the zero matrix.

(d) If $A$ is $n \times n$, then $A^n$ is the zero matrix.

**89.** $3 \ln 4 - \frac{1}{3}\ln(x^2 + 3) = \ln 4^3 - \ln(x^2 + 3)^{1/3}$

$= \ln\left[\dfrac{64}{(x^2 + 3)^{1/3}}\right]$

**91.** $\frac{1}{2}[2 \ln(x + 5) + \ln x - \ln(x - 8)] = \ln(x + 5) + \ln x^{1/2} - \ln(x - 8)^{1/2}$

$= \ln\left[\dfrac{(x + 5)\sqrt{x}}{\sqrt{x - 8}}\right]$

## Section 5.6    The Inverse of a Square Matrix

■ You should be able to find the inverse, if it exists, of a square matrix.

(a) Write the $n \times 2n$ matrix that consists of the given matrix $A$ on the left and the $n \times n$ identity matrix $I$ on the right to obtain $[A \;\vdots\; I]$. Note that we separate the matrices $A$ and $I$ by a dotted line. We call this process **adjoining** the matrices $A$ and $I$.

(b) If possible, row reduce $A$ to $I$ using elementary row operations on the *entire* matrix $[A \;\vdots\; I]$. The result will be the matrix $[I \;\vdots\; A^{-1}]$. If this is not possible, then $A$ is not invertible.

(c) Check your work by multiplying to see that $AA^{-1} = I = A^{-1}A$.

■ You should be able to use inverse matrices to solve systems of equation.

■ You should be able to find inverses using a graphing utility.

**Solutions to Odd-Numbered Exercises**

**1.** $AB = \begin{bmatrix} 2 & 1 \\ 5 & 3 \end{bmatrix} \begin{bmatrix} 3 & -1 \\ -5 & 2 \end{bmatrix} = \begin{bmatrix} 2(3) + 1(-5) & 2(-1) + 1(2) \\ 5(3) + 3(-5) & 5(-1) + 3(2) \end{bmatrix} = \begin{bmatrix} 1 & 0 \\ 0 & 1 \end{bmatrix}$

$BA = \begin{bmatrix} 3 & -1 \\ -5 & 2 \end{bmatrix} \begin{bmatrix} 2 & 1 \\ 5 & 3 \end{bmatrix} = \begin{bmatrix} 3(2) + (-1)(5) & 3(1) + (-1)(3) \\ -5(2) + 2(5) & -5(1) + 2(3) \end{bmatrix} = \begin{bmatrix} 1 & 0 \\ 0 & 1 \end{bmatrix}$

**3.** $AB = \begin{bmatrix} 1 & 2 \\ 3 & 4 \end{bmatrix}\begin{bmatrix} -2 & 1 \\ \frac{3}{2} & -\frac{1}{2} \end{bmatrix} = \begin{bmatrix} -2+3 & 1-1 \\ -6+6 & 3-2 \end{bmatrix} = \begin{bmatrix} 1 & 0 \\ 0 & 1 \end{bmatrix}$

$BA = \begin{bmatrix} -2 & 1 \\ \frac{3}{2} & -\frac{1}{2} \end{bmatrix}\begin{bmatrix} 1 & 2 \\ 3 & 4 \end{bmatrix} = \begin{bmatrix} -2+3 & -4+4 \\ \frac{3}{2}-\frac{3}{2} & 3-2 \end{bmatrix} = \begin{bmatrix} 1 & 0 \\ 0 & 1 \end{bmatrix}$

**5.** $AB = \begin{bmatrix} 2 & -17 & 11 \\ -1 & 11 & -7 \\ 0 & 3 & -2 \end{bmatrix}\begin{bmatrix} 1 & 1 & 2 \\ 2 & 4 & -3 \\ 3 & 6 & -5 \end{bmatrix}$

$= \begin{bmatrix} 2-34+33 & 2-68+66 & 4+51-55 \\ -1+22-21 & -1+44-42 & -2-33+35 \\ 6-6 & 12-12 & -9+10 \end{bmatrix} = \begin{bmatrix} 1 & 0 & 0 \\ 0 & 1 & 0 \\ 0 & 0 & 1 \end{bmatrix}$

$BA = \begin{bmatrix} 1 & 1 & 2 \\ 2 & 4 & -3 \\ 3 & 6 & -5 \end{bmatrix}\begin{bmatrix} 2 & -17 & 11 \\ -1 & 11 & -7 \\ 0 & 3 & -2 \end{bmatrix} = \begin{bmatrix} 2-1 & -17+11+6 & 11-7-4 \\ 4-4 & -34+44-9 & 22-28+6 \\ 6-6 & -51+66-15 & 33-42+10 \end{bmatrix} = \begin{bmatrix} 1 & 0 & 0 \\ 0 & 1 & 0 \\ 0 & 0 & 1 \end{bmatrix}$

**7.** $AB = \frac{1}{3}\begin{bmatrix} -2 & 2 & 3 \\ 1 & -1 & 0 \\ 0 & 1 & 4 \end{bmatrix}\begin{bmatrix} -4 & -5 & 3 \\ -4 & -8 & 3 \\ 1 & 2 & 0 \end{bmatrix} = \frac{1}{3}\begin{bmatrix} 8-8+3 & 10-16+6 & -6+6 \\ -4+4 & -5+8 & 3-3 \\ -4+4 & -8+8 & 3 \end{bmatrix}$

$= \frac{1}{3}\begin{bmatrix} 3 & 0 & 0 \\ 0 & 3 & 0 \\ 0 & 0 & 3 \end{bmatrix} = \begin{bmatrix} 1 & 0 & 0 \\ 0 & 1 & 0 \\ 0 & 0 & 1 \end{bmatrix}$

$BA = \frac{1}{3}\begin{bmatrix} -4 & -5 & 3 \\ -4 & -8 & 3 \\ 1 & 2 & 0 \end{bmatrix}\begin{bmatrix} -2 & 2 & 3 \\ 1 & -1 & 0 \\ 0 & 1 & 4 \end{bmatrix} = \frac{1}{3}\begin{bmatrix} 8-5 & -8+5+3 & -12+12 \\ 8-8 & -8+8+3 & -12+12 \\ -2+2 & 2-2 & 3 \end{bmatrix} = \begin{bmatrix} 1 & 0 & 0 \\ 0 & 1 & 0 \\ 0 & 0 & 1 \end{bmatrix}$

**9.** $AB = \begin{bmatrix} -1 & -4 \\ 1 & 2 \end{bmatrix}\begin{bmatrix} 1 & 2 \\ -\frac{1}{2} & -\frac{1}{2} \end{bmatrix} = \begin{bmatrix} 1 & 0 \\ 0 & 1 \end{bmatrix}; BA = \begin{bmatrix} 1 & 0 \\ 0 & 1 \end{bmatrix}$

**11.** $AB = \begin{bmatrix} 1.6 & 2 \\ -3.5 & -4.5 \end{bmatrix}\begin{bmatrix} 22.5 & 10 \\ -17.5 & -8 \end{bmatrix} = \begin{bmatrix} 1 & 0 \\ 0 & 1 \end{bmatrix}; BA = \begin{bmatrix} 1 & 0 \\ 0 & 1 \end{bmatrix}$

**13.** $[A \ \vdots \ I] = \begin{bmatrix} 2 & 0 & \vdots & 1 & 0 \\ 0 & 3 & \vdots & 0 & 1 \end{bmatrix}$

$\begin{matrix} \frac{1}{2}R_1 \to \\ \frac{1}{3}R_2 \to \end{matrix} \begin{bmatrix} 1 & 0 & \vdots & \frac{1}{2} & 0 \\ 0 & 1 & \vdots & 0 & \frac{1}{3} \end{bmatrix} = [I \ \vdots \ A^{-1}]$

$A^{-1} = \begin{bmatrix} \frac{1}{2} & 0 \\ 0 & \frac{1}{3} \end{bmatrix} = \frac{1}{6}\begin{bmatrix} 3 & 0 \\ 0 & 2 \end{bmatrix}$

**15.** $[A \ \vdots \ I] = \begin{bmatrix} 1 & -2 & \vdots & 1 & 0 \\ 2 & -3 & \vdots & 0 & 1 \end{bmatrix}$

$-2R_1 + R_2 \to \begin{bmatrix} 1 & -2 & \vdots & 1 & 0 \\ 0 & 1 & \vdots & -2 & 1 \end{bmatrix}$

$2R_2 + R_1 \to \begin{bmatrix} 1 & 0 & \vdots & -3 & 2 \\ 0 & 1 & \vdots & -2 & 1 \end{bmatrix} = [I \ \vdots \ A^{-1}]$

$A^{-1} = \begin{bmatrix} -3 & 2 \\ -2 & 1 \end{bmatrix}$

**17.** $[A \;\vdots\; I] = \begin{bmatrix} -1 & 1 & \vdots & 1 & 0 \\ -2 & 1 & \vdots & 0 & 1 \end{bmatrix}$

$-2R_1 + R_2 \rightarrow \begin{bmatrix} -1 & 1 & \vdots & 1 & 0 \\ 0 & -1 & \vdots & -2 & 1 \end{bmatrix}$

$R_2 + R_1 \rightarrow \begin{bmatrix} -1 & 0 & \vdots & -1 & 1 \\ 0 & -1 & \vdots & -2 & 1 \end{bmatrix}$

$\begin{matrix} -R_1 \rightarrow \\ -R_2 \rightarrow \end{matrix} \begin{bmatrix} 1 & 0 & \vdots & 1 & -1 \\ 0 & 1 & \vdots & 2 & -1 \end{bmatrix} = [I \;\vdots\; A^{-1}]$

$A^{-1} = \begin{bmatrix} 1 & -1 \\ 2 & -1 \end{bmatrix}$

**19.** $A = \begin{bmatrix} 2 & 7 & 1 \\ -3 & -9 & 2 \end{bmatrix}$

$A$ has no inverse because it is not square.

**21.** $[A \;\vdots\; I] = \begin{bmatrix} 1 & 1 & 1 & \vdots & 1 & 0 & 0 \\ 3 & 5 & 4 & \vdots & 0 & 1 & 0 \\ 3 & 6 & 5 & \vdots & 0 & 0 & 1 \end{bmatrix}$

$\begin{matrix} -3R_1 + R_2 \rightarrow \\ -3R_1 + R_3 \rightarrow \end{matrix} \begin{bmatrix} 1 & 1 & 1 & \vdots & 1 & 0 & 0 \\ 0 & 2 & 1 & \vdots & -3 & 1 & 0 \\ 0 & 3 & 2 & \vdots & -3 & 0 & 1 \end{bmatrix}$

$\begin{matrix} -R_2 + R_1 \rightarrow \\ \frac{1}{2}R_2 \rightarrow \\ -3R_2 + R_3 \rightarrow \end{matrix} \begin{bmatrix} 1 & 0 & \frac{1}{2} & \vdots & \frac{5}{2} & -\frac{1}{2} & 0 \\ 0 & 1 & \frac{1}{2} & \vdots & -\frac{3}{2} & \frac{1}{2} & 0 \\ 0 & 0 & \frac{1}{2} & \vdots & \frac{3}{2} & -\frac{3}{2} & 1 \end{bmatrix}$

$\begin{matrix} -R_3 + R_1 \rightarrow \\ -R_3 + R_2 \rightarrow \\ 2R_3 \rightarrow \end{matrix} \begin{bmatrix} 1 & 0 & 0 & \vdots & 1 & 1 & -1 \\ 0 & 1 & 0 & \vdots & -3 & 2 & -1 \\ 0 & 0 & 1 & \vdots & 3 & -3 & 2 \end{bmatrix}$

$= [I \;\vdots\; A^{-1}]$

$A^{-1} = \begin{bmatrix} 1 & 1 & -1 \\ -3 & 2 & -1 \\ 3 & -3 & 2 \end{bmatrix}$

**23.** $[A \;\vdots\; I] = \begin{bmatrix} -5 & 0 & 0 & \vdots & 1 & 0 & 0 \\ 2 & 0 & 0 & \vdots & 0 & 1 & 0 \\ -1 & 5 & 7 & \vdots & 0 & 0 & 1 \end{bmatrix}$

$\begin{matrix} \left(-\frac{1}{5}\right)R_1 \\ (-2)R_1 + R_2 \end{matrix} \begin{bmatrix} 1 & 0 & 0 & \vdots & -\frac{1}{5} & 0 & 0 \\ 0 & 0 & 0 & \vdots & \frac{2}{5} & 1 & 0 \\ -1 & 5 & 7 & \vdots & 0 & 0 & 1 \end{bmatrix}$

Not invertible   (row of zeros)

$A^1$ does not exist

**25.** Not invertible. $A^{-1}$ does not exist.

**27.** $A = \begin{bmatrix} -\frac{1}{2} & \frac{3}{4} & \frac{1}{4} \\ 1 & 0 & -\frac{3}{2} \\ 0 & -1 & \frac{1}{2} \end{bmatrix}$

$A^{-1} = \begin{bmatrix} -12 & -5 & -9 \\ -4 & -2 & -4 \\ -8 & -4 & -6 \end{bmatrix}$

**29.** $A = \begin{bmatrix} 0.1 & 0.2 & 0.3 \\ -0.3 & 0.2 & 0.2 \\ 0.5 & 0.4 & 0.4 \end{bmatrix}$

$A^{-1} = \frac{5}{11}\begin{bmatrix} 0 & -4 & 2 \\ -22 & 11 & 11 \\ 22 & -6 & -8 \end{bmatrix}$

**31.**  $A = \begin{bmatrix} -1 & 0 & 1 & 0 \\ 0 & 2 & 0 & -1 \\ 2 & 0 & -1 & 0 \\ 0 & -1 & 0 & 1 \end{bmatrix}$

$A^{-1} = \begin{bmatrix} 1 & 0 & 1 & 0 \\ 0 & 1 & 0 & 1 \\ 2 & 0 & 1 & 0 \\ 0 & 1 & 0 & 2 \end{bmatrix}$

**33.**  $\begin{bmatrix} 5 & 1 \\ -2 & -2 \end{bmatrix}^{-1} = \dfrac{1}{5(-2) - (-2)(1)} \begin{bmatrix} -2 & -1 \\ 2 & 5 \end{bmatrix} = \dfrac{1}{-8} \begin{bmatrix} -2 & -1 \\ 2 & 5 \end{bmatrix} = \begin{bmatrix} \frac{1}{4} & \frac{1}{8} \\ -\frac{1}{4} & -\frac{5}{8} \end{bmatrix}$

**35.**  $\begin{bmatrix} \frac{7}{2} & -\frac{3}{4} \\ \frac{1}{5} & \frac{4}{5} \end{bmatrix}^{-1} = \dfrac{1}{\left(\frac{7}{2}\right)\left(\frac{4}{5}\right) - \left(-\frac{3}{4}\right)\left(\frac{1}{5}\right)} \begin{bmatrix} \frac{4}{5} & \frac{3}{4} \\ -\frac{1}{5} & \frac{7}{2} \end{bmatrix}$

$= \dfrac{20}{59} \begin{bmatrix} \frac{4}{5} & \frac{3}{4} \\ -\frac{1}{5} & \frac{7}{2} \end{bmatrix} = \dfrac{1}{59} \begin{bmatrix} 16 & 15 \\ -4 & 70 \end{bmatrix}$

**37.**  $\begin{bmatrix} x \\ y \end{bmatrix} = \begin{bmatrix} -3 & 2 \\ -2 & 1 \end{bmatrix} \begin{bmatrix} 5 \\ 10 \end{bmatrix} = \begin{bmatrix} 5 \\ 0 \end{bmatrix}$

*Answer:* $(5, 0)$

**39.**  $\begin{bmatrix} x \\ y \end{bmatrix} = \begin{bmatrix} -3 & 2 \\ -2 & 1 \end{bmatrix} \begin{bmatrix} 4 \\ 2 \end{bmatrix} \begin{bmatrix} -8 \\ -6 \end{bmatrix}$

*Answer:* $(-8, -6)$

**41.**  $\begin{bmatrix} x \\ y \\ z \end{bmatrix} = \begin{bmatrix} 1 & 1 & -1 \\ -3 & 2 & -1 \\ 3 & -3 & 2 \end{bmatrix} \begin{bmatrix} 0 \\ 5 \\ 2 \end{bmatrix} = \begin{bmatrix} 3 \\ 8 \\ -11 \end{bmatrix}$

*Answer:* $(3, 8, -11)$

**43.**  $\begin{bmatrix} x_1 \\ x_2 \\ x_3 \\ x_4 \end{bmatrix} = \begin{bmatrix} -24 & 7 & 1 & -2 \\ -10 & 3 & 0 & -1 \\ -29 & 7 & 3 & -2 \\ 12 & -3 & -1 & 1 \end{bmatrix} \begin{bmatrix} 0 \\ 1 \\ -1 \\ 2 \end{bmatrix} = \begin{bmatrix} 2 \\ 1 \\ 0 \\ 0 \end{bmatrix}$

*Answer:* $(2, 1, 0, 0)$

**45.**  $A = \begin{bmatrix} 3 & 4 \\ 5 & 3 \end{bmatrix}$

$A^{-1} = \dfrac{1}{9 - 20} \begin{bmatrix} 3 & -4 \\ -5 & 3 \end{bmatrix}$

$\begin{bmatrix} x \\ y \end{bmatrix} = -\dfrac{1}{11} \begin{bmatrix} 3 & -4 \\ -5 & 3 \end{bmatrix} \begin{bmatrix} -2 \\ 4 \end{bmatrix} = -\dfrac{1}{11} \begin{bmatrix} -22 \\ 22 \end{bmatrix}$

$= \begin{bmatrix} 2 \\ -2 \end{bmatrix}$

*Answer:* $(2, -2)$

**47.**  $A = \begin{bmatrix} -0.4 & 0.8 \\ 2 & -4 \end{bmatrix}$

$A^{-1} = \dfrac{1}{1.6 - 1.6} \begin{bmatrix} -4 & -0.8 \\ -2 & -0.4 \end{bmatrix} \Rightarrow A^{-1}$

does not exist.

[The system actually has no solution.]

**49.**  $A = \begin{bmatrix} -\frac{1}{4} & \frac{3}{8} \\ \frac{3}{2} & \frac{3}{4} \end{bmatrix}$

$A^{-1} = \begin{bmatrix} -1 & \frac{1}{2} \\ 2 & \frac{1}{3} \end{bmatrix}$

$\begin{bmatrix} x \\ y \end{bmatrix} = A^{-1}b = \begin{bmatrix} -1 & \frac{1}{2} \\ 2 & \frac{1}{3} \end{bmatrix} \begin{bmatrix} -2 \\ -12 \end{bmatrix} = \begin{bmatrix} -4 \\ -8 \end{bmatrix}$

*Answer:* $(-4, -8)$

**51.** $A = \begin{bmatrix} 4 & -1 & 1 \\ 2 & 2 & 3 \\ 5 & -2 & 6 \end{bmatrix}$

$A^{-1} = \frac{1}{55} \begin{bmatrix} 18 & 4 & -5 \\ 3 & 19 & -10 \\ -14 & 3 & 10 \end{bmatrix}$

$\begin{bmatrix} x \\ y \\ z \end{bmatrix} = \frac{1}{55} \begin{bmatrix} 18 & 4 & -5 \\ 3 & 19 & -10 \\ -14 & 3 & 10 \end{bmatrix} \begin{bmatrix} -5 \\ 10 \\ 1 \end{bmatrix} = \frac{1}{55} \begin{bmatrix} -55 \\ 165 \\ 110 \end{bmatrix}$

$= \begin{bmatrix} -1 \\ 3 \\ 2 \end{bmatrix}$

*Answer:* $(-1, 3, 2)$

**53.** $A = \begin{bmatrix} 5 & -3 & 2 \\ 2 & 2 & -3 \\ -1 & 7 & -8 \end{bmatrix}$

$A^{-1}$ does not exist. [The system actually has an infinite number of solutions of the form

$x = 0.3125t + 0.8125$

$y = 1.1875t + 0.6875$

$z = t$

where $t$ is any real number.]

**55.** $\begin{bmatrix} 7 & -3 & 0 & 2 & \vdots & 41 \\ -2 & 1 & 0 & -1 & \vdots & -13 \\ 4 & 0 & 1 & -2 & \vdots & 12 \\ -1 & 1 & 0 & -1 & \vdots & -8 \end{bmatrix}$ row reduces to $\begin{bmatrix} 1 & 0 & 0 & 0 & \vdots & 5 \\ 0 & 1 & 0 & 0 & \vdots & 0 \\ 0 & 0 & 1 & 0 & \vdots & -2 \\ 0 & 0 & 0 & 1 & \vdots & 3 \end{bmatrix}$

*Solution:* $(5, 0, -2, 3)$

For 57 and 59 use $A = \begin{bmatrix} 1 & 1 & 1 \\ 0.065 & 0.07 & 0.09 \\ 0 & 2 & -1 \end{bmatrix}$. Using the methods of this section, we have $A^{-1} = \frac{1}{11} \begin{bmatrix} 50 & -600 & -4 \\ -13 & 200 & 5 \\ -26 & 400 & -1 \end{bmatrix}$.

**57.** $X = A^{-1}B = \frac{1}{11} \begin{bmatrix} 50 & -600 & -4 \\ -13 & 200 & 5 \\ -26 & 400 & -1 \end{bmatrix} \begin{bmatrix} 25,000 \\ 1900 \\ 0 \end{bmatrix} = \begin{bmatrix} 10,000 \\ 5000 \\ 10,000 \end{bmatrix}$

*Answer:* $10,000 in AAA bonds, $5000 in A-bonds and $10,000 in B-bonds.

**59.** $X = A^{-1}B = \frac{1}{11} \begin{bmatrix} 50 & -600 & -4 \\ -13 & 200 & 5 \\ -26 & 400 & -1 \end{bmatrix} \begin{bmatrix} 65,000 \\ 5050 \\ 0 \end{bmatrix} = \begin{bmatrix} 20,000 \\ 15,000 \\ 30,000 \end{bmatrix}$

*Answer:* $20,000 in AAA bonds, $15,000 in A-bonds and $30,000 in B-bonds.

**61. (a)** $A = \begin{bmatrix} 2 & 0 & 4 \\ 0 & 1 & 4 \\ 1 & 1 & -1 \end{bmatrix}$    $A^{-1} = \frac{1}{14} \begin{bmatrix} 5 & -4 & 4 \\ -4 & 6 & 8 \\ 1 & 2 & -2 \end{bmatrix}$

$\begin{bmatrix} I_1 \\ I_2 \\ I_3 \end{bmatrix} = \frac{1}{14} \begin{bmatrix} 5 & -4 & 4 \\ -4 & 6 & 8 \\ 1 & 2 & -2 \end{bmatrix} \begin{bmatrix} 14 \\ 28 \\ 0 \end{bmatrix} = \begin{bmatrix} -3 \\ 8 \\ 5 \end{bmatrix}$

*Answer:* $I_1 = -3$ amps, $I_2 = 8$ amps, $I_3 = 5$ amps

**—CONTINUED—**

**61.** **—CONTINUED—**

(b) $A = \begin{bmatrix} 2 & 0 & 4 \\ 0 & 1 & 4 \\ 1 & 1 & -1 \end{bmatrix}$    $A^{-1} = \frac{1}{14}\begin{bmatrix} 5 & -4 & 4 \\ -4 & 6 & 8 \\ 1 & 2 & -2 \end{bmatrix}$

$\begin{bmatrix} I_1 \\ I_2 \\ I_3 \end{bmatrix} = \frac{1}{14}\begin{bmatrix} 5 & -4 & 4 \\ -4 & 6 & 8 \\ 1 & 2 & -2 \end{bmatrix}\begin{bmatrix} 10 \\ 10 \\ 0 \end{bmatrix} = \begin{bmatrix} 5/7 \\ 10/7 \\ 15/7 \end{bmatrix}$

*Answer:* $I_1 = 5/7$ amps, $I_2 = 10/7$ amps, $I_3 = 15/7$ amps

**63.** True. $AA^{-1} = A^{-1}A = I$

**65.** $AA^{-1} = \begin{bmatrix} a & b \\ c & d \end{bmatrix}\left(\frac{1}{ad-bc}\right)\begin{bmatrix} d & -b \\ -c & a \end{bmatrix} = \frac{1}{ad-bc}\begin{bmatrix} a & b \\ c & d \end{bmatrix}\begin{bmatrix} d & -b \\ -c & a \end{bmatrix}$

$\qquad = \frac{1}{ad-bc}\begin{bmatrix} ad-bc & 0 \\ 0 & ad-bc \end{bmatrix} = \begin{bmatrix} 1 & 0 \\ 0 & 1 \end{bmatrix}$

$A^{-1}A = \frac{1}{ad-bc}\begin{bmatrix} d & -b \\ -c & a \end{bmatrix}\begin{bmatrix} a & b \\ c & d \end{bmatrix} = \frac{1}{ad-bc}\begin{bmatrix} ad-bc & 0 \\ 0 & ad-bc \end{bmatrix} = \begin{bmatrix} 1 & 0 \\ 0 & 1 \end{bmatrix}$

**67.** $\dfrac{\left(\dfrac{9}{x}\right)}{\left(\dfrac{6}{x}+2\right)} = \dfrac{\left(\dfrac{9}{x}\right)}{\left(\dfrac{6+2x}{x}\right)} = \dfrac{9}{x}\cdot\dfrac{x}{6+2x} = \dfrac{9}{6+2x}, \quad x \neq 0$

**69.** $\dfrac{\dfrac{4}{x^2-9}+\dfrac{2}{x-2}}{\dfrac{1}{x+3}+\dfrac{1}{x-3}} \cdot \dfrac{(x^2-9)(x-2)}{(x^2-9)(x-2)}$

$\qquad = \dfrac{4(x-2)+2(x^2-9)}{(x-3)(x-2)+(x+3)(x-2)}$

$\qquad = \dfrac{2x^2+4x-26}{2x^2-4x}$

$\qquad = \dfrac{x^2+2x-13}{x(x-2)}, \quad x \neq \pm 3$

**71.** $e^{2x}+2e^x-15 = (e^x+5)(e^x-3) = 0 \implies e^x = 3 \implies x = \ln 3 \approx 1.099$

**73.** $7\ln 3x = 12$

$\qquad \ln 3x = \frac{12}{7}$

$\qquad 3x = e^{12/7}$

$\qquad x = \frac{1}{3}e^{12/7} \approx 1.851$

## Section 5.7    The Determinant of a Square Matrix

- You should be able to determine the determinant of a matrix of order $2 \times 2$ by using the products of the diagonals.
- You should be able to use expansion by cofactors to find the determinant of a matrix of order 3 or greater.
- The determinant of a triangular matrix equals the product of the entries on the main diagonal.
- You should be able to calculate determinants using a graphing utility.

**Solutions to Odd-Numbered Exercises**

**1.** $|4| = 4$

**3.** $\begin{vmatrix} 8 & 4 \\ 2 & 3 \end{vmatrix} = 8(3) - 4(2) = 24 - 8 = 16$

**5.** $\begin{vmatrix} 6 & 2 \\ -5 & 3 \end{vmatrix} = 6(3) - (2)(-5) = 18 + 10 = 28$

**7.** $\begin{vmatrix} -7 & 6 \\ \frac{1}{2} & 3 \end{vmatrix} = -7(3) - 6\left(\frac{1}{2}\right) = -21 - 3 = -24$

**9.** $\begin{vmatrix} 2 & -1 & 0 \\ 4 & 2 & 1 \\ 4 & 2 & 1 \end{vmatrix} = 2\begin{vmatrix} 2 & 1 \\ 2 & 1 \end{vmatrix} - 4\begin{vmatrix} -1 & 0 \\ 2 & 1 \end{vmatrix} + 4\begin{vmatrix} -1 & 0 \\ 2 & 1 \end{vmatrix} = 2(0) - 4(-1) + 4(-1) = 0$

**11.** $\begin{vmatrix} -1 & 2 & -5 \\ 0 & 3 & 4 \\ 0 & 0 & 3 \end{vmatrix} = (-1)(3)(3) = -9$    (Upper Triangular)

**13.** $\begin{vmatrix} 0.3 & 0.2 & 0.2 \\ 0.2 & 0.2 & 0.2 \\ -0.4 & 0.4 & 0.3 \end{vmatrix} = -0.002$

**15.** $\begin{bmatrix} 3 & 4 \\ 2 & -5 \end{bmatrix}$

(a) $M_{11} = -5$

$M_{12} = 2$

$M_{21} = 4$

$M_{22} = 3$

(b) $C_{11} = M_{11} = -5$

$C_{12} = -M_{12} = -2$

$C_{21} = -M_{21} = -4$

$C_{22} = M_{22} = 3$

**17.** $\begin{bmatrix} 3 & -2 & 8 \\ 3 & 2 & -6 \\ -1 & 3 & 6 \end{bmatrix}$

(a) $M_{11} = \begin{vmatrix} 2 & -6 \\ 3 & 6 \end{vmatrix} = 12 + 18 = 30$

$M_{12} = \begin{vmatrix} 3 & -6 \\ -1 & 6 \end{vmatrix} = 18 - 6 = 12$

$M_{13} = \begin{vmatrix} 3 & 2 \\ -1 & 3 \end{vmatrix} = 9 + 2 = 11$

$M_{21} = \begin{vmatrix} -2 & 8 \\ 3 & 6 \end{vmatrix} = -12 - 24 = -36$

$M_{22} = \begin{vmatrix} 3 & 8 \\ -1 & 6 \end{vmatrix} = 18 + 8 = 26$

$M_{23} = \begin{vmatrix} 3 & -2 \\ -1 & 3 \end{vmatrix} = 9 - 2 = 7$

$M_{31} = \begin{vmatrix} -2 & 8 \\ 2 & -6 \end{vmatrix} = 12 - 16 = -4$

$M_{32} = \begin{vmatrix} 3 & 8 \\ 3 & -6 \end{vmatrix} = -18 - 24 = -42$

$M_{33} = \begin{vmatrix} 3 & -2 \\ 3 & 2 \end{vmatrix} = 6 + 6 = 12$

(b) $C_{11} = (-1)^2 M_{11} = \ \ 30$

$C_{12} = (-1)^3 M_{12} = -12$

$C_{13} = (-1)^4 M_{13} = \ \ 11$

$C_{21} = (-1)^3 M_{21} = \ \ 36$

$C_{22} = (-1)^4 M_{22} = \ \ 26$

$C_{23} = (-1)^5 M_{23} = \ \ -7$

$C_{31} = (-1)^4 M_{31} = \ \ -4$

$C_{32} = (-1)^5 M_{32} = \ \ 42$

$C_{33} = (-1)^6 M_{33} = \ \ 12$

**19.** (a) $\begin{vmatrix} -3 & 2 & 1 \\ 4 & 5 & 6 \\ 2 & -3 & 1 \end{vmatrix} = -3 \begin{vmatrix} 5 & 6 \\ -3 & 1 \end{vmatrix} - 2 \begin{vmatrix} 4 & 6 \\ 2 & 1 \end{vmatrix} + \begin{vmatrix} 4 & 5 \\ 2 & -3 \end{vmatrix} = -3(23) - 2(-8) - 22 = -75$

(b) $\begin{vmatrix} -3 & 2 & 1 \\ 4 & 5 & 6 \\ 2 & -3 & 1 \end{vmatrix} = -2 \begin{vmatrix} 4 & 6 \\ 2 & 1 \end{vmatrix} + 5 \begin{vmatrix} -3 & 1 \\ 2 & 1 \end{vmatrix} + 3 \begin{vmatrix} -3 & 1 \\ 4 & 6 \end{vmatrix} = -2(-8) + 5(-5) + 3(-22) = -75$

**21.** (a) $\begin{vmatrix} 6 & 0 & -3 & 5 \\ 4 & 13 & 6 & -8 \\ -1 & 0 & 7 & 4 \\ 8 & 6 & 0 & 2 \end{vmatrix} = -4 \begin{vmatrix} 0 & -3 & 5 \\ 0 & 7 & 4 \\ 6 & 0 & 2 \end{vmatrix} + 13 \begin{vmatrix} 6 & -3 & 5 \\ -1 & 7 & 4 \\ 8 & 0 & 2 \end{vmatrix} - 6 \begin{vmatrix} 6 & 0 & 5 \\ -1 & 0 & 4 \\ 8 & 6 & 2 \end{vmatrix} - 8 \begin{vmatrix} 6 & 0 & -3 \\ -1 & 0 & 7 \\ 8 & 6 & 0 \end{vmatrix}$

$= -4(-282) + 13(-298) - 6(-174) - 8(-234) = 170$

(b) $\begin{vmatrix} 6 & 0 & -3 & 5 \\ 4 & 13 & 6 & -8 \\ -1 & 0 & 7 & 4 \\ 8 & 6 & 0 & 2 \end{vmatrix} = 0 \begin{vmatrix} 4 & 6 & -8 \\ -1 & 7 & 4 \\ 8 & 0 & 2 \end{vmatrix} + 13 \begin{vmatrix} 6 & -3 & 5 \\ -1 & 7 & 4 \\ 8 & 0 & 2 \end{vmatrix} + 0 \begin{vmatrix} 6 & -3 & 5 \\ 4 & 6 & -8 \\ 8 & 0 & 2 \end{vmatrix} + 6 \begin{vmatrix} 6 & -3 & 5 \\ 4 & 6 & -8 \\ -1 & 7 & 4 \end{vmatrix}$

$= 0 + 13(-298) + 0 + 6(674) = 170$

**23.** Expand by Column 3.

$$\begin{vmatrix} 1 & 4 & -2 \\ 3 & 2 & 0 \\ -1 & 4 & 3 \end{vmatrix} = -2\begin{vmatrix} 3 & 2 \\ -1 & 4 \end{vmatrix} + 3\begin{vmatrix} 1 & 4 \\ 3 & 2 \end{vmatrix} = -2(14) + 3(-10) = -58$$

**25.** $\begin{vmatrix} 2 & 4 & 6 \\ 0 & 3 & 1 \\ 0 & 0 & -5 \end{vmatrix} = (2)(3)(-5) = -30$  (Upper Triangular)

**27.** Expand by Column 3.

$$\begin{vmatrix} 2 & 6 & 6 & 2 \\ 2 & 7 & 3 & 6 \\ 1 & 5 & 0 & 1 \\ 3 & 7 & 0 & 7 \end{vmatrix} = 6\begin{vmatrix} 2 & 7 & 6 \\ 1 & 5 & 1 \\ 3 & 7 & 7 \end{vmatrix} - 3\begin{vmatrix} 2 & 6 & 2 \\ 1 & 5 & 1 \\ 3 & 7 & 7 \end{vmatrix} = 6(-20) - 3(16) = -168$$

**29.** Expand by Column 2.

$$\begin{vmatrix} 3 & 2 & 4 & -1 & 5 \\ -2 & 0 & 1 & 3 & 2 \\ 1 & 0 & 0 & 4 & 0 \\ 6 & 0 & 2 & -1 & 0 \\ 3 & 0 & 5 & 1 & 0 \end{vmatrix} = -2\begin{vmatrix} -2 & 1 & 3 & 2 \\ 1 & 0 & 4 & 0 \\ 6 & 2 & -1 & 0 \\ 3 & 5 & 1 & 0 \end{vmatrix} = (-2)(-2)\begin{vmatrix} 1 & 0 & 4 \\ 6 & 2 & -1 \\ 3 & 5 & 1 \end{vmatrix} = 4(103) = 412$$

**31.** $\begin{vmatrix} 4 & 0 & 0 & 0 \\ 6 & -5 & 0 & 0 \\ 1 & 3 & 1 & 0 \\ 1 & -2 & 7 & 3 \end{vmatrix} = (4)(-5)(1)(3) = -60$   (Lower Triangular)

**33.** $\det(A) = (-6)(-1)(-7)(-2)(-2) = -168$   (Upper Triangular)

**35.** $\begin{vmatrix} 1 & -1 & 8 & 4 \\ 2 & 6 & 0 & -4 \\ 2 & 0 & 2 & 6 \\ 0 & 2 & 8 & 0 \end{vmatrix} = -336$

**37.** $\begin{vmatrix} 3 & -2 & 4 & 3 & 1 \\ -1 & 0 & 2 & 1 & 0 \\ 5 & -1 & 0 & 3 & 2 \\ 4 & 7 & -8 & 0 & 0 \\ 1 & 2 & 3 & 0 & 2 \end{vmatrix} = 410$

**39.** (a) $\begin{vmatrix} -1 & 0 \\ 0 & 3 \end{vmatrix} = -3$

(b) $\begin{vmatrix} 2 & 0 \\ 0 & -1 \end{vmatrix} = -2$

(c) $\begin{bmatrix} -1 & 0 \\ 0 & 3 \end{bmatrix}\begin{bmatrix} 2 & 0 \\ 0 & -1 \end{bmatrix} = \begin{bmatrix} -2 & 0 \\ 0 & -3 \end{bmatrix}$

(d) $\begin{vmatrix} -2 & 0 \\ 0 & -3 \end{vmatrix} = 6$   [Note: $|AB| = |A|\,|B|$]

**41. (a)** $\begin{vmatrix} -1 & 2 & 1 \\ 1 & 0 & 1 \\ 0 & 1 & 0 \end{vmatrix} = 2$

**(b)** $\begin{vmatrix} -1 & 0 & 0 \\ 0 & 2 & 0 \\ 0 & 0 & 3 \end{vmatrix} = -6$

**(c)** $\begin{bmatrix} -1 & 2 & 1 \\ 1 & 0 & 1 \\ 0 & 1 & 0 \end{bmatrix} \begin{bmatrix} -1 & 0 & 0 \\ 0 & 2 & 0 \\ 0 & 0 & 3 \end{bmatrix} = \begin{bmatrix} 1 & 4 & 3 \\ -1 & 0 & 3 \\ 0 & 2 & 0 \end{bmatrix}$

**(d)** $\begin{vmatrix} 1 & 4 & 3 \\ -1 & 0 & 3 \\ 0 & 2 & 0 \end{vmatrix} = -12$    [Note: $|AB| = |A|\,|B|$]

**43. (a)** $|A| = -25$

**(b)** $|B| = -220$

**(c)** $AB = \begin{bmatrix} -7 & -16 & -1 & -28 \\ -4 & -14 & -11 & 8 \\ 13 & 4 & 4 & -4 \\ -2 & 3 & 2 & 2 \end{bmatrix}$

**(d)** $|AB| = 5500$    [Note: $|AB| = |A|\,|B|$]

**45.** $\begin{vmatrix} w & x \\ y & z \end{vmatrix} = wz - xy$

$-\begin{vmatrix} y & z \\ w & x \end{vmatrix} = -(xy - wz) = wz - xy$

Thus, $\begin{vmatrix} w & x \\ y & z \end{vmatrix} = -\begin{vmatrix} y & z \\ w & x \end{vmatrix}$.

**47.** $\begin{vmatrix} w & x \\ y & z \end{vmatrix} = wz - xy$

$\begin{vmatrix} w & x + cw \\ y & z + cy \end{vmatrix} = w(z + cy) - y(x + cw) = wz - xy$

Thus, $\begin{vmatrix} w & x \\ y & z \end{vmatrix} = \begin{vmatrix} w & x + cw \\ y & z + cy \end{vmatrix}$.

**49.** $\begin{vmatrix} 1 & x & x^2 \\ 1 & y & y^2 \\ 1 & z & z^2 \end{vmatrix} = \begin{vmatrix} y & y^2 \\ z & z^2 \end{vmatrix} - \begin{vmatrix} x & x^2 \\ z & z^2 \end{vmatrix} + \begin{vmatrix} x & x^2 \\ y & y^2 \end{vmatrix}$

$= (yz^2 - y^2z) - (xz^2 - x^2z) + (xy^2 - x^2y)$

$= yz^2 - xz^2 - y^2z + x^2z + xy(y - x)$

$= z^2(y - x) - z(y^2 - x^2) + xy(y - x)$

$= z^2(y - x) - z(y - x)(y + x) + xy(y - x)$

$= (y - x)[z^2 - z(y + x) + xy]$

$= (y - x)[z^2 - zy - zx + xy]$

$= (y - x)[z^2 - zx - zy + xy]$

$= (y - x)[z(z - x) - y(z - x)]$

$= (y - x)(z - x)(z - y)$

**51.**    $\begin{vmatrix} x + 3 & 2 \\ 1 & x + 2 \end{vmatrix} = 0$

$(x + 3)(x + 2) - 2 = 0$

$x^2 + 5x + 4 = 0$

$(x + 4)(x + 1) = 0$

$x = -4, -1$

**53.**    $\begin{vmatrix} 4u & -1 \\ -1 & 2v \end{vmatrix} = 8uv - 1$

**55.**    $\begin{vmatrix} e^{2x} & e^{3x} \\ 2e^{2x} & 3e^{3x} \end{vmatrix} = 3e^{5x} - 2e^{5x} = e^{5x}$

**57.**    $\begin{vmatrix} x & \ln x \\ 1 & \dfrac{1}{x} \end{vmatrix} = 1 - \ln x$

**59.**    True. Expand along the row of zeros.

**61.**    Let $A = \begin{bmatrix} 1 & 3 \\ -2 & 4 \end{bmatrix}$ and $B = \begin{bmatrix} -4 & 0 \\ 3 & 5 \end{bmatrix}$.

$|A| = \begin{vmatrix} 1 & 3 \\ -2 & 4 \end{vmatrix} = 10, \quad |B| = \begin{vmatrix} -4 & 0 \\ 3 & 5 \end{vmatrix} = -20$

$A + B = \begin{bmatrix} -3 & 3 \\ 1 & 9 \end{bmatrix}, \quad |A + B| = \begin{vmatrix} -3 & 3 \\ 1 & 9 \end{vmatrix} = -30$

Thus, $|A + B| \neq |A| + |B|$. Your answer may differ, depending on how you choose $A$ and $B$.

**63.**    (a)  Columns 2 and 3 are interchanged.

(b)  Rows 1 and 3 are interchanged.

**65.**    (a)  5 is factored out of the first row of $A$.

(b)  4 and 3 are factored out of columns 2 and 3.

**67.**    $x^2 - 3x + 2 = (x - 2)(x - 1)$

**69.**    $4y^2 - 12y + 9 = (2y - 3)^2$

**71.**    $3x - 10y = 46$

$x + \quad y = -2$

$y = -x - 2$

$3x - 10(-x - 2) = 46$

$13x = 26$

$x = 2$

$y = -2 - 2 = -4$

*Solution:* $(2, -4)$

# Section 5.8    Applications of Matrices and Determinants

■ You should be able to find the area of a triangle with vertices $(x_1, y_1)$, $(x_2, y_2)$, and $(x_3, y_3)$.

$$\text{Area} = \pm\frac{1}{2}\begin{vmatrix} x_1 & y_1 & 1 \\ x_2 & y_2 & 1 \\ x_3 & y_3 & 1 \end{vmatrix}$$

The $\pm$ symbol indicates that the appropriate sign should be chosen so that the area is positive.

■ You should be able to test to see if three points, $(x_1, y_1)$, $(x_2, y_2)$, and $(x_3, y_3)$, are collinear.

$$\begin{vmatrix} x_1 & y_1 & 1 \\ x_2 & y_2 & 1 \\ x_3 & y_3 & 1 \end{vmatrix} = 0, \text{ if and only if they are collinear.}$$

■ You should be able to use Cramer's Rule to solve a system of linear equations.

■ Now you should be able to solve a system of linear equations by substitution, elimination, elementary row operations on an augmented matrix, using the inverse matrix, or Cramer's Rule.

■ You should be able to encode and decode messages by using an invertible $n \times n$ matrix.

**Solutions to Odd-Numbered Exercises**

**1.** Vertices: $(-2, -3)$, $(2, -3)$, $(0, 4)$

$$\frac{1}{2}\begin{vmatrix} -2 & -3 & 1 \\ 2 & -3 & 1 \\ 0 & 4 & 1 \end{vmatrix} = \frac{1}{2}\left(-2\begin{vmatrix} -3 & 1 \\ 4 & 1 \end{vmatrix} - 2\begin{vmatrix} -3 & 1 \\ 4 & 1 \end{vmatrix}\right) = \frac{1}{2}(14 + 14).\ \text{Area} = 14 \text{ square units}$$

**3.** Vertices: $(-2, 4)$, $(2, 3)$, $(-1, 5)$

$$\frac{1}{2}\begin{vmatrix} -2 & 4 & 1 \\ 2 & 3 & 1 \\ -1 & 5 & 1 \end{vmatrix} = \frac{1}{2}\left[-2\begin{vmatrix} 3 & 1 \\ 5 & 1 \end{vmatrix} - 4\begin{vmatrix} 2 & 1 \\ -1 & 1 \end{vmatrix} + \begin{vmatrix} 2 & 3 \\ -1 & 5 \end{vmatrix}\right]$$

$$= \frac{1}{2}\left[-2(-2) - 4(3) + 13\right] = \frac{1}{2}(5) = \frac{5}{2}$$

Area $= \frac{5}{2}$ square units

**5.** Vertices: $\left(0, \frac{1}{2}\right)$, $\left(\frac{5}{2}, 0\right)$, $(4, 3)$

$$\frac{1}{2}\begin{vmatrix} 0 & \frac{1}{2} & 1 \\ \frac{5}{2} & 0 & 1 \\ 4 & 3 & 1 \end{vmatrix} = \frac{1}{2}\left[-\frac{1}{2}\left(-\frac{3}{2}\right) + \frac{15}{2}\right] = \frac{1}{2}\left[\frac{33}{4}\right] = \frac{33}{8}.$$

Area $= \frac{33}{8}$ square units

**7.** $4 = \pm\dfrac{1}{2}\begin{vmatrix} -5 & 1 & 1 \\ 0 & 2 & 1 \\ -2 & x & 1 \end{vmatrix}$

$\pm 8 = -5\begin{vmatrix} 2 & 1 \\ x & 1 \end{vmatrix} - 2\begin{vmatrix} 1 & 1 \\ 2 & 1 \end{vmatrix}$

$\pm 8 = -5(2 - x) - 2(-1)$

$\pm 8 = 5x - 8$

$x = \dfrac{8 \pm 8}{5}$

$x = \dfrac{16}{5}$  OR  $x = 0$

**9.** Points: $(3, -1)$, $(0, -3)$, $(12, 5)$

$\begin{vmatrix} 3 & -1 & 1 \\ 0 & -3 & 1 \\ 12 & 5 & 1 \end{vmatrix} = 3(-8) + 12(2) = 0$

The points are collinear.

**11.** Points: $\left(2, -\frac{1}{2}\right)$, $(-4, 4)$, $(6, -3)$

$\begin{vmatrix} 2 & -\frac{1}{2} & 1 \\ -4 & 4 & 1 \\ 6 & -3 & 1 \end{vmatrix} = 2(7) + \frac{1}{2}(-10) + 1(-12) = -3 \neq 0$

The points are not collinear.

**13.** $\begin{vmatrix} 2 & -5 & 1 \\ 4 & x & 1 \\ 5 & -2 & 1 \end{vmatrix} = 0$

$2\begin{vmatrix} x & 1 \\ -2 & 1 \end{vmatrix} + 5\begin{vmatrix} 4 & 1 \\ 5 & 1 \end{vmatrix} + \begin{vmatrix} 4 & x \\ 5 & -2 \end{vmatrix} = 0$

$2(x + 2) + 5(-1) + (-8 - 5x) = 0$

$-3x - 9 = 0$

$x = -3$

**15.** $\begin{cases} -7x + 11y = -1 \\ \phantom{-}3x - \phantom{1}9y = \phantom{-}9 \end{cases}$

$x = \dfrac{\begin{vmatrix} -1 & 11 \\ 9 & -9 \end{vmatrix}}{\begin{vmatrix} -7 & 11 \\ 3 & -9 \end{vmatrix}} = \dfrac{-90}{30} = -3$

$y = \dfrac{\begin{vmatrix} -7 & -1 \\ 3 & 9 \end{vmatrix}}{\begin{vmatrix} -7 & 11 \\ 3 & -9 \end{vmatrix}} = \dfrac{-60}{30} = -2$

*Answer:* $(-3, -2)$

**17.** $\begin{cases} 3x + 2y = -2 \\ 6x + 4y = \phantom{-}4 \end{cases}$

$\begin{vmatrix} 3 & 2 \\ 6 & 4 \end{vmatrix} = 12 - 17 = 0$

Cramer's rule cannot be used.

(In fact, the system is inconsistent)

**19.** $\begin{cases} -0.4x + 0.8y = 1.6 \\ \phantom{-}0.2x + 0.3y = 2.2 \end{cases}$

$D = \begin{vmatrix} -0.4 & 0.8 \\ 0.2 & 0.3 \end{vmatrix} = -0.28$

$x = \dfrac{\begin{vmatrix} 1.6 & 0.8 \\ 2.2 & 0.3 \end{vmatrix}}{-0.28} = \dfrac{-1.28}{-0.28} = \dfrac{32}{7}$

$y = \dfrac{\begin{vmatrix} -0.4 & 1.6 \\ 0.2 & 2.2 \end{vmatrix}}{-0.28} = \dfrac{-1.20}{-0.28} = \dfrac{30}{7}$

*Answer:* $\left(\dfrac{32}{7}, \dfrac{30}{7}\right)$

**21.** $\begin{cases} 4x - y + z = -5 \\ 2x + 2y + 3z = 10 \\ 5x - 2y + 6z = 1 \end{cases}$    $D = \begin{vmatrix} 4 & -1 & 1 \\ 2 & 2 & 3 \\ 5 & -2 & 6 \end{vmatrix} = 55$

$x = \dfrac{\begin{vmatrix} -5 & -1 & 1 \\ 10 & 2 & 3 \\ 1 & -2 & 6 \end{vmatrix}}{55} = \dfrac{-55}{55} = -1, \quad y = \dfrac{\begin{vmatrix} 4 & -5 & 1 \\ 2 & 10 & 3 \\ 5 & 1 & 6 \end{vmatrix}}{55} = \dfrac{165}{55} = 3, \quad z = \dfrac{\begin{vmatrix} 4 & -1 & -5 \\ 2 & 2 & 10 \\ 5 & -2 & 1 \end{vmatrix}}{55} = \dfrac{110}{55} = 2$

Answer: $(-1, 3, 2)$

**23.** $\begin{cases} 3x + 3y + 5z = 1 \\ 3x + 5y + 9z = 2 \\ 5x + 9y + 17z = 4 \end{cases}$    $D = \begin{vmatrix} 3 & 3 & 5 \\ 3 & 5 & 9 \\ 5 & 9 & 17 \end{vmatrix} = 4$

$x = \dfrac{\begin{vmatrix} 1 & 3 & 5 \\ 2 & 5 & 9 \\ 4 & 9 & 17 \end{vmatrix}}{4} = 0, \quad y = \dfrac{\begin{vmatrix} 3 & 1 & 5 \\ 3 & 2 & 9 \\ 5 & 4 & 17 \end{vmatrix}}{4} = -\dfrac{1}{2}, \quad z = \dfrac{\begin{vmatrix} 3 & 3 & 1 \\ 3 & 5 & 2 \\ 5 & 9 & 4 \end{vmatrix}}{4} = \dfrac{1}{2}$

Answer: $\left(0, -\dfrac{1}{2}, \dfrac{1}{2}\right)$

**25.** Vertices: $\overset{A}{(0, 25)}, \overset{B}{(10, 0)}, \overset{C}{(28, 5)}$

$\dfrac{1}{2}\begin{vmatrix} 0 & 25 & 1 \\ 10 & 0 & 1 \\ 28 & 5 & 1 \end{vmatrix} = 250.$ Area $= 250$ square miles

**27.** The uncoded row matrices are the rows of the $6 \times 3$ matrix on the left.

$\begin{matrix} C & A & L \\ L & & M \\ E & & T \\ O & M & O \\ R & R & O \\ W & & \end{matrix} \begin{bmatrix} 3 & 1 & 12 \\ 12 & 0 & 13 \\ 5 & 0 & 20 \\ 15 & 13 & 15 \\ 18 & 18 & 15 \\ 23 & 0 & 0 \end{bmatrix} \begin{bmatrix} 1 & -1 & 0 \\ 1 & 0 & -1 \\ -6 & 2 & 3 \end{bmatrix} = \begin{bmatrix} -68 & 21 & 35 \\ -66 & 14 & 39 \\ -115 & 35 & 60 \\ -62 & 15 & 32 \\ -54 & 12 & 27 \\ 23 & -23 & 0 \end{bmatrix}$

Answer: $[-68, 21, 35], [-66, 14, 39], [-115, 35, 60]$

$[-62, 15, 32], [-54, 12, 27], [23, -23, 0]$

**29.** $\begin{matrix} G & O & N \\ E & & F \\ I & S & H \\ I & N & G \end{matrix} \begin{bmatrix} 7 & 15 & 14 \\ 5 & 0 & 6 \\ 9 & 19 & 8 \\ 9 & 14 & 7 \end{bmatrix} \begin{bmatrix} 1 & 2 & 2 \\ 3 & 7 & 9 \\ -1 & -4 & -7 \end{bmatrix} = \begin{bmatrix} 38 & 63 & 51 \\ -1 & -14 & -32 \\ 58 & 119 & 133 \\ 44 & 88 & 95 \end{bmatrix}$

Cryptogram: 38  63  51  −1  −14  −32  58  119  133  44  88  95

**31.** $A^{-1} = \begin{bmatrix} 1 & 2 \\ 3 & 5 \end{bmatrix}^{-1} = \begin{bmatrix} -5 & 2 \\ 3 & -1 \end{bmatrix}$

$\begin{bmatrix} 11 & 21 \\ 64 & 112 \\ 25 & 50 \\ 29 & 53 \\ 23 & 46 \\ 40 & 75 \\ 55 & 92 \end{bmatrix} \begin{bmatrix} -5 & 2 \\ 3 & -1 \end{bmatrix} = $

| | | | |
|---|---|---|---|
| 8 | 1 | H | A |
| 16 | 16 | P | P |
| 25 | 0 | Y | |
| 14 | 5 | N | E |
| 23 | 0 | W | |
| 25 | 5 | Y | E |
| 1 | 18 | A | R |

Message: HAPPY NEW YEAR

**33.** $A^{-1} = \begin{bmatrix} 1 & 2 & 2 \\ 3 & 7 & 9 \\ -1 & -4 & -7 \end{bmatrix}^{-1} = \begin{bmatrix} -13 & 6 & 4 \\ 12 & -5 & -3 \\ -5 & 2 & 1 \end{bmatrix}$

$\begin{bmatrix} 16 & -1 & -48 \\ 5 & -20 & -65 \\ 8 & 4 & -14 \\ 41 & 83 & 89 \\ 76 & 177 & 227 \end{bmatrix} \begin{bmatrix} -13 & 6 & 4 \\ 12 & -5 & -3 \\ -5 & 2 & 1 \end{bmatrix} = \begin{bmatrix} 20 & 5 & 19 \\ 20 & 0 & 15 \\ 14 & 0 & 6 \\ 18 & 9 & 4 \\ 1 & 25 & 0 \end{bmatrix}$

| T | E | S |
|---|---|---|
| T | | O |
| N | | F |
| R | I | D |
| A | Y | |

Message: TEST ON FRIDAY

**35.** True. Cramer's Rule requires that the determinant of the coefficient matrix be nonzero.

**37.** Answers will vary.

**39.**
$$y - 5 = \frac{5 - 3}{-1 - 7}(x + 1) = \frac{-1}{4}(x + 1)$$
$$4y - 20 = -x - 1$$
$$4y + x = 19$$
$$x + 4y - 19 = 0$$

**41.**
$$y + 3 = \frac{-3 + 1}{3 - 10}(x - 3) = \frac{2}{7}(x - 3)$$
$$7y + 21 = 2x - 6$$
$$7y - 2x = -27$$
$$2x - 7y - 27 = 0$$

**43.** $f(x) = \dfrac{2x^2}{x^2 + 4}$. Horizontal asymptote: $y = 2$

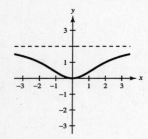

# Review Exercises for Chapter 5

## Solutions to Odd-Numbered Exercises

**1.** $\begin{cases} x + y = 2 \implies & y = 2 - x \\ x - y = 0 \implies & x - (2 - x) = 0 \end{cases}$

$$2x - 2 = 0$$
$$x = 1$$
$$y = 2 - 1 = 1$$

*Answer:* $(1, 1)$

**3.** $\begin{cases} x^2 - y^2 = 9 \\ x - y = 1 \implies & x = y + 1 \end{cases}$

$$(y + 1)^2 - y^2 = 9$$
$$2y + 1 = 9$$
$$y = 4$$
$$x = 5$$

*Answer:* $(5, 4)$

**5.** $\begin{cases} y = 2x^2 \\ y = x^4 - 2x^2 \implies & 2x^2 = x^4 - 2x^2 \end{cases}$

$$0 = x^4 - 4x^2$$
$$0 = x^2(x^2 - 4)$$
$$0 = x^2(x + 2)(x - 2)$$
$$x = 0, x = -2, x = 2$$
$$y = 0, y = 8, y = 8$$

*Answers:* $(0, 0), (-2, 8), (2, 8)$

**7.** $\begin{cases} 5x + 6y = 7 \implies & y_1 = \dfrac{1}{6}(7 - 5x) \\ -x - 4y = 0 \implies & y_2 = -\dfrac{x}{4} \end{cases}$

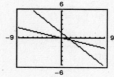

Point of intersection:
$\left(2, -\dfrac{1}{2}\right)$

**9.** $\begin{cases} y^2 - 2y + x = 0 \implies (y - 1)^2 = 1 - x \implies y = 1 \pm \sqrt{1 - x} \\ x + y = 0 \implies \qquad y = -x \end{cases}$

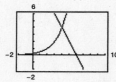

Points of intersection: $(0, 0)$ and $(-3, 3)$

**11.** $\begin{cases} y = 2(6 - x) \\ y = 2^{x-2} \end{cases}$

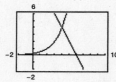

Point of intersection: $(4, 4)$

**13.** Revenue $= 4.95x$

Cost $= 2.85x + 10,000$

Break even when Revenue = Cost

$$4.95x = 2.85x + 10,000$$
$$2.10x = 10,000$$
$$x \approx 4762 \text{ units}$$

**15.** $\begin{cases} 2l + 2w = 480 \\ l = 1.50w \end{cases}$

$2(1.50w) + 2w = 480$

$5w = 480$

$w = 96$

$l = 144$

The dimensions are $96 \times 144$ meters.

**17.** $\begin{cases} 2x - y = 2 \implies 16x - 8y = 16 \\ 6x + 8y = 39 \implies \underline{6x + 8y = 39} \end{cases}$

$22x \qquad = 55$

$x = \frac{55}{22} = \frac{5}{2}$

$y = 3$

*Answer:* $\left(\frac{5}{2}, 3\right)$

**19.** $\begin{cases} 1/5x + 3/10y = 7/50 \implies 20x + 30y = 14 \implies \quad 20x + 30y = \quad 14 \\ 2/5x + 1/2y \ = 1/5 \implies 4x + 5y = 2 \implies \underline{-20x - 25y = -10} \end{cases}$

$5y = \quad 4$

$y = \quad \frac{4}{5}$

$x = \quad -\frac{1}{2}$

*Answer:* $\left(-\frac{1}{2}, \frac{4}{5}\right)$ or $(-0.5, 0.8)$

**21.** $\begin{cases} 3x - 2y = 0 \implies 3x - 2y = 0 \\ 3x + 2(y + 5) = 10 \implies \underline{3x + 2y = 0} \end{cases}$

$6x \qquad = 0$

$x = 0$

$y = 0$

Solution: $(0, 0)$

**23.** $\begin{cases} 1.25x - 2y = 3.5 \implies \quad 5x - 8y = \quad 14 \\ 5x - 8y = 14 \implies \underline{-5x + 8y = -14} \end{cases}$

$0 = \quad 0$

Infinite number of solutions

Let $y = a$, then $5x - 8a = 14 \implies x = \frac{14}{5} + \frac{8}{5}a$.

Solution: $\left(\frac{14}{5} + \frac{8}{5}a, a\right)$

**25.** $\begin{cases} 3x + 2y = 0 \implies y = -\frac{3}{2}x \\ x - y = 4 \implies y = x - 4 \end{cases}$

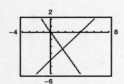

Consistent.

*Answer:* $(1.6, -2.4)$

**27.** $\begin{cases} \frac{1}{4}x - \frac{1}{5}y = 2 \implies y = \frac{5}{4}x - 10 \\ -5x + 4y = 8 \implies y = \frac{1}{4}(8 + 5x) = \frac{5}{4}x + 2 \end{cases}$

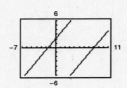

Inconsistent. Lines are parallel.

**29.** $\begin{cases} 2x - 2y = 8 \implies y = x - 4 \\ 4x - 1.5y = -5.5 \implies y = \frac{8}{3}x + \frac{11}{3} \end{cases}$

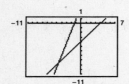

*Answer:* $(-4.6, -8.6)$

**31.** Demand = Supply

$37 - 0.0002x = 22 + 0.00001x$

$15 = 0.00021x$

$x = \frac{500,000}{7}, p = \frac{159}{7}$

Point of equilibrium $\left(\frac{500,000}{7}, \frac{159}{7}\right)$

**33.** Let $x$ = speed of the slower plane.

Let $y$ = speed of the faster plane.

Then, distance of first plane + distance of second plane = 275 miles.

(rate of first plane)(time) + (rate of second plane)(time) = 275 miles

$$\begin{cases} x\left(\frac{40}{60}\right) + y\left(\frac{40}{60}\right) = 275 \\ \qquad\qquad\quad y = x + 25 \end{cases}$$

$$\tfrac{2}{3}x + \tfrac{2}{3}(x + 25) = 275$$

$$4x + 50 = 825$$

$$4x = 775$$

$$x = 193.75 \text{ mph}$$

$$y = x + 25 = 218.75 \text{ mph}$$

**35.** $\begin{cases} x - 4y + 3z = 3 & \text{Equation 1} \\ \quad -y + z = -1 & \text{Equation 2} \\ \qquad\quad z = -5 & \text{Equation 3} \end{cases}$

Substitute $z = -5$ into Equation 2: $-y + (-5) = -1 \implies -y = 4 \implies y = -4$

Substitute $z = -5$ and $y = -4$ into Equation 1:

$$x - 4(-4) + 3(-5) = 3$$

$$x = -16 + 15 + 3$$

$$x = 2$$

*Answer:* $(2, -4, -5)$

**37.** $\begin{cases} x + 3y - z = 13 \\ 2x \qquad - 5z = 23 \\ 4x - y - 2z = 14 \end{cases}$

$\begin{cases} x + 3y - z = 13 \\ \quad - 6y - 3z = -3 \\ \quad - 13y + 2z = -38 \end{cases}$

$\begin{cases} x + 3y - z = 13 \\ \quad - 6y - 3z = -3 \\ \qquad\quad \tfrac{17}{2}z = -\tfrac{63}{2} \end{cases}$

$$\tfrac{17}{2}z = -\tfrac{63}{2} \implies z = -\tfrac{63}{17}$$

$$-6y - 3\left(-\tfrac{63}{17}\right) = -3 \implies y = \tfrac{40}{17}$$

$$x + 3\left(\tfrac{40}{17}\right) - \left(-\tfrac{63}{17}\right) = 13 \implies x = \tfrac{38}{17}$$

Solution: $\left(\tfrac{38}{17}, \tfrac{40}{17}, -\tfrac{63}{17}\right)$

**39.** $\begin{cases} x - 2y + z = -6 \\ 2x - 3y \qquad = -7 \\ -x + 3y - 3z = 11 \end{cases}$

$\begin{cases} x - 2y + z = -6 \\ \quad y - 2z = 5 & -2\,\text{Eq.1} + \text{Eq. 2} \\ \quad y - 2z = 5 & \text{Eq. 1} + \text{Eq. 3} \end{cases}$

$\begin{cases} x - 2y + z = -6 \\ \quad y - 2z = 5 \\ \qquad\quad 0 = 0 & -\text{Eq. 2} + \text{Eq. 3} \end{cases}$

Let $z = a$, then $y = 2a + 5$.

$$x - 2(2a + 5) + a = -6$$

$$x - 3a - 10 = -6$$

$$x = 3a + 4$$

Solution: $(3a + 4, 2a + 5, a)$ where $a$ is any real number.

**41.** $\begin{cases} 5x - 12y + 7z = 16 & \text{Equation 1} \\ 3x - 7y + 4z = 9 & \text{Equation 2} \end{cases}$

3 times Eq. 1 and $(-5)$ times Eq. 2:

$\begin{cases} 15x - 36y + 21z = 48 \\ -15x + 35y - 20z = -45 \end{cases}$

Adding, $-y + z = 3 \implies y = z - 3$.

$5x - 12(z - 3) + 7z = 16$

$5x - 5z + 36 = 16$

$5x = 5z - 20$

$x = z - 4$

Let $z = a$, then $x = a - 4$ and $y = a - 3$.

Solution: $(a - 4, a - 3, a)$ where $a$ is any real number.

**45.** $\dfrac{4 - x}{x^2 + 6x + 8} = \dfrac{A}{x + 2} + \dfrac{B}{x + 4}$

$4 - x = A(x + 4) + B(x + 2) = (A + B)x + (4A + 2B)$

$\begin{cases} A + B = -1 \implies A = 3 \\ 4A + 2B = 4 \implies B = -4 \end{cases}$

$\dfrac{4 - x}{x^2 + 6x + 8} = \dfrac{3}{x + 2} - \dfrac{4}{x + 4}$

**47.** $\dfrac{x^2 + 2x}{x^3 - x^2 + x - 1} = \dfrac{A}{x - 1} + \dfrac{Bx + C}{x^2 + 1}$

$x^2 + 2x = A(x^2 + 1) + (Bx + C)(x - 1) = (A + B)x^2 + (C - B)x + (A - C)$

$\begin{cases} A + B = 1 \implies A = \frac{3}{2} \\ -B + C = 2 \implies B = -\frac{1}{2} \\ A - C = 0 \implies C = \frac{3}{2} \end{cases}$

$\dfrac{x^2 + 2x}{x^3 - x^2 + x - 1} = \dfrac{3/2}{x - 1} + \dfrac{-(1/2)x + 3/2}{x^2 + 1} = \dfrac{1}{2}\left( \dfrac{3}{x - 1} - \dfrac{x - 3}{x^2 + 1} \right)$

**49.** $y = ax^2 + bx + c$ through $(0, -5)$, $(1, -2)$, and $(2, 5)$

$(0, -5)$: $-5 = \phantom{a + b} + c$

$\begin{cases} (1, -2): -2 = a + b + c \implies a + b = 3 \\ (2, 5): 5 = 4a + 2b + c \implies 2a + b = 5 \end{cases}$

$\begin{cases} 2a + b = 5 \\ -a - b = -3 \end{cases}$

$\phantom{2}a \phantom{+ b} = 2$

$\phantom{2a +} b = 1$

The equation of the parabola is $y = 2x^2 + x - 5$.

**43.** Plane: $2x - 4y + z = 8$

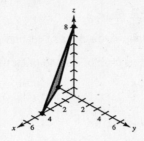

4 points on plane: $(4, 0, 0)$, $(0, -2, 0)$, $(0, 0, 8)$, $(1, 0, 6)$

**51.** Let $x$ = gallons of spray X

Let $y$ = gallons of spray Y

Let $z$ = gallons of spray Z

$$\begin{cases} \text{Chemical A: } \frac{1}{5}x + \phantom{y+} \frac{1}{3}z = 6 & \text{Equation 1} \\ \text{Chemical B: } \frac{2}{5}x + \phantom{y+} \frac{1}{3}z = 8 & \text{Equation 2} \\ \text{Cehmical C: } \frac{2}{5}x + y + \frac{1}{3}z = 13 & \text{Equation 3} \end{cases}$$

Subtracting Eq. 2 − Eq. 1 gives $\frac{1}{5}x = 2 \implies x = 10$.

Then $z = 12$ and $y = 5$.

*Answer:* 10 gallons of spray X

5 gallons of spray Y

12 gallons of spray Z

**53.** Order $3 \times 1$

**55.** Order $1 \times 1$

**57.** $\begin{bmatrix} 3 & -10 & \vdots & 15 \\ 5 & 4 & \vdots & 22 \end{bmatrix}$

**59.** $\begin{bmatrix} 8 & -7 & 4 & \vdots & 12 \\ 3 & -5 & 2 & \vdots & 20 \\ 5 & 3 & -3 & \vdots & 26 \end{bmatrix}$

**61.** $\begin{bmatrix} 5 & 1 & 7 & \vdots & -9 \\ 4 & 2 & 0 & \vdots & 10 \\ 9 & 4 & 2 & \vdots & 3 \end{bmatrix}$
$\quad\begin{aligned} 5x + y + 7z &= -9 \\ 4x + 2y \phantom{{}+ 7z} &= 10 \\ 9x + 4y + 2z &= 3 \end{aligned}$

**63.**
$$\begin{bmatrix} 0 & 1 & 1 \\ 1 & 2 & 3 \\ 2 & 2 & 2 \end{bmatrix}$$

$$\begin{matrix} R_1 + R_2 \to \\ -R_1 + R_2 \to \\ -2R_1 + R_3 \to \end{matrix} \begin{bmatrix} 1 & 3 & 4 \\ 0 & -1 & -1 \\ 0 & -4 & -6 \end{bmatrix}$$

$$\begin{matrix} 3R_2 + R_1 \to \\ -R_2 \to \\ -4R_2 + R_3 \to \end{matrix} \begin{bmatrix} 1 & 0 & 1 \\ 0 & 1 & 1 \\ 0 & 0 & -2 \end{bmatrix}$$

$$\begin{matrix} -R_3 + R_1 \to \\ -R_3 + R_2 \to \\ -\frac{1}{2}R_3 \to \end{matrix} \begin{bmatrix} 1 & 0 & 0 \\ 0 & 1 & 0 \\ 0 & 0 & 1 \end{bmatrix}$$

**65.** $\begin{bmatrix} 3 & -2 & 1 & 0 \\ 4 & -3 & 0 & 1 \end{bmatrix} \Rightarrow \begin{bmatrix} 1 & 0 & 3 & -2 \\ 0 & 1 & 4 & -3 \end{bmatrix}$

**67.** $\begin{bmatrix} 1 & 3 & 4 \\ 0 & 1 & 1 \\ 2 & 4 & 6 \end{bmatrix} \Rightarrow \begin{bmatrix} 1 & 0 & 1 \\ 0 & 1 & 1 \\ 0 & 0 & 0 \end{bmatrix}$

**69.**
$$\begin{bmatrix} 5 & 4 & \vdots & 2 \\ -1 & 1 & \vdots & -22 \end{bmatrix}$$

$$\begin{matrix} 4R_2 + R_1 \to \\ R_1 + R_2 \to \end{matrix} \begin{bmatrix} 1 & 8 & \vdots & -86 \\ 0 & 9 & \vdots & -108 \end{bmatrix}$$

$9y = -108$

$y = -12$

$x = -8(-12) - 86 = 10$

*Answer:* $(10, -12)$

**71.**
$$\begin{bmatrix} 2 & 1 & \vdots & 0.3 \\ 3 & -1 & \vdots & -1.3 \end{bmatrix}$$

$$-R_1 + R_2 \rightarrow \begin{bmatrix} 2 & 1 & \vdots & 0.3 \\ 1 & -2 & \vdots & -1.6 \end{bmatrix}$$

$$\begin{bmatrix} 1 & -2 & \vdots & -1.6 \\ 2 & 1 & \vdots & 0.3 \end{bmatrix}$$

$$-2R_1 + R_2 \rightarrow \begin{bmatrix} 1 & -2 & \vdots & -1.6 \\ 0 & 5 & \vdots & 3.5 \end{bmatrix}$$

$5y = 3.5 \implies y = 0.7$

$x = 2(0.7) - 1.6 = -0.2$

$x = -0.2, \; y = 0.7$

**73.**
$$\begin{bmatrix} 2 & 3 & 3 & \vdots & 3 \\ 6 & 6 & 12 & \vdots & 13 \\ 12 & 9 & -1 & \vdots & 2 \end{bmatrix}$$

$$\begin{matrix} -3R_1 + R_2 \rightarrow \\ -6R_1 + R_3 \rightarrow \end{matrix} \begin{bmatrix} 2 & 3 & 3 & \vdots & 3 \\ 0 & -3 & 3 & \vdots & 4 \\ 0 & -9 & -19 & \vdots & -16 \end{bmatrix}$$

$$\begin{matrix} R_2 + R_1 \rightarrow \\ \\ -3R_2 + R_3 \rightarrow \end{matrix} \begin{bmatrix} 2 & 0 & 6 & \vdots & 7 \\ 0 & -3 & 3 & \vdots & 4 \\ 0 & 0 & -28 & \vdots & -28 \end{bmatrix}$$

$$\begin{matrix} \frac{1}{2}R_1 \rightarrow \\ -\frac{1}{3}R_2 \rightarrow \\ -\frac{1}{28}R_3 \rightarrow \end{matrix} \begin{bmatrix} 1 & 0 & 3 & \vdots & \frac{7}{2} \\ 0 & 1 & -1 & \vdots & -\frac{4}{3} \\ 0 & 0 & 1 & \vdots & 1 \end{bmatrix}$$

$z = 1$

$y - 1 = -\frac{4}{3} \implies y = -\frac{1}{3}$

$x + 3(1) = \frac{7}{2} \implies x = \frac{1}{2}$

*Answer:* $\left( \frac{1}{2}, -\frac{1}{3}, 1 \right)$

**75.**
$$\begin{bmatrix} 3 & 21 & -29 & \vdots & -1 \\ 2 & 15 & -21 & \vdots & 0 \end{bmatrix}$$

$$-R_2 + R_1 \rightarrow \begin{bmatrix} 1 & 6 & -8 & \vdots & -1 \\ 2 & 15 & -21 & \vdots & 0 \end{bmatrix}$$

$$-2R_1 + R_2 \rightarrow \begin{bmatrix} 1 & 6 & -8 & \vdots & -1 \\ 0 & 3 & -5 & \vdots & 2 \end{bmatrix}$$

$$-2R_2 + R_1 \rightarrow \begin{bmatrix} 1 & 0 & 2 & \vdots & -5 \\ 0 & 3 & -5 & \vdots & 2 \end{bmatrix}$$

Let $z = a$, any real number.

$3y - 5a = 2 \implies y = \frac{5}{3}a + \frac{2}{3}$

$x + 2a = -5 \implies x = -2a - 5$

*Answer:* $\left( -2a - 5, \frac{5}{3}a + \frac{2}{3}, a \right)$

**77.**
$$\begin{bmatrix} -1 & 1 & 2 & \vdots & 1 \\ 2 & 3 & 1 & \vdots & -2 \\ 5 & 4 & 2 & \vdots & 4 \end{bmatrix}$$

$$\begin{matrix} -R_1 \rightarrow \\ 2R_1 + R_2 \rightarrow \\ 5R_1 + R_3 \rightarrow \end{matrix} \begin{bmatrix} 1 & -1 & -2 & \vdots & -1 \\ 0 & 5 & 5 & \vdots & 0 \\ 0 & 9 & 12 & \vdots & 9 \end{bmatrix}$$

$$\begin{matrix} \\ \frac{1}{5}R_2 \rightarrow \\ \\ \end{matrix} \begin{bmatrix} 1 & -1 & -2 & \vdots & -1 \\ 0 & 1 & 1 & \vdots & 0 \\ 0 & 9 & 12 & \vdots & 9 \end{bmatrix}$$

$$\begin{matrix} R_2 + R_1 \rightarrow \\ \\ -9R_2 + R_3 \rightarrow \end{matrix} \begin{bmatrix} 1 & 0 & -1 & \vdots & -1 \\ 0 & 1 & 1 & \vdots & 0 \\ 0 & 0 & 3 & \vdots & 9 \end{bmatrix}$$

$$\begin{matrix} \\ \\ \frac{1}{3}R_3 \rightarrow \end{matrix} \begin{bmatrix} 1 & 0 & -1 & \vdots & -1 \\ 0 & 1 & 1 & \vdots & 0 \\ 0 & 0 & 1 & \vdots & 3 \end{bmatrix}$$

$$\begin{matrix} R_3 + R_1 \rightarrow \\ -R_3 + R_2 \rightarrow \\ \end{matrix} \begin{bmatrix} 1 & 0 & 0 & \vdots & 2 \\ 0 & 1 & 0 & \vdots & -3 \\ 0 & 0 & 1 & \vdots & 3 \end{bmatrix}$$

$x = 2, \; y = -3, \; z = 3$

*Answer:* $(2, -3, 3)$

**79.**
$$\begin{bmatrix} 2 & -1 & 9 & \vdots & -8 \\ -1 & -3 & 4 & \vdots & -15 \\ 5 & 2 & -1 & \vdots & 17 \end{bmatrix}$$

$$\begin{array}{c} R_2 \\ R_1 \end{array} \begin{bmatrix} 1 & 3 & -4 & \vdots & 15 \\ 2 & -1 & 9 & \vdots & -8 \\ 5 & 2 & -1 & \vdots & 17 \end{bmatrix}$$

$$\begin{array}{c} \\ -2R_1 + R_2 \rightarrow \\ -5R_1 + R_3 \rightarrow \end{array} \begin{bmatrix} 1 & 3 & -4 & \vdots & 15 \\ 0 & -7 & 17 & \vdots & -38 \\ 0 & -13 & 19 & \vdots & -58 \end{bmatrix}$$

$$\begin{array}{c} \\ \\ R_3 - \frac{13}{7}R_2 \rightarrow \end{array} \begin{bmatrix} 1 & 3 & -4 & \vdots & 15 \\ 0 & -7 & 17 & \vdots & -38 \\ 0 & 0 & -\frac{88}{7} & \vdots & \frac{88}{7} \end{bmatrix} \text{reduces to} \begin{bmatrix} 1 & 0 & 0 & \vdots & 2 \\ 0 & 1 & 0 & \vdots & 3 \\ 0 & 0 & 1 & \vdots & -1 \end{bmatrix}$$

*Answer:* $(2, 3, -1)$

**81.** $\begin{bmatrix} 1 & 2 & -1 & \vdots & 7 \\ 0 & -1 & -1 & \vdots & 4 \\ 4 & 0 & -1 & \vdots & 16 \end{bmatrix}$ reduces to $\begin{bmatrix} 1 & 0 & 0 & \vdots & 3 \\ 0 & 1 & 0 & \vdots & 0 \\ 0 & 0 & 1 & \vdots & -4 \end{bmatrix}$

*Answer:* $(3, 0, -4)$

**83.** $\begin{bmatrix} 3 & -1 & 5 & -2 & \vdots & -44 \\ 1 & 6 & 4 & -1 & \vdots & 1 \\ 5 & -1 & 1 & 3 & \vdots & -15 \\ 0 & 4 & -1 & -8 & \vdots & 58 \end{bmatrix}$ reduces to $\begin{bmatrix} 1 & 0 & 0 & 0 & \vdots & 2 \\ 0 & 1 & 0 & 0 & \vdots & 6 \\ 0 & 0 & 1 & 0 & \vdots & -10 \\ 0 & 0 & 0 & 1 & \vdots & -3 \end{bmatrix}$

*Answer:* $(2, 6, -10, -3)$

**85.** $x = 12$
$y = -7$

**87.** $x + 3 = 5x - 1 \implies x = 1$
$\quad -4y = -44 \quad \implies y = 11$
$\quad y + 5 = 16 \quad \implies y = 11$
$\quad 6x = 6 \quad \implies x = 1$

*Answer:* $x = 1, y = 11$

**89.** (a) $A + B = \begin{bmatrix} 7 & 3 \\ -1 & 5 \end{bmatrix} + \begin{bmatrix} 10 & -20 \\ 14 & -3 \end{bmatrix} = \begin{bmatrix} 17 & -17 \\ 13 & 2 \end{bmatrix}$

(b) $A - B = \begin{bmatrix} -3 & 23 \\ -15 & 8 \end{bmatrix}$

(c) $4A = \begin{bmatrix} 28 & 12 \\ -4 & 20 \end{bmatrix}$

(d) $A + 3B = \begin{bmatrix} 7 & 3 \\ -1 & 5 \end{bmatrix} + \begin{bmatrix} 30 & -60 \\ 42 & -9 \end{bmatrix} = \begin{bmatrix} 37 & -57 \\ 41 & -4 \end{bmatrix}$

**91.** (a) $A + B = \begin{bmatrix} 6 & 0 & 7 \\ 5 & -1 & 2 \\ 3 & 2 & 3 \end{bmatrix} + \begin{bmatrix} 0 & 5 & 1 \\ -4 & 8 & 6 \\ 2 & -1 & 1 \end{bmatrix} = \begin{bmatrix} 6 & 5 & 8 \\ 1 & 7 & 8 \\ 5 & 1 & 4 \end{bmatrix}$

(b) $A - B = \begin{bmatrix} 6 & -5 & 6 \\ 9 & -9 & -4 \\ 1 & 3 & 2 \end{bmatrix}$   (c) $4A = \begin{bmatrix} 24 & 0 & 28 \\ 20 & -4 & 8 \\ 12 & 8 & 12 \end{bmatrix}$

(d) $A + 3B = \begin{bmatrix} 6 & 0 & 7 \\ 5 & -1 & 2 \\ 3 & 2 & 3 \end{bmatrix} + \begin{bmatrix} 0 & 15 & 3 \\ -12 & 24 & 18 \\ 6 & -3 & 3 \end{bmatrix} = \begin{bmatrix} 6 & 15 & 10 \\ -7 & 23 & 20 \\ 9 & -1 & 6 \end{bmatrix}$

**93.** $\begin{bmatrix} 2 & 1 & 0 \\ 0 & 5 & -4 \end{bmatrix} - 3 \begin{bmatrix} 5 & 3 & -6 \\ 0 & -2 & 5 \end{bmatrix} = \begin{bmatrix} 2 & 1 & 0 \\ 0 & 5 & -4 \end{bmatrix} - \begin{bmatrix} 15 & 9 & -18 \\ 0 & -6 & 15 \end{bmatrix}$

$$= \begin{bmatrix} -13 & -8 & 18 \\ 0 & 11 & -19 \end{bmatrix}$$

**95.** $-\begin{bmatrix} 8 & -1 & 8 \\ -2 & 4 & 12 \\ 0 & -6 & 0 \end{bmatrix} - 5 \begin{bmatrix} -2 & 0 & -4 \\ 3 & -1 & 1 \\ 6 & 12 & -8 \end{bmatrix} = \begin{bmatrix} -8 & 1 & -8 \\ 2 & -4 & -12 \\ 0 & 6 & 0 \end{bmatrix} + \begin{bmatrix} 10 & 0 & 20 \\ -15 & 5 & -5 \\ -30 & -60 & 40 \end{bmatrix}$

$$= \begin{bmatrix} 2 & 1 & 12 \\ -13 & 1 & -17 \\ -30 & -54 & 40 \end{bmatrix}$$

**97.** $3 \begin{bmatrix} 8 & -2 & 5 \\ 1 & 3 & -1 \end{bmatrix} + 6 \begin{bmatrix} 4 & -2 & -3 \\ 2 & 7 & 6 \end{bmatrix} = \begin{bmatrix} 48 & -18 & -3 \\ 15 & 51 & 33 \end{bmatrix}$

**99.** $X = 3A - 2B = 3 \begin{bmatrix} -4 & 0 \\ 1 & -5 \\ -3 & 2 \end{bmatrix} - 2 \begin{bmatrix} 1 & 2 \\ -2 & 1 \\ 4 & 4 \end{bmatrix} = \begin{bmatrix} -14 & -4 \\ 7 & -17 \\ -17 & -2 \end{bmatrix}$

**101.** $X = \frac{1}{3}[B - 2A] = \frac{1}{3} \left( \begin{bmatrix} 1 & 2 \\ -2 & 1 \\ 4 & 4 \end{bmatrix} - 2 \begin{bmatrix} -4 & 0 \\ 1 & -5 \\ -3 & 2 \end{bmatrix} \right) = \frac{1}{3} \begin{bmatrix} 9 & 2 \\ -4 & 11 \\ 10 & 0 \end{bmatrix}$

**103.** $\begin{bmatrix} 1 & 2 \\ 5 & -4 \\ 6 & 0 \end{bmatrix} \begin{bmatrix} 6 & -2 & 8 \\ 4 & 0 & 0 \end{bmatrix} = \begin{bmatrix} 1(6) + 2(4) & 1(-2) + 2(0) & 1(8) + 2(0) \\ 5(6) + (-4)(4) & 5(-2) + (-4)(0) & 5(8) + (-4)(0) \\ 6(6) + (0)(4) & 6(-2) + (0)(0) & 6(8) + (0)(0) \end{bmatrix} = \begin{bmatrix} 14 & -2 & 8 \\ 14 & -10 & 40 \\ 36 & -12 & 48 \end{bmatrix}$

**105.** $AB = \begin{bmatrix} 3 & -2 & 0 \\ 1 & 4 & 9 \end{bmatrix} \begin{bmatrix} 7 & 0 \\ 5 & 3 \\ -1 & 3 \end{bmatrix} = \begin{bmatrix} 11 & -6 \\ 18 & 39 \end{bmatrix}$   **107.** $\begin{bmatrix} 4 & 1 \\ 11 & -7 \\ 12 & 3 \end{bmatrix} \begin{bmatrix} 3 & -5 & 6 \\ 2 & -2 & -2 \end{bmatrix} = \begin{bmatrix} 14 & -22 & 22 \\ 19 & -41 & 80 \\ 42 & -66 & 66 \end{bmatrix}$

**109.** $\begin{bmatrix} 2 & 1 \\ 6 & 0 \end{bmatrix} \left( \begin{bmatrix} 4 & 2 \\ -3 & 1 \end{bmatrix} + \begin{bmatrix} -2 & 4 \\ 0 & 4 \end{bmatrix} \right) = \begin{bmatrix} 2 & 1 \\ 6 & 0 \end{bmatrix} \begin{bmatrix} 2 & 6 \\ -3 & 5 \end{bmatrix}$

$$= \begin{bmatrix} 2(2) + 1(-3) & 2(6) + 1(5) \\ 6(2) + 0 & 6(6) + 0 \end{bmatrix}$$

$$= \begin{bmatrix} 1 & 17 \\ 12 & 36 \end{bmatrix}$$

**111.** Increase by 20% corresponds to multiplication by 1.2:

$$1.2\begin{bmatrix} 80 & 70 & 90 & 40 \\ 50 & 30 & 80 & 20 \\ 90 & 60 & 100 & 50 \end{bmatrix} = \begin{bmatrix} 96 & 84 & 108 & 48 \\ 60 & 36 & 96 & 24 \\ 108 & 72 & 120 & 60 \end{bmatrix}$$

**113.** $AB = \begin{bmatrix} -4 & -1 \\ 7 & 2 \end{bmatrix}\begin{bmatrix} -2 & -1 \\ 7 & 4 \end{bmatrix} = \begin{bmatrix} 1 & 0 \\ 0 & 1 \end{bmatrix}; BA = I_2$

**115.** $\begin{bmatrix} -6 & 5 & \vdots & 1 & 0 \\ -5 & 4 & \vdots & 0 & 1 \end{bmatrix}$ row reduces to $\begin{bmatrix} 1 & 0 & \vdots & 4 & -5 \\ 0 & 1 & \vdots & 5 & -6 \end{bmatrix}$

$\begin{bmatrix} -6 & 5 \\ -5 & 4 \end{bmatrix}^{-1} = \begin{bmatrix} 4 & -5 \\ 5 & -6 \end{bmatrix}$

**117.** $\begin{bmatrix} -1 & -2 & -2 & \vdots & 1 & 0 & 0 \\ 3 & 7 & 9 & \vdots & 0 & 1 & 0 \\ 1 & 4 & 7 & \vdots & 0 & 0 & 1 \end{bmatrix}$ row reduces to $\begin{bmatrix} 1 & 0 & 0 & \vdots & 13 & 6 & -4 \\ 0 & 1 & 0 & \vdots & -12 & -5 & 3 \\ 0 & 0 & 1 & \vdots & 5 & 2 & -1 \end{bmatrix}$

$\begin{bmatrix} -1 & -2 & -2 \\ 3 & 7 & 9 \\ 1 & 4 & 7 \end{bmatrix}^{-1} = \begin{bmatrix} 13 & 6 & -4 \\ -12 & -5 & 3 \\ 5 & 2 & -1 \end{bmatrix}$

**119.** $\begin{bmatrix} 2 & 6 \\ 3 & -6 \end{bmatrix}^{-1} = \begin{bmatrix} \frac{1}{5} & \frac{1}{5} \\ \frac{1}{10} & -\frac{1}{15} \end{bmatrix}$

**121.** $\begin{bmatrix} 2 & 0 & 3 \\ -1 & 1 & 1 \\ 2 & -2 & 1 \end{bmatrix}^{-1} = \begin{bmatrix} \frac{1}{2} & -1 & -\frac{1}{2} \\ \frac{1}{2} & -\frac{2}{3} & -\frac{5}{6} \\ 0 & \frac{2}{3} & \frac{1}{3} \end{bmatrix}$

**123.** $\begin{bmatrix} -7 & 2 \\ -8 & 2 \end{bmatrix}^{-1} = \frac{1}{(-7)(2)-(2)(-8)}\begin{bmatrix} 2 & -2 \\ 8 & -7 \end{bmatrix} = \begin{bmatrix} 1 & -1 \\ 4 & -\frac{7}{2} \end{bmatrix}$

**125.** $\begin{bmatrix} -1 & 20 \\ \frac{3}{10} & -6 \end{bmatrix}^{-1} = \frac{1}{(-1)(-6)-(20)\left(\frac{3}{10}\right)}\begin{bmatrix} -6 & -20 \\ -\frac{3}{10} & -1 \end{bmatrix} = \frac{1}{0}\begin{bmatrix} -6 & -20 \\ -\frac{3}{10} & -1 \end{bmatrix}$

Inverse does not exist.

**127.** $\begin{bmatrix} -1 & 4 \\ 2 & -7 \end{bmatrix}^{-1} = \begin{bmatrix} 7 & 4 \\ 2 & 1 \end{bmatrix}$

$\begin{bmatrix} x \\ y \end{bmatrix} = \begin{bmatrix} 7 & 4 \\ 2 & 1 \end{bmatrix}\begin{bmatrix} 8 \\ -5 \end{bmatrix} = \begin{bmatrix} 36 \\ 11 \end{bmatrix}$

*Answer:* (36, 11)

**129.** $\begin{bmatrix} 3 & 2 & -1 \\ 1 & -1 & 2 \\ 5 & 1 & 1 \end{bmatrix}^{-1} = \begin{bmatrix} -1 & -1 & 1 \\ 3 & \frac{8}{3} & -\frac{7}{3} \\ 2 & \frac{7}{3} & -\frac{5}{3} \end{bmatrix}$

$\begin{bmatrix} x \\ y \\ z \end{bmatrix} = \begin{bmatrix} -1 & -1 & 1 \\ 3 & \frac{8}{3} & -\frac{7}{3} \\ 2 & \frac{7}{3} & -\frac{5}{3} \end{bmatrix}\begin{bmatrix} 6 \\ -1 \\ 7 \end{bmatrix} = \begin{bmatrix} 2 \\ -1 \\ -2 \end{bmatrix}$

*Answer:* (2, -1, -2)

**131.** $\begin{bmatrix} -2 & 1 & 2 \\ -1 & -4 & 1 \\ 0 & -1 & -1 \end{bmatrix}^{-1} = \begin{bmatrix} -\frac{5}{9} & \frac{1}{9} & -1 \\ \frac{1}{9} & -\frac{2}{9} & 0 \\ -\frac{1}{9} & \frac{2}{9} & -1 \end{bmatrix}$

$\begin{bmatrix} x \\ y \\ z \end{bmatrix} = \begin{bmatrix} -\frac{5}{9} & \frac{1}{9} & -1 \\ \frac{1}{9} & -\frac{2}{9} & 0 \\ -\frac{1}{9} & \frac{2}{9} & -1 \end{bmatrix} \begin{bmatrix} -13 \\ -11 \\ 0 \end{bmatrix} = \begin{bmatrix} 6 \\ 1 \\ -1 \end{bmatrix}$

*Answer:* $(6, 1, -1)$

**133.** $\begin{cases} x + 2y = -1 \\ 3x + 4y = -5 \end{cases}$

$\begin{bmatrix} 1 & 2 \\ 3 & 4 \end{bmatrix}^{-1} = \begin{bmatrix} -2 & 1 \\ \frac{3}{2} & -\frac{1}{2} \end{bmatrix} \Rightarrow \begin{bmatrix} x \\ y \end{bmatrix} = \begin{bmatrix} -2 & 1 \\ \frac{3}{2} & -\frac{1}{2} \end{bmatrix} \begin{bmatrix} -1 \\ -5 \end{bmatrix} = \begin{bmatrix} -3 \\ 1 \end{bmatrix}$

$x = -3, y = 1$

*Answer:* $(-3, 1)$

**135.** $\begin{cases} -3x - 3y - 4z = 2 \\ y + z = -1 \\ 4x + 3y + 4z = -1 \end{cases}$

$\begin{bmatrix} -3 & -3 & -4 \\ 0 & 1 & 1 \\ 4 & 3 & 4 \end{bmatrix}^{-1} = \begin{bmatrix} 1 & 0 & 1 \\ 4 & 4 & 3 \\ 4 & -3 & -3 \end{bmatrix} \Rightarrow \begin{bmatrix} x \\ y \\ z \end{bmatrix} = \begin{bmatrix} 1 & 0 & 1 \\ 4 & 4 & 3 \\ -4 & -3 & -3 \end{bmatrix} \begin{bmatrix} 2 \\ -1 \\ -1 \end{bmatrix} = \begin{bmatrix} 1 \\ 1 \\ -2 \end{bmatrix}$

$x = 1, y = 1, z = -2$

*Answer:* $(1, 1, -2)$

**137.** $\begin{vmatrix} 8 & 5 \\ 2 & -4 \end{vmatrix} = 8(-4) - 2(5) = -42$

**139.** $\begin{vmatrix} 50 & -30 \\ 10 & 5 \end{vmatrix} = 50(5) - (-30)(10) = 550$

**141.** $A = \begin{bmatrix} 2 & -1 \\ 7 & 4 \end{bmatrix}$

Minors:    $M_{11} = 4$        $M_{21} = -1$

$M_{12} = 7$        $M_{22} = 2$

Cofactors:  $C_{11} = 4$        $C_{21} = 1$

$C_{12} = -7$        $C_{22} = 2$

**143.** $A = \begin{bmatrix} 3 & 2 & -1 \\ -2 & 5 & 0 \\ 1 & 8 & 6 \end{bmatrix}$

Minors: $M_{11} = \begin{vmatrix} 5 & 0 \\ 8 & 6 \end{vmatrix} = 30, M_{12} = \begin{vmatrix} -2 & 0 \\ 1 & 6 \end{vmatrix} = -12, M_{13} = \begin{vmatrix} -2 & 5 \\ 1 & 8 \end{vmatrix} = -21$

$M_{21} = \begin{vmatrix} 2 & -1 \\ 8 & 6 \end{vmatrix} = 20, M_{22} = \begin{vmatrix} 3 & -1 \\ 1 & 6 \end{vmatrix} = 19, M_{23} = \begin{vmatrix} 3 & 2 \\ 1 & 8 \end{vmatrix} = 22$

$M_{31} = \begin{vmatrix} 2 & -1 \\ 5 & 0 \end{vmatrix} = 5, M_{32} = \begin{vmatrix} 3 & -1 \\ -2 & 0 \end{vmatrix} = -2, M_{33} = \begin{vmatrix} 3 & 2 \\ -2 & 5 \end{vmatrix} = 19$

Cofactors: $C_{11} = 30, C_{12} = 12, C_{13} = -21$

$C_{21} = -20, C_{22} = 19, C_{23} = -22$

$C_{31} = 5, C_{32} = 2, C_{33} = 19$

**145.** $\begin{vmatrix} -2 & 4 & 1 \\ -6 & 0 & 2 \\ 5 & 3 & 4 \end{vmatrix} = 6\begin{vmatrix} 4 & 1 \\ 3 & 4 \end{vmatrix} - 2\begin{vmatrix} -2 & 4 \\ 5 & 3 \end{vmatrix} = 6(13) - 2(-26) = 130$

**147.** $\begin{vmatrix} 1 & 0 & -2 \\ 0 & 1 & 0 \\ -2 & 0 & 1 \end{vmatrix} = 1\begin{vmatrix} 1 & -2 \\ -2 & 1 \end{vmatrix} = 1(1) - (-2)(-2) = 1 - 4 = -3$

**149.** $\begin{vmatrix} 3 & 0 & -4 & 0 \\ 0 & 8 & 1 & 2 \\ 6 & 1 & 8 & 2 \\ 0 & 3 & -4 & 1 \end{vmatrix} = 3\begin{vmatrix} 8 & 1 & 2 \\ 1 & 8 & 2 \\ 3 & -4 & 1 \end{vmatrix} + (-4)\begin{vmatrix} 0 & 8 & 2 \\ 6 & 1 & 2 \\ 0 & 3 & 1 \end{vmatrix}$ (Expansion along Row 1)

$= 3[8(8 - (-8)) - 1(1 - 6) + 2(-4 - 24)] - 4[0 - 6(8 - 6) + 0]$

$= 3[128 + 5 - 56] - 4[-12]$

$= 279$

**151.** $\det(A) = 8(-1)(4)(3) = -96$ (Upper Triangular)

**153.** $(1, 0), (5, 0), (5, 8)$

$\frac{1}{2}\begin{vmatrix} 1 & 0 & 1 \\ 5 & 0 & 1 \\ 5 & 8 & 1 \end{vmatrix} = \frac{1}{2}(32) = 16$

Area = 16 square units

**155.** $\frac{1}{2}\begin{vmatrix} \frac{1}{2} & 1 & 1 \\ 2 & -\frac{5}{2} & 1 \\ \frac{3}{2} & 1 & 1 \end{vmatrix} = \frac{1}{2}(\frac{7}{2}) = \frac{7}{4}$

Area = $\frac{7}{4}$ square units

**157.** $\begin{vmatrix} -1 & 7 & 1 \\ 2 & 5 & 1 \\ 4 & 1 & 1 \end{vmatrix} = -8$

The points are not collinear.

**159.** $x = \dfrac{\begin{vmatrix} 5 & 2 \\ 1 & 1 \end{vmatrix}}{\begin{vmatrix} 1 & 2 \\ -1 & 1 \end{vmatrix}} = \dfrac{3}{3} = 1$

$y = \dfrac{\begin{vmatrix} 1 & 5 \\ -1 & 1 \end{vmatrix}}{\begin{vmatrix} 1 & 2 \\ -1 & 1 \end{vmatrix}} = \dfrac{6}{3} = 2$

*Answer:* $(1, 2)$

**161.** $x = \dfrac{\begin{vmatrix} 6 & -2 \\ -23 & 3 \end{vmatrix}}{\begin{vmatrix} 5 & -2 \\ -11 & 3 \end{vmatrix}} = \dfrac{-28}{-7} = 4$

$y = \dfrac{\begin{vmatrix} 5 & 6 \\ -11 & -23 \end{vmatrix}}{\begin{vmatrix} 5 & -2 \\ -11 & 3 \end{vmatrix}} = \dfrac{-49}{-7} = 7$

*Answer:* $(4, 7)$

**163.** $x = \dfrac{\begin{vmatrix} -11 & 3 & -5 \\ -3 & -1 & 1 \\ 15 & -4 & 6 \end{vmatrix}}{\begin{vmatrix} -2 & 3 & -5 \\ 4 & -1 & 1 \\ -1 & -4 & 6 \end{vmatrix}} = \dfrac{-14}{14} = -1$

$y = \dfrac{\begin{vmatrix} -2 & -11 & -5 \\ 4 & -3 & 1 \\ -1 & 15 & 6 \end{vmatrix}}{14} = \dfrac{56}{14} = 4$

$z = \dfrac{\begin{vmatrix} -2 & 3 & -11 \\ 4 & -1 & -3 \\ -1 & -4 & 15 \end{vmatrix}}{14} = \dfrac{70}{14} = 5$

*Answer:* $(-1, 4, 5)$

**165.** $\begin{vmatrix} 1 & -3 & 2 \\ 2 & 2 & -3 \\ 1 & -7 & 8 \end{vmatrix} = 20$

$|A_1| = 0$

$|A_2| = -48$

$|A_3| = -52$

$x = 0,\ y = -\dfrac{48}{20} = -2.4,\ z = -\dfrac{52}{20} = -2.6$

*Solution:* $(0, -2.4, -2.6)$

**167.** L O O K _ O U T _ B E L O W _

$[12 \quad 15 \quad 15][11 \quad 0 \quad 15][21 \quad 20 \quad 0][2 \quad 5 \quad 12][15 \quad 23 \quad 0]$

$[12 \quad 15 \quad 15]A = [-21 \quad 6 \quad 0]$

$[11 \quad 0 \quad 15]A = [-68 \quad 8 \quad 45]$

$[21 \quad 20 \quad 0]A = [102 \quad -42 \quad -60]$

$[2 \quad 5 \quad 12]A = [-53 \quad 20 \quad 21]$

$[15 \quad 23 \quad 0]A = [99 \quad -30 \quad -69]$

Cryptogram: $-21 \ 6 \ 0 \ -68 \ 8 \ 45 \ 102 \ -42 \ -60$

$-53 \ 20 \ 21 \ 99 \ -30 \ -69$

**169.** $\begin{bmatrix} -5 & 11 & -2 \\ 370 & -265 & 225 \\ -57 & 48 & -33 \\ 32 & -15 & 20 \\ 245 & -171 & 147 \end{bmatrix} \begin{bmatrix} -1 & 2 & -3 \\ 2 & 1 & 0 \\ 4 & -2 & 5 \end{bmatrix} = \begin{bmatrix} 19 & 5 & 5 \\ 0 & 25 & 15 \\ 21 & 0 & 6 \\ 18 & 9 & 4 \\ 1 & 25 & 0 \end{bmatrix} \begin{array}{ccc} S & E & E \\ \_ & Y & O \\ U & \_ & F \\ R & I & D \\ A & Y & \_ \end{array}$   Message: SEE YOU FRIDAY

**171.** False.

**173.** The row operations on matrices are equivalent to the operations used in the method of elimination.

# Chapter 5   Practice Test

**For Exercises 1–3, solve the given system by the method of substitution.**

**1.** $x + y = 1$
   $3x - y = 15$

**2.** $x - 3y = -3$
   $x^2 + 6y = 5$

**3.** $x + y + z = 6$
   $2x - y + 3z = 0$
   $5x + 2y - z = -3$

**4.** Find the two numbers whose sum is 110 and product is 2800.

**5.** Find the dimensions of a rectangle if its perimeter is 170 feet and its area is 2800 square feet.

**For Exercises 6–7, solve the linear system by elimination.**

**6.** $2x + 15y = 4$
   $x - 3y = 23$

**7.** $x + y = 2$
   $38x - 19y = 7$

**8.** Use a graphing utility to graph the two equations. Use the graph to approximate the solution of the system. Verify your answer analytically.

$0.4x + 0.5y = 0.112$

$0.3x - 0.7y = -0.131$

**9.** Herbert invests \$17,000 in two funds that pay 11% and 13% simple interest, respectively. If he receives \$2080 in yearly interest, how much is invested in each fund?

**10.** Find the least squares regression line for the points $(4, 3)$, $(1, 1)$, $(-1, -2)$, and $(-2, -1)$.

**For Exercises 11–13, solve the system of equations.**

**11.** $x + y = -2$
   $2x - y + z = 11$
   $4y - 3z = -20$

**12.** $4x - y + 5z = 4$
   $2x + y - z = 0$
   $2x + 4y + 8z = 0$

**13.** $3x + 2y - z = 5$
   $6x - y + 5z = 2$

**14.** Find the equation of the parabola $y = ax^2 + bx + c$ passing through the points $(0, -1)$, $(1, 4)$ and $(2, 13)$.

**15.** Find the position equation $s = \frac{1}{2}at^2 + v_0t + s_0$ given that $s = 12$ feet after 1 second, $s = 5$ feet after 2 seconds, and $s = 4$ after 3 seconds.

**16.** Write the matrix in reduced row-echelon form.

$$\begin{bmatrix} 1 & -2 & 4 \\ 3 & -5 & 9 \end{bmatrix}$$

**For Exercises 17–19, use matrices to solve the system of equations.**

**17.** $3x + 5y = 3$
   $2x - y = -11$

**18.** $2x + 3y = -3$
   $3x + 2y = 8$
   $x + y = 1$

**19.** $x + 3z = -5$
   $2x + y = 0$
   $3x + y - z = 3$

**20.** Multiply $\begin{bmatrix} 1 & 4 & 5 \\ 2 & 0 & -3 \end{bmatrix}\begin{bmatrix} 1 & 6 \\ 0 & -7 \\ -1 & 2 \end{bmatrix}$

**21.** Given $A = \begin{bmatrix} 9 & 1 \\ -4 & 8 \end{bmatrix}$ and $B = \begin{bmatrix} 6 & -2 \\ 3 & 5 \end{bmatrix}$, find $3A - 5B$.

**22.** Find $f(A)$:

$$f(x) = x^2 - 7x + 8, \quad A = \begin{bmatrix} 3 & 0 \\ 7 & 1 \end{bmatrix}$$

**23.** True or false:

$(A + B)(A + 3B) = A^2 + 4AB + 3B^2$ where $A$ and $B$ are matrices.

(Assume that $A^2$, $AB$, and $B^2$ exist.)

**For Exercises 24 and 25, find the inverse of the matrix, if it exists.**

**24.** $\begin{bmatrix} 1 & 2 \\ 3 & 5 \end{bmatrix}$

**25.** $\begin{bmatrix} 1 & 1 & 1 \\ 3 & 6 & 5 \\ 6 & 10 & 8 \end{bmatrix}$

**26.** Use an inverse matrix to solve the systems.

(a)  $x + 2y = 4$        (b)  $x + 2y = 3$
     $3x + 5y = 1$             $3x + 5y = -2$

**For Exercises 27 and 28, find the determinant of the matrix.**

**27.** $\begin{bmatrix} 6 & -1 \\ 3 & 4 \end{bmatrix}$

**28.** $\begin{bmatrix} 1 & 3 & -1 \\ 5 & 9 & 0 \\ 6 & 2 & -5 \end{bmatrix}$

**29.** Use a graphing utility to find the determinant of the matrix.

$\begin{bmatrix} 1 & 4 & 2 & 3 \\ 0 & 1 & -2 & 0 \\ 3 & 5 & -1 & 1 \\ 2 & 0 & 6 & 1 \end{bmatrix}$

**30.** Evaluate $\begin{vmatrix} 6 & 4 & 3 & 0 & 6 \\ 0 & 5 & 1 & 4 & 8 \\ 0 & 0 & 2 & 7 & 3 \\ 0 & 0 & 0 & 9 & 2 \\ 0 & 0 & 0 & 0 & 1 \end{vmatrix}$.

**31.** Use a determinant to find the area of the triangle with vertices $(0, 7)$, $(5, 0)$, and $(3, 9)$.

**32.** Use a determinant to find the equation of the line through $(2, 7)$ and $(-1, 4)$.

**For Exercises 33–35, use Cramer's Rule to find the indicated value.**

**33.** Find $x$.

$6x - 7y = 4$
$2x + 5y = 11$

**34.** Find $z$.

$3x \qquad + z = 1$
$\quad\ y + 4z = 3$
$x - y \qquad = 2$

**35.** Find $y$.

$721.4x - 29.1y = 33.77$
$45.9x + 105.6y = 19.85$

# CHAPTER 6
# Sequences, Series and Probability

# CHAPTER 6
## Sequences, Series, and Probability

### Section 6.1    Sequences and Series

---

■ Given the general $n$th term in a sequence, you should be able to find, or list, some of the terms.

■ You should be able to find an expression for the $n$th term of a sequence.

■ You should be able to use and evaluate factorials.

■ You should be able to use sigma notation for a sum.

---

**Solutions to Odd-Numbered Exercises**

**1.** $a_n = 2n + 5$

$a_1 = 2(1) + 5 = 7$

$a_2 = 2(2) + 5 = 9$

$a_3 = 2(3) + 5 = 11$

$a_4 = 2(4) + 5 = 13$

$a_5 = 2(5) + 5 = 15$

**3.** $a_n = 2^n$

$a_1 = 2^1 = 2$

$a_2 = 2^2 = 4$

$a_3 = 2^3 = 8$

$a_4 = 2^4 = 16$

$a_5 = 2^5 = 32$

**5.** $a_n = \left(-\frac{1}{2}\right)^n$

$a_1 = \left(-\frac{1}{2}\right)^1 = -\frac{1}{2}$

$a_2 = \left(-\frac{1}{2}\right)^2 = \frac{1}{4}$

$a_3 = \left(-\frac{1}{2}\right)^3 = -\frac{1}{8}$

$a_4 = \left(-\frac{1}{2}\right)^4 = \frac{1}{16}$

$a_5 = \left(-\frac{1}{2}\right)^5 = -\frac{1}{32}$

**7.** $a_n = \dfrac{n+1}{n}$

$a_1 = \dfrac{1+1}{1} = 2$

$a_2 = \dfrac{3}{2}$

$a_3 = \dfrac{4}{3}$

$a_4 = \dfrac{5}{4}$

$a_5 = \dfrac{6}{5}$

**9.** $a_n = \dfrac{n}{n^2+1}$

$a_1 = \dfrac{1}{1^2+1} = \dfrac{1}{2}$

$a_2 = \dfrac{2}{2^2+1} = \dfrac{2}{5}$

$a_3 = \dfrac{3}{3^2+1} = \dfrac{3}{10}$

$a_4 = \dfrac{4}{4^2+1} = \dfrac{4}{17}$

$a_5 = \dfrac{5}{5^2+1} = \dfrac{5}{26}$

**11.** $a_n = \dfrac{1+(-1)^n}{n}$

$a_1 = 0$

$a_2 = \dfrac{2}{2} = 1$

$a_3 = 0$

$a_4 = \dfrac{2}{4} = \dfrac{1}{2}$

$a_5 = 0$

**13.** $a_n = 1 - \dfrac{1}{2^n}$

$a_1 = 1 - \dfrac{1}{2^1} = \dfrac{1}{2}$

$a_2 = 1 - \dfrac{1}{2^2} = 1 - \dfrac{1}{4} = \dfrac{3}{4}$

$a_3 = 1 - \dfrac{1}{2^3} = \dfrac{7}{8}$

$a_4 = 1 - \dfrac{1}{2^4} = \dfrac{15}{16}$

$a_5 = 1 - \dfrac{1}{2^5} = \dfrac{31}{32}$

**15.** $a_n = \dfrac{1}{n^{3/2}}$

$a_1 = \dfrac{1}{1} = 1$

$a_2 = \dfrac{1}{2^{3/2}}$

$a_3 = \dfrac{1}{3^{3/2}}$

$a_4 = \dfrac{1}{4^{3/2}} = \dfrac{1}{8}$

$a_5 = \dfrac{1}{5^{3/2}}$

**17.** $a_n = \dfrac{(-1)^n}{n^2}$

$a_1 = \dfrac{-1}{1} = -1$

$a_2 = \dfrac{1}{4}$

$a_3 = \dfrac{-1}{9}$

$a_4 = \dfrac{1}{16}$

$a_5 = \dfrac{-1}{25}$

**19.** $a_n = (2n-1)(2n+1)$

$a_1 = (1)(3) = 3$

$a_2 = (3)(5) = 15$

$a_3 = (5)(7) = 35$

$a_4 = (7)(9) = 63$

$a_5 = (9)(11) = 99$

**21.** $a_{25} = (-1)^{25}[3(25) - 2] = -73$

**23.** $a_{10} = \dfrac{10^2}{10^2 + 1} = \dfrac{100}{101}$

**25.** $a_n = \dfrac{2}{3}n$

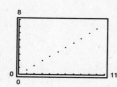

**27.** $a_n = 16(-0.5)^{n-1}$

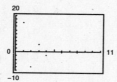

**29.** $a_n = \dfrac{2n}{n+1}$

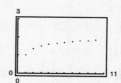

**31.** $a_n = 2(3n - 1) + 5$

| $n$ | 1 | 2 | 3 | 4 | 5 | 6 | 7 | 8 | 9 | 10 |
|---|---|---|---|---|---|---|---|---|---|---|
| $a_n$ | 9 | 15 | 21 | 27 | 33 | 39 | 45 | 51 | 57 | 63 |

**33.** $a_n = 1 + \dfrac{n+1}{n}$

| $n$ | 1 | 2 | 3 | 4 | 5 | 6 | 7 | 8 | 9 | 10 |
|---|---|---|---|---|---|---|---|---|---|---|
| $a_n$ | 3 | 2.5 | 2.33 | 2.25 | 2.2 | 2.17 | 2.14 | 2.13 | 2.11 | 2.1 |

**35.** $a_n = \dfrac{8}{n+1}$

$a_n \to 0$ as $n \to \infty$

$a_1 = 4$, $a_{10} = \dfrac{8}{11}$

Matches graph (c).

**37.** $a_n = 4(0.5)^{n-1}$

$a_n \to 0$ as $n \to \infty$

$a_1 = 4$, $a_{10} \approx 0.008$

Matches graph (d).

**39.** $1, 4, 7, 10, 13, \ldots$

$a_n = 1 + (n-1)3 = 3n - 2$

**41.** $0, 3, 8, 15, 24, \ldots$

$a_n = n^2 - 1$

**43.** $\dfrac{2}{3}, \dfrac{3}{4}, \dfrac{4}{5}, \dfrac{5}{6}, \dfrac{6}{7}, \ldots$

$a_n = \dfrac{n+1}{n+2}$

**45.** $\dfrac{1}{2}, \dfrac{-1}{4}, \dfrac{1}{8}, \dfrac{-1}{16}, \ldots$

$a_n = \dfrac{(-1)^{n+1}}{2^n}$

**47.** $1 + \dfrac{1}{1}, 1 + \dfrac{1}{2}, 1 + \dfrac{1}{3}, 1 + \dfrac{1}{4}, 1 + \dfrac{1}{5}, \ldots$

$a_n = 1 + \dfrac{1}{n}$

**49.** $1, \dfrac{1}{2}, \dfrac{1}{6}, \dfrac{1}{24}, \dfrac{1}{120}, \ldots$

$a_n = \dfrac{1}{n!}$

**51.** $1, 3, 1, 3, 1, 3, \ldots$

$a_n = 2 + (-1)^n$

**53.** $a_1 = 28$ and $a_{k+1} = a_k - 4$

$a_1 = 28$

$a_2 = a_1 - 4 = 28 - 4 = 24$

$a_3 = a_2 - 4 = 24 - 4 = 20$

$a_4 = a_3 - 4 = 20 - 4 = 16$

$a_5 = a_4 - 4 = 16 - 4 = 12$

**55.** $a_1 = 3$ and $a_{k+1} = 2(a_k - 1)$

$a_1 = 3$

$a_2 = 2(a_1 - 1) = 2(3 - 1) = 4$

$a_3 = 2(a_2 - 1) = 2(4 - 1) = 6$

$a_4 = 2(a_3 - 1) = 2(6 - 1) = 10$

$a_5 = 2(a_4 - 1) = 2(10 - 1) = 18$

**57.** $a_1 = 6$ and $a_{k+1} = a_k + 2$

$a_1 = 6$

$a_2 = a_1 + 2 = 6 + 2 = 8$

$a_3 = a_2 + 2 = 8 + 2 = 10$

$a_4 = a_3 + 2 = 10 + 2 = 12$

$a_5 = a_4 + 2 = 12 + 2 = 14$

In general, $a_n = 2n + 4$.

**59.** $a_1 = 81$ and $a_{k+1} = \dfrac{1}{3} a_k$

$a_1 = 81$

$a_2 = \dfrac{1}{3} a_1 = \dfrac{1}{3}(81) = 27$

$a_3 = \dfrac{1}{3} a_2 = \dfrac{1}{3}(27) = 9$

$a_4 = \dfrac{1}{3} a_3 = \dfrac{1}{3}(9) = 3$

$a_5 = \dfrac{1}{3} a_4 = \dfrac{1}{3}(3) = 1$

In general, $a_n = 81\left(\dfrac{1}{3}\right)^{n-1} = 81(3)\left(\dfrac{1}{3}\right)^n = \dfrac{243}{3^n}$.

**61.** $a_n = \dfrac{1}{n!}$

$a_0 = \dfrac{1}{0!} = 1$

$a_1 = \dfrac{1}{1!} = 1$

$a_2 = \dfrac{1}{2}$

$a_3 = \dfrac{1}{3!} = \dfrac{1}{6}$

$a_4 = \dfrac{1}{4!} = \dfrac{1}{24}$

**63.** $a_n = \dfrac{n!}{2n + 1}$

$a_0 = \dfrac{0!}{1} = 1$

$a_1 = \dfrac{1!}{2 + 1} = \dfrac{1}{3}$

$a_2 = \dfrac{2!}{4 + 1} = \dfrac{2}{5}$

$a_3 = \dfrac{3!}{6 + 1} = \dfrac{6}{7}$

$a_4 = \dfrac{4!}{8 + 1} = \dfrac{24}{9} = \dfrac{8}{3}$

**65.** $a_n = \dfrac{(-1)^{2n}}{(2n)!}$

$a_0 = \dfrac{(-1)^0}{0!} = 1$

$a_1 = \dfrac{(-1)^2}{2!} = \dfrac{1}{2}$

$a_2 = \dfrac{(-1)^4}{4!} = \dfrac{1}{24}$

$a_3 = \dfrac{(-1)^6}{6!} = \dfrac{1}{720}$

$a_4 = \dfrac{(-1)^8}{8!} = \dfrac{1}{40,320}$

**67.** $\dfrac{2!}{4!} = \dfrac{2!}{4 \cdot 3 \cdot 2!} = \dfrac{1}{12}$

**69.** $\dfrac{12!}{4!8!} = \dfrac{12 \cdot 11 \cdot 10 \cdot 9 \cdot 8!}{4!8!} = \dfrac{12 \cdot 11 \cdot 10 \cdot 9}{4 \cdot 3 \cdot 2} = 495$

**71.** $\dfrac{(n + 1)!}{n!} = \dfrac{(n + 1)n!}{n!} = n + 1$

**73.** $\dfrac{(2n - 1)!}{(2n + 1)!} = \dfrac{(2n - 1)!}{(2n + 1)(2n)(2n - 1)!}$

$= \dfrac{1}{2n(2n + 1)}$

**75.** $\displaystyle\sum_{i=1}^{5} (2i + 1) = (2 + 1) + (4 + 1) + (6 + 1) + (8 + 1) + (10 + 1) = 35$

**77.** $\displaystyle\sum_{k=1}^{4} 10 = 10 + 10 + 10 + 10 = 40$

**79.** $\displaystyle\sum_{i=0}^{4} i^2 = 0^2 + 1^2 + 2^2 + 3^2 + 4^2 = 30$

**81.** $\displaystyle\sum_{k=0}^{3} \dfrac{1}{k^2 + 1} = \dfrac{1}{1} + \dfrac{1}{1 + 1} + \dfrac{1}{4 + 1} + \dfrac{1}{9 + 1} = \dfrac{9}{5}$

**83.** $\displaystyle\sum_{i=1}^{4} [(i - 1)^2 + (i + 1)^3] = [(0)^2 + (2)^3] + [(1)^2 + (3)^3] + [(2)^2 + (4)^3] + [(3)^2 + (5)^3] = 238$

**85.** $\displaystyle\sum_{i=1}^{4} 2^i = 2^1 + 2^2 + 2^3 + 2^4 = 30$

**87.** $\displaystyle\sum_{j=1}^{6} (24 - 3j) = 81$

**89.** $\displaystyle\sum_{k=0}^{4} \dfrac{(-1)^k}{k + 1} = \dfrac{47}{60}$

**91.** $\dfrac{1}{3(1)} + \dfrac{1}{3(2)} + \dfrac{1}{3(3)} + \cdots + \dfrac{1}{3(9)} = \displaystyle\sum_{i=1}^{9} \dfrac{1}{3i} \approx 0.94299$

**93.** $\left[2\left(\dfrac{1}{8}\right) + 3\right] + \left[2\left(\dfrac{2}{8}\right) + 3\right] + \left[2\left(\dfrac{3}{8}\right) + 3\right] + \cdots + \left[2\left(\dfrac{8}{8}\right) + 3\right] = \displaystyle\sum_{i=1}^{8} \left[2\left(\dfrac{i}{8}\right) + 3\right] = 33$

**95.** $3 - 9 + 27 - 81 + 243 - 729 = \displaystyle\sum_{i=1}^{6} (-1)^{i+1} 3^i = -546$

**97.** $\dfrac{1}{1^2} - \dfrac{1}{2^2} + \dfrac{1}{3^2} - \dfrac{1}{4^2} + \cdots + -\dfrac{1}{20^2} = \displaystyle\sum_{i=1}^{20} \dfrac{(-1)^{i+1}}{i^2} \approx 0.82128$

**99.** $\dfrac{1}{4} + \dfrac{3}{8} + \dfrac{7}{16} + \dfrac{15}{32} + \dfrac{31}{64} = \displaystyle\sum_{i=1}^{5} \dfrac{2^i - 1}{2^{i+1}} = \dfrac{129}{64} = 2.015625$

**101.** $\displaystyle\sum_{i=1}^{4} 5\left(\dfrac{1}{2}\right)^i = 4.6875 = \dfrac{75}{16}$

**103.** $\displaystyle\sum_{n=1}^{3} 4\left(-\dfrac{1}{2}\right)^n = -1.5 = -\dfrac{3}{2}$

**105.** $\displaystyle\sum_{i=1}^{\infty} 6\left(\dfrac{1}{10}\right)^i = 6[0.1 + 0.01 + 0.001 + \cdots]$

$\qquad\qquad = 6[0.111\ldots]$

$\qquad\qquad = 0.666\ldots$

$\qquad\qquad = \dfrac{2}{3}$

**107.** $\displaystyle\sum_{k=1}^{\infty} \left(\dfrac{1}{10}\right)^k = 0.1 + 0.11 + 0.111 + \cdots$

$\qquad\qquad = 0.11111$

$\qquad\qquad = \dfrac{1}{9}$

**109.** $A_n = 5000\left(1 + \dfrac{0.03}{4}\right)^n, \quad n = 1, 2, 3, \ldots$

    (a)   $A_1 = 5000\left(1 + \dfrac{0.03}{4}\right)^1 = \$5037.50$

$\qquad A_2 \approx \$5075.28$

$\qquad A_3 \approx \$5113.35$

$\qquad A_4 \approx \$5151.70$

$\qquad A_5 \approx \$5190.33$

$\qquad A_6 \approx \$5229.26$

$\qquad A_7 \approx \$5268.48$

$\qquad A_8 \approx \$5307.99$

    (b)   $A_{40} \approx \$6741.74$

**111.** $a_n = 1.37n^2 + 3.1n + 698, \quad n = 3, 4, \ldots, 11$

    (a)   $a_3 = 1.37(3)^2 + 3.1(3) + 698 = 719.63 \approx 720$

$\qquad a_4 \approx 732$

$\qquad a_5 \approx 748$

$\qquad a_6 \approx 766$

$\qquad a_7 \approx 787$

$\qquad a_8 \approx 810$

$\qquad a_9 \approx 837$

$\qquad a_{10} \approx 866$

$\qquad a_{11} \approx 898$

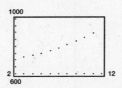

    (b)   The enrollment seems to be growing.

**113.** $a_n = 2.151n^2 + 235.9$,  $n = 4, 5, \ldots, 12$

$\qquad a_4 + a_5 + \cdots + a_{12} = 270.316 + \cdots + 545.644$

$$\approx \$3491.1 \text{ million}$$

**115.** True

**117.** $a_0 = 1$, $a_1 = 1$, $a_{k+2} = a_{k+1} + a_k$

| | | | |
|---|---|---|---|
| $a_0 = 1$ | $b_0 = \dfrac{1}{1} = 1$ | $a_6 = 8 + 5 = 13$ | $b_6 = \dfrac{21}{13}$ |
| $a_1 = 1$ | $b_1 = \dfrac{2}{1} = 2$ | $a_7 = 13 + 8 = 21$ | $b_7 = \dfrac{34}{21}$ |
| $a_2 = 1 + 1 = 2$ | $b_2 = \dfrac{3}{2}$ | $a_8 = 21 + 13 = 34$ | $b_8 = \dfrac{55}{34}$ |
| $a_3 = 2 + 1 = 3$ | $b_3 = \dfrac{5}{3}$ | $a_9 = 34 + 21 = 55$ | $b_9 = \dfrac{89}{55}$ |
| $a_4 = 3 + 2 = 5$ | $b_4 = \dfrac{8}{5}$ | $a_{10} = 55 + 34 = 89$ | |
| $a_5 = 5 + 3 = 8$ | $b_5 = \dfrac{13}{8}$ | $a_{11} = 89 + 55 = 144$ | |

**119.** $a_n = \dfrac{x^n}{n!}$

$a_1 = \dfrac{x}{1} = x$

$a_2 = \dfrac{x^2}{2!} = \dfrac{x^2}{2}$

$a_3 = \dfrac{x^3}{3!} = \dfrac{x^3}{6}$

$a_4 = \dfrac{x^4}{4!} = \dfrac{x^4}{24}$

$a_5 = \dfrac{x^5}{5!} = \dfrac{x^5}{120}$

**121.** $a_n = \dfrac{(-1)^n x^{2n}}{(2n)!}$

$a_1 = \dfrac{-x^2}{2}$

$a_2 = \dfrac{x^4}{4!} = \dfrac{x^4}{24}$

$a_3 = \dfrac{-x^6}{6!} = \dfrac{-x^6}{720}$

$a_4 = \dfrac{x^8}{8!} = \dfrac{x^8}{40,320}$

$a_5 = \dfrac{-x^{10}}{10!} = \dfrac{-x^{10}}{3,628,800}$

**123.** (a) $A - B = \begin{bmatrix} 8 & 1 \\ -3 & 7 \end{bmatrix}$

(b) $2B - 3A = \begin{bmatrix} -22 & -7 \\ 3 & -18 \end{bmatrix}$

(c) $AB = \begin{bmatrix} 18 & 9 \\ 18 & 0 \end{bmatrix}$

(d) $BA = \begin{bmatrix} 0 & 6 \\ 27 & 18 \end{bmatrix}$

**125.** (a) $A - B = \begin{bmatrix} -3 & -7 & 4 \\ 4 & 4 & 1 \\ 1 & 4 & 3 \end{bmatrix}$

(b) $2B - 3A = \begin{bmatrix} 8 & 17 & -14 \\ -12 & -13 & -9 \\ -3 & -15 & -10 \end{bmatrix}$

(c) $AB = \begin{bmatrix} -2 & 7 & -16 \\ 4 & 42 & 45 \\ 1 & 23 & 48 \end{bmatrix}$

(d) $BA = \begin{bmatrix} 16 & 31 & 42 \\ 10 & 47 & 31 \\ 13 & 22 & 25 \end{bmatrix}$

## Section 6.2   Arithmetic Sequences and Partial Sums

- You should be able to recognize an arithmetic sequence, find its common difference, and find its $n$th term.
- You should be able to find the $n$th partial sum of an arithmetic sequence with common difference $d$ using the formula

$$S_n = \frac{n}{2}(a_1 + a_n).$$

**Solutions to Odd-Numbered Exercises**

**1.** $10, 8, 6, 4, 2, \ldots$

Arithmetic sequence, $d = -2$

**3.** $3, \frac{5}{2}, 2, \frac{3}{2}, 1, \ldots$

Arithmetic sequence, $d = -\frac{1}{2}$

**5.** $-24, -16, -8, 0, 8$

Arithmetic sequence, $d = 8$

**7.** $3.7, 4.3, 4.9, 5.5, 6.1, \ldots$

Arithmetic sequence, $d = 0.6$

**9.** $a_n = 8 + 13n$

$21, 34, 47, 60, 73$

Arithmetic sequence, $d = 13$

**11.** $a_n = \dfrac{1}{n+1}$

$\dfrac{1}{2}, \dfrac{1}{3}, \dfrac{1}{4}, \dfrac{1}{5}, \dfrac{1}{6}$

Not an arithmetic sequence

**13.** $a_n = 150 - 7n$

$143, 136, 129, 122, 115$

Arithmetic sequence, $d = -7$

**15.** $a_n = 3 + \dfrac{(-1)^n 2}{n}$

$1, \ 4, \ \dfrac{7}{3}, \ \dfrac{7}{2}, \ \dfrac{13}{5}$

Not an arithmetic sequence

**17.** $a_1 = 1, \ d = 3$

$a_n = a_1 + (n-1)d = 1 + (n-1)(3) = 3n - 2$

**19.** $a_1 = 100, \ d = -8$

$a_n = a_1 + (n-1)d = 100 + (n-1)(-8) = 108 - 8n$

**21.** $4, \frac{3}{2}, -1, -\frac{7}{2}, \ldots$

$d = -\frac{5}{2}$

$a_n = a_1 + (n-1)d = 4 + (n-1)\left(-\frac{5}{2}\right) = \frac{13}{2} - \frac{5}{2}n$

**23.** $a_1 = 5, \ a_4 = 15$

$a_4 = a_1 + 3d \implies 15 = 5 + 3d \implies d = \frac{10}{3}$

$a_n = a_1 + (n-1)d = 5 + (n-1)\left(\frac{10}{3}\right) = \frac{10}{3}n + \frac{5}{3}$

**25.** $a_3 = 94$, $a_6 = 85$

$a_6 = a_3 + 3d \Rightarrow 85 = 94 + 3d \Rightarrow d = -3$

$a_1 = a_3 - 2d \Rightarrow a_1 = 94 - 2(-3) = 100$

$a_n = a_1 + (n-1)d = 100 + (n-1)(-3) = 103 - 3n$

**27.** $a_1 = 5$, $d = 6$

$a_1 = 5$

$a_2 = 5 + 6 = 11$

$a_3 = 11 + 6 = 17$

$a_4 = 17 + 6 = 23$

$a_5 = 23 + 6 = 29$

**29.** $a_1 = -10$, $d = -12$

$a_1 = -10$

$a_2 = -10 - 12 = -22$

$a_3 = -22 - 12 = -34$

$a_4 = -34 - 12 = -46$

$a_5 = -46 - 12 = -58$

**31.** $a_8 = 26$, $a_{12} = 42$

$26 = a_8 = a_1 + (n-1)d = a_1 + 7d$

$42 = a_{12} = a_1 + (n-1)d = a_1 + 11d$

*Answer:* $d = 4$, $a_1 = -2$

$a_1 = -2$

$a_2 = -2 + 4 = 2$

$a_3 = 2 + 4 = 6$

$a_4 = 6 + 4 = 10$

$a_5 = 10 + 4 = 14$

**33.** $a_3 = 19$, $a_{15} = -1.7$

$a_{15} = a_3 + 12d$

$-1.7 = 19 + 12d \Rightarrow d = -1.725$

$a_3 = a_1 + 2d \Rightarrow 19 = a_1 + 2(-1.725)$

$\Rightarrow a_1 = 22.45$

$a_2 = a_1 - 1.725 = 20.725$

$a_3 = 19$

$a_4 = 19 - 1.725 = 17.275$

$a_5 = 17.275 - 1.725 = 15.55$

**35.** $a_1 = 15$, $a_{k+1} = a_k + 4$

$a_2 = a_1 + 4 = 15 + 4 = 19$

$a_3 = 19 + 4 = 23$

$a_4 = 23 + 4 = 27$

$a_5 = 27 + 4 = 31$

$d = 4$, $a_n = 11 + 4n$

**37.** $a_1 = \frac{7}{2}$, $a_{k+1} = a_k - \frac{1}{4}$

$a_2 = \frac{7}{2} - \frac{1}{4} = \frac{13}{4}$

$a_3 = \frac{13}{4} - \frac{1}{4} = \frac{12}{4} = 3$

$a_4 = \frac{12}{4} - \frac{1}{4} = \frac{11}{4}$

$a_5 = \frac{11}{4} - \frac{1}{4} = \frac{10}{4} = \frac{5}{2}$

$d = -\frac{1}{4}$, $a_n = \frac{15}{4} - \frac{1}{4}n$

**39.** $a_1 = 5$, $a_2 = 11 \Rightarrow d = 6$

$a_{10} = a_1 + 9d = 5 + 9(6) = 59$

**41.** $a_1 = 4.2$, $a_2 = 6.6 \Rightarrow d = 2.4$

$a_7 = a_1 + 6d = 4.2 + 6(2.4) = 18.6$

**43.** $a_n = 15 - \frac{3}{2}n$

**45.** $a_n = 0.5n + 4$

**47.** $a_n = 4n - 5$

| $n$ | 1 | 2 | 3 | 4 | 5 | 6 | 7 | 8 | 9 | 10 |
|-----|----|---|---|----|----|----|----|----|----|----|
| $a_n$ | $-1$ | 3 | 7 | 11 | 15 | 19 | 23 | 27 | 31 | 35 |

**49.** $a_n = 20 - \frac{3}{4}n$

| $n$ | 1 | 2 | 3 | 4 | 5 | 6 | 7 | 8 | 9 | 10 |
|-----|---|---|---|---|---|---|---|---|---|----|
| $a_n$ | 19.25 | 18.5 | 17.75 | 17 | 16.25 | 15.5 | 14.75 | 14 | 13.25 | 12.5 |

**51.** $a_n = 1.5 + 0.005n$

| $n$ | 1 | 2 | 3 | 4 | 5 | 6 | 7 | 8 | 9 | 10 |
|-----|---|---|---|---|---|---|---|---|---|----|
| $a_n$ | 1.505 | 1.51 | 1.515 | 1.52 | 1.525 | 1.53 | 1.535 | 1.54 | 1.545 | 1.55 |

**53.** $8, 20, 32, 44, \ldots n = 10$

$a_1 = 8, a_2 = 20 \Rightarrow d = 12$

$a_{10} = a_1 + 9d = 8 + 9(12) = 116$

$S_{10} = \frac{n}{2}[a_1 + a_{10}] = \frac{10}{2}[8 + 116] = 620$

**55.** $a_1 = 0.5, a_2 = 1.3 \Rightarrow d = 0.8$

$a_{10} = a_1 + 9d = 0.5 + 9(0.8) = 7.7$

$S_{10} = \frac{10}{2}(a_1 + a_{10}) = 5(0.5 + 7.7) = 41$

**57.** $a_1 = 100, a_{25} = 220$

$S_{25} = \frac{25}{2}(a_1 + a_{25}) = 12.5(100 + 220) = 4000$

**59.** $a_1 = 1, a_{100} = 199, n = 100$

$\sum_{n=1}^{100} (2n - 1) = \frac{100}{2}(1 + 199) = 10,000$

**61.** $a_1 = 1, a_{50} = 50, n = 50$

$\sum_{n=1}^{50} n = \frac{50}{2}(1 + 50) = 1275$

**63.** $a_1 = 5, a_{100} = 500, n = 100$

$\sum_{n=1}^{100} 5n = \frac{100}{2}(5 + 500) = 25,250$

**65.** $\sum_{n=11}^{30} n - \sum_{n=1}^{10} n = \frac{20}{2}(11 + 30) - \frac{10}{2}(1 + 10) = 410 - 55 = 355$

**67.** $\sum_{n=1}^{500} (n + 8) = \frac{500}{2}[9 + 508] = 129,250$

**69.** $a_1 = 7, a_{20} = 45, n = 20$

$\sum_{n=1}^{20} (2n + 5) = \frac{20}{2}(7 + 45) = 520$

**71.** $a_n = \frac{n + 4}{2} = 2 + \frac{n}{2}$

$a_1 = \frac{5}{2}, a_{100} = 52, d = 2$

$S_{100} = \frac{100}{2}\left(\frac{5}{2} + 52\right) = 2725$

**73.** $a_1 = \frac{742}{3}, a_{60} = 90, n = 60$

$\sum_{i=1}^{60} \left(250 - \frac{8}{3}i\right) = \frac{60}{2}\left(\frac{742}{3} + 90\right) = 10,120$

**75.** (a) $a_1 = 32{,}500, \ d = 1500$

$$a_6 = a_1 + 5d = 32{,}500 + 5(1500) = \$40{,}000$$

(b) $S_6 = \frac{6}{2}[32{,}500 + 40{,}000] = \$217{,}500$

(c) 1st year: \$32,500; 2nd year: \$34,000;
3rd year: \$35,500; 4th year: \$37,000;
5th year: \$38,500; 6th year: \$40,000

$$32{,}500 + 34{,}000 + 35{,}500 + 37{,}000 +$$

$$38{,}500 + 40{,}000 = \$217{,}500$$

**77.** $a_1 = 14, a_{18} = 31$

$$S_{18} = \frac{18}{2}(14 + 31) = 405 \text{ bricks}$$

**79.** $a_1 = 25, \ a_2 = 25 + 2 = 27, \text{ etc.} \implies d = 2 \text{ and } n = 15$

$$a_{15} = 25 + (15 - 1)(2)$$

$$S_{15} = \frac{15}{2}(25 + 53) = \frac{15}{2} \cdot 78 = 585 \text{ seats}$$

**81.** $a_1 = 20{,}000$

$$a_2 = 20{,}000 + 5000 = 25{,}000$$

$$d = 5000$$

$$a_5 = 20{,}000 + 4(5000) = 40{,}000$$

$$S_5 = \frac{5}{2}(20{,}000 + 40{,}000) = 150{,}000$$

**83.** True. Given $a_1$ and $a_2$, you know $d = a_2 - a_1$. Thus, $a_n = a_1 + (n - 1)d$.

**85.** $a_1 = x$      $a_6 = 11x$

$a_2 = x + 2x = 3x$      $a_7 = 13x$

$a_3 = 3x + 2x = 5x$      $a_8 = 15x$

$a_4 = 7x$      $a_9 = 17x$

$a_5 = 9x$      $a_{10} = 19x$

**87.** $S_{20} = \frac{20}{2}(a_1 + [a_1 + 19(3)]) = 650$

$$650 = 10(2a_1 + 57)$$

$$65 = 2a_1 + 57$$

$$a_1 = 4$$

**89.** (a) $1 + 3 = 4$

$$1 + 3 + 5 = 9$$

$$1 + 3 + 5 + 7 = 16$$

$$1 + 3 + 5 + 7 + 9 = 25$$

$$1 + 3 + 5 + 7 + 9 + 11 = 36$$

(b) $S_n = n^2$

$$S_7 = 1 + 3 + 5 + 7 + 9 + 11 + 13 = 49 = 7^2$$

(c) $S_n = \frac{n}{2}[1 + (2n - 1)] = \frac{n}{2}(2n) = n^2$

**91.** $\begin{bmatrix} 2 & -1 & 7 & : & -10 \\ 3 & 2 & -4 & : & 17 \\ 6 & -5 & 1 & : & -20 \end{bmatrix}$ row reduces to $\begin{bmatrix} 1 & 0 & 0 & : & 1 \\ 0 & 1 & 0 & : & 5 \\ 0 & 0 & 1 & : & -1 \end{bmatrix}$

*Answer:* $(1, 5, -1)$

**93.** $\begin{vmatrix} 0 & 0 & 1 \\ 4 & -3 & 1 \\ 2 & 6 & 1 \end{vmatrix} = 30$

Area $= \frac{1}{2}(30) = 15$ square units

# Section 6.3    Geometric Sequences and Series

---

■ You should be able to identify a geometric sequence, find its common ratio, and find the $n$th term.

■ You should be able to find the $n$th partial sum of a geometric sequence with common ratio $r$ using the formula.

$$S_n = a_1 \left( \frac{1 - r^n}{1 - r} \right)$$

■ You should know that if $|r| < 1$, then

$$\sum_{n=1}^{\infty} a_1 r^{n-1} = \frac{a_1}{1 - r}.$$

---

**Solutions to Odd-Numbered Exercises**

**1.** 5, 15, 45, 135, . . .

Geometric sequence, $r = 3$

**3.** 6, 18, 30, 42, . . .

Not a geometric sequence

(Note:  It is an arithmetic sequence with $d = 12$.)

**5.** $1, -\frac{1}{2}, \frac{1}{4}, -\frac{1}{8}, \ldots$

Geometric sequence, $r = -\frac{1}{2}$

**7.** $\frac{1}{8}, \frac{1}{4}, \frac{1}{2}, 1, \ldots$

Geometric sequence, $r = 2$

$$a_n = a_1 r^{n-1}$$

**9.** $1, \frac{1}{2}, \frac{1}{3}, \frac{1}{4}, \ldots$

Not a geometric sequence

**11.** $a_1 = 6, r = 3$

$a_2 = 6(3) = 18$

$a_3 = 18(3) = 54$

$a_4 = 54(3) = 162$

$a_5 = 162(3) = 486$

**13.** $a_1 = 1, r = \frac{1}{2}$

$a_1 = 1$

$a_2 = 1\left(\frac{1}{2}\right) = \frac{1}{2}$

$a_3 = \frac{1}{2}\left(\frac{1}{2}\right) = \frac{1}{4}$

$a_4 = \frac{1}{4}\left(\frac{1}{2}\right) = \frac{1}{8}$

$a_5 = \frac{1}{8}\left(\frac{1}{2}\right) = \frac{1}{16}$

$$a_n = 1\left(\frac{1}{2}\right)^{n-1}$$

**15.** $a_1 = 5, r = -\frac{1}{10}$

$a_1 = 5$

$a_2 = 5\left(-\frac{1}{10}\right) = -\frac{1}{2}$

$a_3 = \left(-\frac{1}{2}\right)\left(-\frac{1}{10}\right) = \frac{1}{20}$

$a_4 = \frac{1}{20}\left(-\frac{1}{10}\right) = -\frac{1}{200}$

$a_5 = \left(-\frac{1}{200}\right)\left(-\frac{1}{10}\right) = \frac{1}{2000}$

**17.** $a_1 = 1, r = e$

$a_1 = 1$

$a_2 = 1(e) = e$

$a_3 = (e)(e) = e^2$

$a_4 = (e^2)(e) = e^3$

$a_5 = (e^3)(e) = e^4$

**19.** $a_1 = 64$, $a_{k+1} = \frac{1}{2}a_k$

$a_1 = 64$

$a_2 = \frac{1}{2}(64) = 32$

$a_3 = \frac{1}{2}(32) = 16$

$a_4 = \frac{1}{2}(16) = 8$

$a_5 = \frac{1}{2}(8) = 4$

$r = \frac{1}{2}$, $a_n = 64\left(\frac{1}{2}\right)^{n-1} = 128\left(\frac{1}{2}\right)^n$

**21.** $a_1 = 9$, $a_{k+1} = 2a_k$

$a_2 = 2(9) = 18$

$a_3 = 2(18) = 36$

$a_4 = 2(36) = 72$

$a_5 = 2(72) = 144$

$r = 2$

$a_n = \left(\frac{9}{2}\right)2^n = 9(2^{n-1})$

**23.** $a_1 = 6$, $a_{k+1} = -\frac{3}{2}a_k$

$a_1 = 6$

$a_2 = -\frac{3}{2}(6) = -9$

$a_3 = -\frac{3}{2}(-9) = \frac{27}{2}$

$a_4 = -\frac{3}{2}\left(\frac{27}{2}\right) = -\frac{81}{4}$

$a_5 = -\frac{3}{2}\left(-\frac{81}{4}\right) = \frac{243}{8}$

$r = -\frac{3}{2}$, $a_n = 6\left(-\frac{3}{2}\right)^{n-1}$

$\quad = -4\left(-\frac{3}{2}\right)^n$

**25.** $a_1 = 4$, $r = \frac{1}{2}$, $n = 10$

$a_n = a_1 r^{n-1}$

$a_{10} = 4\left(\frac{1}{2}\right)^9 = \left(\frac{1}{2}\right)^7 = \frac{1}{128}$

**27.** $a_1 = 6$, $r = -\frac{1}{3}$, $n = 12$

$a_n = a_1 r^{n-1}$

$a_{12} = 6\left(-\frac{1}{3}\right)^{11} = \frac{-2}{3^{10}}$

**29.** $a_1 = 500$, $r = 1.02$, $n = 14$

$a_n = a_1 r^{n-1}$

$a_{14} = 500(1.02)^{13} \approx 646.8$

**31.** $a_2 = a_1 r = -18 \implies a_1 = \frac{-18}{r}$

$a_5 = a_1 r^4 = (a_1 r)r^3 = -18r^3 = \frac{2}{3} \implies r = -\frac{1}{3}$

$a_1 = \frac{-18}{r} = \frac{-18}{-1/3} = 54$

$a_6 = a_1 r^5 = 54\left(\frac{-1}{3}\right)^5 = \frac{-54}{243} = -\frac{2}{9}$

**33.** $r = \frac{21}{7} = 3$

$a_9 = a_1 r^{9-1} = 7(3)^8 = 45{,}927$

**35.** $r = \frac{30}{5} = 6$

$a_{10} = a_1 r^{10-1} = 5(6)^9 = 50{,}388{,}480$

**37.** $a_n = 12(-0.75)^{n-1}$

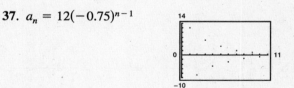

**39.** $a_n = 2(1.3)^{n-1}$

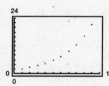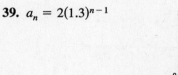

**41.** $8, -4, 2, -1, \frac{1}{2}$

$S_1 = 8$

$S_2 = 8 + (-4) = 4$

$S_3 = 8 + (-4) + 2 = 6$

$S_4 = 8 + (-4) + 2 + (-1) = 5$

**43.** $\displaystyle\sum_{n=1}^{\infty} 16\left(-\frac{1}{2}\right)^{n-1}$

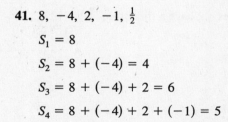

| $n$ | 1 | 2 | 3 | 4 | 5 | 6 | 7 | 8 | 9 | 10 |
|---|---|---|---|---|---|---|---|---|---|---|
| $S_n$ | 16 | 24 | 28 | 30 | 31 | 31.5 | 31.75 | 31.875 | 31.9375 | 31.96875 |

**45.** $\sum_{n=1}^{9} 2^{n-1} \Rightarrow a_1 = 1, \; r = 2$

$S_9 = \dfrac{1(1 - 2^9)}{1 - 2} = 511$

**47.** $\sum_{i=1}^{7} 64\left(-\dfrac{1}{2}\right)^{i-1} \Rightarrow a_1 = 64, \; r = -\dfrac{1}{2}$

$S_7 = 64\left[\dfrac{1 - (-1/2)^7}{1 - (-1/2)}\right] = \dfrac{128}{3}\left[1 - \left(-\dfrac{1}{2}\right)^7\right] = 43$

**49.** $\sum_{n=0}^{20} 3\left(\dfrac{3}{2}\right)^n = \sum_{n=1}^{21} 3\left(\dfrac{3}{2}\right)^{n-1} \Rightarrow a_1 = 3, \; r = \dfrac{3}{2}$

$S_{21} = 3\left[\dfrac{1 - (3/2)^{21}}{1 - (3/2)}\right] = -6\left[1 - \left(\dfrac{3}{2}\right)^{21}\right] \approx 29{,}921.31$

**51.** $\sum_{i=1}^{10} 8\left(-\dfrac{1}{4}\right)^{i-1} \Rightarrow a_1 = 8, \; r = -\dfrac{1}{4}$

$S_{10} = 8\left[\dfrac{1 - (-1/4)^{10}}{1 - (-1/4)}\right] = \dfrac{32}{5}\left[1 - \left(-\dfrac{1}{4}\right)^{10}\right] \approx 6.4$

**53.** $\sum_{n=0}^{5} 300(1.06)^n = \sum_{n=1}^{6} 300(1.06)^{n-1} \Rightarrow a_1 = 300, \; r = 1.06$

$S_6 = 300\left[\dfrac{1 - (1.06)^6}{1 - 1.06}\right] \approx 2092.60$

**55.** $5 + 15 + 45 + \cdots + 3645$

$r = 3$ and $3645 = 5(3)^{n-1} \Rightarrow n = 7$

Thus, the sum can be written as $\sum_{n=1}^{7} 5(3)^{n-1}$.

**57.** $2 - \frac{1}{2} + \frac{1}{8} - \cdots + \frac{1}{2048}$

$r = -\frac{1}{4}$ and $\frac{1}{2048} = 2\left(-\frac{1}{4}\right)^{n-1} \Rightarrow n = 7$

$\sum_{n=1}^{7} 2\left(-\frac{1}{4}\right)^{n-1}$

**59.** $a_1 = 1, \; r = \dfrac{1}{2}$

$\sum_{n=0}^{\infty} \left(\dfrac{1}{2}\right)^n = \dfrac{a_1}{1 - r} = \dfrac{1}{1 - (1/2)} = 2$

**61.** $a_1 = 1, \; r = -\dfrac{1}{2}$

$\sum_{n=0}^{\infty} \left(-\dfrac{1}{2}\right)^n = \sum_{n=1}^{\infty} \left(-\dfrac{1}{2}\right)^{n-1} = \dfrac{a_1}{1 - r} = \dfrac{1}{1 - (-1/2)} = \dfrac{2}{3}$

**63.** $\sum_{n=1}^{\infty} 2\left(\frac{7}{3}\right)^{n-1}$ does not have a finite sum $\left(\frac{7}{3} > 1\right)$.

**65.** $a_1 = 1, \, r = 0.4$

$\sum_{n=0}^{\infty} (0.4)^n = \dfrac{a_1}{1 - r} = \dfrac{1}{1 - 0.4} = \dfrac{1}{0.6} = \dfrac{10}{6} = \dfrac{5}{3}$

**67.** $a = -3, \, r = 0.9$

$\sum_{n=0}^{\infty} -3(0.9)^n = \dfrac{a_1}{1 - r} = \dfrac{-3}{1 - 0.9} = \dfrac{-3}{0.1} = -30$

**69.** $8 + 6 + \dfrac{9}{2} + \dfrac{27}{8} + \cdots = \sum_{n=0}^{\infty} 8\left(\dfrac{3}{4}\right)^n$

$= \dfrac{8}{1 - 3/4} = 32$

**71.** $3 - 1 + \dfrac{1}{3} - \dfrac{1}{9} + \cdots = \displaystyle\sum_{n=0}^{\infty} 3\left(-\dfrac{1}{3}\right)^n = \dfrac{a_1}{1 - r} = \dfrac{3}{1 - \left(-\dfrac{1}{3}\right)} = 3\left(\dfrac{3}{4}\right) = \dfrac{9}{4}$

**73.** $0.\overline{36} = \displaystyle\sum_{n=0}^{\infty} 0.36(0.01)^n = \dfrac{0.36}{1 - 0.01} = \dfrac{0.36}{0.99} = \dfrac{36}{99} = \dfrac{4}{11}$

**75.** $0.3\overline{18} = 0.3 + \displaystyle\sum_{n=0}^{\infty} 0.018(0.01)^n = \dfrac{3}{10} + \dfrac{0.018}{1 - 0.01}$

$\qquad\qquad = \dfrac{3}{10} + \dfrac{0.018}{0.99} = \dfrac{3}{10} + \dfrac{18}{990} = \dfrac{3}{10} + \dfrac{2}{110}$

$\qquad\qquad = \dfrac{35}{110} = \dfrac{7}{22}$

**77.** $A = P\left(1 + \dfrac{r}{n}\right)^{nt} = 1000\left(1 + \dfrac{0.03}{n}\right)^{n(10)}$

(a) $n = 1$: $A = 1000(1 + 0.03)^{10} \approx 1343.92$

(b) $n = 2$: $A = 1000\left(1 + \dfrac{0.03}{2}\right)^{2(10)} \approx 1346.86$

(c) $n = 4$: $A = 1000\left(1 + \dfrac{0.03}{4}\right)^{4(10)} \approx 1348.35$

(d) $n = 12$: $A = 1000\left(1 + \dfrac{0.03}{12}\right)^{12(10)} \approx 1349.35$

(e) $n = 365$: $A = 1000\left(1 + \dfrac{0.03}{365}\right)^{365(10)} \approx 1349.84$

**79.** $A = \displaystyle\sum_{n=1}^{60} 100\left(1 + \dfrac{0.06}{12}\right)^n = 100\left(1 + \dfrac{0.06}{12}\right) \cdot \dfrac{\left[1 - \left(1 + \dfrac{0.06}{12}\right)^{60}\right]}{\left[1 - \left(1 + \dfrac{0.06}{12}\right)\right]} \approx \$7011.89$

**81.** Let $N = 12t$ be the total number of deposits.

$A = P\left(1 + \dfrac{r}{12}\right) + P\left(1 + \dfrac{r}{12}\right)^2 + \cdots + P\left(1 + \dfrac{r}{12}\right)^N$

$\quad = \left(1 + \dfrac{r}{12}\right)\left[P + P\left(1 + \dfrac{r}{12}\right) + \cdots + P\left(1 + \dfrac{r}{12}\right)^{N-1}\right]$

$\quad = P\left(1 + \dfrac{r}{12}\right)\displaystyle\sum_{n=1}^{N}\left(1 + \dfrac{r}{12}\right)^{n-1}$

$\quad = P\left(1 + \dfrac{r}{12}\right)\dfrac{1 - \left(1 + \dfrac{r}{12}\right)^N}{1 - \left(1 + \dfrac{r}{12}\right)}$

$\quad = P\left(1 + \dfrac{r}{12}\right)\left(-\dfrac{12}{r}\right)\left[1 - \left(1 + \dfrac{r}{12}\right)^N\right]$

$\quad = P\left(\dfrac{12}{r} + 1\right)\left[-1 + \left(1 + \dfrac{r}{12}\right)^N\right]$

$\quad = P\left[\left(1 + \dfrac{r}{12}\right)^N - 1\right]\left(1 + \dfrac{12}{r}\right)$

$\quad = P\left[\left(1 + \dfrac{r}{12}\right)^{12t} - 1\right]\left(1 + \dfrac{12}{r}\right)$

**83.** $P = \$50$, $r = 7\%$, $t = 20$ years

(a) Compounded monthly: $A = 50\left[\left(1 + \dfrac{0.07}{12}\right)^{12(20)} - 1\right]\left(1 + \dfrac{12}{0.07}\right) \approx \$26,198.27$

(b) Compounded continuously: $A = \dfrac{50e^{0.07/12}(e^{0.07(20)} - 1)}{e^{0.07/12} - 1} \approx \$26,263.88$

**85.** $P = 100$, $r = 5\% = 0.05$, $t = 40$

(a) Compounded monthly: $A = 100\left[\left(1 + \dfrac{0.05}{12}\right)^{12(40)} - 1\right]\left(1 + \dfrac{12}{0.05}\right) \approx \$153,237.86$

(b) Compounded continuously: $A = \dfrac{100e^{0.05/12}(e^{0.05(40)} - 1)}{e^{0.05/12} - 1} \approx \$153,657.02$

**87.** First shaded area: $\dfrac{16^2}{4}$ 　　　　　　　Second shaded area: $\dfrac{16^2}{4} + \dfrac{1}{2} \cdot \dfrac{16^2}{4}$

Third shaded area: $\dfrac{16^2}{4} + \dfrac{1}{2}\dfrac{16^2}{4} + \dfrac{1}{4}\dfrac{16^2}{4}$, etc

Total area of shaded region:

$$\dfrac{16^2}{4}\sum_{n=0}^{5}\left(\dfrac{1}{2}\right)^n = 64\left[\dfrac{1 - \left(\dfrac{1}{2}\right)^6}{1 - \dfrac{1}{2}}\right] = 128\left(1 - \left(\dfrac{1}{2}\right)^6\right) = 126 \text{ square units}$$

**89.** $a_n = 3343(1.013)^n$

(a) The population is growing at 0.013 or 1.3% per year.

(b) In 2010, $n = 20$ and $a_{20} \approx 4328$ or 4,328,000 people.

(c) $a_n = 4100$ when $n = 15.8$ or during 2005.

**91.** $S_n = \displaystyle\sum_{i=1}^{n} 0.01(2)^{i-1}$

$S_{29} = \$5,368,709.11$

$S_{30} = \$10,737,418.23$

$S_{31} = \$21,474,836.47$

**93.** False. See definition page 516.

**95.** $a_1 = 3$, $r = \dfrac{x}{2}$

$a_2 = 3\left(\dfrac{x}{2}\right) = \dfrac{3x}{2}$

$a_3 = \dfrac{3x}{2}\left(\dfrac{x}{2}\right) = \dfrac{3x^2}{4}$

$a_4 = \dfrac{3x^2}{4}\left(\dfrac{x}{2}\right) = \dfrac{3x^3}{8}$

$a_5 = \dfrac{3x^3}{8}\left(\dfrac{x}{2}\right) = \dfrac{3x^4}{16}$

**97.** $a_1 = 100$, $r = e^x$, $n = 9$

$a_n = a_1 r^{n-1}$

$a_9 = 100(e^x)^8 = 100e^{8x}$

**99.** (a) $f(x) = 6\left[\dfrac{1 - 0.5^x}{1 - 0.5}\right]$

$$\sum_{n=0}^{\infty} 6\left(\frac{1}{2}\right)^n = \dfrac{6}{1 - \dfrac{1}{2}} = 12$$

The horizontal asymptote of $f(x)$ is $y = 12$. This corresponds to the sum of the series.

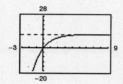

(b) $f(x) = 2\left[\dfrac{1 - 0.8^x}{1 - 0.8}\right]$

$$\sum_{n=0}^{\infty} 2\left(\frac{4}{5}\right)^n = \dfrac{2}{1 - \dfrac{4}{5}} = 10$$

The horizontal asymptote of $f(x)$ is $y = 10$. This corresponds to the sum of the series.

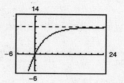

**101.** To use the first two terms of a geometric series to find the $n$th term, first divide the second term by the first term to obtain the constant ratio. The $n$th term is the first term multiplied by the common ratio raised to the $(n - 1)$ power.

$$r = \frac{a_2}{a_1},\, a_n = a_1 r^{n-1}$$

**103.** $\text{Time} = \dfrac{\text{Distance}}{\text{Speed}} = \dfrac{200}{50} + \dfrac{200}{42} = 200\left[\dfrac{92}{2100}\right] \text{ hours}$

$\text{Speed} = \dfrac{\text{Distance}}{\text{Time}} = \dfrac{400}{200\left[\frac{92}{2100}\right]} = \dfrac{2(2100)}{92} \approx 45.65 \text{ mph}$

**105.** $\det\begin{bmatrix} 4 & -1 \\ 6 & 2 \end{bmatrix} = 8 - (-6) = 14$

**107.** $\det\begin{bmatrix} -1 & 3 & 4 \\ -2 & 8 & 0 \\ 2 & 5 & -1 \end{bmatrix}$

$= 4(-10 - 16) - 1(-8 + 6)$

$= -104 + 2 = -102$

## Section 6.4   Mathematical Induction

> ■ You should be sure that you understand the principle of mathematical induction. If $P_n$ is a statement involving the positive integer $n$, where $P_1$ is true and the truth of $P_k$ implies the truth of $P_{k+1}$, then $P_n$ is true for all positive integers $n$.
>
> ■ You should be able to verify (by induction) the formulas for the sums of powers of integers and be able to use these formulas.
>
> ■ You should be able to work with finite differences.

**Solutions to Odd-Numbered Exercises**

**1.** $P_k = \dfrac{5}{k(k+1)}$

$P_{k+1} = \dfrac{5}{(k+1)[(k+1)+1]} = \dfrac{5}{(k+1)(k+2)}$

**3.** $P_k = \dfrac{3(2k+1)}{k-1}$

$P_{k+1} = \dfrac{3(2(k+1)+1)}{(k+1)-1} = \dfrac{3(2k+3)}{k}$

**5.** $P_k = 1 + 6 + 11 + \cdots + [5(k-1) - 4] + [5k - 4]$

$P_{k+1} = 1 + 6 + 11 + \cdots + [5k - 4] + [5(k+1) - 4]$

$\qquad = 1 + 6 + 11 + \cdots + [5k - 4] + [5k + 1]$

**7.** 1. When $n = 1$, $S_1 = 2 = 1(1+1)$.

   2. Assume that

     $S_k = 2 + 4 + 6 + 8 + \cdots + 2k = k(k+1)$.

     Then,

     $S_{k+1} = 2 + 4 + 6 + 8 + \cdots + 2k + 2(k+1)$

     $\qquad = S_k + 2(k+1) = k(k+1) + 2(k+1) = (k+1)(k+2)$.

We conclude by mathematical induction that the formula is valid for all positive integer values of $n$.

**9.** 1. When $n = 1$, $S_1 = 3 = \dfrac{1}{2}(5(1) + 1)$

   2. Assume that $S_k = 3 + 8 + 13 + \cdots + (5k - 2) = \dfrac{k}{2}(5k + 1)$

     Then: $S_{k+1} = 3 + 8 + 13 + \cdots + (5k - 2) + [5(k+1) - 2]$

         $= S_k + [5k + 3] = \dfrac{k}{2}(5k + 1) + 5k + 3$

         $= \dfrac{1}{2}[5k^2 + 11k + 6] = \dfrac{1}{2}(k + 1)(5k + 6)$

         $= \dfrac{1}{2}(k + 1)(5(k + 1) + 1)$

We conclude by mathematical induction that the formula is valid for all positive integers $n$.

**11.** 1. When $n = 1$, $S_1 = 1 = 2^1 - 1$.

    2. Assume that

$$S_k = 1 + 2 + 2^2 + 2^3 + \cdots + 2^{k-1} = 2^k - 1.$$

    Then,

$$S_{k+1} = 1 + 2 + 2^2 + 2^3 + \cdots + 2^{k-1} + 2^k$$
$$= S_k + 2^k = 2^k - 1 + 2^k = 2(2^k) - 1 = 2^{k+1} - 1.$$

Therefore, by mathematical induction, the formula is valid for all positive integer values of $n$.

**13.** 1. When $n = 1$, $S_1 = 1 = \dfrac{1(1 + 1)}{2}$.

    2. Assume that

$$S_k = 1 + 2 + 3 + 4 + \cdots + k = \frac{k(k + 1)}{2}.$$

    Then,

$$S_{k+1} = 1 + 2 + 3 + 4 + \cdots + k + (k + 1)$$
$$= S_k + (k + 1) = \frac{k(k + 1)}{2} + \frac{2(k + 1)}{2} = \frac{(k + 1)(k + 2)}{2}.$$

Therefore, we conclude that this formula holds for all positive integer values of $n$.

**15.** 1. When $n = 1$,

$$S_1 = 1^4 = \frac{1(1 + 1)(2 \cdot 1 + 1)(3 \cdot 1^2 + 3 \cdot 1 - 1)}{30}.$$

    2. Assume that $S_k = \displaystyle\sum_{i=1}^{k} i^4 = \dfrac{k(k + 1)(2k + 1)(3k^2 + 3k - 1)}{30}$.

    Then, $S_{k+1} = S_k + (k + 1)^4$

$$= \frac{k(k + 1)(2k + 1)(3k^2 + 3k - 1)}{30} + (k + 1)^4 = \frac{k(k + 1)(2k + 1)(3k^2 + 3k - 1) + 30(k + 1)^4}{30}$$

$$= \frac{(k + 1)[k(2k + 1)(3k^2 + 3k - 1) + 30(k + 1)^3]}{30} = \frac{(k + 1)(6k^4 + 39k^3 + 91k^2 + 89k + 30)}{30}$$

$$= \frac{(k + 1)(k + 2)(2k + 3)(3k^2 + 9k + 5)}{30} = \frac{(k + 1)(k + 2)(2(k + 1) + 1)(3(k + 1)^2 + 3(k + 1) - 1)}{30}.$$

Therefore, we conclude that this formula holds for all positive integer values of $n$.

**17.** 1. When $n = 1$, $S_1 = 2 = \dfrac{1(2)(3)}{3}$.

2. Assume that $S_k = 1(2) + 2(3) + 3(4) + \cdots + k(k + 1) = \dfrac{k(k + 1)(k + 2)}{3}$.

Then,

$$S_{k+1} = 1(2) + 2(3) + 3(4) + \cdots + k(k + 1) + (k + 1)(k + 2)$$

$$= S_k + (k + 1)(k + 2) = \frac{k(k + 1)(k + 2)}{3} + \frac{3(k + 1)(k + 2)}{3}$$

$$= \frac{(k + 1)(k + 2)(k + 3)}{3}.$$

Thus, this formula is valid for all positive integer values of $n$.

**19.** $\displaystyle\sum_{n=1}^{10} n^3 = \frac{10^2(10 + 1)^2}{4} = 3025$

**21.** $\displaystyle\sum_{n=1}^{6} (n^2 - n) = \sum_{n=1}^{6} n^2 - \sum_{n=1}^{6} n$

$$= \frac{6(6 + 1)(2(6) + 1)}{6} - \frac{6(6 + 1)}{2}$$

$$= 91 - 21 = 70$$

**23.** 1. When $n = 4$, $4! = 24$ and $2^4 = 16$, thus $4! > 2^4$.

2. Assume $k! > 2^k$, $k > 4$. Then, $(k + 1)! = k!(k + 1) > 2^k(2)$ since $k + 1 > 2$. Thus, $(k + 1)! > 2^{k+1}$.

Therefore, by mathematical induction, the formula is valid for all integers $n$ such that $n \geq 4$.

**25.** 1. When $n = 2$, $\dfrac{1}{\sqrt{1}} + \dfrac{1}{\sqrt{2}} \approx 1.707$ and $\sqrt{2} \approx 1.414$, thus $\dfrac{1}{\sqrt{1}} + \dfrac{1}{\sqrt{2}} > \sqrt{2}$.

2. Assume $\dfrac{1}{\sqrt{1}} + \dfrac{1}{\sqrt{2}} + \dfrac{1}{\sqrt{3}} + \cdots + \dfrac{1}{\sqrt{k}} > \sqrt{k}, k > 2$.

Then, $\dfrac{1}{\sqrt{1}} + \dfrac{1}{\sqrt{2}} + \dfrac{1}{\sqrt{3}} + \cdots + \dfrac{1}{\sqrt{k}} + \dfrac{1}{\sqrt{k + 1}} > \sqrt{k} + \dfrac{1}{\sqrt{k + 1}}$.

Now we need to show that $\sqrt{k} + \dfrac{1}{\sqrt{k + 1}} > \sqrt{k + 1}, k > 2$.

This is true because

$$\sqrt{k(k + 1)} > k$$

$$\sqrt{k(k + 1)} + 1 > k + 1$$

$$\frac{\sqrt{k(k + 1)} + 1}{\sqrt{k + 1}} > \frac{k + 1}{\sqrt{k + 1}}$$

$$\sqrt{k} + \frac{1}{\sqrt{k + 1}} > \sqrt{k + 1}.$$

Therefore, $\dfrac{1}{\sqrt{1}} + \dfrac{1}{\sqrt{2}} + \dfrac{1}{\sqrt{3}} + \cdots + \dfrac{1}{\sqrt{k}} + \dfrac{1}{\sqrt{k + 1}} > \sqrt{k + 1}$.

Therefore, by mathematical induction, the formula is valid for all integers n such that $n \geq 2$.

**27.** 1. When $n = 1, 1 + a \geq a$ since $1 > 0$.

2. Assume $(1 + a)^k \geq ka$

Then $(1 + a)^{k+1} = (1 + a)^k(1 + a) \geq ka(1 + a)$

$$= ka + ka^2 \geq ka + a \quad \text{(because } a > 1\text{)}$$

$$= (k + 1)a$$

Therefore, by mathematical induction, the inequality is valid for all integers $n \geq 1$.

**29.** 1. When $n = 1, (ab)^1 = a^1b^1 = ab$.

2. Assume that $(ab)^k = a^kb^k$.

Then, $(ab)^{k+1} = (ab)^k(ab)$

$$= a^kb^kab$$

$$= a^{k+1}b^{k+1}.$$

Thus, $(ab)^n = a^nb^n$.

**31.** 1. When $n = 1, (x_1)^{-1} = x_1^{-1}$.

2. Assume that

$$(x_1x_2x_3 \cdots x_k)^{-1} = x_1^{-1}x_2^{-1}x_3^{-1} \cdots x_k^{-1}.$$

Then,

$$(x_1x_2x_3 \cdots x_kx_{k+1})^{-1} = [(x_1x_2x_3 \cdots x_k)x_{k+1}]^{-1}$$

$$= (x_1x_2x_3 \cdots x_k)^{-1}x_{k+1}^{-1}$$

$$= x_1^{-1}x_2^{-1}x_3^{-1} \cdots x_k^{-1}x_{k+1}^{-1}.$$

Thus, the formula is valid.

**33.** 1. When $n = 1, x(y_1) = xy_1$.

2. Assume that $x(y_1 + y_2 + \cdots + y_k) = xy_1 + xy_2 + \cdots + xy_k$.

Then,

$$xy_1 + xy_2 + \cdots + xy_k + xy_{k+1} = x(y_1 + y_2 + \cdots + y_k) + xy_{k+1}$$

$$= x[(y_1 + y_2 + \cdots + y_k) + y_{k+1}]$$

$$= x(y_1 + y_2 + \cdots + y_k + y_{k+1}).$$

Hence, the formula holds.

**35.** 1. When $n = 1, [1^3 + 3(1)^2 + 2(1)] = 6$ and 3 is a factor.

2. Assume that 3 is a factor of $(k^3 + 3k^2 + 2k)$. Then,

$$[(k + 1)^3 + 3(k + 1)^2 + 2(k + 1)] = k^3 + 3k^2 + 3k + 1 + 3k^2 + 6k + 3 + 2k + 2$$

$$= (k^3 + 3k^2 + 2k) + (3k^2 + 9k + 6)$$

$$= (k^3 + 3k^2 + 2k) + 3(k^2 + 3k + 2).$$

Since 3 is a factor of $(k^3 + 3k^2 + 2k)$ by our assumption, and 3 is a factor of $3(k^2 + 3k + 2)$ then 3 is a factor of the whole sum.

Thus, 3 is a factor of $(n^3 + 3n^2 + 2n)$ for every positive integer $n$.

**37.** $a_1 = 0, a_n = a_{n-1} + 3$

$a_1 = 0$

$a_2 = a_1 + 3 = 0 + 3 = 3$

$a_3 = a_2 + 3 = 3 + 3 = 6$

$a_4 = a_3 + 3 = 6 + 3 = 9$

$a_5 = a_4 + 3 = 9 + 3 = 12$

$a_n$:    0    3    6    9    12

First differences:    3    3    3    3

Second differences:    0    0    0

Since the first differences are equal, the sequence has a linear model.

**39.** $a_1 = 3, a_n = a_{n-1} - n$

$a_1 = 3$

$a_2 = a_1 - 2 = 3 - 2 = 1$

$a_3 = a_2 - 3 = 1 - 3 = -2$

$a_4 = a_3 - 4 = -2 - 4 = -6$

$a_5 = a_4 - 5 = -6 - 5 = -11$

$a_n$:    3    1    -2    -6    -11

First differences:    -2    -3    -4    -5

Second differences:    -1    -1    -1

Since the second differences are all the same, the sequence has a quadratic model.

**41.** $a_0 = 0, a_n = a_{n-1} + n$

$a_0 = 0$

$a_1 = a_0 + 1 = 0 + 1 = 1$

$a_2 = a_1 + 2 = 1 + 2 = 3$

$a_3 = a_2 + 3 = 3 + 3 = 6$

$a_4 = a_3 + 4 = 6 + 4 = 10$

$a_n$:    0    1    3    6    10

First differences:    1    2    3    4

Second differences:    1    1    1

Since the second differences are equal, the sequence has a quadratic model.

**43.** $a_1 = 2, a_n = a_{n-1} + 2$

$a_1 = 2$

$a_2 = a_1 + 2 = 2 + 2 = 4$

$a_3 = a_2 + 2 = 4 + 2 = 6$

$a_4 = a_3 + 2 = 6 + 2 = 8$

$a_5 = a_4 + 2 = 8 + 2 = 10$

$a_n$:    2    4    6    8    10

First differences:    2    2    2    2

Second differences:    0    0    0

Since the first differences are equal, the sequence has a linear model.

**45.** $a_0 = 3, a_1 = 3, a_4 = 15$

Let $a_n = an^2 + bn + c$. Thus

$a_0 = a(0)^2 + b(0) + c = 3 \implies c = 3$

$a_1 = a(1)^2 + b(1) + c = 3 \implies a + b + c = 3$

$\phantom{a_1 = a(1)^2 + b(1) + c = 3 \implies} a + b \phantom{+ c} = 0$

$a_4 = a(4)^2 + b(4) + c = 15 \implies 16a + 4b + c = 15$

$\phantom{a_4 = a(4)^2 + b(4) + c = 15 \implies} 16a + 4b \phantom{+ c} = 12$

$\phantom{a_4 = a(4)^2 + b(4) + c = 15 \implies} 4a + b \phantom{+ 4} = 3$

By elimination:  $-a - b = 0$

$\phantom{By elimination:  } \underline{4a + b = 3}$

$\phantom{By elimination:  } 3a \phantom{+ b} = 3$

$\phantom{By elimination:  3} a = 1 \implies b = -1$

Thus, $a_n = n^2 - n + 3$.

**47.** $a_0 = -3, a_2 = 1, a_4 = 9$

Let $a_n = an^2 + bn + c$. Then

$a_0 = a(0)^2 + b(0) + c = -3 \implies c = -3$

$a_2 = a(2)^2 + b(2) + c = 1 \implies 4a + 2b + c = 1$

$\phantom{a_2 = a(2)^2 + b(2) + c = 1 \implies} 4a + 2b = 4$

$\phantom{a_2 = a(2)^2 + b(2) + c = 1 \implies} 2a + b = 2$

$a_4 = a(4)^2 + b(4) + c = 9 \implies 16a + 4b + c = 9$

$\phantom{a_4 = a(4)^2 + b(4) + c = 9 \implies} 16a + 4b = 12$

$\phantom{a_4 = a(4)^2 + b(4) + c = 9 \implies} 4a + b = 3$

By elimination: $-2a - b = -2$

$$\frac{4a + b = 3}{2a \quad = 1}$$

$a = \frac{1}{2} \implies b = 1$

Thus, $a_n = \frac{1}{2}n^2 + n - 3$.

**49.** (a) 33.3  35.5  38.0  40.1  41.9  44.2  46.1

First difference: 2.2  2.5  2.1  1.8  2.3  1.9

(b) Yes, the data is approximately linear because the first differences are nearly constant.

$a_n = 2.13n + 22.8$

(c)

| $n$ | 5 | 6 | 7 | 8 | 9 | 10 | 11 |
|---|---|---|---|---|---|---|---|
| $a_n$ | 33.5 | 35.6 | 37.7 | 39.8 | 42.0 | 44.1 | 46.2 |

The model is a good fit.

(d) For 2006, $n = 16$ and $a_{16} \approx 56.9$ or %56,900.

**51.** False. $P_1$ might not even be defined.

**53.** False. It has $n - 2$ second differences.

**55.** $(2x^2 - 1)^2 = 4x^4 - 4x^2 + 1$

**57.** $(5 - 4x)^3 = -64x^3 + 240x^2 - 300x + 125$

**59.**
$$-4 \begin{array}{|rrrr} 1 & 1 & -10 & 8 \\ & -4 & 12 & -8 \\ \hline 1 & -3 & 2 & 0 \end{array}$$
$x^2 - 3x + 2$

**61.**
$$-5 \begin{array}{|rrrr} 4 & 11 & -43 & 10 \\ & -20 & 45 & -10 \\ \hline 4 & -9 & 2 & 0 \end{array}$$
$4x^2 - 9x + 2$

**63.** $3\sqrt{-27} - \sqrt{-12} = 3\sqrt{3 \cdot 3 \cdot (-3)} - \sqrt{2 \cdot 2(-3)}$

$\phantom{3\sqrt{-27} - \sqrt{-12}} = 9\sqrt{3}i - 2\sqrt{3}i = 7\sqrt{3}i$

**65.** $10\left(\sqrt[3]{64} - 2\sqrt[3]{-16}\right) = 10\left(4 - 2^2\sqrt[3]{-2}\right)$

$\phantom{10\left(\sqrt[3]{64} - 2\sqrt[3]{-16}\right)} = 40 - 40\sqrt[3]{-2}$

$\phantom{10\left(\sqrt[3]{64} - 2\sqrt[3]{-16}\right)} = 40\left(1 + \sqrt[3]{2}\right)$

## Section 6.5    The Binomial Theorem

---

■  You should be able to use the Binomial Theorem

$$(x + y)^n = x^n + nx^{n-1}y + \frac{n(n-1)}{2!}x^{n-2}y^2 + \cdots + {}_nC_r\,x^{n-r}y^r + \cdots + y^n$$

where ${}_nC_r = \dfrac{n!}{(n-r)!r!}$, to expand $(x + y)^n$.

■  You should be able to use Pascal's Triangle.

---

**Solutions to Odd-Numbered Exercises**

**1.** ${}_7C_5 = \dfrac{7!}{2!5!} = \dfrac{7 \cdot 6 \cdot 5!}{2 \cdot 5!} = \dfrac{42}{2} = 21$

**3.** $\dbinom{12}{0} = {}_{12}C_0 = \dfrac{12!}{0!12!} = 1$

**5.** ${}_{20}C_{15} = \dfrac{20!}{15!5!} = \dfrac{20 \cdot 19 \cdot 18 \cdot 17 \cdot 16}{5 \cdot 4 \cdot 3 \cdot 2 \cdot 1} = 15{,}504$

**7.** ${}_{14}C_1 = \dfrac{14!}{13!1!} = \dfrac{14 \cdot 13!}{13!} = 14$

**9.** $\dbinom{100}{98} = {}_{100}C_{98} = \dfrac{100!}{98!2!} = \dfrac{100 \cdot 99}{2 \cdot 1} = 4950$

**11.** ${}_{32}C_{28} = 35{,}960$

**13.** ${}_{22}C_9 = 497{,}420$

**15.** ${}_{41}C_{36} = 749{,}398$

**17.**
```
            1
          1   1
        1   2   1
      1   3   3   1
    1   4   6   4   1
  1   5  10  10   5   1
 1  6  15  20  15   6   1
1  7  21  35 (35) 21  7   1
```

${}_7C_4 = 35$, the 5th entry in the 7th row.

**19.**
```
             1
           1   1
         1   2   1
       1   3   3   1
     1   4   6   4   1
   1   5  10  10   5   1
 1   6  15  20  15   6   1
1  7  21  35  35  21  7   1
1  8  28  56  70 (56) 28  8   1
```

${}_8C_5 = 56$, the 6th entry in the 8th row.

**21.** $(x + 2)^4 = {}_4C_0x^4 + {}_4C_1x^3(2) + {}_4C_2x^2(2)^2 + {}_4C_3x(2)^3 + {}_4C_4(2)^4$

$\qquad\quad = x^4 + 8x^3 + 24x^2 + 32x + 16$

**23.** $(a + 3)^3 = {}_3C_0a^3 + {}_3C_1a^2(3) + {}_3C_2a(3)^2 + {}_3C_3(3)^3$

$\qquad\qquad = a^3 + 3a^2(3) + 3a(3)^2 + (3)^3$

$\qquad\qquad = a^3 + 9a^2 + 27a + 27$

**25.** $(y - 2)^4 = {}_4C_0y^4 - {}_4C_1y^3(2) + {}_4C_2y^2(2)^2 - {}_4C_3y(2)^3 + {}_4C_4(2)^4$

$\qquad\qquad = y^4 - 4y^3(2) + 6y^2(4) - 4y(8) + 16$

$\qquad\qquad = y^4 - 8y^3 + 24y^2 - 32y + 16$

**27.** $(x + y)^5 = {}_5C_0x^5 + {}_5C_1x^4y + {}_5C_2x^3y^2 + {}_5C_3x^2y^3 + {}_5C_4xy^4 + {}_5C_5y^5$

$\qquad = x^5 + 5x^4y + 10x^3y^2 + 10x^2y^3 + 5xy^4 + y^5$

**29.** $(3r + 2s)^6 = {}_6C_0(3r)^6 + {}_6C_1(3r)^5(2s) + {}_6C_2(3r)^4(2s)^2 + {}_6C_3(3r)^3(2s)^3 + {}_6C_4(3r)^2(2s)^4 + {}_6C_5(3r)(2s)^5 + {}_6C_6(2s)^6$

$\qquad = 729r^6 + 2916r^5s + 4860r^4s^2 + 4320r^3s^3 + 2160r^2s^4 + 576rs^5 + 64s^6$

**31.** $(x - y)^5 = {}_5C_0x^5 - {}_5C_1x^4y + {}_5C_2x^3y^2 - {}_5C_3x^2y^3 - {}_5C_4xy^4 - {}_5C_5y^5$

$\qquad = x^5 - 5x^4y + 10x^3y^2 - 10x^2y^3 + 5xy^4 - y^5$

**33.** $(1 - 4x)^3 = {}_3C_01^3 - {}_3C_11^2(4x) + {}_3C_21(4x)^2 - {}_3C_3(4x)^3$

$\qquad = 1 - 3(4x) + 3(4x)^2 - (4x)^3$

$\qquad = 1 - 12x + 48x^2 - 64x^3$

**35.** $(x^2 + y^2)^4 = {}_4C_0(x^2)^4 + {}_4C_1(x^2)^3(y^2) + {}_4C_2(x^2)^2(y^2)^2 + {}_4C_3(x^2)(y^2)^3 + {}_4C_4(y^2)^4$

$\qquad = x^8 + 4x^6y^2 + 6x^4y^4 + 4x^2y^6 + y^8$

**37.** $\left(\dfrac{1}{x} + y\right)^5 = {}_5C_0\left(\dfrac{1}{x}\right)^5 + {}_5C_1\left(\dfrac{1}{x}\right)^4y + {}_5C_2\left(\dfrac{1}{x}\right)^3y^2 + {}_5C_3\left(\dfrac{1}{x}\right)^2y^3 + {}_5C_4\left(\dfrac{1}{x}\right)y^4 + {}_5C_5y^5$

$\qquad = \dfrac{1}{x^5} + \dfrac{5y}{x^4} + \dfrac{10y^2}{x^3} + \dfrac{10y^3}{x^2} + \dfrac{5y^4}{x} + y^5$

**39.** $2(x - 3)^4 + 5(x - 3)^2 = 2[x^4 - 4(x^3)(3) + 6(x^2)(3^2) - 4(x)(3^3) + 3^4] + 5[x^2 - 2(x)(3) + 3^2]$

$\qquad = 2(x^4 - 12x^3 + 54x^2 - 108x + 81) + 5(x^2 - 6x + 9)$

$\qquad = 2x^4 - 24x^3 + 113x^2 - 246x + 207$

**41.** $-3(x - 2)^3 - 4(x + 1)^6$

$\qquad = [-3x^3 + 18x^2 - 36x + 24] - [4x^6 + 24x^5 + 60x^4 + 80x^3 + 60x^2 + 24x + 4]$

$\qquad = -4x^6 - 24x^5 - 60x^4 - 83x^3 - 42x^2 - 60x + 20$

**43.** 5th Row of Pascal's Triangle:   1    5    10    10    5    1

$\quad (3t - s)^5 = 1(3t)^5 + 5(3t)^4(-s) + 10(3t)^3(-s)^2 + 10(3t)^2(-s)^3 + 5(3t)(-s)^4 + 1(-s)^5$

$\qquad = 243t^5 - 405t^4s + 270t^3s^2 - 90t^2s^3 + 15ts^4 - s^5$

**45.** 5th row of Pascal's Triangle:   1    5    10    10    5    1

$\quad (x + 2y)^5 = (1)x^5 + 5x^4 2y + 10x^3(2y)^2 + 10x^2(2y)^3 + 5x(2y)^4 + (2y)^5$

$\qquad = x^5 + 10x^4y + 40x^3y^2 + 80x^2y^3 + 80xy^4 + 32y^5$

**47.** $(x + 8)^{10}, n = 4$

$\quad {}_{10}C_3x^{10-3}(8)^3 = 120x^7(512) = 61{,}440x^7$

**49.** $(x - 6y)^5, n = 3$

$\quad {}_5C_2x^{5-2}(-6y)^2 = 10x^3(36)y^2 = 360x^3y^2$

**51.** $(4x + 3y)^9, n = 8$

$$_9C_7(4x)^{9-7}(3y)^7 = 36(16)x^2(3^7)y^7$$
$$= 1,259,712x^2y^7$$

**53.** $(10x - 3y)^{12}, n = 9$

$$_{12}C_8(10x)^{12-8}(-3y)^8 = 495(10^4)(3^8)x^4y^8$$
$$= 32,476,950,000x^4y^8$$

**55.** The term involving $x^4$ in the expansion of $(x + 3)^{12}$ is $_{12}C_8x^4(3)^8 = 495x^4(3)^8 = 3,247,695x^4$.
The coefficient is 3,247,695.

**57.** The term involving $x^8 y^2$ in the expansion of $(x - 2y)^{10}$ is

$$_{10}C_2x^8(-2y)^2 = \frac{10!}{2!8!} \cdot 4x^8y^2 = 180x^8y^2.$$ The coefficient is 180.

**59.** The term involving $x^6y^3$ in $(3x - 2y)^9$ is $_9C_3(3x)^6(-2y)^3 = 84(3)^6(-2)^3x^6y^3 = -489,888x^6y^3$.
The coefficient is $-489,888$.

**61.** The coefficient of $x^8 y^6 = (x^2)^4y^6$ in the expansion of $(x^2 + y)^{10}$ is $_{10}C_6 = 210$.

**63.** $\left(\sqrt{x} + 5\right)^4 = \left(\sqrt{x}\right)^4 + 4\left(\sqrt{x}\right)^3(5) + 6\left(\sqrt{x}\right)^2(5)^2 + 4\left(\sqrt{x}\right)(5^3) + 5^4$

$$= x^2 + 20x\sqrt{x} + 150x + 500\sqrt{x} + 625$$
$$= x^2 + 20x^{3/2} + 150x + 500x^{1/2} + 625$$

**65.** $(x^{2/3} - y^{1/3})^3 = (x^{2/3})^3 - 3(x^{2/3})^2 (y^{1/3}) + 3(x^{2/3}) (y^{1/3})^2 - (y^{1/3})^3$

$$= x^2 - 3x^{4/3}y^{1/3} + 3x^{2/3}y^{2/3} - y$$

**67.** $\dfrac{f(x + h) - f(x)}{h} = \dfrac{(x + h)^3 - x^3}{h}$

$$= \frac{x^3 + 3x^2h + 3xh^2 + h^3 - x^3}{h}$$
$$= \frac{h(3x^2 + 3xh + h^2)}{h}$$
$$= 3x^2 + 3xh + h^2, \ h \neq 0$$

**69.** $\dfrac{f(x + h) - f(x)}{h} = \dfrac{\sqrt{x + h} - \sqrt{x}}{h}$

$$= \frac{\sqrt{x + h} - \sqrt{x}}{h} \cdot \frac{\sqrt{x + h} + \sqrt{x}}{\sqrt{x + h} + \sqrt{x}}$$
$$= \frac{(x + h) - x}{h\left[\sqrt{x + h} + \sqrt{x}\right]}$$
$$= \frac{1}{\sqrt{x + h} + \sqrt{x}}, h \neq 0$$

**71.** $(1 + i)^4 = {}_4C_0 1^4 + {}_4C_1(1)^3i + {}_4C_2(1)^2i^2 + {}_4C_3 1 \cdot i^3 + {}_4C_4i^4$

$$= 1 + 4i - 6 - 4i + 1$$
$$= -4$$

**73.** $(2 - 3i)^6 = {}_6C_0 2^6 - {}_6C_1 2^5(3i) + {}_6C_2 2^4(3i)^2 - {}_6C_3 2^3(3i)^3 + {}_6C_4 2^2(3i)^4 - {}_6C_5 2(3i)^5 + {}_6C_6(3i)^6$

$$= 64 - 576i - 2160 + 4320i + 4860 - 2916i - 729$$
$$= 2035 + 828i$$

**75.** $\left(-\dfrac{1}{2} + \dfrac{\sqrt{3}}{2}i\right)^3 = \dfrac{1}{8}\left(-1 + \sqrt{3}i\right)^3$

$$= \dfrac{1}{8}\left[(-1)^3 + 3(-1)^2\left(\sqrt{3}i\right) + 3(-1)\left(\sqrt{3}i\right)^2 + \left(\sqrt{3}i\right)^3\right]$$

$$= \dfrac{1}{8}\left[-1 + 3\sqrt{3}i + 9 - 3\sqrt{3}i\right]$$

$$= 1$$

**77.** $(1.02)^8 = (1 + 0.02)^8 = 1 + 8(0.02) + 28(0.02)^2 + 56(0.02)^3 + 70(0.02)^4 + 56(0.02)^5$

$$+ 28(0.02)^6 + 8(0.02)^7 + (0.02)^8$$

$$= 1 + 0.16 + 0.0112 + 0.000448 + \cdots \approx 1.172$$

**79.** $(2.99)^{12} = (3 - 0.01)^{12}$

$$= 3^{12} - 12(3)^{11}(0.01) + 66(3)^{10}(0.01)^2 - 220(3)^9(0.01)^3 + 495(3)^8(0.01)^4$$

$$- 792(3)^7(0.01)^5 + 924(3)^6(0.01)^6 - 792(3)^5(0.01)^7 + 495(3)^4(0.01)^8$$

$$- 220(3)^3(0.01)^9 + 66(3)^2(0.01)^{10} - 12(3)(0.01)^{11} + (0.01)^{12}$$

$$\approx 510{,}568.785$$

**81.** $f(x) = x^3 - 4x$

$g(x) = f(x + 3)$

$\quad = (x + 3)^3 - 4(x + 3)$

$\quad = x^3 + 9x^2 + 27x + 27 - 4x - 12$

$\quad = x^3 + 9x^2 + 23x + 15$

$g$ is shifted 3 units to the left

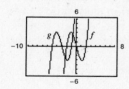

**83.** $f(x) = (1 - x)^3$

$g(x) = 1 - 3x$

$h(x) = 1 - 3x + 3x^2$

$p(x) = 1 - 3x + 3x^2 - x^3$

Since $p(x)$ is the expansion of $f(x)$, they have the same graph.

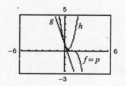

**85.** $_7C_4\left(\dfrac{1}{2}\right)^4\left(\dfrac{1}{2}\right)^3 = 35\left(\dfrac{1}{16}\right)\left(\dfrac{1}{8}\right) \approx 0.273$

**87.** $_8C_4\left(\dfrac{1}{3}\right)^4\left(\dfrac{2}{3}\right)^4 = 70\left(\dfrac{1}{81}\right)\left(\dfrac{16}{81}\right) \approx 0.171$

**89.** $f(t) = 0.018t^2 + 5.15t + 41.6 \quad 5 \le t \le 20$

(a) $g(t) = f(t + 10)$

$\quad = 0.018(t + 10)^2 + 5.15(t + 10) + 41.6 \quad -5 \le t \le 10$

$\quad = 0.018(t^2 + 20t + 100) + 5.15t + 51.5 + 41.6$

$\quad = 0.018t^2 + 5.51t + 94.9$

(b)

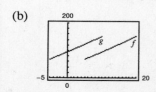

**91.** False. The $x^4y^8$ term is

$$_{12}C_4x^4(-2y)^8 = 495x^4(-2)^8y^8 = 126,720x^4y^8$$

[Note 7920 is the coefficient of $x^8y^4$]

**93.** Answers will vary. See page 536.

**95.** The expansions of $(x + y)^n$ and $(x - y)^n$ are almost the same except that the signs of the terms in the expansion of $(x - y)^n$ alternate from positive to negative.

**97.**
$$_nC_{n-r} = \frac{n!}{[n - (n - r)]!(n - r)!}$$
$$= \frac{n!}{r!(n - r)!}$$
$$= \frac{n!}{(n - r)!r!}$$
$$= {_nC_r}$$

**99.**
$$_nC_r + {_nC_{r-1}} = \frac{n!}{(n - r)!r!} + \frac{n!}{(n - r + 1)!(r - 1)!}$$
$$= \frac{n!(n - r + 1)}{(n - r)!r!(n - r + 1)} + \frac{n!}{(n - r + 1)!(r - 1)!}\frac{r}{r}$$
$$= \frac{n!(n - r + 1)}{(n - r + 1)!r!} + \frac{n!r}{(n - r + 1)!r!}$$
$$= \frac{n!(n - r + 1 + r)}{(n - r + 1)!r!}$$
$$= \frac{n!(n + 1)}{(n - r + 1)!r!}$$
$$= \frac{(n + 1)!}{(n + 1 - r)!r!} = {_{n+1}C_r}$$

**101.** $g(x) = f(x) + 8$

$g(x)$ is shifted 8 units up from $f(x)$.

**103.** $g(x) = f(-x)$

$g(x)$ is the reflection of $f(x)$ in the $y$-axis.

**105.**
$$\begin{bmatrix} -6 & 5 \\ -5 & 4 \end{bmatrix}^{-1} = \frac{1}{-24 + 25}\begin{bmatrix} 4 & -5 \\ 5 & -6 \end{bmatrix} = \begin{bmatrix} 4 & -5 \\ 5 & -6 \end{bmatrix}$$

## Section 6.6    Counting Principles

■  You should know The Fundamental Counting Principle.

■  $_nP_r = \dfrac{n!}{(n - r)!}$ is the number of permutations of $n$ elements taken $r$ at a time.

■  Given a set of $n$ objects that has $n_1$ of one kind, $n_2$ of a second kind, and so on, the number of distinguishable permutations is

$$\frac{n!}{n_1!n_2! \cdots n_k!}$$

■  $_nC_r = \dfrac{n!}{(n - r)!r!}$ is the number of combinations of $n$ elements taken $r$ at a time.

**Solutions to Odd-Numbered Exercises**

**1.** Odd integers: 1, 3, 5, 7, 9, 11
   6 ways

**3.** Prime integers: 2, 3, 5, 7, 11
   5 ways

**5.** Divisible by 4: 4, 8, 12
   3 ways

**7.** Sum is 8: $1 + 7, 2 + 6, 3 + 5, 4 + 4, 5 + 3, 6 + 2, 7 + 1$
   7 ways

**9.** Amplifiers: 4 choices
   Compact disc players: 6 choices
   Speakers: 5 choices
   Total: $4 \cdot 6 \cdot 5 = 120$ ways

**11.** Chemist: 3 choices
   Statistician: 8 choices
   Total: $3 \cdot 8 = 24$ ways

**13.** $2^{10} = 1024$ ways

**15.** 1st Position: 2 choices
   2nd Position: 3 choices
   3rd Position: 2 choices
   4th Position: 1 choice
   Total: $2 \cdot 3 \cdot 2 \cdot 1 = 12$ ways

   Label the four people A, B, C, and D and suppose
   that A and B are willing to take the first position.
   The twelve combinations are as follows.

   | ABCD | BACD |
   |------|------|
   | ABDC | BADC |
   | ACBD | BCAD |
   | ACDB | BCDA |
   | ADBC | BDAC |
   | ADCB | BDCA |

**17.** $10 \cdot 10 \cdot 10 \cdot 26 \cdot 26 \cdot 26 = 17{,}576{,}000$ license plates

**19.** (a) $9 \cdot 10 \cdot 10 = 900$
   (c) $9 \cdot 10 \cdot 2 = 180$

   (b) $9 \cdot 9 \cdot 8 = 648$
   (d) $10 \cdot 10 \cdot 10 - 400 = 600$

**21.** $2(8 \cdot 10 \cdot 10)(10 \cdot 10 \cdot 10 \cdot 10) = 16{,}000{,}000$ numbers

**23.** (a) $6 \cdot 5 \cdot 4 \cdot 3 \cdot 2 \cdot 1 = 720$
   (b) $6 \cdot 1 \cdot 4 \cdot 1 \cdot 2 \cdot 1 = 48$

**25.** $_nP_r = \dfrac{n!}{(n-r)!}$

   So, $_4P_4 = \dfrac{4!}{0!} = 4! = 24.$

**27.** $_8P_3 = \dfrac{8!}{5!} = 8 \cdot 7 \cdot 6 = 336$

**29.** $_5P_4 = \dfrac{5!}{1!} = 120$

**31.**    $14 \cdot {}_nP_3 = {}_{n+2}P_4$   Note $n \geq 3$ for this to be defined.

$$14\left[\frac{n!}{(n-3)!}\right] = \frac{(n+2)!}{(n-2)!}$$

$14n(n-1)(n-2) = (n+2)(n+1)n(n-1)$ (We can divide here by $n(n-1)$ since $n \neq 0, n \neq 1$.)

$$14n - 28 = n^2 + 3n + 2$$
$$0 = n^2 - 11n + 30$$
$$0 = (n-5)(n-6)$$
$$n = 5 \quad \text{or} \quad n = 6$$

**33.** ${}_{20}P_6 = 27,907,200$ 　　　　　**35.** ${}_{120}P_4 = 197,149,680$ 　　　　　**37.** ${}_{20}C_4 = 4845$

**39.** $5! = 120$ ways 　　　　　**41.** ${}_{12}P_4 = \dfrac{12!}{8!} = 12 \cdot 11 \cdot 10 \cdot 9 = 11,880$ ways

**43.** $\dfrac{7!}{2!1!3!1!} = \dfrac{7!}{2!3!} = 420$ 　　　　　**45.** $\dfrac{7!}{2!1!1!1!1!1!} = \dfrac{7!}{2!} = 7 \cdot 6 \cdot 5 \cdot 4 \cdot 3 = 2520$

**47.** (a)

| | | | |
|---|---|---|---|
| ABCD | BACD | CABD | DABC |
| ABDC | BADC | CADB | DACB |
| ACBD | BCAD | CBAD | DBAC |
| ACDB | BCDA | CBDA | DBCA |
| ADBC | BDAC | CDAB | DCAB |
| ADCB | BDCA | CDBA | DCBA |

(b)  ABCD
　　 ACBD
　　 DBCA
　　 DCBA

**49.** ${}_{20}C_4 = 4845$ groups 　　　　　**51.** ${}_{49}C_6 = 13,983,816$ ways 　　　　　**53.** ${}_9C_2 = 36$ lines

**55.** Select type of card for three of a kind: ${}_{13}C_1$

Select three of four cards for three of a kind: ${}_4C_3$

Select type of card for pair: ${}_{12}C_1$

Select two of four cards for pair: ${}_4C_2$

${}_{13}C_1 \cdot {}_4C_3 \cdot {}_{12}C_1 \cdot {}_4C_2 = 13 \cdot 4 \cdot 12 \cdot 6 = 3744$ ways to get a full house

**57.** (a) ${}_{12}C_4 = 495$ ways

(b) $({}_5C_2)({}_7C_2) = (10)(21) = 210$ ways

**59.** $({}_7C_1)({}_{12}C_3)({}_{20}C_2) = 7 \cdot 220 \cdot 190$
$$= 292,600 \text{ ways}$$

**61.** ${}_5C_2 - 5 = 10 - 5 = 5$ diagonals

**63.** ${}_8C_2 - 8 = 28 - 8 = 20$ diagonals

**65.** False. This is an example of a combination.

**67.** ${}_{100}P_{80} \approx 3.836 \times 10^{139}$.

This number is too large for some calculators to evaluate.

**69.** ${}_nC_r = {}_nC_{n-r} = \dfrac{n!}{r!(n-r)!}$

**71.** $_nP_{n-1} = \dfrac{n!}{(n-(n-1))!} = \dfrac{n!}{1!} = \dfrac{n!}{0!} = {_nP_n}$

**73.** $_nC_{n-1} = \dfrac{n!}{[n-(n-1)]!(n-1)!} = \dfrac{n!}{(1)!(n-1)!} = \dfrac{n!}{(n-1)!1!} = {_nC_1}$

**75.** From the graph of $y = \sqrt{x-3} - x + 6$, you see that there is one zero, $x \approx 8.303$. Analytically,

$$\sqrt{x-3} = x - 6$$

$$x - 3 = x^2 - 12x + 36$$

$$0 = x^2 - 13x + 39.$$

By the Quadratic Formula, $x = \dfrac{13 \pm \sqrt{(-13)^2 - 4(39)}}{2} = \dfrac{13 \pm \sqrt{13}}{2}$.

Selecting the larger solution, $x = \dfrac{13 + \sqrt{13}}{2} \approx 8.303$. (The other solution is extraneous)

**77.** $\log_2(x-3) = 5$

$$2^5 = x - 3$$

$$2^5 + 3 = x$$

$$x = 35$$

**79.** $x = \dfrac{\begin{vmatrix} -14 & 3 \\ 2 & -2 \end{vmatrix}}{\begin{vmatrix} -5 & 3 \\ 7 & -2 \end{vmatrix}} = \dfrac{22}{-11} = -2$

$y = \dfrac{\begin{vmatrix} -5 & -14 \\ 7 & 2 \end{vmatrix}}{\begin{vmatrix} -5 & 3 \\ 7 & -2 \end{vmatrix}} = \dfrac{88}{-11} = -8$

*Answer:* $(-2, -8)$

**81.** $x = \dfrac{\begin{vmatrix} -1 & -4 \\ -4 & 5 \end{vmatrix}}{\begin{vmatrix} -3 & -4 \\ 9 & 5 \end{vmatrix}} = \dfrac{-21}{21} = -1$

$y = \dfrac{\begin{vmatrix} -3 & -1 \\ 9 & -4 \end{vmatrix}}{\begin{vmatrix} -3 & -4 \\ 9 & 5 \end{vmatrix}} = \dfrac{21}{21} = 1$

*Answer:* $(-1, 1)$

## Section 6.7    Probability

You should know the following basic principles of probability.

- If an event $E$ has $n(E)$ equally likely outcomes and its sample space has $n(S)$ equally likely outcomes, then the probability of event $E$ is

$$P(E) = \frac{n(E)}{n(S)}, \text{ where } 0 \le P(E) \le 1.$$

- If $A$ and $B$ are mutually exclusive events, then $P(A \cup B) = P(A) + P(B)$.

  If $A$ and $B$ are not mutually exclusive events, then $P(A \cup B) = P(A) + P(B) - P(A \cap B)$.

- If $A$ and $B$ are independent events, then the probability that both $A$ and $B$ will occur is $P(A)P(B)$.

- The probability of the complement of an event $A$ is $P(A') = 1 - P(A)$.

**Solutions to Odd-Numbered Exercises**

**1.** $\{(H, 1), (H, 2), (H, 3), (H, 4), (H, 5), (H, 6),$
$(T, 1), (T, 2), (T, 3), (T, 4), (T, 5), (T, 6)\}$

**3.** $\{ABC, ACB, BAC, BCA, CAB, CBA\}$

**5.** $\{(A, B), (A, C), (A, D), (A, E), (B, C), (B, D), (B, E), (C, D), (C, E), (D, E)\}$

**7.** $E = \{HTT, THT, TTH\}$
$$P(E) = \frac{n(E)}{n(S)} = \frac{3}{8}$$

**9.** $E = \{HHH, HHT, HTH, HTT, THH, THT, TTH\}$
$$P(E) = \frac{n(E)}{n(S)} = \frac{7}{8}$$

**11.** $E = \{K, K, K, K, Q, Q, Q, Q, J, J, J, J\}$
$$P(E) = \frac{n(E)}{n(S)} = \frac{12}{52} = \frac{3}{13}$$

**13.** $E = \{K, K, Q, Q, J, J\}$
$$P(E) = \frac{n(E)}{n(S)} = \frac{6}{52} = \frac{3}{26}$$

**15.** $E = \{(1, 4), (2, 3), (3, 2), (4, 1)\}$
$$P(E) = \frac{n(E)}{n(S)} = \frac{4}{36} = \frac{1}{9}$$

**17.** not $E = \{(5, 6), (6, 5), (6, 6)\}$
$$n(E) = n(S) - n(\text{not } E) = 36 - 3 = 33$$
$$P(E) = \frac{n(E)}{n(S)} = \frac{33}{36} = \frac{11}{12}$$

**19.** $P(E) = \dfrac{_3C_2}{_6C_2} = \dfrac{3}{15} = \dfrac{1}{5}$

**21.** $P(E) = \dfrac{_4C_2}{_6C_2} = \dfrac{6}{15} = \dfrac{2}{5}$

**23.** $P(E') = 1 - P(E) = 1 - 0.8 = 0.2$

**25.** $P(E') = 1 - P(E) = 1 - \frac{1}{3} = \frac{2}{3}$

**27.** $P(E) = 1 - P(E') = 1 - p = 1 - 0.12 = 0.88$

**29.** $P(E) = 1 - P(E') = 1 - \frac{13}{20} = \frac{7}{20}$

**31.** (a) $0.18(6.7) \approx 1.2$ million

(b) $\dfrac{0.42}{1.0} = 0.42$

(c) $\dfrac{0.21}{1.0} = 0.21$

(d) $\dfrac{0.21 + 0.01}{1.0} = 0.22$

**33.** (a) $\dfrac{34}{100} = 0.34$

(b) $\dfrac{45}{100} = 0.45$

(c) $\dfrac{23}{100} = 0.23$

**35.** (a) $\dfrac{672}{1254}$

(b) $\dfrac{582}{1254}$

(c) $\dfrac{672 - 124}{1254} = \dfrac{548}{1254}$

**37.** $p + p + 2p = 1$
$$p = 0.25$$
Taylor: $0.50 = \dfrac{1}{2}$

Moore: $0.25 = \dfrac{1}{4}$

Perez: $0.25 = \dfrac{1}{4}$

**39.** (a) $\dfrac{_{15}C_{10}}{_{20}C_{10}} = \dfrac{3003}{184{,}756} = \dfrac{21}{1292} \approx 0.016$

(b) $\dfrac{_{15}C_8 \cdot {}_5C_2}{_{20}C_{10}} = \dfrac{64{,}350}{184{,}756} = \dfrac{225}{646} \approx 0.348$

(c) $\dfrac{_{15}C_9 \cdot {}_5C_1}{_{20}C_{10}} + \dfrac{_{15}C_{10}}{_{20}C_{10}} = \dfrac{25{,}025 + 3003}{184{,}756} = \dfrac{28{,}028}{184{,}756} = \dfrac{49}{323} \approx 0.152$

**41.** (a) $\dfrac{1}{_5P_5} = \dfrac{1}{120}$

(b) $\dfrac{1}{_4P_4} = \dfrac{1}{24}$

**43.** (a) $\dfrac{20}{52} = \dfrac{5}{13}$

(b) $\dfrac{13 + 13}{52} = \dfrac{1}{2}$

(c) $\dfrac{4 + 12}{52} = \dfrac{4}{13}$

**45.** (a) $\dfrac{_9C_4}{_{12}C_4} = \dfrac{126}{495} = \dfrac{14}{55}$   (4 good units)

(b) $\dfrac{({}_9C_2)\,({}_3C_2)}{_{12}C_4} = \dfrac{108}{495} = \dfrac{12}{55}$   (2 good units)

(c) $\dfrac{({}_9C_3)({}_3C_1)}{_{12}C_4} = \dfrac{252}{495} = \dfrac{28}{55}$   (3 good units)

At least 2 good units: $\dfrac{12}{55} + \dfrac{28}{55} + \dfrac{14}{55} = \dfrac{54}{55}$

**47.** $(0.32)^2 = 0.1024$

**49.** (a) $P(SS) = (0.985)^2 \approx 0.9702$

(b) $P(S) = 1 - P(FF) = 1 - (0.015)^2 \approx 0.9998$

(c) $P(FF) = (0.015)^2 \approx 0.0002$

**51.** (a) $\left(\dfrac{1}{5}\right)^6 = \dfrac{1}{15{,}625}$

(b) $\left(\dfrac{4}{5}\right)^6 = \dfrac{4096}{15{,}625} = 0.262144$

(c) $1 - 0.262144 = 0.737856 = \dfrac{11{,}529}{15{,}625}$

**53.** $1 - \dfrac{(45)^2}{(60)^2} = 1 - \left(\dfrac{45}{60}\right)^2 = 1 - \left(\dfrac{3}{4}\right)^2 = 1 - \dfrac{9}{16} = \dfrac{7}{16}$

**55.** True. $P(E) + P(E') = 1$

**57.** (a) As you consider successive people with distinct birthdays, the probabilities must decrease to take into account the birth dates already used. Since the birth dates of people are independent events, multiply the respective probabilities of distinct birthdays.

(b) $\dfrac{365}{365} \cdot \dfrac{364}{365} \cdot \dfrac{363}{365} \cdot \dfrac{362}{365}$

(c) $P_1 = \dfrac{365}{365} = 1$

$P_2 = \dfrac{365}{365} \cdot \dfrac{364}{365} = \dfrac{364}{365} P_1 = \dfrac{365 - (2 - 1)}{365} P_1$

$P_3 = \dfrac{365}{365} \cdot \dfrac{364}{365} \cdot \dfrac{363}{365} = \dfrac{363}{365} P_2 = \dfrac{365 - (3 - 1)}{365} P_2$

$P_n = \dfrac{365}{365} \cdot \dfrac{364}{365} \cdot \dfrac{363}{365} \cdot \ldots \cdot \dfrac{365 - (n - 1)}{365} = \dfrac{365 - (n - 1)}{365} P_{n-1}$

(d) $Q_n$ is the probability that the birthdays are *not* distinct which is equivalent to at least 2 people having the same birthday.

(e)

| $n$ | 10 | 15 | 20 | 23 | 30 | 40 | 50 |
|-----|------|------|------|------|------|------|------|
| $P_n$ | 0.88 | 0.75 | 0.59 | 0.49 | 0.29 | 0.11 | 0.03 |
| $Q_n$ | 0.12 | 0.25 | 0.41 | (0.51) | 0.71 | 0.89 | 0.97 |

(f) 23, See the chart above.

**59.** $\dfrac{2}{x - 5} = 4$

$2 = 4(x - 5) = 4x - 20$

$4x = 22$

$x = \dfrac{11}{2}$

**61.** $\dfrac{3}{x - 2} + \dfrac{x}{x + 2} = 1$

$3(x + 2) + x(x - 2) = (x - 2)(x + 2)$

$3x + 6 + x^2 - 2x = x^2 - 4$

$x = -10$

**63.** $e^x + 7 = 35$

$e^x = 28$

$x = \ln(28) \approx 3.332$

**65.** $4 \ln 6x = 16$

$\ln 6x = 4$

$e^4 = 6x$

$x = \tfrac{1}{6} e^4 \approx 9.10$

**67.** $_5P_3 = \dfrac{5!}{(5 - 3)!} = \dfrac{120}{2} = 60$

**69.** $_{11}P_8 = \dfrac{11!}{(11 - 8)!} = \dfrac{11!}{3!} = 6,652,800$

**71.** $_6C_2 = \dfrac{6!}{4!2!} = \dfrac{6 \cdot 5 \cdot 4!}{4!2} = 15$

**73.** $_{11}C_8 = \dfrac{11!}{8!3!} = \dfrac{11 \cdot 10 \cdot 9 \cdot 8!}{8!6} = 165$

# Review Exercises for Chapter 6

**Solutions to Odd-Numbered Exercises**

**1.** $a_n = 2 + \dfrac{6}{n}$

$a_1 = 2 + \dfrac{6}{1} = 8$

$a_2 = 2 + \dfrac{6}{2} = 5$

$a_3 = 2 + \dfrac{6}{3} = 4$

$a_4 = 2 + \dfrac{6}{4} = \dfrac{7}{2}$

$a_5 = 2 + \dfrac{6}{5} = \dfrac{16}{5}$

**3.** $a_n = \dfrac{72}{n!}$

$a_1 = \dfrac{72}{1!} = 72$

$a_2 = \dfrac{72}{2!} = 36$

$a_3 = \dfrac{72}{3!} = 12$

$a_4 = \dfrac{72}{4!} = 3$

$a_5 = \dfrac{72}{5!} = \dfrac{3}{5}$

**5.** $a_n = \dfrac{3}{2}n$

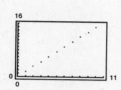

**7.** $a_n = 4(0.4)^{n-1}$

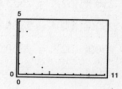

**9.** $\dfrac{18!}{20!} = \dfrac{18!}{20 \cdot 19 \cdot 18!} = \dfrac{1}{20 \cdot 19} = \dfrac{1}{380}$

**11.** $\dfrac{2!5!}{6!} = \dfrac{2 \cdot 5!}{6 \cdot 5!} = \dfrac{2}{6} = \dfrac{1}{3}$

**13.** $\displaystyle\sum_{i=1}^{6} 5 = 6(5) = 30$

**15.** $\displaystyle\sum_{j=1}^{4} \dfrac{6}{j^2} = \dfrac{6}{1^2} + \dfrac{6}{2^2} + \dfrac{6}{3^2} + \dfrac{6}{4^2} = 6 + \dfrac{3}{2} + \dfrac{2}{3} + \dfrac{3}{8} = \dfrac{205}{24}$

**17.** $\displaystyle\sum_{k=1}^{10} 2k^3 = 2(1)^3 + 2(2)^3 + 2(3)^3 + \cdots + 2(10)^3 = 6050$

**19.** $\displaystyle\sum_{n=0}^{10} (n^2 + 3) = \sum_{n=0}^{10} n^2 + \sum_{n=0}^{10} 3 = \dfrac{10(11)(21)}{6} + 11(3) = 418$

**21.** $\dfrac{1}{2(1)} + \dfrac{1}{2(2)} + \dfrac{1}{2(3)} + \cdots + \dfrac{1}{2(20)} = \displaystyle\sum_{k=1}^{20} \dfrac{1}{2k} \approx 1.799$

**23.** $\dfrac{1}{2} + \dfrac{2}{3} + \dfrac{3}{4} + \cdots + \dfrac{9}{10} = \displaystyle\sum_{k=1}^{9} \dfrac{k}{k+1} \approx 7.071$

**25.** (a) $\displaystyle\sum_{k=1}^{4}\frac{5}{10^k} = \frac{5}{10} + \frac{5}{100} + \frac{5}{1000} + \frac{5}{10,000} = .5 + .05 + .005 + .0005 = .5555 = \frac{1111}{2000}$

(b) $\displaystyle\sum_{k=1}^{\infty}\frac{5}{10^k} = \frac{5}{10}\sum_{k=0}^{\infty}\frac{1}{10^k} = \frac{5}{10}\cdot\frac{1}{1-\dfrac{1}{10}} = \frac{5}{10}\cdot\frac{10}{9} = \frac{5}{9}$

**27.** (a) $\displaystyle\sum_{k=1}^{4}2\left(\frac{1}{100}\right)^k = 2\left(\frac{1}{100}\right) + 2\left(\frac{1}{100}\right)^2 + 2\left(\frac{1}{100}\right)^3 + 2\left(\frac{1}{100}\right)^4$

$$= 0.02 + 0.0002 + 0.000002 + 0.00000002$$

$$= 0.02020202$$

(b) $\displaystyle\sum_{k=1}^{\infty}2\left(\frac{1}{100}\right)^k = \frac{2}{100}\sum_{k=0}^{\infty}\left(\frac{1}{100}\right)^k = \frac{2}{100}\cdot\frac{1}{1-\left(\dfrac{1}{100}\right)} = \frac{2}{100}\cdot\frac{100}{99} = \frac{2}{99}$

**29.** (a) $a_1 = 2500\left(1 + \dfrac{0.08}{4}\right)^1 = 2500(1.02) = 2550$     (b) $a_{40} = 2500(1.02)^{40} = 5520.10$

$a_2 = 2500(1.02)^2 = 2601$

$a_3 = 2500(1.02)^3 = 2653.02$

$a_4 = 2500(1.02)^4 = 2706.08$

$a_5 = 2500(1.02)^5 = 2760.20$

$a_6 = 2500(1.02)^6 = 2815.41$

$a_7 = 2500(1.02)^7 = 2871.71$

$a_8 = 2500(1.02)^8 = 2929.15$

**31.** Yes. $d = 3 - 5 = -2$

**33.** Yes. $d = 1 - \frac{1}{2} = \frac{1}{2}$

**35.** $a_1 = 3, d = 4$

$a_1 = 3$

$a_2 = 3 + 4 = 7$

$a_3 = 7 + 4 = 11$

$a_4 = 11 + 4 = 15$

$a_5 = 15 + 4 = 19$

**37.** $a_4 = 10 \quad a_{10} = 28$

$a_{10} = a_4 + 6d$

$28 = 10 + 6d$

$18 = 6d$

$3 = d$

$a_1 = a_4 - 3d$

$a_1 = 10 - 3(3)$

$a_1 = 1$

$a_2 = 1 + 3 = 4$

$a_3 = 4 + 3 = 7$

$a_4 = 7 + 3 = 10$

$a_5 = 10 + 3 = 13$

**39.** $a_1 = 35, a_{k+1} = a_k - 3$

$a_1 = 35$

$a_2 = a_1 - 3 = 35 - 3 = 32$

$a_3 = a_2 - 3 = 32 - 3 = 29$

$a_4 = a_3 - 3 = 29 - 3 = 26$

$a_5 = a_4 - 3 = 26 - 3 = 23$

$a_n = 35 + (n-1)(-3) = 38 - 3n, d = -3$

**41.** $a_1 = 9, a_{k+1} = a_k + 7$

$a_1 = 9$

$a_2 = a_1 + 7 = 9 + 7 = 16$

$a_3 = a_2 + 7 = 16 + 7 = 23$

$a_4 = a_3 + 7 = 23 + 7 = 30$

$a_5 = a_4 + 7 = 30 + 7 = 37$

$a_n = 9 + (n-1)(7) = 2 + 7n, d = 7$

**43.** $a_n = 100 + (n-1)(-3) = 103 - 3n$

$$\sum_{n=1}^{20}(103 - 3n) = \sum_{n=1}^{20}103 - 3\sum_{n=1}^{20}n = 20(103) - 3\left[\frac{(20)(21)}{2}\right] = 1430$$

**45.** $\displaystyle\sum_{j=1}^{10}(2j-3) = 2\sum_{j=1}^{10}j - \sum_{j=1}^{10}3 = 2\left[\frac{10(11)}{2}\right] - 10(3) = 80$

**47.** $\displaystyle\sum_{k=1}^{11}\left(\frac{2}{3}k + 4\right) = \frac{2}{3}\sum_{k=1}^{11}k + \sum_{k=1}^{11}4 = \frac{2}{3}\cdot\frac{(11)(12)}{2} + 11(4) = 88$

**49.** $\displaystyle\sum_{k=1}^{100}5k = 5\left[\frac{(100)(101)}{2}\right] = 25{,}250$

**51.** (a) $34{,}000 + 4(2250) = \$43{,}000$

(b) $\displaystyle\sum_{k=1}^{5}[34{,}000 + (k-1)(2250)] = \sum_{k=1}^{5}(31{,}750 + 2250k) = \$192{,}500$

**53.** $5, 10, 20, 40$

Geometric: $r = 2$

**55.** $\frac{1}{2}, \frac{2}{3}, \frac{3}{4}, \frac{4}{5}$

Not geometric: $\frac{1}{2}r = \frac{2}{3} \Rightarrow r = \frac{4}{3}$

$\frac{2}{3}r = \frac{3}{4} \Rightarrow r = \frac{9}{8}$

**57.** $a_1 = 4, r = -\frac{1}{4}$

$a_1 = 4$

$a_2 = 4\left(-\frac{1}{4}\right) = -1$

$a_3 = -1\left(-\frac{1}{4}\right) = \frac{1}{4}$

$a_4 = \frac{1}{4}\left(-\frac{1}{4}\right) = -\frac{1}{16}$

$a_5 = -\frac{1}{16}\left(-\frac{1}{4}\right) = \frac{1}{64}$

**59.** $a_1 = 9, a_3 = 4$

$a_3 = a_1 r^2$

$4 = 9r^2$

$\frac{4}{9} = r^2 \Rightarrow r = \pm\frac{2}{3}$

$a_1 = 9 \qquad\qquad a_1 = 9$

$a_2 = 9\left(\frac{2}{3}\right) = 6 \qquad a_2 = 9\left(-\frac{2}{3}\right) = -6$

$a_3 = 6\left(\frac{2}{3}\right) = 4$  OR  $a_3 = -6\left(-\frac{2}{3}\right) = 4$

$a_4 = 4\left(\frac{2}{3}\right) = \frac{8}{3} \qquad a_4 = 4\left(-\frac{2}{3}\right) = -\frac{8}{3}$

$a_5 = \frac{8}{3}\left(\frac{2}{3}\right) = \frac{16}{9} \qquad a_5 = -\frac{8}{3}\left(-\frac{2}{3}\right) = \frac{16}{9}$

**61.** $a_1 = 120, a_{k+1} = \frac{1}{3}a_k$

$a_1 = 120$

$a_2 = \frac{1}{3}(120) = 40$

$a_3 = \frac{1}{3}(40) = \frac{40}{3}$

$a_4 = \frac{1}{3}\left(\frac{40}{3}\right) = \frac{40}{9}$

$a_5 = \frac{1}{3}\left(\frac{40}{9}\right) = \frac{40}{27}$

$a_n = 120\left(\frac{1}{3}\right)^{n-1}, r = \frac{1}{3}$

**63.** $a_1 = 25, \ a_{k+1} = -\frac{3}{5}a_k$

$a_1 = 25$

$a_2 = -\frac{3}{5}(25) = -15$

$a_3 = -\frac{3}{5}(-15) = 9$

$a_4 = -\frac{3}{5}(9) = -\frac{27}{5}$

$a_5 = -\frac{3}{5}\left(-\frac{27}{5}\right) = \frac{81}{25}$

$a_n = 25\left(-\frac{3}{5}\right)^{n-1}, r = -\frac{3}{5}$

**65.** $a_2 = a_1 r$

$-8 = 16r$

$-\frac{1}{2} = r$

$a_n = 16\left(-\frac{1}{2}\right)^{n-1}$

$\sum_{n=1}^{20} 16\left(-\frac{1}{2}\right)^{n-1} = 16\left[\frac{1 - (-1/2)^{20}}{1 - (-1/2)}\right] \approx 10.67$

**67.** $a_1 = 100, r = 1.05$

$a_n = 100(1.05)^{n-1}$

$\sum_{n=1}^{20} 100(1.05)^{n-1} = 100\left[\frac{1 - (1.05)^{20}}{1 - 1.05}\right] \approx 3306.60$

**69.** $\sum_{i=1}^{7} 2^{i-1} = \frac{1 - 2^7}{1 - 2} = 127$

**71.** $\sum_{n=1}^{7} (-4)^{n-1} = \frac{1 - (-4)^7}{1 - (-4)} = 3277$

**73.** $\sum_{n=0}^{4} 250(1.02)^n = 250\left(\frac{1 - 1.02^5}{1 - 1.02}\right) = 1301.01004$

**75.** $\sum_{i=1}^{10} 10\left(\frac{3}{5}\right)^{i-1} \approx 24.849$

**77.** $\sum_{i=1}^{\infty} \left(\frac{7}{8}\right)^{i-1} = \frac{1}{1 - 7/8} = 8$

**79.** $\sum_{k=1}^{\infty} 4\left(\frac{2}{3}\right)^{k-1} = \frac{4}{1 - 2/3} = 12$

**81.** (a) $a_t = 120,000(0.7)^t$

(b) $a_5 = 120,000(0.7)^5 = \$20,168.40$

**83.** 1. When $n = 1, 2 = \frac{1}{2}(5(1) - 1)$

2. Assume that $S_k = 2 + 7 + \cdots + (5k - 3) = \frac{k}{2}(5k - 1)$

Then, $S_{k+1} = 2 + 7 + \cdots + (5k - 3) + [5(k + 1) - 3]$

$= S_k + 5k + 2$

$= \frac{k}{2}(5k - 1) + 5k + 2$

$= \frac{1}{2}[5k^2 + 9k + 4]$

$= \frac{1}{2}[(5k + 4)(k + 1)]$

$= \frac{k + 1}{2}(5(k + 1) - 1)$

Therefore, by mathematical induction, the formula is true for all positive integers $n$.

**85.** 1. When $n = 1$, $a = a\left(\dfrac{1 - r}{1 - r}\right)$.

2. Assume that

$$S_k = \sum_{i=0}^{k-1} ar^i = \frac{a(1 - r^k)}{1 - r}.$$

Then,

$$S_{k+1} = \sum_{i=0}^{k} ar^i = \sum_{i=0}^{k-1} ar^i + ar^k = \frac{a(1 - r^k)}{1 - r} + ar^k$$

$$= \frac{a(1 - r^k + r^k - r^{k+1})}{1 - r} = \frac{a(1 - r^{k+1})}{1 - r}.$$

Therefore, by mathematical induction, the formula is valid for all positive integer values of $n$.

**87.** $\displaystyle\sum_{n=1}^{30} n = \frac{30(31)}{2} = 465$

**89.** $\displaystyle\sum_{n=1}^{7} (n^4 - n) = \sum_{n=1}^{7} n^4 - \sum_{n=1}^{7} n$

$$= \frac{7(8)(15)[3(7)^2 + 3(7) - 1]}{30} - \frac{7(8)}{2}$$

$$= \frac{840(167)}{30} - 28$$

$$= 4676 - 28 = 4648$$

**91.** $a_1 = f(1) = 5$

$a_2 = a_1 + 5 = 5 + 5 = 10$

$a_3 = a_2 + 5 = 15$

$a_4 = a_3 + 5 = 20$

$a_5 = a_4 + 5 = 25$

| $n$: | 1 | 2 | 3 | 4 | 5 |
|---|---|---|---|---|---|
| $a_n$: | 5 | 10 | 15 | 20 | 25 |

First differences: 5  5  5  5

Second difference: 0  0  0

Linear model: $a_n = 5n$

**93.** $a_1 = f(1) = 16$

$a_2 = a_1 - 1 = 16 - 1 = 15$

$a_3 = a_2 - 1 = 15 - 1 = 14$

$a_4 = 14 - 1 = 13$

$a_5 = 13 - 1 = 12$

| $n$: | 1 | 2 | 3 | 4 | 5 |
|---|---|---|---|---|---|
| $a_n$: | 16 | 15 | 14 | 13 | 12 |

First differences: $-1$  $-1$  $-1$  $-1$

Second difference: 0  0  0

Linear model: $a_n = 17 - n$

**95.** $_{10}C_8 = 45$

**97.** $\dbinom{9}{4} = {}_9C_4 = 126$

**99.** 4th number in 6$^{\text{th}}$ row is $_6C_3 = 20$

**101.** 5th number in 8$^{\text{th}}$ row is $\dbinom{8}{4} = {}_8C_4 = 70$

**103.** $(x + 5)^4 = x^4 + 4x^3(5) + 6x^2(5^2) + 4x(5^3) + 5^4$

$$= x^4 + 20x^3 + 150x^2 + 500x + 625$$

**105.** $(a - 4b)^5 = a^5 - 5a^4(4b) + 10a^3(4b)^2 - 10a^2(4b)^3 + 5a(4b)^4 - (4b)^5$

$$= a^5 - 20a^4b + 160a^3b^2 - 640a^2b^3 + 1280ab^4 - 1024b^5$$

**107.** $(7 + 2i)^4 = 7^4 + 4(7)^3(2i) + 6(7)^2(2i)^2 + 4(7)(2i)^3 + (2i)^4$

$$= 2401 + 2744i - 1176 - 224i + 16$$

$$= 1241 + 2520i$$

**109.** $E = \{(1, 11), (2, 10), (3, 9), (4, 8), (5, 7), (7, 5), (8, 4), (9, 3), (10, 2), (11, 1)\}$

$n(E) = 10$

**111.** $(4)(6)(2) = 48$ schedules

**113.** $\dfrac{8!}{2!2!2!1!1!} = \dfrac{8!}{8} = 7! = 5040$ permutations

**115.** $10! = 3{,}628{,}800$ ways

**117.** $_{20}C_{15} = 15{,}504$ ways

**119.** $\dfrac{10}{10} \cdot \dfrac{1}{9} = \dfrac{1}{9}$

**121.** (a) $\dfrac{208}{500} = 0.416$

(b) $\dfrac{400}{500} = 0.8$

(c) $\dfrac{37}{500} = 0.074$

**123.** $P(2 \text{ pairs}) = \dfrac{(_{13}C_2)(_4C_2)(_4C_2)(_{44}C_1)}{(_{52}C_5)} = 0.0475$

**125.** True.

$$\dfrac{(n + 2)!}{n!} = \dfrac{(n + 2)(n + 1)n!}{n!} = (n + 2)(n + 1)$$

**127.** Answers will vary. See pages 507 and 516.

**129.** (a) arithmetic-linear model

(b) geometric model

**131.** Answers will vary. See page 509. To define a sequence recursively, you need to be given one or more of the first few terms. All other terms are defined using previous terms.

**133.** If $n$ is even, the expansion are the same. If $n$ is odd, the expansion of $(-x + y)^n$ is the negative of that of $(x - y)^n$.

# Chapter 6   Practice Test

1. Write out the first five terms of the sequence $a_n = \dfrac{2n}{(n+2)!}$.

2. Write an expression for the $n$th term of the sequence $\left\{\frac{4}{3},\ \frac{5}{9},\ \frac{6}{27},\ \frac{7}{81},\ \frac{8}{243},\ \ldots\right\}$.

3. Find the sum $\displaystyle\sum_{i=1}^{6}(2i-1)$.

4. Write out the first five terms of the arithmetic sequence where $a_1 = 23$ and $d = -2$.

5. Find $a_{50}$ for the arithmetic sequence with $a_1 = 12$, $d = 3$, and $n = 50$.

6. Find the sum of the first 200 positive integers.

7. Write out the first five terms of the geometric sequence with $a_1 = 7$ and $r = 2$.

8. Evaluate $\displaystyle\sum_{n=0}^{9} 6\left(\frac{2}{3}\right)^n$.

9. Evaluate $\displaystyle\sum_{n=0}^{\infty} (0.03)^n$.

10. Use mathematical induction to prove that $1 + 2 + 3 + 4 + \cdots + n = \dfrac{n(n+1)}{2}$.

11. Use mathematical induction to prove that $n! > 2^n$, $n \geq 4$.

12. Evaluate $_{13}C_4$. Verify with a graphing utility.

13. Expand $(x+3)^5$.

14. Find the term involving $x^7$ in $(x-2)^{12}$.

15. Evaluate $_{30}P_4$.

16. How many ways can six people sit at a table with six chairs?

17. Twelve cars run in a race. How many different ways can they come in first, second, and third place? (Assume that there are no ties.)

18. Two six-sided dice are tossed. Find the probability that the total of the two dice is less than 5.

19. Two cards are selected at random from a deck of 52 playing cards without replacement. Find the probability that the first card is a King and the second card is a black ten.

20. A manufacturer has determined that for every 1000 units it produces, 3 will be faulty. What is the probability that an order of 50 units will have one or more faulty units?

# C H A P T E R   7
## Conics and Parametric Equations

---

# C H A P T E R   7
# Conics and Parametric Equations

## Section 7.1    Conics

You should know the following basic definitions of conic sections.

- A parabola is the set of all points $(x, y)$ that are equidistant from a fixed line (directrix) and a fixed point (focus) not on the line.

    (a) Standard equation with Vertex $(0, 0)$ and Directrix $y = -p$ (vertical axis): $x^2 = 4py$

    (b) Standard equation with Vertex $(0, 0)$ and Directrix $x = -p$ (horizontal axis): $y^2 = 4px$

    (c) The focus lies on the axis $p$ units (directed distance) from the vertex.

- An ellipse is the set of all points $(x, y)$ the sum of whose distances from two distinct fixed points (foci) is constant.

    (a) Standard Equation of an Ellipse with Center $(0, 0)$, Major Axis Length $2a$, and Minor Axis Length $2b$:

    1. Horizontal Major Axis: $\dfrac{x^2}{a^2} + \dfrac{y^2}{b^2} = 1$      2. Vertical Major Axis: $\dfrac{x^2}{b^2} + \dfrac{y^2}{a^2} = 1$

    (b) The foci lie on the major axis, $c$ units from the center, where $a$, $b$, and $c$ are related by the equation $c^2 = a^2 - b^2$.

    (c) The vertices and endpoints of the minor axis are:

    1. Horizontal Axis: $(\pm a, 0)$ and $(0, \pm b)$      2. Vertical Axis: $(0, \pm a)$ and $(\pm b, 0)$

- A hyperbola is the set of all points $(x, y)$ the difference of whose distances from two distinct fixed points (foci) is constant.

    (a) Standard Equation of Hyperbola with Center $(0, 0)$

    1. Horizontal Transverse Axis: $\dfrac{x^2}{a^2} - \dfrac{y^2}{b^2} = 1$      2. Vertical Transverse Axis: $\dfrac{y^2}{a^2} - \dfrac{x^2}{b^2} = 1$

    (b) The vertices and foci are $a$ and $c$ units from the center and $b^2 = c^2 - a^2$.

    (c) The asymptotes of the hyperbola are:

    1. Horizontal Transverse Axis: $y = \pm \dfrac{b}{a}x$      2. Vertical Transverse Axis: $y = \pm \dfrac{a}{b}x$

**Solutions to Odd-Numbered Exercises**

**1.** $x^2 + y^2 = 36$

**3.** Diameter $= \dfrac{10}{7} \implies$ radius $= \dfrac{5}{7}$

$$x^2 + y^2 = \left(\tfrac{5}{7}\right)^2$$
$$x^2 + y^2 = \tfrac{25}{49}$$

**5.** $y = \frac{1}{2}x^2$

$x^2 = 2y = 4\left(\frac{1}{2}\right)y; \; p = \frac{1}{2}$

Vertex: $(0, 0)$

Focus: $\left(0, \frac{1}{2}\right)$

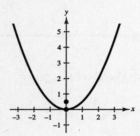

**7.** $y^2 = -6x$

$y^2 = 4\left(-\frac{3}{2}\right)x; \; p = -\frac{3}{2}$

Vertex: $(0, 0)$

Focus: $\left(-\frac{3}{2}, 0\right)$

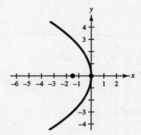

**9.** $x^2 + 8y = 0$

$x^2 = 4(-2)y; \; p = -2$

Vertex: $(0, 0)$

Focus: $(0, -2)$

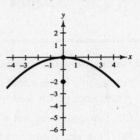

**11.** Focus: $\left(0, -\frac{3}{2}\right)$

$x^2 = 4\left(-\frac{3}{2}\right)y$

$x^2 = -6y$

**13.** Focus: $(-2, 0)$

$y^2 = 4(-2)x$

$y^2 = -8x$

**15.** Directrix: $y = -1$

$x^2 = 4(1)y$

$x^2 = 4y$

**17.** Directrix: $x = 3$

$y^2 = 4(-3)x$

$y^2 = -12x$

**19.** $y^2 = 4px$

$6^2 = 4p(4)$

$36 = 16p$

$p = \frac{9}{4}$

$y^2 = 4\left(\frac{9}{4}\right)x$

$y^2 = 9x$

**21.** $x^2 = 4py$

$3^2 = 4p(6)$

$9 = 24p$

$\frac{3}{8} = p$

$x^2 = 4\left(\frac{3}{8}\right)y$

$x^2 = \frac{3}{2}y \quad \text{OR} \quad y = \frac{2}{3}x^2$

Focus: $\left(0, \frac{3}{8}\right)$

**23.**     $y^2 = 4px$

$(-3)^2 = 4p(5)$

$9 = 20p$

$\frac{9}{20} = p$

$y^2 = 4\left(\frac{9}{20}\right)x \text{ or } y^2 = \frac{9}{5}x \text{ or } x = \frac{5}{9}y^2$

Focus: $\left(\frac{9}{20}, 0\right)$

**25.** $y^2 - 8x = 0 \quad$ and $\quad x - y + 2 = 0$

$\qquad y^2 = 8x \qquad\qquad\qquad y_3 = x + 2$

$\qquad y_1 = \sqrt{8x}$

$\qquad y_2 = -\sqrt{8x}$

The point of tangency is $(2, 4)$.

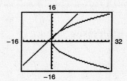

**27.** $\dfrac{x^2}{25} + \dfrac{y^2}{16} = 1$

Horizontal major axis

$a = 5, \; b = 4$

Center: $(0, 0)$

Vertices: $(\pm 5, 0)$

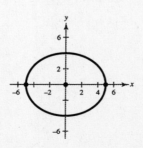

**29.** $\dfrac{x^2}{\frac{25}{9}} + \dfrac{y^2}{\frac{16}{9}} = 1$

Horizontal major axis

$a = \dfrac{5}{3}, b = \dfrac{4}{3}$

Center: $(0, 0)$

Vertices: $\left(\pm \dfrac{5}{3}, 0\right)$

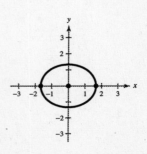

**31.** $\dfrac{x^2}{9} + \dfrac{y^2}{5} = 1$

Horizontal major axis

$a = 3,\ b = \sqrt{5}$

Center: $(0, 0)$

Vertices: $(\pm 3, 0)$

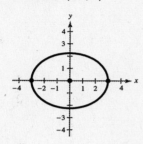

**33.** $4x^2 + y^2 = 1$

$\dfrac{x^2}{\left(\frac{1}{2}\right)^2} + \dfrac{y^2}{1^2} = 1$

Vertical major axis

$a = 1,\ b = \dfrac{1}{2}$

Center: $(0, 0)$

Vertices: $(0, \pm 1)$

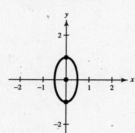

**35.** $5x^2 + 3y^2 = 15$

$3y^2 = 15 - 5x^2$

$y^2 = 5 - \dfrac{5}{3}x^2$

$y = \pm\sqrt{5 - \dfrac{5}{3}x^2}$

$y_1 = \sqrt{5 - \dfrac{5}{3}x^2}$

$y_2 = -\sqrt{5 - \dfrac{5}{3}x^2}$

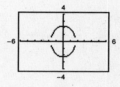

**37.** $6x^2 + y^2 = 36$

$y^2 = 36 - 6x^2$

$y = \pm\sqrt{36 - 6x^2}$

$y_1 = \sqrt{36 - 6x^2}$

$y_2 = -\sqrt{36 - 6x^2}$

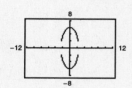

**39.** Vertices: $(0, \pm 2) \implies a = 2$

Minor axis of length $2 \implies b = 1$

Vertical major axis

$\dfrac{x^2}{b^2} + \dfrac{y^2}{a^2} = 1$

$\dfrac{x^2}{1} + \dfrac{y^2}{4} = 1$

**41.** Vertices: $(\pm 2, 0) \implies a = 2$

Minor axis of length $3 \implies b = \dfrac{3}{2}$

Horizontal major axis

$\dfrac{x^2}{a^2} + \dfrac{y^2}{b^2} = 1$

$\dfrac{x^2}{4} + \dfrac{4y^2}{9} = 1$

$\dfrac{x^2}{4} + \dfrac{y^2}{\left(\frac{9}{4}\right)} = 1$

**43.** Vertices: $(\pm 5, 0) \implies a = 5$

Foci: $(\pm 2, 0) \implies b = \sqrt{5^2 - 2^2} = \sqrt{21}$

Horizontal major axis

$\dfrac{x^2}{a^2} + \dfrac{y^2}{b^2} = 1$

$\dfrac{x^2}{25} + \dfrac{y^2}{21} = 1$

**45.** Foci: $(\pm 5, 0) \implies c = 5$

Major axis of length 12 $\implies a = 6$

$b = \sqrt{6^2 - 5^2} = \sqrt{11}$

Horizontal major axis

$$\frac{x^2}{a^2} + \frac{y^2}{b^2} = 1$$

$$\frac{x^2}{36} + \frac{y^2}{11} = 1$$

**47.** Vertices: $(0, \pm 5) \implies a = 5$

Vertical major axis

$$\frac{x^2}{b^2} + \frac{y^2}{25} = 1$$

$$\frac{(4)^2}{b^2} + \frac{(2)^2}{25} = 1$$

$$\frac{400}{21} = b^2$$

$$\frac{21x^2}{400} + \frac{y^2}{25} = 1$$

**49.** $x^2 - y^2 = 1$

$a = 1, \ b = 1$

Center: $(0, 0)$

Vertices: $(\pm 1, 0)$

Asymptotes: $y = \pm x$

$c^2 = a^2 + b^2 = 2 \implies$ foci: $\left( \pm \sqrt{2}, 0 \right)$

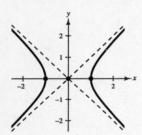

**51.** $\dfrac{y^2}{1} - \dfrac{x^2}{9} = 1$

$a = \ , b = 3$

Center: $(0, 0)$

Vertices: $(0, \pm 1)$

Asymptotes: $y = \pm \dfrac{1}{3} x$

$c^2 = a^2 + b^2 = 10 \implies$ foci: $\left( 0, \pm \sqrt{10} \right)$

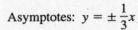

**53.** $\dfrac{y^2}{25} - \dfrac{x^2}{144} = 1$

$a = 5, \ b = 12$

Center: $(0, 0)$

Vertices: $(0, \pm 5)$

Asymptotes: $y = \pm \dfrac{5}{12} x$

$c^2 = a^2 + b^2 = 169 \implies$ foci: $(0, \pm 13)$

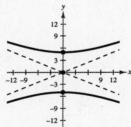

**55.** $4y^2 - x^2 = 1$

$$\frac{y^2}{\left(\frac{1}{2}\right)^2} - \frac{x^2}{1^2} = 1$$

$a = \dfrac{1}{2}, b = 1$

Center: $(0, 0)$

Vertices: $\left( 0, \pm \dfrac{1}{2} \right)$

Asymptotes: $y = \pm \dfrac{1}{2} x$

$c^2 = a^2 + b^2 = \dfrac{1}{4} + 1 = \dfrac{5}{4} \implies$

Foci: $\left( 0, \pm \dfrac{\sqrt{5}}{2} \right)$

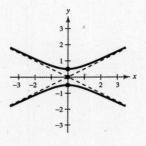

**57.** $2x^2 - 3y^2 = 6$

$$y^2 = \frac{1}{3}(2x^2 - 6)$$

$$y = \pm\sqrt{\frac{2}{3}(x^2 - 3)}$$

Asymptotes: $y = \pm\sqrt{\frac{2}{3}}x$

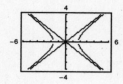

**59.** $4y^2 - 6x^2 = 12$

$$4y^2 = 6x^2 + 12$$

$$y^2 = \frac{3}{2}x^2 + 3$$

$$y = \pm\sqrt{\frac{3}{2}x^2 + 3}$$

$$a = \sqrt{3}, b = \sqrt{2}$$

Asymptotes: $y = \pm\sqrt{\frac{3}{2}}x$

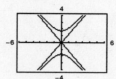

**61.** Vertices: $(0, \pm 2) \implies a = 2$

Foci: $(0, \pm 4) \implies c = 4$

$b^2 = c^2 - a^2 = 12$

Vertical transverse axis

$$\frac{y^2}{a^2} - \frac{x^2}{b^2} = 1$$

$$\frac{y^2}{4} - \frac{x^2}{12} = 1$$

**63.** Vertices: $(\pm 1, 0) \implies a = 1$

Asymptotes: $y = \pm 3x$

Horizontal transverse axis

$$3 = \frac{b}{a} = \frac{b}{1} \implies b = 3$$

$$\frac{x^2}{a^2} - \frac{y^2}{b^2} = 1$$

$$\frac{x^2}{1} - \frac{y^2}{9} = 1$$

**65.** Foci: $(0, \pm 8) \implies c = 8$

Asymptotes: $y = \pm 4x$

Vertical transverse axis

$$4 = \frac{a}{b} \implies a = 4b$$

$$16b^2 + b^2 = (8)^2$$

$$b^2 = \frac{64}{17} \implies a^2 = \frac{1024}{17}$$

$$\frac{y^2}{a^2} - \frac{x^2}{b^2} = 1$$

$$\frac{17y^2}{1024} - \frac{17x^2}{64} = 1$$

**67.** Vertices: $(0, \pm 3) \implies a = 3$

Vertical transverse axis

$$\frac{y^2}{9} - \frac{x^2}{b^2} = 1$$

$$\frac{5^2}{9} - \frac{(-2)^2}{b^2} = 1$$

$$b^2 = \frac{9}{4}$$

$$\frac{y^2}{9} - \frac{x^2}{\frac{9}{4}} = 1$$

**69.** $x^2 = 2y$

Parabola opening upward

Matches (f).

**71.** $9x^2 + y^2 = 9$

$$\frac{x^2}{1} + \frac{y^2}{9} = 1$$

Ellipse with vertical major axis
Matches (g).

**73.** $9x^2 - y^2 = 9$

$$\frac{x^2}{1} - \frac{y^2}{9} = 1$$

Hyperbola with horizontal transverse axis

Matches (d).

**75.** $x^2 + y^2 = 16$

Circle with center $(0, 0)$ and radius 4.

Matches (b).

**77.** $4x^2 + 4y^2 - 16 = 0$

$x^2 + y^2 = 4$   Circle

**79.** $3y^2 - 6x = 0$

$3y^2 = 6x$

$y^2 = 2x$

$y^2 = 4(\frac{1}{2})x$   Parabola

**81.** $4x^2 + y^2 - 16 = 0$

$4x^2 + y^2 = 16$

$\dfrac{x^2}{4} + \dfrac{y^2}{16} = 1$   Ellipse

**83.** $x^2 = 4py, \ p = 3.5$

$x^2 = 4(3.5)y$

$x^2 = 14y$

$y = \dfrac{1}{14}x^2$

**85.** (a)   $x^2 = 4py$

$32^2 = 4p\left(\dfrac{1}{12}\right)$

$1024 = \dfrac{1}{3}p$

$3072 = p$

$x^2 = 4(3072)y$

$y = \dfrac{x^2}{12{,}288}$

(b)   $\dfrac{1}{24} = \dfrac{x^2}{12{,}288}$

$\dfrac{12{,}288}{24} = x^2$

$512 = x^2$

$x \approx 22.6 \text{ feet}$

**87.** (a)

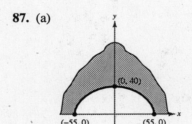

(b)   Vertices: $(\pm 55, 0) \implies a = 55$

height at center: $40 \implies b = 40$

Horizontal major axis

$\dfrac{x^2}{a^2} + \dfrac{y^2}{b^2} = 1$

$\dfrac{x^2}{3025} + \dfrac{y^2}{1600} = 1, \quad y \geq 0$

(c)   For $x = 50, \ \dfrac{50^2}{3025} + \dfrac{y^2}{1600} = 1$

$y^2 = 1600\left(1 - \dfrac{50^2}{3025}\right)$

$y^2 \approx 277.686$

$y \approx 16.66$

The height five feet from the edge of the tunnel is approximately 16.66 feet.

**89.** $\dfrac{x^2}{4} + \dfrac{y^2}{1} = 1$

$a = 2, \ b = 1, \ c = \sqrt{3}$

Points on the ellipse: $(\pm 2, 0), (0, \pm 1)$

Length of latus recta: $\dfrac{2b^2}{a} = 1$

Additional points: $\left(\sqrt{3}, \pm\dfrac{1}{2}\right), \left(-\sqrt{3}, \pm\dfrac{1}{2}\right)$

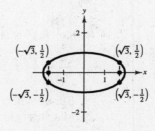

**91.** $9x^2 + 4y^2 = 36$

$$\frac{x^2}{4} + \frac{y^2}{9} = 1$$

Points on the ellipse: $(\pm 2, 0), (0, \pm 3)$

Length of latus recta: $\dfrac{2b^2}{a} = \dfrac{2 \cdot 2^2}{3} = \dfrac{8}{3}$

Additional points: $\left(\pm\dfrac{4}{3}, -\sqrt{5}\right), \left(\pm\dfrac{4}{3}, \sqrt{5}\right)$

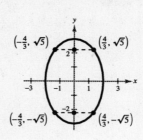

**93.** $r = 186,000$ miles per second

$t_1 = t_2 - 0.001$

$d_1 = \sqrt{(150 - x)^2 + (0 - 75)^2}$

$\phantom{d_1} = \sqrt{x^2 - 300x + 28,125}$

$d_2 = \sqrt{(-150 - x)^2 + (0 - 75)^2}$

$\phantom{d_2} = \sqrt{x^2 - 300x + 28,125}$

$$\frac{\sqrt{x^2 - 300x + 28,125}}{186,000} = \frac{\sqrt{x^2 + 300x + 28,125}}{186,000} - 0.001$$

$$\sqrt{x^2 - 300x + 28,125} = \sqrt{x^2 + 300x + 28,125} - 186$$

$$x^2 - 300x + 28,125 = x^2 + 300x + 28,125 - 372\sqrt{x^2 + 300x + 28,125} + 34,596$$

$$372\sqrt{x^2 + 300x + 28,125} = 600x + 34,596$$

$$138,384(x^2 + 300x + 28,125) = 360,000x^2 + 41,515,200x + 1,196,883,216$$

$$2,695,166,784 = 221,616x^2$$

$$x = 110.3 \text{ miles}$$

**95.** False. The equation represents a hyperbola.

**97.** False. A parabola cannot intersect its directrix nor focus.

**99.** $y - y_1 = \dfrac{x_1}{2p}(x - x_1)$

(a) The slope is $\dfrac{x_1}{2p}$.

(b)  $y - 1 = \dfrac{2}{2(1)}(x - 2)$

$x - y - 1 = 0$

$x + y + 1 = 0$

$y - 3 = \dfrac{6}{2(3)}(x - 6)$

$x - y - 3 = 0$

$x + y + 3 = 0$

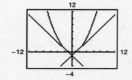

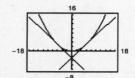

**—CONTINUED—**

**99. (b)** —CONTINUED—

$$y - 2 = \frac{4}{2(2)}(x - 4)$$

$$y - 4 = \frac{8}{2(4)}(x - 8)$$

$$x - y - 2 = 0 \qquad\qquad\qquad x - y - 4 = 0$$

$$x + y + 2 = 0 \qquad\qquad\qquad x + y + 4 = 0$$

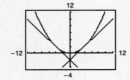

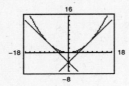

**101.** Since $x$ is negative, $x = -\frac{3}{2}\sqrt{4 - y^2}$ represents the left half of the ellipse.

**103.** Since $y$ is negative, $y = -\frac{2}{3}\sqrt{x^2 - 9}$ represents the bottom half of the hyperbola.

**105.** No it is not an ellipse. The exponent on $y$ is not 2.

**107.** $\dfrac{x^2}{a^2} + \dfrac{y^2}{4^2} = 1, \ 1 \leq a \leq 8$ The shape changes from an ellipse with a vertical major axis of length 8 and a minor axis of length 2 to a circle with a diameter of 8 and then to an ellipse with a horizontal major axis of length 16 and a minor axis of length 8.

**109.** Let $(x, y)$ be such that the sum of the distance from $(c, 0)$ and $(-c, 0)$ is $2a$.
(Note that this is only deriving the standard form for the ellipse with horizontal major axis.)

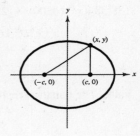

$$2a = \sqrt{(x - c)^2 + y^2} + \sqrt{(x + c)^2 + y^2}$$

$$2a - \sqrt{(x + c)^2 + y^2} = \sqrt{(x - c)^2 + y^2}$$

$$4a^2 - 4a\sqrt{(x + c)^2 + y^2} + (x + c)^2 + y^2 = (x - c)^2 + y^2$$

$$4a^2 + 4cx = 4a\sqrt{(x + c)^2 + y^2}$$

$$a^2 + cx = a\sqrt{(x + c)^2 + y^2}$$

$$a^4 + 2a^2cx + c^2x^2 = a^2(x^2 + 2cx + c^2 + y^2)$$

$$a^4 + c^2x^2 = a^2x^2 + a^2c^2 + a^2y^2$$

$$a^2(a^2 - c^2) = (a^2 - c^2)x^2 + a^2y^2$$

Let $b^2 = a^2 - c^2$. Then we have

$$a^2b^2 = b^2x^2 + a^2y^2 \implies 1 = \frac{x^2}{a^2} + \frac{y^2}{b^2}.$$

**111.** $12x^2 + 7x - 10 = (3x - 2)(4x + 5)$

**113.** $12z^4 + 17z^3 + 5z^2 = z^2(12z^2 + 17z + 5) = z^2(z + 1)(12z + 5)$

**115.** (Many answers are possible.)

$$(x - 0)(x - 3)(x - 4) = x(x^2 - 7x + 12) = x^3 - 7x^2 + 12x$$

**117.** (Many answers are possible.)

$$(x + 3)(x - 1 - \sqrt{2})(x - 1 + \sqrt{2}) = (x + 3)(x^2 - 2x - 1)$$
$$= x^3 + x^2 - 7x - 3$$

**119.** $2x^3 - 3x^2 + 50x - 75 = \left(x - \frac{3}{2}\right)(2x^2 + 50)$

$$= \left(x - \frac{3}{2}\right)2(x^2 + 25)$$

$$= (2x - 3)(x + 5i)(x - 5i)$$

Zeros: $\frac{3}{2}, 5i, -5i$

**121.** $11! = 39,916,800$ ways

## Section 7.2    Translation of Conics

You should know the following basic facts about conic sections.

■ Parabola with Vertex $(h, k)$

   (a) Vertex Axis

      1. Standard equation: $(x - h)^2 = 4p(y - k)$

      2. Focus: $(h, k + p)$

      3. Directrix: $y = k - p$

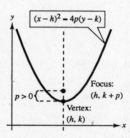

   (b) Horizontal Axis

      1. Standard equation: $(y - k)^2 = 4p(x - h)$

      2. Focus: $(h + p, k)$

      3. Directrix: $x = h - p$

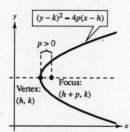

■ Circle with Center $(h, k)$ and Radius $r$

   Standard equation: $(x - h)^2 + (y - k)^2 = r^2$

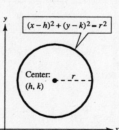

—CONTINUED—

**—CONTINUED—**

■ Ellipse with Center $(h, k)$

    (a) Horizontal Major Axis:

        1. Standard equation:

$$\frac{(x-h)^2}{a^2} + \frac{(y-k)^2}{b^2} = 1$$

        2. Vertices: $(h \pm a, k)$

        3. Foci: $(h \pm c, k)$

        4. $c^2 = a^2 - b^2$

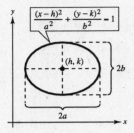

    (b) Vertical Major Axis:

        1. Standard equation:

$$\frac{(x-h)^2}{b^2} + \frac{(y-k)^2}{a^2} = 1$$

        2. Vertices: $(h, k \pm a)$

        3. Foci: $(h, k \pm c)$

        4. $c^2 = a^2 - b^2$

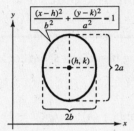

■ Hyperbola with Center $(h, k)$

    (a) Horizontal Transverse Axis:

        1. Standard equation:

$$\frac{(x-h)^2}{a^2} - \frac{(y-k)^2}{b^2} = 1$$

        2. Vertices: $(h \pm a, k)$

        3. Foci: $(h \pm c, k)$

        4. Asymptotes: $y - k = \pm\dfrac{b}{a}(x - h)$

        5. $c^2 = a^2 + b^2$

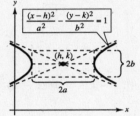

    (b) Vertical Transverse Axis:

        1. Standard equation:

$$\frac{(y-h)^2}{a^2} - \frac{(x-k)^2}{b^2} = 1$$

        2. Vertices: $(h, k \pm a)$

        3. Foci: $(h, k \pm c)$

        4. Asymptotes: $y - k = \pm\dfrac{a}{b}(x - h)$

        5. $c^2 = a^2 + b^2$

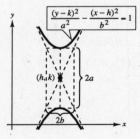

### Solutions to Odd-Numbered Exercises

**1.** The circle of radius 2 is translated 2 units to the left and 1 unit upward.

**3.** The hyperbola is translated 1 unit to the right and 3 units downward.

**5.** The ellipse is translated 1 unit to the right and 2 units downward.

**7.** $x^2 + y^2 = 49$

Center: $(0, 0)$

Radius: $7$

**9.** $(x + 2)^2 + (y - 7)^2 = 16$

Center: $(-2, 7)$

Radius: $4$

**11.** $(x - 1)^2 + y^2 = 15$

Center: $(1, 0)$

Radius: $\sqrt{15}$

**13.** $(x^2 - 2x + 1) + (y^2 + 6y + 9) = -9 + 1 + 9$

$\qquad (x - 1)^2 + (y + 3)^2 = 1$

Center: $(1, -3)$

Radius: $1$

**15.** $4\left(x^2 + 3x + \frac{9}{4}\right) + 4(y^2 - 6y + 9) = -41 + 9 + 36$

$\qquad 4\left(x + \frac{3}{2}\right)^2 + 4(y - 3)^2 = 4$

$\qquad \left(x + \frac{3}{2}\right)^2 + (y - 3)^2 = 1$

Center: $\left(-\frac{3}{2}, 3\right)$

Radius: $1$

**17.** $(x + 1)^2 + 8(y + 2) = 0$

$(x + 1)^2 = 4(-2)(y + 2); \ p = -2$

Vertex: $(-1, -2)$

Focus: $(-1, -2 - 2) = (-1, -4)$

Directrix: $y = 0$

**19.** $\left(y - \frac{1}{2}\right)^2 = 2(x - 5)$

$\left(y - \frac{1}{2}\right)^2 = 4\left(\frac{1}{2}\right)(x - 5); \ p = \frac{1}{2}$

Vertex: $\left(5, \frac{1}{2}\right)$

Focus: $\left(\frac{11}{2}, \frac{1}{2}\right)$

Directrix: $x = \frac{9}{2}$

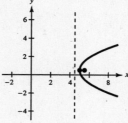

**21.** $y = \frac{1}{4}(x^2 - 2x + 5)$

$y = \frac{1}{4}(x - 1)^2 + 1$

$4(y - 1) = (x - 1)^2; \ p = 1$

Vertex: $(1, 1)$

Focus: $(1, 2)$

Directrix: $y = 0$

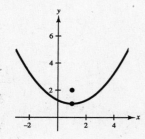

**23.** $y^2 + 6y + 8x + 25 = 0$

$\qquad (y + 3)^2 = 4(-2)(x + 2); \ p = -2$

Vertex: $(-2, -3)$

Focus: $(-4, -3)$

Directrix: $x = 0$

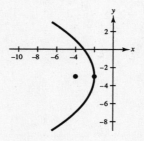

**25.** Vertex: (3, 1) and opens downward

Passes through (2, 0) and (4, 0)

$(x - 3)^2 = 4p(y - 1)$

$(2 - 3)^2 = 4p(0 - 1)$

$1 = -4p$

$(x - 3)^2 = -(y - 1)$

$(x - 3)^2 = 4\left(-\frac{1}{4}\right)(y - 1)$

**27.** Vertex: (3, 2)

Focus: (1, 2)

Horizontal axis

$p = 1 - 3 = -2$

$(y - 2)^2 = 4(-2)(x - 3)$

$(y - 2)^2 = -8(x - 3)$

**29.** Vertex: (0, 4)

Directrix: $y = 2$

Vertical axis

$p = 4 - 2 = 2$

$(x - 0)^2 = 4(2)(y - 4)$

$x^2 = 8(y - 4)$

**31.** Focus: (2, 2)

Directrix: $x = -2$

Horizontal axis

Vertex: (0, 2)

$p = 2 - 0 = 2$

$(y - 2)^2 = 4(2)(x - 0)$

$(y - 2)^2 = 8x$

**33.** $\dfrac{(x - 1)^2}{9} + \dfrac{(y - 3)^2}{25} = 1$

$a = 5, b = 3, c = \sqrt{a^2 - b^2} = 4$

Center: (1, 3)

Foci: (1, 7), (1, −1)

Vertices: (1, 8), (1, −2)

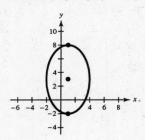

**35.** $(x + 2)^2 + \dfrac{(y - 4)^2}{\frac{1}{4}} = 1$

$a = 1, b = \dfrac{1}{2}, c = \sqrt{a^2 - b^2} = \dfrac{\sqrt{3}}{2}$

Center: (−2, 4)

Foci: $\left(-2 + \dfrac{\sqrt{3}}{2}, 4\right), \left(-2 - \dfrac{\sqrt{3}}{2}, 4\right)$

Vertices: (−3, 4), (−1, 4)

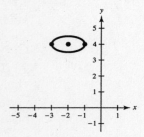

**37.** $9x^2 + 4y^2 + 36x - 24y + 36 = 0$

$9(x^2 + 4x + 4) + 4(y^2 - 6y + 9) = 36$

$\dfrac{(x + 2)^2}{4} + \dfrac{(y - 3)^2}{9} = 1$

$a = 3, b = 2, c = \sqrt{a^2 - b^2} = \sqrt{5}$

Vertical major axis

Center: (−2, 3)

Foci: $\left(-2, 3 - \sqrt{5}\right), \left(-2, 3 + \sqrt{5}\right)$

Vertices: (−2, 0), (−2, 6)

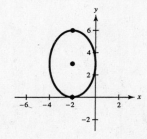

**39.**
$$16x^2 + 25y^2 - 32x + 50y + 16 = 0$$
$$16(x^2 - 2x + 1) + 25(y^2 + 2y + 1) = 25$$
$$\frac{(x-1)^2}{25/16} + (y+1)^2 = 1$$
$$a = \frac{5}{4}, \ b = 1, \ c = \sqrt{a^2 - b^2} = \frac{3}{4}$$

Horizontal major axis

Center: $(1, -1)$

Foci: $\left(\frac{1}{4}, -1\right), \left(\frac{7}{4}, -1\right)$

Vertices: $\left(-\frac{1}{4}, -1\right), \left(\frac{9}{4}, -1\right)$

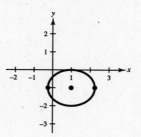

**41.** Center: $(2, 2)$

Horizontal major axis

$a = 3, \ b = 2$

$$\frac{(x-2)^2}{9} + \frac{(y-2)^2}{4} = 1$$

**43.** Vertices: $(0, 2), \ (4, 2)$

Minor axis of length 2

$$a = \frac{4-0}{2} = 2, \ b = \frac{2}{2} = 1$$

Center: $(2, 2)$

Horizontal major axis

$$\frac{(x-2)^2}{4} + \frac{(y-2)^2}{1} = 1$$

**45.** Foci: $(0, 0), \ (0, 8)$

Major axis of length 16

$a = 8, \ c = 4, \ b^2 = 48$

Center: $(0, 4)$

Vertical major axis

$$\frac{x^2}{48} + \frac{(y-4)^2}{64} = 1$$

**47.** Vertices: $(3, 1), \ (3, 9)$

Minor axis of length 6

$a = 4, \ b = 3$

Center: $(3, 5)$

Vertical major axis

$$\frac{(x-3)^2}{9} + \frac{(y-5)^2}{16} = 1$$

**49.** Center: $(0, 4)$

$a = 2c$

Vertices: $(-4, 4), \ (4, 4)$

$a = 4, \ c = 2, \ b^2 = 12$

Horizontal major axis

$$\frac{x^2}{16} + \frac{(y-4)^2}{12} = 1$$

**51.** $\dfrac{(x+1)^2}{4} - \dfrac{(y-2)^2}{1} = 1$

$a = 2, b = 1, c = \sqrt{a^2 + b^2} = \sqrt{5}$

Center: $(-1, 2)$ Horizontal transverse axis

Vertices: $(1, 2), (-3, 2)$

Foci: $\left(-1 + \sqrt{5}, 2\right), \left(-1 - \sqrt{5}, 2\right)$

Asymptotes: $y - 2 = \pm\dfrac{1}{2}(x + 1)$

$$y = 2 \pm \frac{1}{2}(x + 1)$$

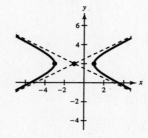

**53.** $(y + 6)^2 - (x - 2)^2 = 1$

$a = 1, b = 1, c = \sqrt{a^2 + b^2} = \sqrt{2}$

Center: $(2, -6)$

Vertical transverse axis

Vertices: $(2, -5), (2, -7)$

Foci: $\left(2, -6 \pm \sqrt{2}\right)$

Asymptotes: $y + 6 = \pm(x - 2)$

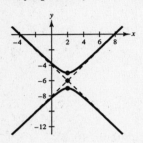

**55.**     $9x^2 - y^2 - 36x - 6y + 18 = 0$

$9(x^2 - 4x + 4) - (y^2 + 6y + 9) = 9$

$$(x - 2)^2 - \frac{(y + 3)^2}{9} = 1$$

$a = 1, b = 3, c = \sqrt{a^2 + b^2} = \sqrt{10}$

Center: $(2, -3)$

Horizontal transverse axis

Vertices: $(1, -3), (3, -3)$

Foci: $\left(2 \pm \sqrt{10}, -3\right)$

Asymptotes: $y = \pm 3(x - 2) - 3$

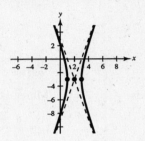

**57.**     $x^2 - 9y^2 + 2x - 54y - 80 = 0$

$(x^2 + 2x + 1) - 9(y^2 + 6y + 9) = 0$

$9(y + 3)^2 = (x + 1)^2$

$$y = \pm \frac{1}{3}(x + 1) - 3$$

The graph of this equation is two lines intersecting at $(-1, -3)$.

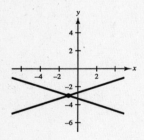

**59.** $9(y^2 + 2y + 1) - 4(x^2 - 2x + 1) = -41 + 9 - 4$

$9(y + 1)^2 - 4(x - 1)^2 = -36$

$$\frac{(x - 1)^2}{9} - \frac{(y + 1)^2}{4} = 1$$

$a = 3, b = 2, c = \sqrt{a^2 + b^2} = \sqrt{13}$

Center: $(1, -1)$

Horizontal transverse axis

Vertices: $(4, -1), (-2, -1)$

Foci: $\left(1 \pm \sqrt{13}, -1\right)$

Asymptotes: $y + 1 = \pm \frac{2}{3}(x - 1)$

$$y = \pm \frac{2}{3}(x - 1) - 1$$

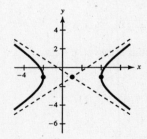

**61.** Vertices: $(0, 0)$, $(0, 2)$

Passes through $\left(\sqrt{3}, 3\right)$

Center: $(0, 1)$

Vertical transverse axis

$$a = 1$$

$$\frac{(y - 1)^2}{1} - \frac{x^2}{b^2} = 1$$

$$\frac{(3 - 1)^2}{1} - \frac{\left(\sqrt{3}\right)^2}{b^2} = 1$$

$$b^2 = 1$$

$$(y - 1)^2 - x^2 = 1$$

**63.** Vertices: $(2, 0)$, $(6, 0)$

Foci: $(0, 0)$, $(8, 0)$

Center: $(4, 0)$

Horizontal transverse axis

$$a = 2, \ c = 4, \ b^2 = c^2 - a^2 = 12$$

$$\frac{(x - 4)^2}{4} - \frac{y^2}{12} = 1$$

**65.** Vertices: $(4, 1)$, $(4, 9)$

Foci: $(4, 0)$, $(4, 10)$

Center: $(4, 5)$

Vertical transverse axis

$$a = 4, \ c = 5, \ b^2 = c^2 - a^2 = 9$$

$$\frac{(y - 5)^2}{16} - \frac{(x - 4)^2}{9} = 1$$

**67.** Vertices: $(2, 3)$, $(2, -3)$

Passes through the point $(0, 5)$

Center: $(2, 0)$

Vertical transverse axis

$$a = 3$$

$$\frac{y^2}{9} - \frac{(x - 2)^2}{b^2} = 1$$

$$\frac{5^2}{9} - \frac{(x - 2)^2}{b^2} = 1$$

$$b^2 = \frac{9}{4}$$

$$\frac{y^2}{9} - \frac{4(x - 2)^2}{9} = 1$$

**69.** $x^2 + y^2 - 6x + 4y + 9 = 0$

$(x - 3)^2 + (y + 2)^2 = 4$

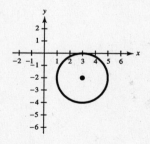

Circle

**71.** $4x^2 - y^2 - 4x - 3 = 0$

$$4\left(x - \tfrac{1}{2}\right)^2 - y^2 = 4$$

$$\left(x - \tfrac{1}{2}\right)^2 - \frac{y^2}{4} = 1$$

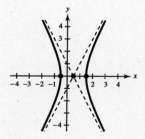

Hyperbola

**73.** $4x^2 + 3y^2 + 8x - 24y + 51 = 0$

$4(x + 1)^2 + 3(y - 4)^2 = 1$

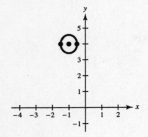

Ellipse

**75.** $25x^2 - 10x - 200y - 119 = 0$

$$25\left(x - \tfrac{1}{5}\right)^2 = 200\left(y + \tfrac{3}{5}\right)$$

$$\left(x - \tfrac{1}{5}\right)^2 = 8\left(y + \tfrac{3}{5}\right)$$

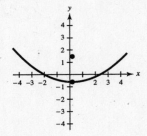

Parabola

**77.** (a) $V = 17,500\sqrt{2} \text{ mi/hr}$

$\approx 24,750 \text{ mi/hr}$

(b) $p = -4100, (h, k) = (0, 4100)$

$(x - 0)^2 = 4(-4100)(y - 4100)$

$x^2 = -16,400(y - 4100)$

**79.** $-12.5(y - 7.125) = (x - 6.25)^2$

$-12.5y + 89.0625 = x^2 - 12.5x + 39.0625$

$y = -0.08x^2 + x + 4$

(a)

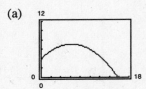

(b) The highest point is at $(6.25, 7.125)$. The distance is the $x$-intercept of $\approx 15.69$ feet.

**81.** Vertices: $(\pm 5, 0)$

$e = \dfrac{c}{a} = \dfrac{3}{5}$

$a = 5, \; c = 3, \; b = 4$

$\dfrac{x^2}{25} + \dfrac{y^2}{16} = 1$

**83.** $a = 3.67 \times 10^9$

$e = \dfrac{c}{a} = 0.249$

$c = 913,830,000$

Smallest distance: $a - c = 2,756,170,000$ miles

Greatest distance: $a + c = 4,583,830,000$ miles

**85.** $a = 100, b = 85$

Center: $(0, 85)$

$\dfrac{x^2}{100^2} + \dfrac{(y - 85)^2}{85^2} = 1$

**87.** $3(x^2 - 6x + 9) + 2(y^2 - 8y + 16) = -58 + 27 + 32$

$3(x - 3)^2 + 2(y - 4)^2 = 1$

True

**89.** $(y - 3)^2 = 6(x + 1)$

For the upper half of the parabola,

$$y - 3 = +\sqrt{6(x + 1)}$$
$$y = \sqrt{6(x + 1)} + 3.$$

**91.** $\dfrac{(x - 3)^2}{9} + \dfrac{y^2}{4} = 1$

$$\frac{(x - 3)^2}{9} = 1 - \frac{y^2}{4}$$

$$(x - 3)^2 = 9\left(\frac{4 - y^2}{4}\right)$$

$$x - 3 = \pm\frac{3}{2}\sqrt{4 - y^2}$$

$$x = 3 \pm \frac{3}{2}\sqrt{4 - y^2}$$

$$x = \frac{3}{2}\left(2 \pm \sqrt{4 - y^2}\right)$$

The right half of the ellipse is

$$x = \tfrac{3}{2}\left(2 + \sqrt{4 - y^2}\right).$$

**93.** $\dfrac{x^2}{a^2} + \dfrac{y^2}{b^2} = 1$

(a) Let $e = \dfrac{c}{a}$ be the eccentricity. Then,

$$c^2 = e^2 a^2 \implies a^2 - b^2 = e^2 a^2 \implies b^2 = a^2 - e^2 a^2 = a^2(1 - e^2).$$

So if the center of the ellipse is $(h, k)$,

$$\frac{(x - h)^2}{a^2} + \frac{(y - k)^2}{a^2(1 - e^2)} = 1.$$

(b) Solve for $y$: $\dfrac{(y - k)^2}{a^2(1 - e^2)} = 1 - \dfrac{(x - h)^2}{a^2} = \dfrac{a^2 - (x - h)^2}{a^2}$

$$(y - k)^2 = (1 - e^2)[a^2 - (x - h)^2]$$

$$y = k \pm \sqrt{(1 - e^2)[a^2 - (x - h)^2]}$$

$h = 2$, $k = 3$, $a = 2$, graph:

$$y_1 = 3 + \sqrt{(1 - e^2)[4 - (x - 2)^2]}$$
$$y_2 = 3 - \sqrt{(1 - e^2)[4 - (x - 2)^2]}.$$

(c) As $e$ approaches 0, the ellipse becomes more circular.

**95.** $\displaystyle\sum_{n=1}^{9} \frac{1}{6n} \approx 0.4715$

**97.** $\displaystyle\sum_{n=0}^{8} \frac{(-1)^n}{2^{2n}} \approx 0.8$

**99.** $(x - 4)^4 = x^4 - 16x^3 + 96x^2 - 256x + 256$

**101.** $(3x + 1)^5 = 243x^5 + 405x^4 + 270x^3 + 90x^2 + 15x + 1$

# Section 7.3   Parametric Equations

■ If $f$ and $g$ are continuous functions of $t$ on an interval $I$, then the set of ordered pairs $(f(t), g(t))$ is a *plane curve C*. The equations $x = f(t)$ and $y = g(t)$ are *parametric equations* for $C$ and $t$ is the *parameter.*

■ You should be able to graph plane curves with your graphing utility.

■ To eliminate the parameter:

Solve for $t$ in one equation and substitute into the second equation.

■ You should be able to find the parametric equations for a graph.

**Solutions to Odd-Numbered Exercises**

**1.** $x = t$

$y = t + 2$

$y = x + 2$ line

Matches (c).

**3.** $x = \sqrt{t}$

$y = t$

$y = x^2$ parabola, $x \geq 0$

Matches (b).

**5.** $x = \ln t \iff t = e^x$

$y = \frac{1}{2}t - 2$

$y = \frac{1}{2}e^x - 2$

Matches (f).

**7.** $x = \sqrt{t}, y = 2 - t$

(a)

| $t$ | 0 | 1 | 2 | 3 | 4 |
|---|---|---|---|---|---|
| $x$ | 0 | 1 | $\sqrt{2}$ | $\sqrt{3}$ | 2 |
| $y$ | 2 | 1 | 0 | $-1$ | $-2$ |

(b) Graph by hand:

Note: $x \geq 0$

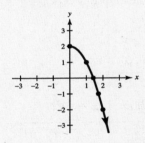

(c)

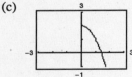

(d) $y = 2 - t = 2 - x^2$ parabola

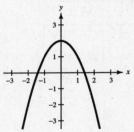

In part (c), $x \geq 0$.

**9.** $x = t, y = -4t$

$y = -4x$

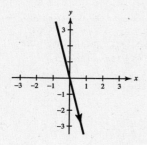

**11.** $x = 3t - 3, y = 2t + 1$

$y = 2\left[\dfrac{x + 3}{3}\right] + 1 = \dfrac{2}{3}x + 3$   line

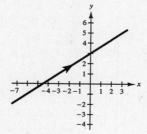

**13** $x = \dfrac{1}{4}t, \; y = t^2$

$y = (4x)^2$

$y = 16x^2$

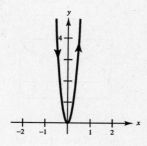

**15.** $x = t + 2, y = t^2$

$y = (x - 2)^2$   parabola

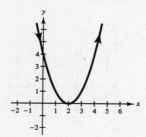

**17.** $x = t + 1, y = \dfrac{t}{t + 1}$

$y = \dfrac{x - 1}{x} = 1 - \dfrac{1}{x}$

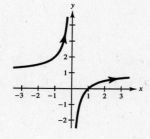

**19.** $x = 2t$

$y = |t - 2|$

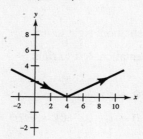

$t = \dfrac{x}{2} \implies y = |t - 2|$

$\qquad = \left|\dfrac{x}{2} - 2\right|$

$\qquad = \dfrac{1}{2}|x - 4|$

**21.** $x = e^{-t} \implies \dfrac{1}{x} = e^t$

$y = e^{3t} \implies y = (e^t)^3$

$y = \left(\dfrac{1}{x}\right)^3$

$y = \dfrac{1}{x^3}, \ x > 0, \ y > 0$

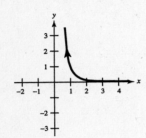

**23.** $x = t^3 \implies x^{1/3} = t$

$y = 3 \ln t \implies y = \ln t^3$

$y = \ln(x^{1/3})^3$

$y = \ln x$

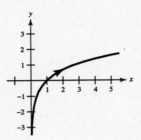

**25.** $x = 12t$

$y = -8t^2 + 32t$

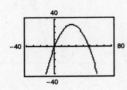

**27.** $x = \dfrac{t}{2}$

$y = \ln(t^2 + 1)$

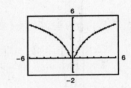

**29.** Each curve represents a portion of the line $y = 2x + 1$.

   (a) Domain: $(-\infty, \infty)$; orientation: left to right

   (b) Domain: all $x \neq 0$; orientation: oscillates, right to left

   (c) Domain: $x > 0$; orientation: right to left

   (d) Domain: $x > 0$; orientation: left to right

**31.** Each curve represents a portion of the graph of $y = x^3 - 1$. The portions are as follows.

   (a) $0 \leq x \leq 1$

   (b) $0 \leq x \leq 3$

   (c) $-2 \leq x \leq 3$

   (d) $-3 \leq x \leq 3$

**33.** $x = x_1 + t(x_2 - x_1) = 0 + t(5 - 0) = 5t$

$y = y_1 + t(y_2 - y_1) = 0 + t(-2 - 0) = -2t$

(Solution not unique.)

**35.** $x = x_1 + t(x_2 - x_1) = -2 + t[3 - (-2)] = 5t - 2$

$y = y_1 + t(y_2 - y_1) = 3 + t(10 - 3) = 7t + 3$

(Solution not unique.)

**37.** (a) $x = x_1 + t(x_2 - x_1) = 3 + t(3 - 3) = 3$

     $y = y_1 + t(y_2 - y_1) = -1 + t[5 - (-1)] = 6t - 1$

     One answer: $x = 3, \ y = 6t - 1, \ 0 \leq t \leq 1$

     Alternative: $x = 3, \ y = t, \ -1 \leq t \leq 5$

   (b) $x = x_1 + t(x_2 - x_1) = 3 + t(3 - 3) = 3$

     $y = y_1 + t(y_2 - y_1) = 5 + t(-1 - 5) = 5 - 6t$

     One answer: $x = 3, \ y = 5 - 6t, 0 \leq t \leq 1$

     Alternative: $x = 3, \ y = -t, -5 \leq t \leq 1$

**39.** Answers not unique

$x = t, y = 4t - 3$

$x = 2t, y = 8t - 3$

**41.** Answers not unique

$x = t, y = \dfrac{1}{t}$

$x = \dfrac{1}{t}, y = t$

**43.** Answers not unique

$x = t, y = t^2 + 4$

$x = -t, y = t^2 + 4$

**45.** Answers not unique

$x = t, y = t^3 + 2t$

$x = 2t, y = 8t^3 + 4t$

**47.** (a)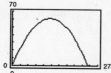

Maximum height: 60.5 ft

Range: 242 ft

(b)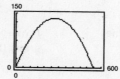

Maximum height: 136.1 ft

Range: 544.5 ft

(c)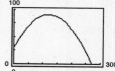

Maximum height: 90.5 ft

Range: 269 ft

**49.** (a) $x = 145t$

$y = 3 + 39t - 16t^2$

When $x = 400 = 145t$, $t = \frac{400}{145}$ and $y = 3 + 39\left(\frac{400}{145}\right) - 16\left(\frac{400}{145}\right)^2 \approx -11$.

Hence, it is not a home run. You can verify this graphically by tracing along the curve.

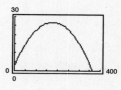

(b) $x = 138t$

$y = 3 + 59t - 16t^2$

When $x = 400 = 138t$, $t = \frac{400}{138}$ and $y \approx 39.6$.

Yes, it is a home run.

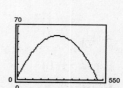

**51.** True.

$x = t$    first set

$y = t^2 + 1 = x^2 + 1$

$x = 3t$    second set

$y = 9t^2 + 1$

$= (3t)^2 + 1 = x^2 + 1$

**53.** The graph is the same, but the orientation is reversed.

**55.** $5x^2 + 8 = 0$

$x^2 = -\dfrac{8}{5}$

$x = \pm\sqrt{\dfrac{8}{5}}i = \pm\dfrac{2}{5}\sqrt{10}\,i$

**57.** $4x^2 + 4x - 11 = 0$

$x = \dfrac{-4 \pm \sqrt{16 + 4(4)(11)}}{8} = \dfrac{-1 \pm \sqrt{12}}{2} = -\dfrac{1}{2} \pm \sqrt{3}$

**59.** $f(-x) = \dfrac{4(-x)^2}{(-x)^2 + 1} = \dfrac{4x^2}{x^2 + 1} = f(x)$

Symmetric about the $y$-axis

Even function

**61.** $y = e^x \neq e^{-x};\ e^{-x} \neq -e^x$

No symmetry

Neither even nor odd

**63.** $\displaystyle\sum_{n=1}^{50} 8n = 8\,\dfrac{50(51)}{2} = 10{,}200$

**65.** $\displaystyle\sum_{n=1}^{40}\left(300 - \dfrac{1}{2}n\right) = 300(40) - \dfrac{1}{2}\,\dfrac{40(41)}{2} = 11{,}590$

# Review Exercises for Chapter 7

**Solutions to Odd-Numbered Exercises**

**1.** $4x^2 + y^2 = 4$

$x^2 + \dfrac{y^2}{2^2} = 1$

Ellipse; matches graph (e).

**3.** $4x^2 - y^2 = 4$

$x^2 - \dfrac{y^2}{2^2} = 1$

Hyperbola; matches graph (c).

**5.** $x^2 - 5y^2 = -5$

$y^2 - \dfrac{x^2}{5} = 1$

Hyperbola; matches graph (f).

**7.** $x^2 + y^2 = 81$

Circle radius: 9

Matches graph (h).

**9.** Vertex: $(0, 0)$:  $y^2 = 4px$

$(1, 2)$ on graph: $2^2 = 4p(1) \implies p = 1$

$y^2 = 4x$

**11.** Vertex: $(0, 0)$

Focus: $(-6, 0)$ Parabola opens to left.

$y^2 = 4px$

$y^2 = 4(-6)x$

$y^2 = -24x$

**13.** $4x - y^2 = 0$

$4x = y^2$

$x = \tfrac{1}{4}y^2$

$P = 1$

Focus: $(1, 0)$

Vertex: $(0, 0)$

**15.** $\tfrac{1}{2}y^2 + 18x = 0$

$y^2 = -36x = 4(-9)x$

Focus: $(-9, 0)$

Vertex: $(0, 0)$

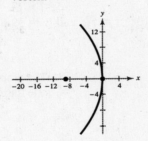

**17.** $y = \dfrac{x^2}{200}$, $-100 \le x \le 100$

Vertex: $(0, 0)$

$x^2 = 200y$

$4p = 200$

$p = 50$

Focus: $(0, 50)$

**19.** Vertices: $(\pm 5, 0) \implies a = 5$

Also, $b = 3$

$\dfrac{x^2}{25} + \dfrac{y^2}{9} = 1$

**21.** Vertices: $(0, \pm 6)$

Passes through $(2, 2)$

Vertical Major axis

Center: $(0, 0)$, $a = 6$

$\dfrac{x^2}{b^2} + \dfrac{y^2}{36} = 1$

$\dfrac{2^2}{b^2} + \dfrac{2^2}{36} = 1$

$\dfrac{4}{b^2} = 1 - \dfrac{1}{9} = \dfrac{8}{9}$

$b^2 = \dfrac{36}{8} = \dfrac{9}{2}$

$\dfrac{x^2}{\frac{9}{2}} + \dfrac{y^2}{36} = 1$

**23.** $\dfrac{x^2}{4} + \dfrac{y^2}{16} = 1$   Ellipse

Center: $(0, 0)$

Vertices: $(0, 4), (0, -4)$

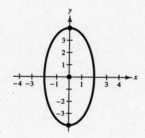

**25.** $6x^2 + 4y^2 = 36$   Ellipse

$\dfrac{x^2}{6} + \dfrac{y^2}{9} = 1$

Center: $(0, 0)$

Vertices: $(0, 3), (0, -3)$

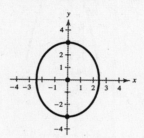

**27.** $a = 5, b = 4, c = \sqrt{a^2 - b^2} = \sqrt{25 - 16} = 3$

The foci should be placed 3 feet on either side of the center and have the same height as the pillars.

**29.** Vertices: $(\pm 1, 0)$

Horizontal transverse axis

Center: $(0, 0)$

$a = 1$

$\pm\dfrac{b}{a} = \pm 2 \implies b = 2$

$\dfrac{x^2}{1} - \dfrac{y^2}{4} = 1$

**31.** Vertices: $(0, \pm 1)$

Foci: $(0, \pm 3)$

Vertical transverse axis

Center: $(0, 0)$

$a = 1, c = 3,$

$b = \sqrt{9 - 1} = \sqrt{8}$

$\dfrac{y^2}{1} - \dfrac{x^2}{8} = 1$

**33.** $\dfrac{y^2}{9} - \dfrac{x^2}{64} = 1$

Vertical transverse axis

$a = 3, b = 8$

$c = \sqrt{9 + 64} = \sqrt{73}$

Asymptotes: $y = \pm\dfrac{3}{8}x$

Center: $(0, 0)$

Vertices: $(0, 3), (0, -3)$

Foci: $\left(0, \pm\sqrt{73}\right)$

**35.** $5y^2 - 4x^2 = 20$

$\dfrac{y^2}{4} - \dfrac{x^2}{5} = 1$

Hyperbola

Vertical transverse axis

Center: $(0, 0)$

Vertices: $(0, \pm 2)$

Foci: $\left(0, \pm\sqrt{41}\right)$

Asymptote: $y = \pm\dfrac{2\sqrt{5}}{5}x$

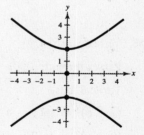

**37.** $x^2 - 6x + 2y + 9 = 0$

$$(x - 3)^2 = -2y$$

Parabola

Vertex: $(3, 0)$

Focus: $\left(3, -\dfrac{1}{2}\right)$

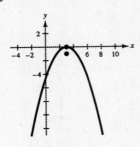

**39.** $x^2 + 9y^2 + 10x - 18y + 25 = 0$

$$(x + 5)^2 + 9(y - 1)^2 = 9$$

$$\dfrac{(x + 5)^2}{9} + (y - 1)^2 = 1$$

Ellipse

Center: $(-5, 1)$

Vertices: $(-8, 1)$,

$\qquad\qquad (-2, 1)$

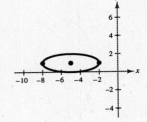

**41.** $\qquad\qquad 4x^2 - 4x - 4y^2 + 8y = 11$

$$4\left(x^2 - x + \dfrac{1}{4}\right) - 4(y^2 - 2y + 1) = 11 + 1 - 4$$

$$4\left(x - \dfrac{1}{2}\right)^2 - 4(y - 1)^2 = 8$$

$$\dfrac{\left(x - \frac{1}{2}\right)^2}{2} - \dfrac{(y - 1)^2}{2} = 1$$

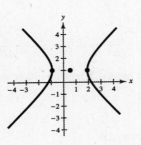

Hyperbola

**43.** $4x^2 + y^2 - 16x + 15 = 0$

$$4(x - 2)^2 + y^2 = 1$$

Ellipse

Center: $(2, 0)$

Vertices: $(2, -1), (2, 1)$

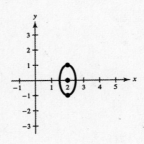

**45.** Vertex: $(-6, 4)$

Passes through $(0, 0)$

Vertical axis

$(x + 6)^2 = 4p(y - 4)$

$(0 + 6)^2 = 4p(0 - 4)$

$\qquad 36 = -16p$

$\qquad -\dfrac{9}{4} = p$

$(x + 6)^2 = 4\left(-\dfrac{9}{4}\right)(y - 4)$

$(x + 6)^2 = -9(y - 4)$

**47.** Vertex: $(4, 2)$

Focus: $(4, 0)$

Vertical axis, $p = -2$

$(x - 4)^2 = 4(-2)(y - 2)$

$(x - 4)^2 = -8(y - 2)$

**49.** Vertex: $(0, 2)$

Horizontal axis

Passes through $(-1, 0)$

$(y - 2)^2 = 4p(x - 0)$

$(0 - 2)^2 = 4p(-1 - 0)$

$\qquad 4 = -4p$

$\qquad -1 = p$

$(y - 2)^2 = 4(-1)(x - 0)$

$(y - 2)^2 = -4x$

**51.** Vertices: $(0, 3)$, $(10, 3)$

Passes through $(5, 0)$

Center: $(5, 3)$

Horizontal major axis

$a = 5$, $b = 3$

$$\frac{(x - 5)^2}{25} + \frac{(y - 3)^2}{9} = 1$$

**53.** Vertices: $(-3, 0)$, $(7, 0)$

Foci: $(0, 0)$, $(4, 0)$

Horizontal major axis

Center: $(2, 0)$

$a = 5$, $c = 2$,

$b = \sqrt{25 - 4} = \sqrt{21}$

$$\frac{(x - h)^2}{a^2} + \frac{(y - k)^2}{b^2} = 1$$

$$\frac{(x - 2)^2}{25} + \frac{y^2}{21} = 1$$

**55.** Horizontal major axis

Center: $(2, 1)$

$a = 2$, $b = 1$

$$\frac{(x - h)^2}{a^2} + \frac{(y - k)^2}{b^2} = 1$$

$$\frac{(x - 2)^2}{4} + \frac{(y - 1)^2}{1} = 1$$

**57.** Center: $(0, 7)$

Horizontal transverse axis length $2a = 12 \implies a = 6$

Vertices: $(\pm 6, 7)$

Asymptotes: $y - 7 = \pm\frac{1}{2}x = \pm\frac{b}{a}x = \pm\frac{b}{6}x \implies b = 3$

$$\frac{x^2}{36} - \frac{(y - 7)^2}{9} = 1$$

**59.** Vertices: $(-10, 3), (6, 3) \implies$ Center: $(-2, 3), a = 8$

Foci: $(-12, 3), (8, 3) \implies c = 10$ and

$$b = \sqrt{c^2 - a^2} = 6.$$

$$\frac{(x + 2)^2}{64} - \frac{(y - 3)^2}{36} = 1$$

**61.** Foci: $(0, 0)$, $(8, 0)$

Asymptotes: $y = \pm 2(x - 4)$

Horizontal transverse axis

Center: $(4, 0) \implies c = 4$

$\frac{b}{a} = 2 \implies b = 2a$

$$a^2 + b^2 = c^2$$

$$a^2 + (2a)^2 = 4^2$$

$$a^2 = \frac{16}{5}$$

$$b^2 = \frac{64}{5}$$

$$\frac{(x - h)^2}{a^2} - \frac{(y - k)^2}{b^2} = 1$$

$$\frac{5(x - 4)^2}{16} - \frac{5y^2}{64} = 1$$

**63.** $x^2 = 4p(y - 12)$

$(4, 10)$ on curve:

$16 = 4p(10 - 12) = -8p \implies p = -2$

$x^2 = 4(-2)(y - 12) = -8y + 96$

$y = (-x^2 + 96)/8$

$y = 0$ if $x^2 = 96 \implies x = 4\sqrt{6} \implies$ width is

$8\sqrt{6}$ meters.

**65.** $a - c = 1.3495 \times 10^9$

$a + c = 1.5045 \times 10^9$

Adding, $2a = 2.854 \times 10^9 \implies a = 1.427 \times 10^9$

Then $c = 1.5045 \times 10^9 - 1.427 \times 10^9 = 0.0775 \times 10^9$

$e = \frac{c}{a} \approx 0.0543$

**67.**

| $t$ | $-2$ | $-1$ | $0$ | $1$ | $2$ | $3$ |
|---|---|---|---|---|---|---|
| $x$ | $-8$ | $-5$ | $-2$ | $1$ | $4$ | $7$ |
| $y$ | $15$ | $11$ | $7$ | $3$ | $-1$ | $-5$ |

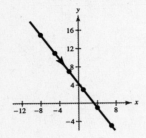

**69.**

| $t$ | $-2$ | $-1$ | $1$ | $2$ | $3$ | $4$ |
|---|---|---|---|---|---|---|
| $x$ | $-3$ | $-6$ | $6$ | $3$ | $2$ | $\frac{3}{2}$ |
| $y$ | $2$ | $3$ | $5$ | $6$ | $7$ | $8$ |

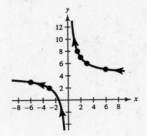

**71.** $x = t^3, y = t^2 + 1$

$t = x^{1/3} \implies y = x^{2/3} + 1, -1 \le x \le 1$

Matches (a)

**73.** $x = (2t)^3 = 8t^3, y = (2t)^2 + 1 = 4t^2 + 1$

$t = \left(\dfrac{x}{8}\right)^{1/3} = \dfrac{1}{2}x^{1/3} \implies y = 4\left(\dfrac{1}{2}x^{1/3}\right)^2 + 1$

$= x^{2/3} + 1, -8 \le x \le 8$

Matches (d)

**75.** $x = 5t - 1, y = 2t + 5$

$t = \frac{1}{5}(x + 1) \implies y = \frac{2}{5}(x + 1) + 5 = \frac{2}{5}x + \frac{27}{5}$
line

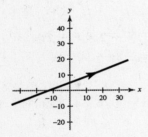

**77.** $x = t^2 + 2, y = 4t^2 - 3$

$t^2 = x - 2 \implies y = 4(x - 2) - 3 = 4x - 11,$

$x \ge 2$

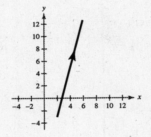

**79.** $x = t^3, y = \frac{1}{2}t^2$

$t = x^{1/3} \implies y = \frac{1}{2}x^{2/3}$

**81.** $x = \sqrt[3]{t}$

$y = t$

$t = x^3 \implies y = t = x^3$

$y = x^3$

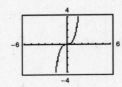

**83.** $x = \dfrac{1}{t}$

$y = t$

$y = t = \dfrac{1}{x}$

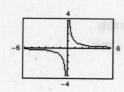

**85.** $x = 2t$

$y = 4t$

$y = 2(2t) = 2x$

line

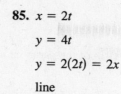

**87.** $x = 1 + 4t$

$y = 2 - 3t$

$t = \dfrac{x - 1}{4} \implies y = 2 - 3\left(\dfrac{x - 1}{4}\right) = 2 - \dfrac{3}{4}x + \dfrac{3}{4}$

$y = \dfrac{11}{4} - \dfrac{3}{4}x$

$3x + 4y - 11 = 0$

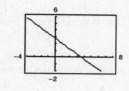

**89.** $x = \dfrac{1}{t}$

$y = t^2$

$y = \left(\dfrac{1}{x}\right)^2$

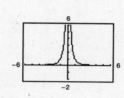

**91.** $x = 3$

$y = t$

Vertical line: $x = 3$

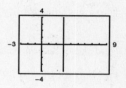

**93.** $y = 6x + 2$

$x = t,\, y = 6t + 2$

$x = -t,\, y = -6t + 2$

Other answers possible

**95.** $y = x^2 + 2$

$x = t,\, y = t^2 + 2$

$x = t + 1,\, y = (t + 1)^2 + 2 = t^2 + 2t + 3$

Other answers possible

**97.** $x = x_1 + t(x_2 - x_1) = 3 + t(8 - 3) = 5t + 3$

$y = y_1 + t(y_2 - y_1) = 5 + t(0 - 0) = 5$

or: $x = t,\ y = 5$

**99.** $x = x_1 + t(x_2 - x_1) = -1 + t[10 - (-1)] = 11t - 1$

$y = y_1 + t(y_2 - y_1) = 6 + t(0 - 6) = -6t + 6$

**101.** False. $\dfrac{x^2}{4} - y^4 = 1$ is not of degree 2 in $y$.

**103.** False. There are an infinite number of parametric equations. Two examples are

$x = t,\, y = 3 - 2t$

$x = t + 1,\, y = 3 - 2(t + 1) = 1 - 2t$

**105.** Answers will vary. See pages 579 and 589. The extended diagonals of the central rectangle are asymptotes of the hyperbola.

# Chapter 7   Practice Test

1. Find the vertex, focus, and directrix of the parabola $x^2 = 20y$.

2. Find the equation of the parabola with vertex $(0, 0)$ and focus $(7, 0)$.

3. Find the center, foci, and vertices of the ellipse $\dfrac{x^2}{144} + \dfrac{y^2}{25} = 1$.

4. Find the equation of the ellipse with foci $(\pm 4, 0)$ and minor axis of length 6.

5. Find the center, vertices, foci, and asymptotes of the hyperbola $\dfrac{y^2}{144} - \dfrac{x^2}{169} = 1$.

6. Find the equation of the hyperbola with vertices $(\pm 4, 0)$ and asymptotes $y = \pm\frac{1}{2}x$.
   Use a graphing utility to graph the curve.

7. Find the equation of the parabola with vertex $(6, -1)$ and focus $(6, 3)$.
   Use a graphing utility to graph the curve.

8. Find the center, foci, and vertices of the ellipse $16x^2 + 9y^2 - 96x + 36y + 36 = 0$.

9. Find the equation of the ellipse with vertices $(-1, 1)$ and $(7, 1)$ and minor axis of length 2.

10. Find the center, vertices, foci, and asymptotes of the hyperbola $4(x + 3)^2 - 9(y - 1)^2 = 1$.

11. Find the equation of the hyperbola with vertices $(3, 4)$ and $(3, -4)$ and foci $(3, 7)$ and $(3, -7)$.

12. Use a graphing utility to sketch the curve represented by the parametric equations.

    $x = 2t + 1$

    $y = -1 - 3t$

    Then eliminate the parameter and write the corresponding rectangular equation.

13. Use a graphing utility to sketch the curve represented by the parametric equations

    $x = 2 \ln t$

    $y = t^3$

14. Describe how the plane curves $x = t, y = t^2$ and $x = t^2, y = t^4$ differ from each other.

# APPENDICES

# APPENDIX C
## Concepts in Statistics

### Appendix C.1    Measures of Central Tendency and Dispersion

Solutions to Odd-Numbered Exercises

**1.** Mean $= \dfrac{5 + 12 + 7 + 14 + 8 + 9 + 7}{7} = \dfrac{62}{7} \approx 8.86$

Median: 8

Mode: 7

**3.** Mean $= \dfrac{5 + 12 + 7 + 24 + 8 + 9 + 7}{7} = \dfrac{72}{7} \approx 10.29$

Median: 8

Mode: 7

**5.** Mean $= \dfrac{5 + 12 + 7 + 14 + 9 + 7}{6} = \dfrac{54}{6} = 9$

Median: $\dfrac{7 + 9}{2} = 8$

Mode: 7

**7.** (a)  The mean is sensitive to extreme values

(b)  Mean: 14.86

Median: 14

Mode: 13

Each is increased by 6.

(c)  Each will increase by $k$.

**9.** Mean $= \dfrac{410 + 260 + 320 + 320 + 460 + 150}{6} = \dfrac{1920}{6} = 320$

Median: 320

Mode: 320

**11.** There are many possible answers. For example: $\{4, 4, 10\}$

**13.** The mean is 76.55 and the median is 82. The median is the best description.

**15.** (a)  Mean $= 12$, $\sigma \approx 2.83$

(b)  Mean $= 20$, $\sigma \approx 2.83$

(c)  Mean $= 12$, $\sigma \approx 1.41$

(d)  Mean $= 9$, $\sigma \approx 1.41$

**17.** $\bar{x} = 6$

$v = 10$

$\sigma \approx 3.16$

**19.** $\bar{x} = 2$

$v = \frac{4}{3}$

$\sigma \approx 1.15$

**21.** $\bar{x} = 4$

$v = 4$

$\sigma \approx 2$

**23.** $\bar{x} = 47$

$v = 226$

$\sigma \approx 15.03$

**25.** $\bar{x} = 6$

$$\sigma = \sqrt{\frac{2^2 + 4^2 + 6^2 + 6^2 + 13^2 + 5^2}{6} - 6^2}$$

$$= \sqrt{\frac{286}{6} - 36}$$

$$= \sqrt{\frac{35}{3}} \approx 3.42$$

**27.** $\bar{x} = 5.8$

$$\sigma = \sqrt{\frac{8.1^2 + 6.9^2 + 3.7^2 + 4.2^2 + 6.1^2}{5} - 5.8^2}$$

$$= \sqrt{2.712} \approx 1.65$$

**29.** $\bar{x} = 12$ and $|x_i - 12| = 8$ for all $x_i$. Hence, $\sigma = 8$.

**31.** The mean will increase by 5. The standard deviation will not change.

**33.** The first histogram has a smaller standard deviation.

**35.** (a) 12, 13, 13, 14, 14, 15, 20, 23, 23

Median: 14

Lower quartile is median of {12, 13, 13} = 13

Upper quartile is median of {15, 20, 23, 23} = 21.5

(b)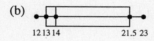

**37.** (a) 46, 47, 47, 48, 48, 49, 50, 51, 52, 53

Median: $\dfrac{48 + 49}{2} = 48.5$

Lower quartile is median of {46, 47, 47, 48, 48} = 47

Upper quartile is median of {49, 50, 51, 52, 53} = 51

(b)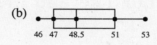

**39.**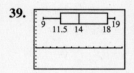

**41.**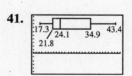

**43.**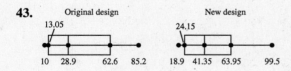

From the plots, you can see that the lifetimes of the units in the new design are greater than the original design. The median increased by over 12 months.

## Appendix C.2   Least Squares Regression

**1.**

| $x$ | $y$ | $xy$ | $x^2$ |
|---|---|---|---|
| $-4$ | 1 | $-4$ | 16 |
| $-3$ | 3 | $-9$ | 9 |
| $-2$ | 4 | $-8$ | 4 |
| $-1$ | 6 | $-6$ | 1 |
| Total | $-10$ | 14 | $-27$ | 30 |

$n = 4$

$4b + (-10)a = 14$

$(-10)b + 30a = -27$

Solving this system, $a = 1.6$ and $b = 7.5$

*Answer:* $y = 1.6x + 7.5$

**3.**

| $x$ | $y$ | $xy$ | $x^2$ |
|---|---|---|---|
| $-3$ | 1 | $-3$ | 9 |
| $-1$ | 2 | $-2$ | 1 |
| 1 | 2 | 2 | 1 |
| 4 | 3 | 12 | 16 |
| Total | 1 | 8 | 9 | 27 |

$n = 4$

$4b + a = 8$

$b + 27a = 9$

Solving this system, $a \approx 0.262$ and $b \approx 1.93$

*Answer:* $y = 0.262x + 1.93$

# A P P E N D I X   D
# Solving Linear Equations and Inequalities

**Solutions to Odd-Numbered Exercises**

**1.** $x + 11 = 15$

$x = 15 - 11$

$x = 4$

**3.** $x - 2 = 5$

$x = 5 + 2$

$x = 7$

**5.** $3x = 12$

$x = \frac{12}{3}$

$x = 4$

**7.** $\frac{x}{5} = 4$

$x = 4(5)$

$x = 20$

**9.** $8x + 7 = 39$

$8x = 32$

$x = 4$

**11.** $24 - 7x = 3$

$-7x = -21$

$x = 3$

**13.** $8x - 5 = 3x + 20$

$5x = 25$

$x = 5$

**15.** $-2(x + 5) = 10$

$-2x - 10 = 10$

$-2x = 20$

$x = -10$

**17.** $2x + 3 = 2x - 2$

$3 = -2$

No solution

**19.** $\frac{3}{2}(x + 5) - \frac{1}{4}(x + 24) = 0$

$\frac{3}{2}(x + 5) = \frac{1}{4}(x + 24)$

$12(x + 5) = 2(x + 24)$

$12x + 60 = 2x + 48$

$10x = -12$

$x = -\frac{12}{10}$

$x = -\frac{6}{5}$

**21.** $0.25x + 0.75(10 - x) = 3$

$25x + 75(10 - x) = 300$

$25x + 750 - 75x = 300$

$-50x = -450$

$x = 9$

**23.** $x + 6 < 8$

$x < 8 - 6$

$x < 2$

**25.** $-x - 8 > -17$

$17 - 8 > x$

$9 > x$

$x < 9$

**27.** $6 + x \le -8$

$x \le -8 - 6$

$x \le -14$

**29.** $\frac{4}{5}x > 8$

$x > \frac{5}{4}(8)$

$x > 10$

**31.** $-\frac{3}{4}x > -3$

$\frac{3}{4}x < 3$

$x < 4$

**33.** $4x < 12$

$x < 3$

**35.** $-11x \le -22$

$11x \ge 22$

$x \ge 2$

**37.** $x - 3(x + 1) \ge 7$

$x - 3x - 3 \ge 7$

$-2x \ge 10$

$x \le -5$

**39.** $7x - 12 < 4x + 6$

$3x < 18$

$x < 6$

**41.** $\frac{3}{4}x - 6 \le x - 7$

$1 \le \frac{1}{4}x$

$4 \le x$

$x \ge 4$

**43.** $3.6x + 11 \ge -3.4$

$3.6x \ge -14.4$

$x \ge \frac{-14.4}{3.6}$

$x \ge -4$

# A P P E N D I X   E
# Systems of Inequalities

## Appendix E.1    Solving Systems of Inequalities

- ■   You should be able to sketch the graph of an inequality in two variables:

  (a)  Replace the inequality with an equal sign and graph the equation. Use a dashed line for $<$ or $>$, a solid line for $\le$ or $\ge$.

  (b)  Test a point in each region formed by the graph. If the point satisfies the inequality, shade the whole region.

### Solutions to Odd-Numbered Exercises

**1.** $x < 2$

Vertical boundary

Matches graph (g).

**3.** $2x + 3y \ge 6$

$y \ge -\frac{2}{3}x + 2$

Line with negative slope

Matches (a).

**5.** $x^2 + y^2 < 9$

Circular boundary

Matches (e).

**7.** $xy > 1$ or $y > \dfrac{1}{x}$

Matches (f).

**9.** $y < 2 - x^2$

Graph the parabola $y = 2 - x^2$. The region lies below the parabola.

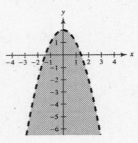

**11.** $x \geq 4$

Using a solid line, graph the vertical line $x = 4$ and shade to the right of this line.

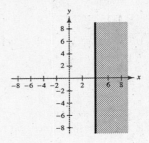

**13.** $y \geq -1$

Using a solid line, graph the horizontal line $y = -1$ and shade above this line.

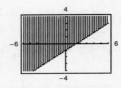

**15.** $2y - x \geq 4$

Using a solid line, graph $2y - x = 4$, and then shade above the line. (Use $(0, 0)$ as a test point.)

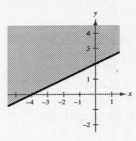

**17.** $y > 3x^2 + 1$

Sketch the parabola $y = 3x^2 + 1$. The region lies above the parabola.

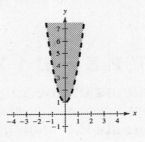

**19.** $(x + 1)^2 + y^2 < 9$

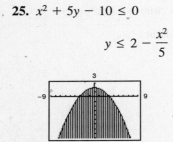

**21.** $y \geq \frac{2}{3}x - 1$

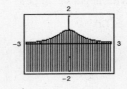

**23.** $y < -3.8x + 1.1$

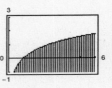

**25.** $x^2 + 5y - 10 \leq 0$

$$y \leq 2 - \dfrac{x^2}{5}$$

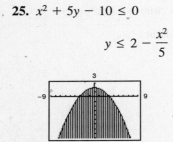

**27.** $y \leq \dfrac{1}{1 + x^2}$

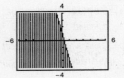

**29.** $y < \ln x$

Using a dashed line, graph $y = \ln x$, and shade to the right of the curve. (Use $(2, 0)$ as a test point.)

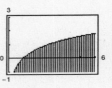

**31.** $y > 3^{-x-4}$

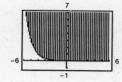

**33.** The line through $(0, 2)$ and $(3, 0)$ is $y = -\frac{2}{3}x + 2$. For the shaded region above the line, we have:

$$y > -\frac{2}{3}x + 2$$

$$3y > -2x + 6$$

$$2x + 3y > 6$$

$$\frac{x}{3} + \frac{y}{2} > 1$$

**35.** The circle shown is $x^2 + y^2 = 9$. For the shaded region inside the circle, we have $x^2 + y^2 \leq 9$.

**37.** (a) $(0, 2)$ is a solution:  $-2(0) + 5(2) \geq 3$

$$2 < 4$$

$$-4(0) + 2(2) < 7$$

(b) $(-6, 4)$ is not a solution:  $4 \not< 4$

(c) $(-8, -2)$ is not a solution:
$-4(-8) + 2(-2) \not< 7$

(d) $(-3, 2)$ is not a solution:  $-4(-3) + 2(2) \not< 7$

**39.** $\begin{cases} x + y \leq 1 \\ -x + y \leq 1 \\ \qquad y \geq 0 \end{cases}$

First, find the points of intersection of each pair of equations.

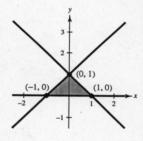

| **Vertex A** | **Vertex B** | **Vertex C** |
|---|---|---|
| $\begin{cases} x + y = 1 \\ -x + y = 1 \end{cases}$ | $\begin{cases} x + y = 1 \\ \qquad y = 0 \end{cases}$ | $\begin{cases} -x + y = 1 \\ \qquad y = 0 \end{cases}$ |
| $(0, 1)$ | $(1, 0)$ | $(-1, 0)$ |

**41.** $\begin{cases} -3x + 2y < 6 \\ \quad x - 4y > -2 \\ \quad 2x + y < 3 \end{cases}$

First, find the points of intersection of each pair of equations.

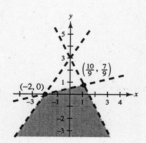

| **Vertex A** | **Vertex B** | **Vertex C** |
|---|---|---|
| $\begin{cases} -3x + 2y = \phantom{-}6 \\ \quad x - 4y = -2 \end{cases}$ | $\begin{cases} -3x + 2y = 6 \\ \quad 2x + \phantom{0}y = 3 \end{cases}$ | $\begin{cases} x - 4y = -2 \\ 2x + \phantom{0}y = \phantom{-}3 \end{cases}$ |
| $(-2, 0)$ | $(0, 3)$ | $\left(\frac{10}{9}, \frac{7}{9}\right)$ |

**43.** $3x + y \leq y^2$

$x - y > 0$

The curves given by $3x + y = y^2$ and $x - y = 0$ intersect as follows:

$3x + x = x^2$

$4x = x^2$

$x = 0, 4$

Intersection points: $(0, 0), (4, 4)$

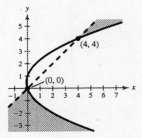

**45.** $2x + y < 2 \implies y < 2 - 2x$

$x + 3y > 2 \implies y > \frac{1}{3}(2 - x)$

$2 - 2x = \frac{1}{3}(2 - x)$

$6 - 6x = (2 - x)$

$4 = 5x$

$x = \frac{4}{5}$

Intersection: $\left(\frac{4}{5}, \frac{2}{5}\right)$

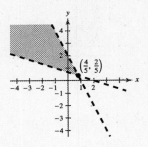

**47.** $\begin{cases} x < y^2 \\ x > y + 2 \end{cases}$

Points of intersection:

$y^2 = y + 2$

$y^2 - y - 2 = 0$

$(y + 1)(y - 2) = 0$

$y = -1, 2$

$(1, -1), (4, 2)$

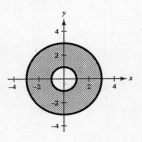

**49.** $\begin{cases} x^2 + y^2 \leq 9 \\ x^2 + y^2 \geq 1 \end{cases}$

There are no points of intersection. The region in common to both inequalities is the region between the circles.

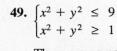

**51.** $\begin{cases} y \leq \sqrt{3x} + 1 \\ y \geq x^2 + 1 \end{cases}$

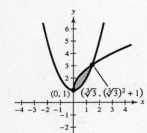

**53.** $\begin{cases} y < x^3 - 2x + 1 \\ y > -2x \\ x \leq 1 \end{cases}$

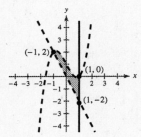

**55.** $\begin{cases} x^2 y \geq 1 \\ 0 < x \leq 4 \\ y \leq 4 \end{cases}$

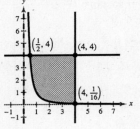

**57.** $\begin{cases} y < -x + 4 \implies \dfrac{x}{4} + \dfrac{y}{4} < 1 \\ x \geq 0 \qquad\qquad\quad x \geq 0 \\ y \geq 0 \qquad\qquad\quad y \geq 0 \end{cases}$

**59.** $(0, 4), (4, 0)$ Line: $y \leq 4 - x$

$(0, 2), (8, 0)$ Line: $y \leq -\frac{1}{4}x + 2$

$x \geq 0, \ y \geq 0$

**61.** $\begin{cases} x \geq 2 \\ x \leq 5 \\ y \geq 1 \\ y \leq 7 \end{cases}$

Thus,
$2 \leq x \leq 5, 1 \leq y \leq 7.$

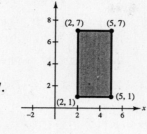

**63.** $(0, 0), (5, 0)$

Line: $y \geq 0$

$(0, 0), (2, 3)$

Line: $y \leq \frac{3}{2}x$

$(2, 3), (5, 0)$

Line: $y \leq -x + 5$

**65.**    Demand = Supply

$50 - 0.5x = 0.125x$

$50 = 0.625x$

$x = 80$

$p = 10$

Point of equilibrium: $(80, 10)$

Consumer surplus $= \frac{1}{2}(40)(80) = 1600$

Producer surplus $= \frac{1}{2}(10)(80) = 400$

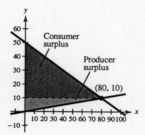

**67.**        Demand = Supply

$300 - 0.0002x = 225 + 0.0005x$

$75 = 0.0007x$

$x = \dfrac{75}{0.0007} = \dfrac{750,000}{7}$

Equilibrium point: $\left(\dfrac{750,000}{7}, \dfrac{1950}{7}\right) \approx (107,142.86, 278.57)$

Consumer surplus: $\dfrac{(107,142.86)(300 - 278.57)}{2} \approx 1,148,036$

Producer surplus: $\dfrac{(107,142.86)(278.57 - 225)}{2} \approx 2,869,822$

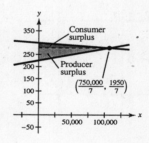

**69.** $x + y \leq 30,000$

$x \geq 7500$

$y \geq 7500$

$x \geq 2y$

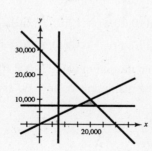

**71. (a)** Let $x$ = number of ounces of food X

Let $y$ = number of ounces of food Y

Calcium: $20x + 10y \geq 280$

Iron: $15x + 10y \geq 160$

Vitamin B: $10x + 20y \geq 180$

$$x \geq 0$$

$$y \geq 0$$

**(b)**

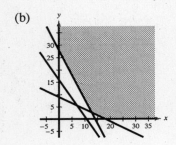

**73. (a)**

| | |
|---|---|
| $xy \geq 500$ | Body-building space |
| $2x + \pi y \geq 125$ | Track (Two semi-circles and two lengths) |
| $x \geq 0$ | Physical constraint |
| $y \geq 0$ | Physical constraint |

**(b)**

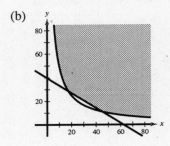

**75.** Area $= 9 \cdot 11 = 99$ square units

True

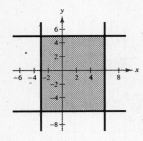

**77.** Test a point on either side of the boundary.

# Appendix E.2   Linear Programming

**Solutions to Odd-Numbered Exercises**

**1.** $z = 3x + 5y$

At $(0, 6)$: $z = 3(0) + 5(6) = 30$

At $(0, 0)$: $z = 3(0) + 5(0) = 0$

At $(6, 0)$: $z = 3(6) + 5(0) = 18$

The minimum value is 0 at $(0, 0)$.

The maximum value is 30 at $(0, 6)$.

**3.** $z = 10x + 7y$

At $(0, 6)$: $z = 10(0) + 7(6) = 42$

At $(0, 0)$: $z = 10(0) + 7(0) = 0$

At $(6, 0)$: $z = 10(6) + 7(0) = 60$

The minimum value is 0 at $(0, 0)$.

The maximum value is 60 at $(6, 0)$.

**5.** $z = 3x + 2y$

$x + 3y = 15 \Rightarrow y = \frac{1}{3}(15 - x)$

$4x + y = 16 \Rightarrow y = (16 - 4x)$

$\frac{1}{3}(15 - x) = 16 - 4x$

$(15 - x) = 48 - 12x$

$11x = 33$

$x = 3$

$y = 4$

At $(0, 0)$: $z = 0$

At $(0, 5)$: $z = 10$

At $(4, 0)$: $z = 12$

At $(3, 4)$: $z = 17$

Minimum at $(0, 0)$ is 0.

Maximum at $(3, 4)$ is 17.

**7.** $z = 5x + 0.5y$

At $(0, 0)$: $z = 0$

At $(0, 5)$: $z = 2.5$

At $(4, 0)$: $z = 20$

At $(3, 4)$: $z = 17$

Minimum at $(0, 0)$ is 0.

Maximum at $(4, 0)$ is 20.

**9.** $z = 10x + 7y$

At $(0, 45)$:   $z = 10(0)\ \ + 7(45) = 315$

At $(30, 45)$: $z = 10(30) + 7(45) = 615$

At $(60, 20)$: $z = 10(60) + 7(20) = 740$

At $(60, 0)$:   $z = 10(60) + 7(0)\ \ = 600$

At $(0, 0)$:     $z = 10(0)\ \ + 7(0)\ \ =\ \ 0$

The minimum value is 0 at $(0, 0)$.

The maximum value is 740 at $(60, 20)$.

**11.** $z = 25x + 30y$

At $(0, 45)$:   $z = 25(0)\ \ + 30(45) = 1350$

At $(30, 45)$: $z = 25(30) + 30(45) = 2100$

At $(60, 20)$: $z = 25(60) + 30(20) = 2100$

At $(60, 0)$:   $z = 25(60) + 30(0)\ \ = 1500$

At $(0, 0)$:     $z = 25(0)\ \ + 30(0)\ \ =\ \ 0$

The minimum value is 0 at $(0, 0)$.

The maximum value is 2100 at any point along the line segment connecting $(30, 45)$ and $(60, 20)$.

**13.** $z = 6x + 10y$

At $(0, 2)$: $z = 6(0) + 10(2) = 20$

At $(5, 0)$: $z = 6(5) + 10(0) = 30$

At $(0, 0)$: $z = 6(0) + 10(0) =\ \ 0$

The minimum value is 0 at $(0, 0)$.

The maximum value is 30 at $(5, 0)$.

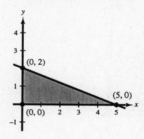

**15.** $z = 3x + 4y$

At $(0, 0)$: $z = 0$

At $(7, 0)$: $z = 21$

At $(0, 10)$: $z = 40$

At $(5, 8)$: $z = 47$

Minimum at $(0, 0)$ is 0.

Maximum at $(5, 8)$ is 47.

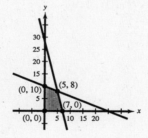

**17.** $z = 4x + y$

At $(36, 0)$: $z = 4(36) + 0 = 144$

At $(40, 0)$: $z = 4(40) + 0 = 160$

At $(24, 8)$: $z = 4(24) + 8 = 104$

The minimum value is 104 at $(24, 8)$.

The maximum value is 160 at $(40, 0)$.

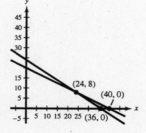

**19.** $z = x + 4y$

At $(36, 0)$: $z = 36 + 4(0) = 36$

At $(40, 0)$: $z = 40 + 4(0) = 40$

At $(24, 8)$: $z = 24 + 4(8) = 56$

The minimum value is 36 at $(36, 0)$.

The maximum value is 56 at $(24, 8)$.

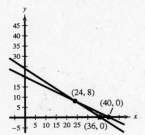

**21.** $z = 2x + 3y$

At $(36, 0)$: $z = 2(36) + 3(0) = 72$

At $(40, 0)$: $z = 2(40) + 3(0) = 80$

At $(24, 8)$: $z = 2(24) + 3(8) = 72$

Minimum at any point on the line segment joining $(36, 0)$ and $(24, 8)$: 72.

Maximum at $(40, 0)$: 80.

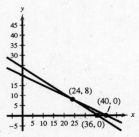

**23.** $z = 2x + y$

(a), (b)

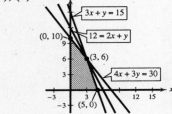

(c) At $(0, 10)$: $z = 2(0) + (10) = 10$

At $(3, 6)$: $z = 2(3) + (6) = 12$

At $(5, 0)$: $z = 2(5) + (0) = 10$

At $(0, 0)$: $z = 2(0) + (0) = 0$

The maximum value is 12 at $(3, 6)$.

**25.** $z = x + y$

(a), (b)

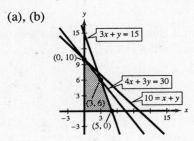

(c) At $(0, 10)$: $z = (0) + (10) = 10$

At $(3, 6)$: $z = (3) + (6) = 9$

At $(5, 0)$: $z = (5) + (0) = 5$

At $(0, 0)$: $z = (0) + (0) = 0$

The maximum value is 10 at $(0, 10)$.

**27.** $-x + y \le 1 \implies y \le x + 1$

$-x + 2y \le 4 \implies y \le \frac{1}{2}x + 2$

Intersection: $(2, 3)$

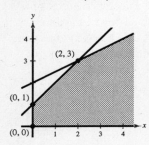

The constraints do not form a closed set of points. Therefore, $z = x + y$ is unbounded.

**29.** $-x + y \le 0 \implies y \le x$

$-3x + y \ge 3 \implies y \ge 3x + 3$

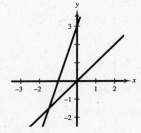

The feasible set is empty.

**31.** Let $x$ = number of audits

Let $y$ = number of tax returns

Constraints:  $100x + 12.5y \leq 800$

$$8x + 2y \leq 96$$

$$x \geq 0$$

$$y \geq 0$$

Objective function: $R = 2000x + 300y$

Vertices of feasible region:  $(0, 0), (8, 0), (0, 48), (4, 32)$

At $(0, 0)$:  $R = 0$

At $(8, 0)$:  $R = 16,000$

At $(0, 48)$:  $R = 14,400$

At $(4, 32)$:  $R = 17,600$

4 audits, 32 tax returns yields maximum revenue of $17,600.

**33.** $x$ = number of bags of Brand X
$y$ = number of bags of Brand Y

Constraints:   $2x + y \geq 12$

$$2x + 9y \geq 36$$

$$2x + 3y \geq 24$$

$$x \geq 0$$

$$y \geq 0$$

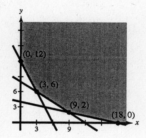

Objective function:  $C = 25x + 20y$

Vertices:  $(0, 12), (3, 6), (9, 2), (18, 0)$

At $(0, 12)$:  $C = 25(0) + 20(12) = 240$

At $(3, 6)$:   $C = 25(3) + 20(6) = 195$

At $(9, 2)$:   $C = 25(9) + 20(2) = 265$

At $(18, 0)$:  $C = 25(18) + 20(0) = 450$

To minimize cost, use three bags of Brand X and six bags of Brand Y for a total cost of $195.

**35.** True, the maximum value is attained at all points in the segment joining these two vertices.

**37.** There are an infinite number of objective functions that would have a maximum at $(0, 4)$. One such objective function is $z = x + 5y$.

**39.** There are an infinite number of objective functions that would have a maximum at $(5, 0)$. One such objective function is $z = 4x + y$.

**41.** Constraints: $x \geq 0, y \geq 0, x + 3y \leq 15, 4x + y \leq 16$

| Vertex | Value of $z = 3x + ty$ |
|--------|------------------------|
| $(0, 0)$ | $z = 0$ |
| $(0, 5)$ | $z = 5t$ |
| $(3, 4)$ | $z = 9 + 4t$ |
| $(4, 0)$ | $z = 12$ |

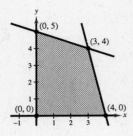

(a) For the maximum value to be at $(0, 5)$, $z = 5t$ must be greater than

$$z = 9 + 4t \quad \text{and} \quad z = 12.$$

$$5t > 9 + 4t \quad \text{and} \quad 5t > 12$$

$$t > 9 \qquad\qquad t > \tfrac{12}{5}$$

Thus, $t > 9$.

(b) For the maximum value to be at $(3, 4)$, $z = 9 + 4t$ must be greater than $z = 5t$ and $z = 12$.

$$9 + 4t > 5t \quad \text{and} \quad 9 + 4t > 12$$

$$9 > t \qquad\qquad t > 3$$

$$\qquad\qquad\qquad t > \tfrac{3}{4}$$

Thus, $\tfrac{3}{4} < t < 9$.

# Chapter 1    Practice Test Solutions

**1.**

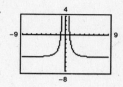

$x$-intercepts: $\pm 0.894$

**2.**

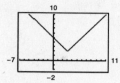

No $x$-intercepts

**3.** $3x - 5y = 15$

Line

$x$-intercept: $(5, 0)$

$y$-intercept: $(0, -3)$

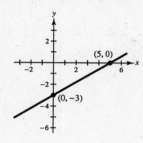

**4.** $y = \sqrt{9 - x}$

Domain: $(-\infty, 9]$

$x$-intercept: $(9, 0)$

$y$-intercept: $(0, 3)$

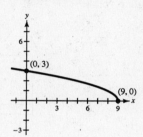

**5.** $5x + 4 = 7x - 8$

$4 + 8 = 7x - 5x$

$12 = 2x$

$x = 6$

**6.**

$$\frac{x}{3} - 5 = \frac{x}{5} + 1$$

$$15\left(\frac{x}{3} - 5\right) = 15\left(\frac{x}{5} + 1\right)$$

$$5x - 75 = 3x + 15$$

$$2x = 90$$

$$x = 45$$

**7.**

$$\frac{3x + 1}{6x - 7} = \frac{2}{5}$$

$$5(3x + 1) = 2(6x - 7)$$

$$15x + 5 = 12x - 14$$

$$3x = -19$$

$$x = -\frac{19}{3}$$

**8.**

$$(x - 3)^2 + 4 = (x + 1)^2$$

$$x^2 - 6x + 9 + 4 = x^2 + 2x + 1$$

$$-8x = -12$$

$$x = \frac{-12}{-8}$$

$$x = \frac{3}{2}$$

**9.** $\text{Slope} = \dfrac{-2 - (-5)}{3 - 4} = \dfrac{3}{-1} = -3$

$y + 2 = -3(x - 3)$

$y + 2 = -3x + 9$

$y + 3x = 7$   or   $y = -3x + 7$

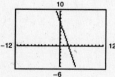

**10.** $y - 5 = -3(x + 1)$

$y - 5 = -3x - 3$

$y + 3x = 2$ or $y = -3x + 2$

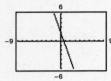

**11.** No, $y$ is not a function of $x$. For example, $(0, 2)$ and $(0, -2)$ both satisfy the equation.

**12.** $f(0) = \dfrac{|0 - 2|}{(0 - 2)} = \dfrac{2}{-2} = -1$

$f(2)$ is not defined.

$f(4) = \dfrac{|4 - 2|}{(4 - 2)} = \dfrac{2}{2} = 1$

**13.** The domain of

$$f(x) = \frac{5}{x^2 - 16}$$

is all $x \neq \pm 4$.

**14.** The domain of $g(t) = \sqrt{4 - t}$ consists of all $t$ satisfying

$$4 - t \geq 0 \text{ or } t \leq 4.$$

**15.**

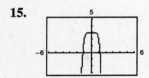

$f(x) = 3 - x^6$ is even.

**16.**

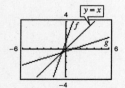

Relative minimum: $(0.577, 3.615)$

Relative maximum: $(-0.577, 4.385)$

**17.** $f(x) = x^3 - 3$ is a vertical shift of 3 units downward of $y = x^3$.

**18.** $f(x) = \sqrt{x - 6}$ is a horizontal shift 6 units to the right of $y = \sqrt{x}$.

**19.** $(g \circ f)(x) = g(f(x))$

$= g(\sqrt{x}) = (\sqrt{x})^2 - 2 = x - 2$

Domain: $x \geq 0$

**20.** $\left(\dfrac{f}{g}\right)(x) = \dfrac{f(x)}{g(x)} = \dfrac{3x^2}{16 - x^4}$

The domain is all $x \neq \pm 2$.

**21.** $(f \circ g)(x) = f\left(\dfrac{x - 1}{3}\right)$

$= 3\left(\dfrac{x - 1}{3}\right) + 1 = (x - 1) + 1 = x$

$(g \circ f)(x) = g(3x + 1) = \dfrac{(3x + 1) - 1}{3} = \dfrac{3x}{3} = x$

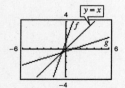

Correction: 

**22.** $y = \sqrt{9 - x^2}, \quad 0 \leq x \leq 3$

$x = \sqrt{9 - y^2}$

$x^2 = 9 - y^2$

$y^2 = 9 - x^2$

$y = \sqrt{9 - x^2}$

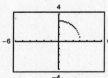

# Chapter 2    Practice Test Solutions

**1.**    $\frac{1}{2}x - \frac{1}{3}(x - 1) = 10$

$3x - 2(x - 1) = 60$

$3x - 2x + 2 = 60$

$x = 58$

**2.**    $(x + 1)^2 - 6 = x^2 + 3x$

$x^2 + 2x + 1 - 6 = x^2 + 3x$

$2x - 5 = 3x$

$x = -5$

**3.**    $A = \dfrac{1}{2}(a + b)h$

$2A = ah + bh$

$2A - bh = ah$

$\dfrac{2A - bh}{h} = a$

**4.** Percent $= \dfrac{301}{4300} = 0.07 = 7\%$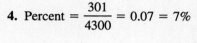

**5.** Let $x$ = number of quarters. Then $53 - x$ = number of nickels.

$25x + 5(53 - x) = 605$

$20x + 265 = 605$

$20x = 340$

$x = 17$ quarters

$53 - x = 36$ nickels

**6.** Let $x$ = amount in $9\frac{1}{2}\%$ fund. Then $15{,}000 - x$ = amount in 11% fund.

$0.095x + 0.11(15{,}000 - x) = 1582.50$

$-0.015x + 1650 = 1582.50$

$-0.015x = -67.5$

$x = \$4500$ at $9\frac{1}{2}\%$

$15{,}000 - x = \$10{,}500$ at 11%

**7.** 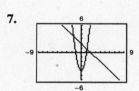    (1.257, 0.743),
(−1.591, 3.591)

**8.** 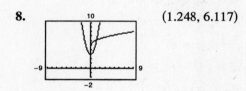    (1.248, 6.117)

**9.** $\dfrac{2}{1 + i} = \dfrac{2}{1 + i} \cdot \dfrac{1 - i}{1 - i} = \dfrac{2 - 2i}{1 + 1} = 1 - i$

**10.** $\dfrac{3 + i}{2} - \dfrac{i + 1}{4} = \dfrac{6 + 2i - i - 1}{4} = \dfrac{5}{4} + \dfrac{i}{4}$

**11.**    $28 + 5x - 3x^2 = 0$

$(4 - x)(7 + 3x) = 0$

$4 - x = 0 \Rightarrow x = 4$

$7 + 3x = 0 \Rightarrow x = -\frac{7}{3}$

**12.** $(x - 2)^2 = 24$

$x - 2 = \pm\sqrt{24}$

$x - 2 = \pm 2\sqrt{6}$

$x = 2 \pm 2\sqrt{6}$

**13.**    $x^2 - 4x - 9 = 0$

$x^2 - 4x + 2^2 = 9 + 2^2$

$(x - 2)^2 = 13$

$x - 2 = \pm\sqrt{13}$

$x = 2 \pm \sqrt{13}$

**14.** $x^2 + 5x - 1 = 0$

$a = 1,\ b = 5,\ c = -1$

$x = \dfrac{-5 \pm \sqrt{(5)^2 - 4(1)(-1)}}{2(1)} = \dfrac{-5 \pm \sqrt{25 + 4}}{2} = \dfrac{-5 \pm \sqrt{29}}{2}$

**15.** $3x^2 - 2x + 4 = 0$

$a = 3, \ b = -2, \ c = 4$

$x = \dfrac{-(-2) \pm \sqrt{(-2)^2 - 4(3)(4)}}{2(3)}$

$= \dfrac{2 \pm \sqrt{4 - 48}}{6}$

$= \dfrac{2 \pm \sqrt{-44}}{6}$

$= \dfrac{2 \pm 2i\sqrt{11}}{6}$

$= \dfrac{1 \pm i\sqrt{11}}{3} = \dfrac{1}{3} \pm \dfrac{\sqrt{11}}{3}i$

**16.**

$60,000 = xy$

$y = \dfrac{60,000}{x}$

$2x + 2y = 1100$

$2x + 2\left(\dfrac{60,000}{x}\right) = 1100$

$x + \dfrac{60,000}{x} = 550$

$x^2 + 60,000 = 550x$

$x^2 - 550x + 60,000 = 0$

$(x - 150)(x - 400) = 0$

$x = 150 \quad \text{or} \quad x = 400$

$y = 400 \qquad y = 150$

Length: 400 feet

Width: 150 feet

**17.**

$x(x + 2) = 624$

$x^2 + 2x - 624 = 0$

$(x - 24)(x + 26) = 0$

$x = 24 \quad \text{or} \quad x = -26, \text{(extraneous solution)}$

$x + 2 = 26$

**18.** $x^3 - 10x^2 + 24x = 0$

$x(x^2 - 10x + 24) = 0$

$x(x - 4)(x - 6) = 0$

$x = 0, \ x = 4, \ x = 6$

**19.** $\sqrt[3]{6 - x} = 4$

$6 - x = 64$

$-x = 58$

$x = -58$

**20.** $(x^2 - 8)^{2/5} = 4$

$x^2 - 8 = \pm 4^{5/2}$

$x^2 - 8 = 32 \quad \text{or} \quad x^2 - 8 = -32$

$x^2 = 40 \qquad\qquad x^2 = -24$

$x = \pm\sqrt{40} \qquad x = \pm\sqrt{-24}$

$x = \pm 2\sqrt{10} \qquad x = \pm 2\sqrt{6}\,i$

**21.** $x^4 - x^2 - 12 = 0$

$(x^2 - 4)(x^2 + 3) = 0$

$x^2 = 4 \quad \text{or} \quad x^2 = -3$

$x^2 = \pm 2 \qquad x = \pm\sqrt{3}\,i$

**22.** $4 - 3x > 16$

$-3x > 12$

$x < -4$

**23.**
$$\left| \frac{x-3}{2} \right| < 5$$

$$-5 < \frac{x-3}{2} < 5$$

$$-10 < x - 3 < 10$$

$$-7 < x < 13$$

**24.**
$$\frac{x+1}{x-3} < 2$$

$$\frac{x+1}{x-3} - 2 < 0$$

$$\frac{x + 1 - 2(x - 3)}{x - 3} < 0$$

$$\frac{7 - x}{x - 3} < 0$$

Critical numbers: $x = 7$ and $x = 3$

Test intervals: $(-\infty, 3), (3, 7), (7, \infty)$

Solution intervals: $(-\infty, 3) \cup (7, \infty)$

**25.** $|3x - 4| \geq 9$

$$3x - 4 \leq -9 \quad \text{or} \quad 3x - 4 \geq 9$$

$$3x \leq -5 \qquad\qquad 3x \geq 13$$

$$x \leq -\frac{5}{3} \qquad\qquad x \geq \frac{13}{3}$$

**26.** $y = 0.882 + 0.912x$

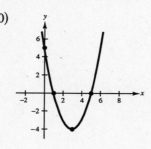

## Chapter 3    Practice Test Solutions

**1.** $x$-intercepts: $(1, 0), (5, 0)$

$y$-intercept: $(0, 5)$

Vertex: $(3, -4)$

**2.** $a = 0.01, b = -90$

$$\frac{-b}{2a} = \frac{90}{2(.01)} = 4500 \text{ units}$$

**3.** Vertex: $(1, 7)$

Opening downward through $(2, 5)$

$y = a(x - 1)^2 + 7$ Standard form

$5 = a(2 - 1)^2 + 7$

$5 = a + 7$

$a = -2$

$y = -2(x - 1)^2 + 7$

$\quad = -2(x^2 - 2x + 1) + 7$

$\quad = -2x^2 + 4x + 5$

**4.** $y = \pm a(x - 2)(3x - 4)$ where $a$ is any real number.

$y = \pm(3x^2 - 10x + 8)$

**5.** Leading coefficient: $-3$

Degree: $5$

Moves down to the right and up to the left.

**6.** $0 = x^5 - 5x^3 + 4x$

$\quad = x(x^4 - 5x^2 + 4)$

$\quad = x(x^2 - 1)(x^2 - 4)$

$\quad = x(x + 1)(x - 1)(x + 2)(x - 2)$

$x = 0, x = \pm 1, x = \pm 2$

**7.** $f(x) = x(x - 3)(x + 2)$

$\quad = x(x^2 - x - 6)$

$\quad = x^3 - x^2 - 6x$

**8.** Intercepts: $(0, 0), \left(\pm 2\sqrt{3}, 0\right)$

Moves up to the right.

Moves down to the left.

| $x$ | $-2$ | $-1$ | $0$ | $1$ | $2$ |
|---|---|---|---|---|---|
| $y$ | $16$ | $11$ | $0$ | $-11$ | $-16$ |

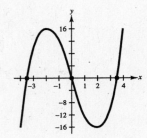

**9.**

$$3x^3 + 9x^2 + 20x + 62 + \frac{176}{x - 3}$$

$$x - 3 \overline{) 3x^4 + 0x^3 - 7x^2 + 2x - 10}$$

$$\underline{3x^4 - 9x^3}$$

$$9x^3 - 7x^2$$

$$\underline{9x^3 - 27x^2}$$

$$20x^2 + 2x$$

$$\underline{20x^2 - 60x}$$

$$62x - 10$$

$$\underline{62x - 186}$$

$$176$$

**10.**

$$x - 2 + \frac{5x - 13}{x^2 + 2x - 1}$$

$$x^2 + 2x - 1 \overline{) x^3 + 0x^2 + 0x - 11}$$

$$\underline{x^3 + 2x^2 - x}$$

$$-2x^2 + x - 11$$

$$\underline{-2x^2 - 4x + 2}$$

$$5x - 13$$

**11.** $-5 \begin{array}{|rrrrrr} 3 & 13 & 0 & 0 & 12 & -1 \\ & -15 & 10 & -50 & 250 & -1310 \\ \hline 3 & -2 & 10 & -50 & 262 & -1311 \end{array}$

$$\frac{3x^5 + 13x^4 + 12x - 1}{x + 5} = 3x^4 - 2x^3 + 10x^2 - 50x + 262 - \frac{1311}{x + 5}$$

**12.** $-6 \begin{array}{|rrrr} 7 & 40 & -12 & 15 \\ & -42 & 12 & 0 \\ \hline 7 & -2 & 0 & 15 \end{array}$

$f(-6) = 15$

**13.** $0 = x^3 - 19x - 30$

Possible rational roots:

$\quad \pm 1, \pm 2, \pm 3, \pm 5, \pm 6, \pm 10, \pm 15, \pm 30$

$-2 \begin{array}{|rrrr} 1 & 0 & -19 & -30 \\ & -2 & 4 & 30 \\ \hline 1 & -2 & -15 & 0 \end{array}$

$-2$ is a zero.

$0 = (x + 2)(x^2 - 2x - 15)$

$0 = (x + 2)(x + 3)(x - 5)$

Zeros: $x = -2, x = -3, x = 5$

**14.** $0 = x^4 + x^3 - 8x^2 - 9x - 9$

Possible rational roots: $\pm 1, \pm 3, \pm 9$

$$
\begin{array}{r|rrrr}
3 & 1 & 1 & -8 & -9 & -9 \\
  &   & 3 & 12 & 12 & 9 \\
\hline
  & 1 & 4 & 4 & 3 & 0
\end{array}
$$

$x = 3$ is a zero.

$0 = (x - 3)(x^3 + 4x^2 + 4x + 3)$

Possible rational roots of $x^3 + 4x^2 + 4x + 3$: $\pm 1, \pm 3$

$$
\begin{array}{r|rrrr}
-3 & 1 & 4 & 4 & 3 \\
   &   & -3 & -3 & -3 \\
\hline
   & 1 & 1 & 1 & 0
\end{array}
$$

$x = -3$ is a zero.

$0 = (x - 3)(x + 3)(x^2 + x + 1)$

The zeros of $x^2 + x + 1$ are $x = \dfrac{-1 \pm \sqrt{3}i}{2}$.

Zeros: $x = 3, x = -3, x = -\dfrac{1}{2} + \dfrac{\sqrt{3}}{2}i, x = -\dfrac{1}{2} - \dfrac{\sqrt{3}}{2}i$

**15.** $0 = 6x^3 - 5x^2 + 4x - 15$

Possible rational roots: $\pm 1, \pm 3, \pm 5, \pm 15, \pm\frac{1}{2}, \pm\frac{3}{2}, \pm\frac{5}{2}, \pm\frac{15}{2}, \pm\frac{1}{3}, \pm\frac{5}{3}, \pm\frac{1}{6}, \pm\frac{5}{6}$

**16.** $0 = x^3 - \frac{20}{3}x^2 + 9x - \frac{10}{3}$

$0 = 3x^3 - 20x^2 + 27x - 10$

Possible rational roots:

$\pm 1, \pm 2, \pm 5, \pm 10, \pm\frac{1}{3}, \pm\frac{2}{3}, \pm\frac{5}{3}, \pm\frac{10}{3}$

$$
\begin{array}{r|rrrr}
1 & 3 & -20 & 27 & -10 \\
  &   & 3 & -17 & 10 \\
\hline
  & 3 & -17 & 10 & 0
\end{array}
$$

$x = 1$ is a zero.

$0 = (x - 1)(3x^2 - 17x + 10)$

$0 = (x - 1)(3x - 2)(x - 5)$

Zeros: $x = 1, x = \frac{2}{3}, x = 5$

**17.** $f(x) = x^4 + x^3 + 3x^2 + 5x - 10$

Possible rational roots: $\pm 1, \pm 2, \pm 5, \pm 10$

$$
\begin{array}{r|rrrrr}
1 & 1 & 1 & 3 & 5 & -10 \\
  &   & 1 & 2 & 5 & 10 \\
\hline
  & 1 & 2 & 5 & 10 & 0
\end{array}
$$

$x = 1$ is a zero.

$$
\begin{array}{r|rrrr}
-2 & 1 & 2 & 5 & 10 \\
   &   & -2 & 0 & -10 \\
\hline
   & 1 & 0 & 5 & 0
\end{array}
$$

$x = -2$ is a zero.

$f(x) = (x - 1)(x + 2)(x^2 + 5)$

$\quad = (x - 1)(x + 2)(x + 5i)(x - 5i)$

**18.** $f(x) = (x - 2)[x - (3 + i)][x - (3 - i)][x - (3 - 2i)][x - (3 + 2i)]$

$\quad = (x - 2)[(x - 3)^2 + 1][(x - 3)^2 + 4]$

$\quad = (x - 2)(x^2 - 6x + 10)(x^2 - 6x + 13)$

$\quad = x^5 - 14x^4 + 83x^3 - 256x^2 + 406x - 260$

**19.** $3i$

| | 1 | 4 | 9 | 36 |
|---|---|---|---|---|
| | | $3i$ | $12i - 9$ | $-36$ |
| | 1 | $4 + 3i$ | $12i$ | 0 |

**20.** $z = \dfrac{kx^2}{\sqrt{y}}$

**21.** $f(x) = \dfrac{x - 1}{2x}$

Vertical asymptote: $x = 0$

Horizontal asymptote: $y = \dfrac{1}{2}$

$x$-intercept: $(1, 0)$

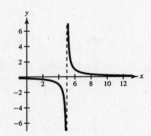

**22.** $f(x) = \dfrac{3x^2 - 4}{x}$

Vertical asymptote: $x = 0$

Slant asymptote: $y = 3x$

$x$-intercepts: $\left(\pm\dfrac{2}{\sqrt{3}}, 0\right)$

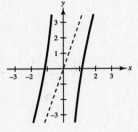

**23.** $y = 8$ is a horizontal asymptote since the degree of the numerator equals the degree of the denominator. There are no vertical asymptotes.

**24.** $x = 1$ is a vertical asymptote.

$$\dfrac{4x^2 - 2x + 7}{x - 1} = 4x + 2 + \dfrac{9}{x - 1}$$

so $y = 4x + 2$ is a slant asymptote.

**25.** $f(x) = \dfrac{x - 5}{(x - 5)^2} = \dfrac{1}{x - 5}$

Vertical asymptote: $x = 5$

Horizontal asymptote: $y = 0$

y-intercept: $\left(0, -\dfrac{1}{5}\right)$

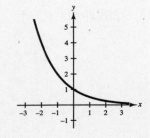

# Chapter 4    Practice Test Solutions

**1.** $x^{3/5} = 8$

$x = 8^{5/3}$

$\quad = \left(\sqrt[3]{8}\right)^5 = 2^5 = 32$

**2.** $3^{x-1} = \frac{1}{81}$

$3^{x-1} = 3^{-4}$

$x - 1 = -4$

$x = -3$

**3.** $f(x) = 2^{-x} = \left(\frac{1}{2}\right)^x$

| $x$ | $-2$ | $-1$ | 0 | 1 | 2 |
|---|---|---|---|---|---|
| $f(x)$ | 4 | 2 | 1 | $\frac{1}{2}$ | $\frac{1}{4}$ |

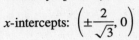

**4.** $g(x) = e^x + 1$

| $x$ | $-2$ | $-1$ | $0$ | $1$ | $2$ |
|------|------|------|-----|------|------|
| $g(x)$ | 1.14 | 1.37 | 2 | 3.72 | 8.39 |

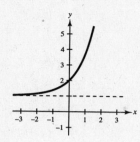

**5.** $A = P\left(1 + \dfrac{r}{n}\right)^{nt}$

(a) $A = 5000\left(1 + \dfrac{0.09}{12}\right)^{12(3)} \approx \$6543.23$

(b) $A = 5000\left(1 + \dfrac{0.09}{4}\right)^{4(3)} \approx \$6530.25$

(c) $A = 5000e^{(0.09)(3)} \approx \$6549.82$

**6.** $7^{-2} = \dfrac{1}{49}$

$\log_7 \dfrac{1}{49} = -2$

**7.** $x - 4 = \log_2 \dfrac{1}{64}$

$2^{x-4} = \dfrac{1}{64}$

$2^{x-4} = 2^{-6}$

$x - 4 = -6$

$x = -2$

**8.** $\log_b \sqrt[4]{\dfrac{8}{25}} = \dfrac{1}{4}\log_b \dfrac{8}{25}$

$= \dfrac{1}{4}[\log_b 8 - \log_b 25]$

$= \dfrac{1}{4}[\log_b 2^3 - \log_b 5^2]$

$= \dfrac{1}{4}[3\log_b 2 - 2\log_b 5]$

$= \dfrac{1}{4}[3(0.3562) - 2(0.8271)]$

$= -0.1464$

**9.** $5\ln x - \dfrac{1}{2}\ln y + 6\ln z = \ln x^5 - \ln \sqrt{y} + \ln z^6 = \ln\left(\dfrac{x^5 z^6}{\sqrt{y}}\right)$

**10.** $\log_9 28 = \dfrac{\log 28}{\log 9} \approx 1.5166$

**11.** $\log_{10} N = 0.6646$

$N = 10^{0.6646} \approx 4.62$

**12.**

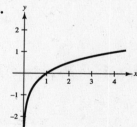

**13.** Domain:

$x^2 - 9 > 0$

$(x + 3)(x - 3) > 0$

$x < -3$ or $x > 3$

**14.**

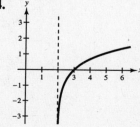

**15.** $\dfrac{\ln x}{\ln y} \neq \ln(x - y)$ since

$\dfrac{\ln x}{\ln y} = \log_y x.$

**16.** $5^x = 41$

$x = \log_5 41$

$= \dfrac{\ln 41}{\ln 5}$

$\approx 2.3074$

**17.** $x - x^2 = \log_5 \frac{1}{25}$

$5^{x-x^2} = \frac{1}{25}$

$5^{x-x^2} = 5^{-2}$

$x - x^2 = -2$

$0 = x^2 - x - 2$

$0 = (x + 1)(x - 2)$

$x = -1 \text{ or } x = 2$

**18.** $\log_2 x + \log_2(x - 3) = 2$

$\log_2[x(x - 3)] = 2$

$x(x - 3) = 2^2$

$x^2 - 3x = 4$

$x^2 - 3x - 4 = 0$

$(x + 1)(x - 4) = 0$

$x = 4$

$x = -1$

(extraneous solution)

**19.**    $\dfrac{e^x + e^{-x}}{3} = 4$

$e^x(e^x + e^{-x}) = 12e^x$

$e^{2x} + 1 = 12e^x$

$e^{2x} - 12e^x + 1 = 0$

$e^x = \dfrac{12 \pm \sqrt{144 - 4}}{2}$

$e^x \approx 11.9161 \quad \text{or} \quad e^x \approx 0.0839$

$x \approx \ln 11.9161 \qquad x \approx \ln 0.0839$

$x \approx 2.4779 \qquad\quad x \approx -2.4779$

**20.**      $A = Pe^{rt}$

$12{,}000 = 6000e^{0.13t}$

$2 = e^{0.13t}$

$\ln 2 = 0.13t$

$\dfrac{\ln 2}{0.13} = t$

$t \approx 5.3319 \text{ yr or 5 yr 4 mo}$

**21.** There are two points of intersection:

$(0.0169, -2.983)$,
$(1.731, 1.647)$

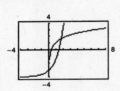

**22.** $y = 1.0597x^{1.9792}$

# Chapter 5    Practice Test Solutions

**1.** $\begin{cases} x + y = 1 \\ 3x - y = 15 \end{cases} \implies y = 3x - 15$

$x + (3x - 15) = 1$

$4x = 16$

$x = 4$

$y = -3$

**2.** $\begin{cases} x - 3y = -3 \implies x = 3y - 3 \\ x^2 + 6y = 5 \end{cases}$

$(3y - 3)^2 + 6y = 5$

$9y^2 - 18y + 9 + 6y = 5$

$9y^2 - 12y + 4 = 0$

$(3y - 2)^2 = 0$

$y = \frac{2}{3}$

$x = -1$

**3.** $\begin{cases} x + y + z = 6 \\ 2x - y + 3z = 0 \\ 5x + 2y - z = -3 \end{cases}$ $\Rightarrow$ $z = 6 - x - y$

$2x - y + 3(6 - x - y) = 0$ $\Rightarrow$ $-x - 4y = -18$

$5x + 2y - (6 - x - y) = -3$ $\Rightarrow$ $6x + 3y = 3$

$$x = 18 - 4y$$

$$6(18 - 4y) + 3y = 3$$

$$-21y = -105$$

$$y = 5$$

$$x = 18 - 4y = -2$$

$$z = 6 - x - y = 3$$

**4.** $\begin{cases} x + y = 110 \\ xy = 2800 \end{cases}$ $\Rightarrow$ $y = 110 - x$

$$x(110 - x) = 2800$$

$$0 = x^2 - 110x + 2800$$

$$0 = (x - 40)(x - 70)$$

$$x = 40 \quad \text{or} \quad x = 70$$

$$y = 70 \qquad y = 40$$

**5.** $\begin{cases} 2x + 2y = 170 \\ xy = 1500 \end{cases}$ $\Rightarrow$ $y = \dfrac{170 - 2x}{2} = 85 - x$

$$x(85 - x) = 1500$$

$$0 = x^2 - 85x + 1500$$

$$0 = (x - 25)(x - 60)$$

$$x = 25 \quad \text{or} \quad x = 60$$

$$y = 60 \qquad y = 25$$

Dimensions: $60' \times 25'$

**6.** $\begin{cases} 2x + 15y = 4 \\ x - 3y = 23 \end{cases}$ $\begin{array}{l} \Rightarrow \quad 2x + 15y = 4 \\ \Rightarrow \quad \underline{5x - 15y = 115} \\ \qquad 7x \qquad = 119 \end{array}$

$$x = 17$$

$$y = \frac{x - 23}{3} = -2$$

**7.** $\begin{cases} x + y = 2 \\ 38x - 19y = 7 \end{cases}$ $\begin{array}{l} \Rightarrow \quad 19x + 19y = 38 \\ \Rightarrow \quad \underline{38x - 19y = 7} \\ \qquad 57x \qquad = 45 \end{array}$

$$x = \frac{45}{57} = \frac{15}{19}$$

$$y = 2 - x$$

$$= \frac{38}{19} - \frac{15}{19}$$

$$= \frac{23}{19}$$

**8.** $y_1 = 2(0.112 - 0.4x)$

$$y_2 = \frac{(0.131 + 0.3x)}{0.7}$$

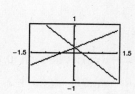

$\begin{cases} 0.4x + 0.5y = 0.112 \\ 0.3x - 0.7y = -0.131 \end{cases}$ $\begin{array}{l} \Rightarrow \quad 0.28x + 0.35y = 0.0784 \\ \Rightarrow \quad \underline{0.15x - 0.35y = -0.0655} \\ \qquad 0.43x \qquad = 0.0129 \end{array}$

$$x = \frac{0.0129}{0.43} = 0.03$$

$$y = (2)(0.112 - 0.4x) = 0.20$$

**9.** Let $x$ = amount in 11% fund and $y$ = amount in 13% fund.

$$\begin{cases} x + y = 17{,}000 \implies y = 17{,}000 - x \\ 0.11x + 0.13y = 2080 \end{cases}$$

$$0.11x + 0.13(17{,}000 - x) = 2080$$

$$-0.02x = -130$$

$$x = \$6500$$

$$y = \$10{,}500$$

**10.** Using a graphing utility, you obtain
$y = 0.7857x - 0.1429$. Analytically, $(4, 3)$,
$(1, 1)$, $(-1, -2)$, $(-2, -1)$.

$$n = 4, \sum_{i=1}^{4} x_i = 2, \sum_{i=1}^{4} y_i = 1, \sum_{i=1}^{4} x_i^2 = 22, \sum_{i=1}^{4} x_i y_i = 17$$

$$4b + 2a = 1 \implies 4b + 2a = 1$$

$$2b + 22a = 17 \implies \underline{-4b - 44a = -34}$$

$$-42a = -33$$

$$a = \tfrac{33}{42} = \tfrac{11}{14}$$

$$b = \tfrac{1}{4}\left(1 - 2\left(\tfrac{33}{42}\right)\right) = -\tfrac{1}{7}$$

$$y = ax + b = \tfrac{11}{14}x - \tfrac{1}{7}$$

**11.**
$$\begin{array}{ll} x + y \phantom{+ z} = -2 & \text{Equation 1} \\ 2x - y + z = 11 & \text{Equation 2} \\ \phantom{2x +} 4y - 3z = -20 & \text{Equation 3} \end{array}$$

$$\begin{cases} x + \phantom{3}y \phantom{+ z} = -2 \\ \phantom{x +} -3y + z = 15 \quad -2\text{Eq.1} + \text{Eq.2} \\ \phantom{x +} 4y - 3z = -20 \end{cases}$$

$$\begin{cases} x + \phantom{3}y \phantom{+ z} = -2 \\ \phantom{x +} -3y + z = 15 \\ \phantom{x +} -5y \phantom{+ z} = 25 \quad 3\text{Eq.2} + \text{Eq.3} \end{cases}$$

*Answer:* $y = -5$
$$x = 3$$
$$z = 0$$

**12.**
$$\begin{array}{ll} 4x - y + 5z = 4 & \text{Equation 1} \\ 2x + y - z = 0 & \text{Equation 2} \\ 2x + 4y + 8z = 0 & \text{Equation 3} \end{array}$$

$$\begin{cases} 4x - \phantom{3}y + 5z = 4 \\ \phantom{4x} -3y + 7z = 4 \quad \text{Eq.1} - 2\text{Eq.2} \\ \phantom{4x +} 3y + 9z = 0 \quad -\text{Eq.2} + \text{Eq.3} \end{cases}$$

$$\begin{cases} 4x - \phantom{3}y + 5z = 4 \\ \phantom{4x} -3y + 7z = 4 \\ \phantom{4x + 3y +} 16z = 4 \quad \text{Eq.2} + \text{Eq.3} \end{cases}$$

*Answer:* $z = \tfrac{1}{4}$
$$y = -\tfrac{3}{4}$$
$$x = \tfrac{1}{2}$$

**13.**
$$\begin{cases} 3x + 2y - z = 5 \implies \phantom{-} 6x + 4y - 2z = 10 \\ 6x - y + 5z = 2 \implies \underline{-6x + y - 5z = -2} \end{cases}$$

$$5y - 7z = 8$$

$$y = \frac{8 + 7z}{5}$$

$$\begin{array}{l} 3x + 2y - \phantom{1}z = 5 \\ \underline{12x - 2y + 10z = 4} \\ 15x \phantom{- 2y} + 9z = 9 \end{array}$$

$$x = \frac{9 - 9z}{15} = \frac{3 - 3z}{5}$$

Let $z = a$, then $x = \dfrac{3 - 3a}{5}$ and $y = \dfrac{8 + 7a}{5}$.

**14.** $y = ax^2 + bx + c$ passes through $(0, -1)$, $(1, 4)$, and $(2, 13)$.

At $(0, -1)$: $-1 = a(0)^2 + b(0) + c \implies c = -1$

At $(1, 4)$: $\phantom{1}4 = a(1)^2 + b(1) - 1 \implies 5 = a + b \implies 5 = a + b$

At $(2, 13)$: $13 = a(2)^2 + b(2) - 1 \implies 14 = 4a + 2b \implies \underline{-7 = -2a - b}$

$$-2 = -a$$

$$a = 2$$

$$b = 3$$

Thus, $y = 2x^2 + 3x - 1$.

**15.** $s = \frac{1}{2}at^2 + v_0t + s_0$ passes through $(1, 12)$, $(2, 5)$, and $(3, 4)$.

At $(1, 12)$: $12 = \frac{1}{2}a + v_0 + s_0 \implies$ $\begin{cases} \frac{1}{2}a + v_0 + s_0 = 12 \\ -a \quad\quad\; + s_0 = 19 \quad 2\text{Eq.1} - \text{Eq.2} \\ -3a \quad\quad + s_0 = 7 \quad\; 3\text{Eq.2} - 2\text{Eq.3} \end{cases}$

At $(2, 5)$: $5 = 2a + 2v_0 + s_0 \implies$

At $(3, 4)$: $4 = \frac{9}{2}a + 3v_0 + s_0 \implies$

$a = 6$

$s_0 = 25$ $\begin{cases} \frac{1}{2}a + v_0 + s_0 = 12 \\ -a \quad\quad\; + s_0 = 19 \\ -2a \quad\quad\quad = -12 \quad -\text{Eq.2} + \text{Eq.3} \end{cases}$

$v_0 = -16$

Thus,

$s = \frac{1}{2}(6)t^2 - 16t + 25 = 3t^2 - 16t + 25.$

**16.** $\begin{bmatrix} 1 & -2 & 4 \\ 3 & -5 & 9 \end{bmatrix}$

$-3R_1 + R_2 \rightarrow \begin{bmatrix} 1 & -2 & 4 \\ 0 & 1 & -3 \end{bmatrix}$

$2R_2 + R_1 \rightarrow \begin{bmatrix} 1 & 0 & -2 \\ 0 & 1 & -3 \end{bmatrix}$

**17.** $3x + 5y = 3$

$\quad\; 2x - y = -11$

$\begin{bmatrix} 3 & 5 & \vdots & 3 \\ 2 & -1 & \vdots & -11 \end{bmatrix}$

$-R_2 + R_1 \rightarrow \begin{bmatrix} 1 & 6 & \vdots & 14 \\ 2 & -1 & \vdots & -11 \end{bmatrix}$

$-2R_1 + R_2 \rightarrow \begin{bmatrix} 1 & 6 & \vdots & 14 \\ 0 & -13 & \vdots & -39 \end{bmatrix}$

$-\frac{1}{13}R_2 \rightarrow \begin{bmatrix} 1 & 6 & \vdots & 14 \\ 0 & 1 & \vdots & 3 \end{bmatrix}$

$-6R_2 + R_1 \rightarrow \begin{bmatrix} 1 & 0 & \vdots & -4 \\ 0 & 1 & \vdots & 3 \end{bmatrix}$

*Answer:* $x = -4, y = 3$

**18.** $\begin{cases} 2x + 3y = -3 \\ 3x + 2y = 8 \\ x + y = 1 \end{cases}$

$\begin{bmatrix} 2 & 3 & \vdots & -3 \\ 3 & 2 & \vdots & 8 \\ 1 & 1 & \vdots & 1 \end{bmatrix}$

$\begin{matrix} R_3 \\ \\ R_1 \end{matrix} \begin{bmatrix} 1 & 1 & \vdots & 1 \\ 3 & 2 & \vdots & 8 \\ 2 & 3 & \vdots & -3 \end{bmatrix}$

$\begin{matrix} -3R_1 + R_2 \rightarrow \\ -2R_1 + R_3 \rightarrow \end{matrix} \begin{bmatrix} 1 & 1 & \vdots & 1 \\ 0 & -1 & \vdots & 5 \\ 0 & 1 & \vdots & -5 \end{bmatrix}$

$\begin{matrix} R_2 + R_1 \rightarrow \\ -R_2 \rightarrow \\ -R_2 + R_3 \rightarrow \end{matrix} \begin{bmatrix} 1 & 0 & \vdots & 6 \\ 0 & 1 & \vdots & -5 \\ 0 & 0 & \vdots & 0 \end{bmatrix}$

*Answer:* $x = 6, y = -5$

**19.** $\begin{cases} x + 3z = -5 \\ 2x + y = 0 \\ 3x + y - z = 3 \end{cases}$

$\begin{bmatrix} 1 & 0 & 3 & \vdots & -5 \\ 2 & 1 & 0 & \vdots & 0 \\ 3 & 1 & -1 & \vdots & 3 \end{bmatrix}$

$\begin{matrix} -2R_1 + R_2 \rightarrow \\ -3R_1 + R_3 \rightarrow \end{matrix} \begin{bmatrix} 1 & 0 & 3 & \vdots & -5 \\ 0 & 1 & -6 & \vdots & 10 \\ 0 & 1 & -10 & \vdots & 18 \end{bmatrix}$

$-R_2 + R_3 \rightarrow \begin{bmatrix} 1 & 0 & 3 & \vdots & -5 \\ 0 & 1 & -6 & \vdots & 10 \\ 0 & 0 & -4 & \vdots & 8 \end{bmatrix}$

$\begin{matrix} -3R_3 + R_1 \rightarrow \\ 6R_3 + R_2 \rightarrow \\ -\frac{1}{4}R_3 \rightarrow \end{matrix} \begin{bmatrix} 1 & 0 & 0 & \vdots & 1 \\ 0 & 1 & 0 & \vdots & -2 \\ 0 & 0 & 1 & \vdots & -2 \end{bmatrix}$

*Answer:* $x = 1, y = -2, z = -2$

**20.** $\begin{bmatrix} 1 & 4 & 5 \\ 2 & 0 & -3 \end{bmatrix} \begin{bmatrix} 1 & 6 \\ 0 & -7 \\ -1 & 2 \end{bmatrix} = \begin{bmatrix} -4 & -12 \\ 5 & 6 \end{bmatrix}$

**21.** $3A - 5B = 3\begin{bmatrix} 9 & 1 \\ -4 & 8 \end{bmatrix} - 5\begin{bmatrix} 6 & -2 \\ 3 & 5 \end{bmatrix}$

$\qquad\qquad = \begin{bmatrix} -3 & 13 \\ -27 & -1 \end{bmatrix}$

**22.** $f(A) = \begin{bmatrix} 3 & 0 \\ 7 & 1 \end{bmatrix}^2 - 7\begin{bmatrix} 3 & 0 \\ 7 & 1 \end{bmatrix} + 8\begin{bmatrix} 1 & 0 \\ 0 & 1 \end{bmatrix}$

$\qquad = \begin{bmatrix} 3 & 0 \\ 7 & 1 \end{bmatrix}\begin{bmatrix} 3 & 0 \\ 7 & 1 \end{bmatrix} - \begin{bmatrix} 21 & 0 \\ 49 & 7 \end{bmatrix} + \begin{bmatrix} 8 & 0 \\ 0 & 8 \end{bmatrix}$

$\qquad = \begin{bmatrix} 9 & 0 \\ 28 & 1 \end{bmatrix} - \begin{bmatrix} 21 & 0 \\ 49 & 7 \end{bmatrix} + \begin{bmatrix} 8 & 0 \\ 0 & 8 \end{bmatrix}$

$\qquad = \begin{bmatrix} -4 & 0 \\ -21 & 2 \end{bmatrix}$

**23.** False.

$\qquad (A + B)(A + 3B) = A(A + 3B) + B(A + 3B)$

$\qquad\qquad\qquad\qquad\quad = A^2 + 3AB + BA + 3B^2$

**24.**

$$\begin{bmatrix} 1 & 2 & \vdots & 1 & 0 \\ 3 & 5 & \vdots & 0 & 1 \end{bmatrix}$$

$-3R_1 + R_2 \rightarrow \begin{bmatrix} 1 & 2 & \vdots & 1 & 0 \\ 0 & -1 & \vdots & -3 & 1 \end{bmatrix}$

$\begin{matrix} 2R_2 + R_1 \rightarrow \\ -R_2 \rightarrow \end{matrix} \begin{bmatrix} 1 & 0 & \vdots & -5 & 2 \\ 0 & 1 & \vdots & 3 & -1 \end{bmatrix}$

$A^{-1} = \begin{bmatrix} -5 & 2 \\ 3 & -1 \end{bmatrix}$

**25.**

$$\begin{bmatrix} 1 & 1 & 1 & \vdots & 1 & 0 & 0 \\ 3 & 6 & 5 & \vdots & 0 & 1 & 0 \\ 6 & 10 & 8 & \vdots & 0 & 0 & 1 \end{bmatrix}$$

$\begin{matrix} -3R_1 + R_2 \rightarrow \\ -6R_1 + R_3 \rightarrow \end{matrix} \begin{bmatrix} 1 & 1 & 1 & \vdots & 1 & 0 & 0 \\ 0 & 3 & 2 & \vdots & -3 & 1 & 0 \\ 0 & 4 & 2 & \vdots & -6 & 0 & 1 \end{bmatrix}$

$\begin{matrix} -\frac{1}{3}R_2 + R_1 \rightarrow \\ \frac{1}{3}R_2 \rightarrow \\ -4R_2 + R_3 \rightarrow \end{matrix} \begin{bmatrix} 1 & 0 & \frac{1}{3} & \vdots & 2 & -\frac{1}{3} & 0 \\ 0 & 1 & \frac{2}{3} & \vdots & -1 & \frac{1}{3} & 0 \\ 0 & 0 & -\frac{2}{3} & \vdots & -2 & -\frac{4}{3} & 1 \end{bmatrix}$

$\begin{matrix} \frac{1}{2}R_3 + R_1 \rightarrow \\ R_3 + R_2 \rightarrow \\ -\frac{3}{2}R_3 \rightarrow \end{matrix} \begin{bmatrix} 1 & 0 & 0 & \vdots & 1 & -1 & \frac{1}{2} \\ 0 & 1 & 0 & \vdots & -3 & -1 & 1 \\ 0 & 0 & 1 & \vdots & 3 & 2 & -\frac{3}{2} \end{bmatrix}$

$A^{-1} = \begin{bmatrix} 1 & -1 & \frac{1}{2} \\ -3 & -1 & 1 \\ 3 & 2 & -\frac{3}{2} \end{bmatrix}$

**26. (a)** $x + 2y = 4$

$\qquad 3x + 5y = 1$

$$\begin{bmatrix} 1 & 2 & \vdots & 1 & 0 \\ 3 & 5 & \vdots & 0 & 1 \end{bmatrix}$$

$-3R_1 + R_2 \rightarrow \begin{bmatrix} 1 & 2 & \vdots & 1 & 0 \\ 0 & -1 & \vdots & -3 & 1 \end{bmatrix}$

$\begin{matrix} -2R_2 + R_1 \rightarrow \\ -R_2 \rightarrow \end{matrix} \begin{bmatrix} 1 & 0 & \vdots & -5 & 2 \\ 0 & 1 & \vdots & 3 & -1 \end{bmatrix}$

$X = A^{-1}B = \begin{bmatrix} -5 & 2 \\ 3 & -1 \end{bmatrix}\begin{bmatrix} 4 \\ 1 \end{bmatrix} = \begin{bmatrix} -18 \\ 11 \end{bmatrix}$

$x = -18, y = 11$

**(b)** $x + 2y = \phantom{-}3$

$\qquad 3x + 5y = -2$

$X = A^{-1}B - \begin{bmatrix} -5 & 2 \\ 3 & -1 \end{bmatrix}\begin{bmatrix} 3 \\ -2 \end{bmatrix} = \begin{bmatrix} -19 \\ 11 \end{bmatrix}$

$x = -19, y = 11$

**27.** $\begin{vmatrix} 6 & -1 \\ 3 & 4 \end{vmatrix} = 24 - (-3) = 27$

**28.** $\begin{vmatrix} 1 & 3 & -1 \\ 5 & 9 & 0 \\ 6 & 2 & -5 \end{vmatrix} = 1(-45) + (-3)(-25) + (-1)(-44)$

$$= 74$$

**29.** $\begin{vmatrix} 1 & 4 & 2 & 3 \\ 0 & 1 & -2 & 0 \\ 3 & 5 & -1 & 1 \\ 2 & 0 & 6 & 1 \end{vmatrix} = -7$

**30.** $\begin{vmatrix} 6 & 4 & 3 & 0 & 6 \\ 0 & 5 & 1 & 4 & 8 \\ 0 & 0 & 2 & 7 & 3 \\ 0 & 0 & 0 & 9 & 2 \\ 0 & 0 & 0 & 0 & 1 \end{vmatrix} = 6(5)(2)(9)(1) = 540$

**31.** Area $= \dfrac{1}{2}\begin{vmatrix} 0 & 7 & 1 \\ 5 & 0 & 1 \\ 3 & 9 & 1 \end{vmatrix} = \dfrac{1}{2}(31)$

$\qquad = 15.5$ square units

**32.** $\begin{vmatrix} x & y & 1 \\ 2 & 7 & 1 \\ -1 & 4 & 1 \end{vmatrix} = 3x - 3y + 15 = 0$

or $\quad x - y + 5 = 0$

**33.** $x = \dfrac{\begin{vmatrix} 4 & -7 \\ 11 & 5 \end{vmatrix}}{\begin{vmatrix} 6 & -7 \\ 2 & 5 \end{vmatrix}} = \dfrac{97}{44}$

**34.** $z = \dfrac{\begin{vmatrix} 3 & 0 & 1 \\ 0 & 1 & 3 \\ 1 & -1 & 2 \end{vmatrix}}{\begin{vmatrix} 3 & 0 & 1 \\ 0 & 1 & 4 \\ 1 & -1 & 0 \end{vmatrix}} = \dfrac{14}{11}$

**35.** $y = \dfrac{\begin{vmatrix} 721.4 & 33.77 \\ 45.9 & 19.85 \end{vmatrix}}{\begin{vmatrix} 721.4 & -29.1 \\ 45.9 & 105.6 \end{vmatrix}}$

$\qquad = \dfrac{12,769.747}{77,515.530} \approx 0.1647$

# Chapter 6    Practice Test Solutions

**1.** $a_n = \dfrac{2n}{(n + 2)!}$

$a_1 = \dfrac{2(1)}{3!} = \dfrac{2}{6} = \dfrac{1}{3}$

$a_2 = \dfrac{2(2)}{4!} = \dfrac{4}{24} = \dfrac{1}{6}$

$a_3 = \dfrac{2(3)}{5!} = \dfrac{6}{120} = \dfrac{1}{20}$

$a_4 = \dfrac{2(4)}{6!} = \dfrac{8}{720} = \dfrac{1}{90}$

$a_5 = \dfrac{2(5)}{7!} = \dfrac{10}{5040} = \dfrac{1}{504}$

Terms: $\dfrac{1}{3}, \dfrac{1}{6}, \dfrac{1}{20}, \dfrac{1}{90}, \dfrac{1}{504}$

**2.** $a_n = \dfrac{n + 3}{3^n}$

**3.** $\displaystyle\sum_{i=1}^{6}(2i - 1) = 1 + 3 + 5 + 7 + 9 + 11 = 36$

**4.** $a_1 = 23$, $d = -2$

$a_2 = a_1 + d = 21$

$a_3 = a_2 + d = 19$

$a_4 = a_3 + d = 17$

$a_5 = a_4 + d = 15$

Terms:  23, 21, 19, 17, 15

**5.** $a_1 = 12$, $d = 3$, $n = 50$

$a_n = a_1 + (n - 1)d$

$a_{50} = 12 + (50 - 1)3 = 159$

**6.** $a_1 = 1$

$a_{200} = 200$

$S_n = \dfrac{n}{2}(a_1 + a_n)$

$S_{200} = \dfrac{200}{2}(1 + 200) = 20{,}100$

**7.** $a_1 = 7$, $r = 2$

$a_2 = a_1 r = 14$

$a_3 = a_1 r^2 = 28$

$a_4 = a_1 r^3 = 56$

$a_5 = a_1 r^4 = 112$

Terms: 7, 14, 28, 56, 112

**8.** $\displaystyle\sum_{n=0}^{9} 6\left(\dfrac{2}{3}\right)^n$, $a_1 = 6$, $r = \dfrac{2}{3}$, $n = 10$

$S_n = \dfrac{a_1(1 - r^n)}{1 - r} = \dfrac{6\left(1 - (2/3)^{10}\right)}{1 - (2/3)} \approx 17.6879$

**9.** $\displaystyle\sum_{n=0}^{\infty} (0.03)^n$, $a_1 = 1$, $r = 0.03$

$S = \dfrac{a_1}{1 - r} = \dfrac{1}{1 - 0.03} = \dfrac{1}{0.97} = \dfrac{100}{97} \approx 1.0309$

**10.** For $n = 1$, $1 = \dfrac{1(1 + 1)}{2}$. Assume that $1 + 2 + 3 = 4 + \cdots + k = \dfrac{k(k + 1)}{2}$. Now for $n = k + 1$,

$$1 + 2 + 3 + 4 + \cdots + k + (k + 1) = \dfrac{k(k + 1)}{2} + k + 1$$

$$= \dfrac{k(k + 1)}{2} + \dfrac{2(k + 1)}{2}$$

$$= \dfrac{(k + 1)(k + 2)}{2}.$$

Thus, $1 + 2 + 3 + 4 + \cdots + n = \dfrac{n(n + 1)}{2}$ for all integers $n \geq 1$.

**11.** For $n = 4$, $4! > 2^4$. Assume that $k! > 2^k$. Then

$(k + 1)! = (k + 1)(k!) > (k + 1)2^k > 2 \cdot 2^k$

$= 2^{k+1}$.

Thus, $n! > 2^n$ for all integers $n \geq 4$.

**12.** $_{13}C_4 = \dfrac{13!}{(13 - 4)!4!} = 715$

**13.** $(x + 3)^5 = x^5 + 5x^4(3) + 10x^3(3)^2 + 10x^2(3)^3 + 5x(3)^4 + (3)^5$

$= x^5 + 15x^4 + 90x^3 + 270x^2 + 405x + 243$

**14.** $_{12}C_5 x^7(-2)^5 = -25{,}344x^7$

**15.** $_{30}P_4 = \dfrac{30!}{(30 - 4)!} = 657{,}720$

**16.** $6! = 720$ ways

**17.** $_{12}P_3 = 1320$

**18.** $P(2) + P(3) + P(4) = \frac{1}{36} + \frac{2}{36} + \frac{3}{36}$
$$= \frac{6}{36} = \frac{1}{6}$$

**19.** $P(K, B10) = \frac{4}{52} \cdot \frac{2}{51} = \frac{2}{663}$

**20.** Let $A$ = probability of no faulty units.

$P(A) = \left(\frac{997}{1000}\right)^{50} \approx 0.8605$

$P(A') = 1 - P(A) \approx 0.1395$

# Chapter 7    Practice Test Solutions

**1.** $(x - 0)^2 = 4(5)(y - 0)$

Vertex: $(0, 0)$

Focus: $(0, 5)$

Directrix: $y = -5$

**2.** $(y - 0)^2 = 4(7)(x - 0)$

$y^2 = 28x$

**3.** $a = 12, b = 5, h = k = 0,$

$c = \sqrt{144 - 25} = \sqrt{119}$

Center: $(0, 0)$

Foci: $\left(\pm\sqrt{119}, 0\right)$

Vertices: $(\pm 12, 0)$

**4.** Center: $(0, 0)$

$c = 4, 2b = 6 \implies b = 3,$

$a = \sqrt{16 + 9} = 5$

$\frac{x^2}{25} + \frac{y^2}{9} = 1$

**5.** $a = 12, b = 13, c = \sqrt{144 + 169} = \sqrt{313}$

Center: $(0, 0)$

Foci: $\left(0, \pm\sqrt{313}\right)$

Vertices: $(0, \pm 12)$

Asymptotes: $y = \pm\frac{12}{13}x$

**6.** Center: $(0, 0)$

$a = 4, \pm\frac{1}{2} = \pm\frac{b}{4} \implies b = 2$

$\frac{x^2}{16} - \frac{y^2}{4} = 1$

$y_1 = 2\sqrt{\frac{x^2}{16} - 1}$

$y_2 = -2\sqrt{\frac{x^2}{16} - 1}$

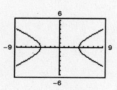

**7.** $p = 4$

$(x - 6)^2 = 4(4)(y + 1)$

$(x - 6)^2 = 16(y + 1)$

$y = \frac{(x - 6)^2}{16} - 1$

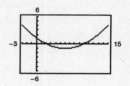

**8.**
$$16x^2 - 96x + 9y^2 + 36y = -36$$
$$16(x^2 - 6x + 9) + 9(y^2 + 4y + 4) = -36 + 144 + 36$$
$$16(x - 3)^2 + 9(y + 2)^2 = 144$$
$$\frac{(x - 3)^2}{9} + \frac{(y + 2)^2}{16} = 1$$

$a = 4, b = 3, c = \sqrt{16 - 9} = \sqrt{7}$

Center: $(3, -2)$

Foci: $\left(3, -2 \pm \sqrt{7}\right)$

Vertices: $(3, -2 \pm 4)$  or  $(3, 2)$ and $(3, -6)$

**9.** Center: $(3, 1)$

$a = 4, 2b = 2 \implies b = 1$

$\frac{(x - 3)^2}{16} + \frac{(y - 1)^2}{1} = 1$

**10.** Center: $(-3, 1)$

Vertices: $\left(-3 \pm \dfrac{1}{2}, 1\right)$ or $\left(-\dfrac{5}{2}, 1\right)$ and $\left(-\dfrac{7}{2}, 1\right)$

Foci: $\left(-3 \pm \dfrac{\sqrt{13}}{6}, 1\right)$

Asymptotes: $y = \pm \dfrac{1/3}{1/2}(x + 3) + 1 = \pm \dfrac{2}{3}(x + 3) + 1$

$a = \dfrac{1}{2}, b = \dfrac{1}{3}, c = \sqrt{\dfrac{1}{4} + \dfrac{1}{9}} = \dfrac{\sqrt{13}}{6}$

$\dfrac{(x + 3)^2}{1/4} - \dfrac{(y - 1)^2}{1/9} = 1$

**11.** Center: $(3, 0)$

$a = 4, c = 7, b = \sqrt{49 - 16} = \sqrt{33}$

$\dfrac{y^2}{16} - \dfrac{(x - 3)^2}{33} = 1$

**12.**

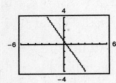

$x = 2t + 1 \implies t = \dfrac{x - 1}{2}$

$y = -1 - 3t = -1 - 3\left(\dfrac{x - 1}{2}\right)$

$= -1 - \dfrac{3}{2}x + \dfrac{3}{2} = -\dfrac{3}{2}x + \dfrac{1}{2}$

**13.**

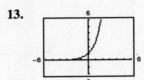

**14.** The first curve consists of the entire parabola $y = x^2$, whereas the second consists of just the branch $x \geq 0$.

# PART II

1. $-\frac{10}{3} \approx -3.3$ and $-|-4| = -4$, hence $-\frac{10}{3} > -|-4|$.

2. $d(-17, 39) = |39 - (-17)| = |39 + 17| = |56| = 56$

3. $(5 - x) + 0 = 5 - x$   Additive Identity Property

4. (a) $27\left(\frac{-2}{3}\right) = -2\left(\frac{27}{3}\right) = -2(9) = -18$   (b) $\frac{5}{18} \div \frac{15}{8} = \frac{5}{18} \cdot \frac{8}{15} = \frac{5 \cdot 2 \cdot 4}{2 \cdot 9 \cdot 3 \cdot 5} = \frac{4}{27}$

   (c) $\left(-\frac{2}{7}\right)^3 = -\frac{2^3}{7^3} = -\frac{8}{343}$   (d) $\left(\frac{3^2}{2}\right)^{-3} = \left(\frac{2}{3^2}\right)^3 = \frac{2^3}{3^6} = \frac{8}{729}$

5. (a) $\sqrt{5} \cdot \sqrt{125} = \sqrt{5} \cdot 5\sqrt{5} = 25$   (b) $\frac{\sqrt{72}}{\sqrt{2}} = \frac{6\sqrt{2}}{\sqrt{2}} = 6$

   (c) $\frac{5.4 \times 10^8}{3 \times 10^3} = \frac{5.4}{3} \times 10^5 = 1.8 \times 10^5$   (d) $(3 \times 10^4)^3 = 3^3 \times (10^4)^3 = 27 \times 10^{12} = 2.7 \times 10^{13}$

6. (a) $3z^2(2z^3)^2 = 3z^2 4 \cdot z^6 = 12z^8$   (b) $(u - 2)^{-4}(u - 2)^{-3} = (u - 2)^{-7} = \frac{1}{(u - 2)^7}$

   (c) $\left(\frac{x^{-2}y^2}{3}\right)^{-1} = \frac{3}{x^{-2}y^2} = \frac{3x^2}{y^2}$

7. (a) $9z\sqrt{8z} - 3\sqrt{2z^3} = 9z \cdot 2\sqrt{2z} - 3z\sqrt{2z} = 15z\sqrt{2z}$

   (b) $-5\sqrt{16y} + 10\sqrt{y} = -5 \cdot 4\sqrt{y} + 10\sqrt{y} = -10\sqrt{y}$

   (c) $\sqrt[3]{\frac{16}{v^5}} = \sqrt[3]{\frac{2^3 \cdot 2}{v^3 v^2}} = \frac{2}{v} \sqrt[3]{\frac{2}{v^2}}$

8. $3 - 2x^5 + 3x^3 - x^4 = -2x^5 - x^4 + 3x^3 + 3$

   Degree: 5

   Leading coefficient: $-2$

9. $(x^2 + 3) - [3x + (8 - x^2)] = x^2 + 3 - 3x - 8 + x^2 = 2x^2 - 3x - 5$

10. $\left(x + \sqrt{5}\right)\left(x - \sqrt{5}\right) = x^2 - 5$

11. $\frac{8x}{x - 3} + \frac{24}{3 - x} = \frac{8x}{x - 3} - \frac{24}{x - 3} = \frac{8x - 24}{x - 3} = \frac{8(x - 3)}{x - 3} = 8, \ x \neq 3$

12. $\left(\frac{2}{x} - \frac{2}{x + 1}\right) \div \frac{4}{x^2 - 1} = \frac{2(x + 1) - 2x}{x(x + 1)} \cdot \frac{(x - 1)(x + 1)}{4} = \frac{2}{x} \cdot \frac{x - 1}{4} = \frac{x - 1}{2x}, \ x \neq \pm 1$

13. $2x^4 - 3x^3 - 2x^2 = x^2(2x^2 - 3x - 2) = x^2(2x + 1)(x - 2)$

**14.** $x^3 + 2x^2 - 4x - 8 = x^2(x + 2) - 4(x + 2) = (x + 2)(x^2 - 4)$

$$= (x + 2)(x + 2)(x - 2)$$

$$= (x + 2)^2(x - 2)$$

**15.** $8x^3 - 27 = (2x)^3 - 3^3 = (2x - 3)((2x)^2 + (2x)(3) + 3^2)$

$$= (2x - 3)(4x^2 + 6x + 9)$$

**16.** (a) $\dfrac{16}{\sqrt[3]{16}} = \dfrac{16}{\sqrt[3]{8 \cdot 2}} = \dfrac{16}{2} \dfrac{1}{2^{1/3}} \cdot \dfrac{2^{2/3}}{2^{2/3}} = 4 \cdot 2^{2/3} = 4\sqrt[3]{4}$

(b) $\dfrac{6}{1 - \sqrt{3}} = \dfrac{6}{1 - \sqrt{3}} \cdot \dfrac{1 + \sqrt{3}}{1 + \sqrt{3}} = \dfrac{6(1 + \sqrt{3})}{1 - 3} = -3\left(1 + \sqrt{3}\right)$

**17.** Shaded region = (area big triangle) − (area small triangle)

$$= \tfrac{1}{2}(3x)\left(\sqrt{3}x\right) - \tfrac{1}{2}(2x)\left(\tfrac{2}{3}\sqrt{3}x\right)$$

$$= \tfrac{1}{2}3\sqrt{3}x^2 - \tfrac{2}{3}\sqrt{3}x^2 = \left(\tfrac{3}{2} - \tfrac{2}{3}\right)\sqrt{3}x^2 = \tfrac{5}{6}\sqrt{3}x^2$$

**18.** Midpoint $= \left(\dfrac{-2 + 6}{2}, \dfrac{5 + 0}{2}\right) = \left(2, \dfrac{5}{2}\right)$

Distance $= \sqrt{(-2 - 6)^2 + (5 - 0)^2}$

$$= \sqrt{64 + 25} = \sqrt{89} \approx 9.43$$

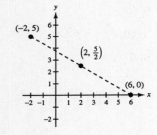

**19.**

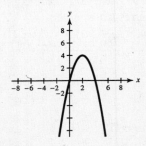

# Chapter 1    Chapter Test

**1.** $y = 4 - \tfrac{3}{4}|x|$

$x = 0 \implies y = 4 \quad (0, 4)$

$y = 0 \implies 4 = \tfrac{3}{4}|x| \implies |x| = \tfrac{16}{3} \implies x = \pm\tfrac{16}{3}$

Intercepts: $(0, 4), \left(\tfrac{16}{3}, 0\right), \left(-\tfrac{16}{3}, 0\right)$

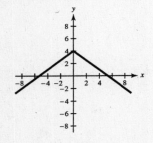

**2.** $y = 4 - (x - 2)^2$

$x = 0 \implies y = 4 - 4 = 0 \quad (0, 0)$

$y = 0 \implies 0 = 4 - (x - 2)^2 \implies (x - 2) = \pm 2$

$x = 0, 4$

Intercepts: $(0, 0), (4, 0)$

**3.** $y = x - x^3$

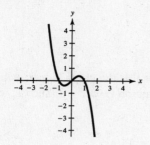

$x = 0 \implies y = 0 \quad (0, 0)$

$y = 0 \implies 0 = x(1 - x)(1 + x)$

Intercepts: $(0, 0), \ (1, 0), \ (-1, 0)$

**4.** $y = -x^3 + 2x - 4$

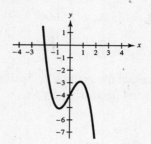

$x = 0 \implies y = -4 \quad (0, -4)$

$y = 0 \implies x = -2 \quad (-2, 0)$

Intercepts: $(0, -4), \ (-2, 0)$

**5.** $y = \sqrt{3 - x}$

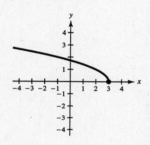

$x = 0 \implies y = \sqrt{3}$

$y = 0 \implies 0 = \sqrt{3 - x} \implies x = 3$

Intercepts: $\left(0, \sqrt{3}\right), \ (3, 0)$

**6.** $y = \frac{1}{2}x\sqrt{x + 3}$

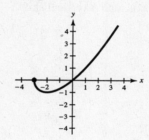

$x = 0 \implies y = 0 \quad (0, 0)$

$y = 0 \implies x = 0, -3 \quad (0, 0), (-3, 0)$

Intercepts: $\left(0, 0\right), \ (-3, 0)$

**7.** $y - (-1) = \frac{3}{2}(x - 3) = \frac{3}{2}x - \frac{9}{2}$

$\qquad y = \frac{3}{2}x - \frac{11}{2}$

Additional points: $(1, -4), \ \left(4, \frac{1}{2}\right), \ (5, 2)$

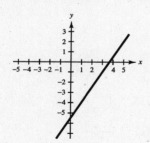

**8.** $5x + 2y = 3$

$\qquad 2y = -5x + 3$

$\qquad y = -\frac{5}{2}x + \frac{3}{2} \quad \text{slope} = -\frac{5}{2}$

(a) parallel line: $\qquad y - 4 = -\frac{5}{2}(x - 0)$

$\qquad\qquad\qquad\qquad\qquad y = -\frac{5}{2}x + 4$

$\qquad\qquad 5x + 2y - 8 = 0$

(b) perpendicular line: $\qquad y - 4 = \frac{2}{5}(x - 0)$

$\qquad\qquad\qquad\qquad\qquad y = \frac{2}{5}x + 4$

$\qquad\qquad 2x - 5y + 20 = 0$

**9.** No, for some $x$ there corresponds more than one value of $y$. For instance, if $x = 1$, $y = \pm 1/\sqrt{3}$.

**10.** $f(x) = |x + 2| - 15$

(a) $f(-8) = |-8 + 2| - 15 = 6 - 15 = -9$

(b) $f(14) = |14 + 2| - 15 = 16 - 15 = 1$

(c) $f(t - 6) = |t - 6 + 2| - 15 = |t - 4| - 15$

**11.** $3 - x \geq 0 \implies$ domain is all $x \leq 3$.

**12.** $C = 5.60x + 24{,}000$

$P = R - C = 99.50x - (5.60x + 24{,}000) = 93.9x - 24{,}000$

**13** $h(x) = \frac{1}{4}x^4 - 2x^2 = \frac{1}{4}x^2(x^2 - 8)$

By graphing $h$, you see that the graph is increasing on $(-2, 0)$ and $(2, \infty)$ and decreasing on $(-\infty, -2)$ and $(0, 2)$.

**14.** $g(t) = |t + 2| - |t - 2|$.

By graphing $g$, you see that the graph is increasing on $(-2, 2)$, and constant on $(-\infty, -2)$ and $(2, \infty)$.

**15.** Relative minimum: $(-3.33, -6.52)$.

Relative maximum: $(0, 12)$

**16.** Relative minimum: $(0.77, 1.81)$.

Relative maximum: $(-0.77, 2.19)$

**17. (a)** Common function $f(x) = x^3$

**(b)** $g$ is obtained from $f$ by a horizontal shift 5 units to the right, a vertical stretch of 2, a reflection in the $x$-axis, and a vertical shift 3 units upward.

**(c)**

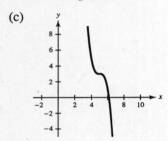

**18. (a)** Common function $f(x) = \sqrt{x}$.

**(b)** $g$ is obtained from $f$ by a reflection in the $y$-axis, and a horizontal shift 7 units to the left.

**(c)**

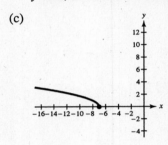

**19. (a)** Common function $f(x) = |x|$.

**(b)** $g(x) = 4|-x| - 7 = 4|x| - 7$ is obtained from $f$ by a vertical stretch of 4 followed by a vertical shift 7 units downward.

**(c)**

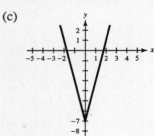

**20. (a)** $(f - g)(x) = x^2 - \sqrt{2 - x}$,   Domain: $x \le 2$

**(b)** $\left(\dfrac{f}{g}\right)(x) = \dfrac{x^2}{\sqrt{2 - x}}$,   Domain: $x < 2$

**(c)** $(f \circ g)(x) = f(\sqrt{2 - x}) = 2 - x$, Domain $x \le 2$

**(d)** $(g \circ f)x = g(x^2) = \sqrt{2 - x^2}$ Domain:
$$-\sqrt{2} \le x \le \sqrt{2}$$

**21.** $f(x) = x^3 + 8$

Yes, $f$ is one-to-one and has an inverse function.

$y = x^3 + 8$

$x = y^3 + 8$

$x - 8 = y^3$

$\sqrt[3]{x - 8} = y$

$f^{-1}(x) = \sqrt[3]{x - 8}$

**22.** $f(x) = x^2 + 6$

No, $f$ is not one-to-one, and does not have an inverse function.

**23.** $f(x) = \dfrac{3x\sqrt{x}}{8}$

Yes, $f$ is one-to-one and has an inverse function.

$y = \frac{3}{8}x^{3/2}$,   $x \ge 0$,   $y \ge 0$

$x = \frac{3}{8}y^{3/2}$,   $y \ge 0$,   $x \ge 0$

$\frac{8}{3}x = y^{3/2}$

$\left(\frac{8}{3}x\right)^{2/3} = y$

$f^{-1}(x) = \left(\frac{8}{3}x\right)^{2/3}$,   $x \ge 0$

## Chapter 2    Chapter Test

**1.** $\dfrac{12}{x} - 7 = -\dfrac{27}{x} + 6$

$\qquad \dfrac{39}{x} = 13$

$\qquad 39 = 13x$

$\qquad 3 = x \implies x = 3$

**2.** $\dfrac{4}{3x - 2} - \dfrac{9x}{3x + 2} = -3$

$\qquad 4(3x + 2) - 9x(3x - 2) = -3(3x - 2)(3x + 2)$

$\qquad 12x + 8 - 27x^2 + 18x = -3(9x^2 - 4)$

$\qquad -27x^2 + 30x + 8 = -27x^2 + 12$

$\qquad\qquad 30x \quad\;\; = 4$

$\qquad\qquad\quad x = \dfrac{2}{15}$

**3.** $(-8 - 3i) + (-1 - 15i) = -9 - 18i$

**4.** $\left(10 + \sqrt{-20}\right) - \left(4 - \sqrt{-14}\right) = 6 + 2\sqrt{5}\,i + \sqrt{14}\,i = 6 + \left(2\sqrt{5} + \sqrt{14}\right)i$

**5.** $(2 + i)(6 - i) = 12 + 6i - 2i + 1 = 13 + 4i$

**6.** $(4 + 3i)^2 - (5 + i)^2 = (16 + 24i - 9) - (25 + 10i - 1) = -17 + 14i$

**7.** $\dfrac{8 + 5i}{6 - i} \cdot \dfrac{6 + i}{6 + i} = \dfrac{48 + 30i + 8i - 5}{36 + 1} = \dfrac{43}{37} + \dfrac{38}{37}i$

**8.** $\dfrac{5i}{2 + i} \cdot \dfrac{2 - i}{2 - i} = \dfrac{10i + 5}{4 + 1} = 1 + 2i$

**9.** $y = 3x^2 + 1$

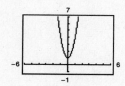

No $x$-intercepts. No real zeros

$3x^2 + 1 = 0$

$\quad 3x^2 = -1$ No real solutions

**10.** $y = 2 + 8x^{-2} = 2 + \dfrac{8}{x^2}$

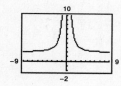

No $x$-intercepts. No real zeros.

**11.** $y = x^3 - 4x^2 + 5x$

$x$-intercept: $(0, 0)$

Real Zero: $x = 0$

$\quad x^3 - 4x^2 + 5x = 0$

$\quad x(x^2 - 4x + 5) = 0$

Since $x^2 - 4x + 5 > 0$ for all $x$, $x = 0$ is the only zero.

**12.** $y = x^3 + x$

$x$-intercept: $(0, 0)$

Real Zero: $x = 0$

$\quad x^3 + x = 0$

$x(x^2 + 1) = 0$

$\qquad x = 0$

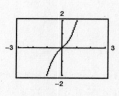

**13.**  $x^2 - 10x + 9 = 0$

$(x - 1)(x - 9) = 0$

$x = 1, 9$

**14.**  $x^2 + 12x - 2 = 0$

$x = \dfrac{-12 \pm \sqrt{12^2 - 4(-2)}}{2} = -6 \pm \sqrt{38}$

**15.**  $4x^2 - 81 = 0$

$4x^2 = 81$

$x^2 = \frac{81}{4}$

$x = \pm \frac{9}{2}$

**16.**  $5x^2 + 14x - 3 = 0$

$(x + 3)(5x - 1) = 0$

$x = -3, \frac{1}{5}$

**17.**  $3x^3 - 4x^2 - 12x + 16 = 0$

$x^2(3x - 4) - 4(3x - 4) = 0$

$(x^2 - 4)(3x - 4) = 0$

$x = 2, -2, \frac{4}{3}$

**18.**  $x + \sqrt{22 - 3x} = 6$

$\sqrt{22 - 3x} = 6 - x$

$22 - 3x = (6 - x)^2$

$22 - 3x = 36 - 12x + x^2$

$x^2 - 9x + 14 = 0$

$(x - 2)(x - 7) = 0$

$x = 2$   ($x = 7$ is extraneous)

**19.**  $(x^2 + 6)^{2/3} = 16$

$x^2 + 6 = 16^{3/2} = 64$

$x^2 = 58$

$x = \pm \sqrt{58} \approx \pm 7.616$

**20.**  $|8x - 1| = 21$

$8x - 1 = 21$   or   $-(8x - 1) = 21$

$8x = 22$   or   $-8x = 20$

$x = \frac{11}{4}$   $\qquad$   $x = -\frac{5}{2}$

**21.**  $-\frac{5}{6} < x - 2 < \frac{1}{8}$

$2 - \frac{5}{6} < x \qquad < 2 + \frac{1}{8}$

$\frac{7}{6} < x \qquad < \frac{17}{8}$

**22.**  $2|x - 8| < 10$

$|x - 8| < 5$

$-5 < x - 8 < 5$

$3 < x \qquad < 13$

**23.**  $\dfrac{3 - 5x}{2 + 3x} < -2$

$\dfrac{3 - 5x}{2 + 3x} + 2 < 0$

$\dfrac{3 - 5x + 2(2 + 3x)}{2 + 3x} < 0$

$\dfrac{x + 7}{2 + 3x} < 0$

Critical numbers: $-7, -\frac{2}{3}$. Checking the three intervals, we obtain $-7 < x < -\frac{2}{3}$

**24.**  $L = 14.8t + 92.4$

$L > 300$

$14.8t + 92.4 > 300$

$14.8t > 207.6$

$t > 14$

The number of lines will exceed 300 in 2004.

# Chapters P–2   Cumulative Test

**1.** $\dfrac{14x^2y^{-3}}{32x^{-1}y^2} = \dfrac{7x^3}{16y^5}, x \neq 0$

**2.** $8\sqrt{60} - 2\sqrt{135} - \sqrt{15} = 16\sqrt{15} - 6\sqrt{15} - \sqrt{15} = 9\sqrt{15}$

**3.** $\sqrt{28x^4y^3} = 2x^2y\sqrt{7y}$

**4.** $4x - [2x + 5(2 - x)] = 4x - [-3x + 10] = -10 + 7x = 7x - 10$

**5.** $(x - 2)(x^2 + x - 3) = x^3 + x^2 - 3x - 2x^2 - 2x + 6$
$$= x^3 - x^2 - 5x + 6$$

**6.** $\dfrac{2}{x + 3} - \dfrac{1}{x + 1} = \dfrac{2(x + 1) - (x + 3)}{(x + 3)(x + 1)} = \dfrac{x - 1}{(x + 3)(x + 1)}$

**7.** $25 - (x - 2)^2 = [5 - (x - 2)][5 + (x - 2)] = (7 - x)(3 + x)$

**8.** $x - 5x^2 - 6x^3 = -x(6x^2 + 5x - 1) = -x(6x - 1)(x + 1)$
$$= x(x + 1)(1 - 6x)$$

**9.** $54 - 16x^3 = 2(27 - 8x^3) = 2(3 - 2x)(9 + 6x + 4x^2)$

**10.** Midpoint: $\left(\dfrac{-7/2 + 13/2}{2}, \dfrac{4 - 8}{2}\right) = (3/2, -2)$

Distance: $\sqrt{(13/2 + 7/2)^2 + (-8 - 4)^2} = \sqrt{100 + 144} = \sqrt{244} = 2\sqrt{61} \approx 15.62$

**11.** $\left(x + \frac{1}{2}\right)^2 + (y + 8)^2 = \left(\frac{5}{4}\right)^2 = \frac{25}{16}$

**12.** $x - 3y + 12 = 0$

$-3y = -x - 12$

$y = \dfrac{x}{3} + 4$

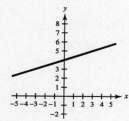

**13.** $y = x^2 - 9$

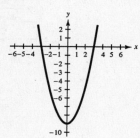

**14.** $y = \sqrt{4 - x}$

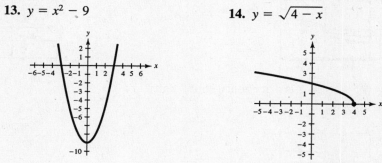

**15.** (a) $y - 8 = \dfrac{-6 - 8}{\frac{1}{2} + 5}(x + 5) = \dfrac{-14}{\frac{11}{2}}(x + 5) = -\dfrac{28}{11}(x + 5)$

$11y - 88 = -28x - 140$

$11y + 28x + 52 = 0$

(b) Three other points: $\left(0, -\frac{52}{11}\right), \left(-\frac{52}{28}, 0\right), \left(1, -\frac{80}{11}\right)$

**16.** (a) $y - 1 = -2\left(x + \frac{1}{2}\right)$

$y - 1 = -2x - 1$

$y = -2x$

$y + 2x = 0$

(b) Three additional points:

$(0, 0), (1, -2). (2, -4)$

**17.** (a) Vertical line: $x = -\frac{3}{7}$   or   $x + \frac{3}{7} = 0$

(b) Three additional points;

$\left(-\frac{3}{7}, 0\right), \left(-\frac{3}{7}, 1\right), \left(-\frac{3}{7}, 2\right)$

**18.** (a) $f(6) = \dfrac{6}{6 - 2} = \dfrac{3}{2}$

(b) $f(2)$ is undefined (division by zero).

(c) $f(s + 2) = \dfrac{s + 2}{(s + 2) - 2} = \dfrac{s + 2}{s}$

**19.** (a) $f\left(-\frac{5}{3}\right) = 3\left(-\frac{5}{3}\right) - 8 = -13$

(b) $f(-1) = 3(-1)^2 + 9(-1) - 8$

$= 3 - 9 - 8 = -14$

(c) $f(0) = 3(0)^2 + 9(0) - 8 = -8$

**20.** No, for some $x$ there corresponds two values of $y$.

**21.**

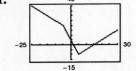

Decreasing on $(-\infty, 5)$, increasing on $(5, \infty)$

**22.** (a) $r(x) = \frac{1}{2}\sqrt[3]{x}$ is a vertical shrink of $y = \sqrt[3]{x}$

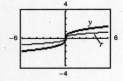

(b) $h(x) = \sqrt[3]{x} + 2$ is a vertical shift 2 units upward.

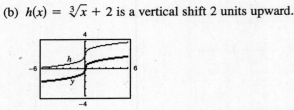

(c) $g(x) = \sqrt[3]{x + 2}$ is a horizontal shift 2 units to the left.

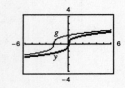

**23.** $(f + g)(-4) = f(-4) + g(-4) = [-(-4)^2 + 3(-4) - 10] + [4(-4) + 1]$

$= -38 - 15 = -53$

**24.** $(g - f)\left(\frac{3}{4}\right) = \left[4\left(\frac{3}{4}\right) + 1\right] - \left[-\left(\frac{3}{4}\right)^2 + 3\left(\frac{3}{4}\right) - 10\right]$

$= 4 - (-8.3125) = 12.3125 = \frac{197}{16}$

**25.** $(g \circ f)(-2) = g(f(-2)) = g(-20) = 4(-20) + 1 = -79$

**26.** $(fg)(-1) = f(-1)g(-1) = (-14)(-3) = 42$

**27.** Yes, $h(x) = 5x - 2$ has an inverse function

$y = 5x - 2$

$x = 5y - 2$

$x + 2 = 5y$

$\dfrac{x + 2}{5} = y$

$h^{-1}(x) = \dfrac{x + 2}{5}$

**28.**

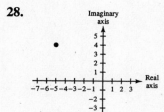

**29.**

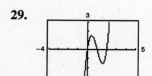

$x$-intercepts: $0, 1, 2$

$4x^3 - 12x^2 + 8x = 0$

$4x(x^2 - 3x + 2) = 0$

$4x(x - 2)(x - 1) = 0$

$x$-intercepts: $0, 1, 2$

**30.**

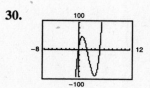

$x$-intercepts: $0, 2, 5$

$12x^3 - 84x^2 + 120x = 0$

$12x(x^2 - 7x + 10) = 0$

$12x(x - 2)(x - 5) = 0$

$x$-intercepts: $0, 2, 5$

**31.**

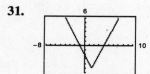

$x$-intercepts: $-1, 4$

$|2x - 3| - 5 = 0$

$|2x - 3| = 5$

$2x - 3 = 5$   or   $2x - 3 = -5$

$x = 4$   or   $x = -1$

$x$-intercepts: $-1, 4$

**32.**

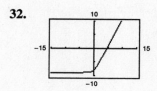

$x$-intercept: $4.444$

$\sqrt{x^2 + 1} = 9 - x$

$x^2 + 1 = (9 - x)^2$

$x^2 + 1 = 81 - 18x + x^2$

$18x = 80$

$x = \dfrac{40}{9} = 4.4\overline{4}$

**33.** $Z = \sqrt{R^2 - X^2}$

$Z^2 = R^2 - X^2$

$X^2 = R^2 - Z^2$

$X = \pm\sqrt{R^2 - Z^2}$

**34.** $L = \dfrac{k}{3\pi r^2 p}$

$pL = \dfrac{k}{3\pi r^2}$

$p = \dfrac{k}{3\pi r^2 L}$

**35.** $\dfrac{x}{5} - 6 \le \dfrac{-x}{2} + 6$

$\dfrac{x}{2} + \dfrac{x}{5} \le 12$

$\dfrac{7x}{10} \le 12$

$x \le \dfrac{120}{7} \approx 17.143$

**36.** $2x^2 + x \ge 15$

$2x^2 + x - 15 \ge 0$

$(2x - 5)(x + 3) \ge 0$

Critical numbers: $-3, \dfrac{5}{2}$

Test Intervals: $(-\infty, -3), \left(-3, \dfrac{5}{2}\right), \left(\dfrac{5}{2}, \infty\right)$

you obtain $(-\infty, -3] \cup \left[\dfrac{5}{2}, \infty\right)$

**37.** $|7 + 8x| > 5$

$7 + 8x > 5 \quad$ or $\quad 7 + 8x < -5$

$8x > -2 \quad$ or $\qquad 8x < -12$

$x > -\dfrac{1}{4} \quad$ or $\qquad x < -\dfrac{3}{2}$

**38.** $\dfrac{2(x - 2)}{x + 1} \le 0$

Critical numbers: $x = -1, x = 2$

Test intervals: $(-\infty, -1), (-1, 2), (2, \infty)$

Is $\dfrac{2(x - 2)}{x + 1} \le 0$?

$-1 < x \le 2$

**39.** $V = \dfrac{4}{3}\pi r^3 \implies r = \sqrt[3]{\dfrac{3}{4\pi} V}$

$= \sqrt[3]{\dfrac{3}{4\pi}(370.7)}$

$\approx 4.456$ inches

**40.** (a) Let $x$ and $y$ be the lengths of the sides $2x + 2y = 546 \implies y = 273 - x$

$A = xy = x(273 - x)$

(b)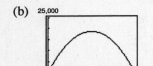

Domain: $0 < x < 273$

(c) If $A = 15000$, then $x = 76.23$ or $196.77$

Dimensions in feet:

$76.23 \times 196.77 \quad$ or $\quad 196.77 \times 76.23$

**41.** (a) $R = 129.25t - 389.27$

(b)
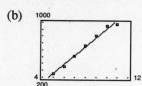

(c) The revenue is increasing about 129 million per year.

(d) For 2007, $t = 17$ and $R \approx 1808$ million

(e)

| Year | Data | Model |
|------|------|-------|
| 1995 | 253.4 | 257 |
| 1996 | 360.1 | 386 |
| 1997 | 508.8 | 515 |
| 1998 | 669.8 | 645 |
| 1999 | 805.3 | 774 |
| 2000 | 944.7 | 903 |
| 2001 | 971.2 | 1032 |

## Chapter 3    Chapter Test

**1.** (a) $g(x) = 6 - x^2$ is a reflection in the $x$-axis followed by a vertical shift 6 units upward.

(b) $g(x) = \left(x - \frac{3}{2}\right)^2$ is a horizontal shift $\frac{3}{2}$ units to the right.

**2.** $y = x^2 + 4x + 3 = x^2 + 4x + 4 - 1 = (x + 2)^2 - 1$

Vertex: $(-2, -1)$

$x = 0 \implies y = 3$

$y = 0 \implies x^2 + 4x + 3 = 0 \implies (x + 3)(x + 1) = 0 \implies x = -1, -3$

Intercepts: $(0, 3), \ (-1, 0), \ (-3, 0)$

**3.** Let $y = a(x - h)^2 + k$. The vertex $(3, -6)$ implies that $y = a(x - 3)^2 - 6$. For $(0, 3)$ you obtain

$3 = a(0 - 3)^2 - 6 = 9a - 6 \implies a = 1$.

Thus, $y = (x - 3)^2 - 6 = x^2 - 6x + 3$.

**4.** (a) $y = -\frac{1}{20}x^2 + 3x + 5 = -\frac{1}{20}(x^2 - 60x + 900) + 5 + 45$

$$= -\frac{1}{20}(x - 30)^2 + 50$$

Maximum height: $y = 50$ feet

(b) The term 5 determines the height at which the ball was thrown. Changing the constant term results in a vertical shift of the graph and therefore changes the maximum height.

**5.**
$$\begin{array}{r} 3x \phantom{ + 0x^2} \\ x^2 + 1 \overline{) 3x^3 + 0x^2 + 4x - 1} \\ \underline{3x^3 \phantom{ + 0x^2} + 3x} \\ x - 1 \end{array}$$

$$3x + \frac{x - 1}{x^2 + 1}$$

**6.**
$$\begin{array}{r|rrrrr} 2 & 2 & 0 & -5 & 0 & -3 \\ & & 4 & 8 & 6 & 12 \\ \hline & 2 & 4 & 3 & 6 & 9 \end{array}$$

$$2x^3 + 4x^2 + 3x + 6 + \frac{9}{x - 2}$$

**7.** Possible rational zeros:

$\pm 24, \pm 12, \pm 8, \pm 6, \pm 4, \pm 3, \pm 2, \pm 1, \pm \frac{3}{2}, \pm \frac{1}{2}$

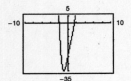

Rational zeros: $-2, \frac{3}{2}$

**8.** Possible rational zeros:

$\pm 2, \pm 1, \pm \frac{2}{3}, \pm \frac{1}{3}$

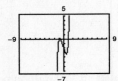

Rational zeros: $\pm 1, -\frac{2}{3}$

**9.** Real zeros: $1.380, -0.819$

**10.** Real zeros: $-1.414, -0.667, 1.414$

**11.** $(x - 0)(x - 3)(x - (3 + i))(x - (3 - i))$

$x(x - 3)(x^2 - 6x + 10)$

$x^4 - 9x^3 + 28x^2 - 30x$

**12.** $\left(x - (1 + \sqrt{3}i)\right)\left(x - (1 - \sqrt{3}i)\right)(x - 2)(x - 2)$

$(x^2 - 2x + 4)(x^2 - 4x + 4)$

$x^4 - 6x^3 + 16x^2 - 24x + 16$

**13.** $(x - 0)(x + 5)(x - 1 - i)(x - 1 + i)$

$(x^2 + 5x)((x - 1)^2 + 1)$

$(x^2 + 5x)(x^2 - 2x + 2)$

$x^4 + 3x^3 - 8x^2 + 10x$

**14.**

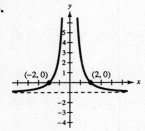

Vertical asymptote: $x = 0$

Intercepts: $(2, 0), (-2, 0)$

Symmetry: $y$-axis

Horizontal asymptote: $y = -1$

**15.** $g(x) = \dfrac{x^2 + 2}{x - 1} = x + 1 + \dfrac{3}{x - 1}$

Vertical asymptote: $x = 1$

Intercept: $(0, -2)$

Slant asymptote: $y = x + 1$

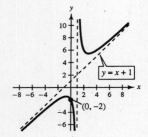

**16.** $f(x) = \dfrac{2x^2 + 9}{5x^2 + 2}$

Horizontal asymptote: $y = \dfrac{2}{5}$

$y$-axis symmetry

Intercept: $\left(0, \dfrac{9}{2}\right)$

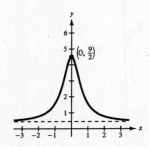

**17.** (a)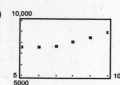

(b) $C = 62.5536t^2 - 654.875t + 9269.1429$

(c)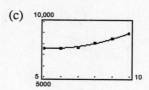

(d) $C = 15,000$ when $t \approx 16.1$, or 2006.

## Chapter 4    Chapter Test

**1.** $12.4^{2.79} \approx 1123.690$

**2.** $4^{3\pi/2} \approx 687.291$

**3.** $e^{-7/10} \approx 0.497$

**4.** $e^{3.1} \approx 22.198$

**5.** $f(x) = 10^{-x}$

| $x$ | $-3$ | $-1$ | 0 | 1 | 2 | 4 |
|---|---|---|---|---|---|---|
| $f(x)$ | 1000 | 10 | 1 | 0.1 | 0.01 | 0.0001 |

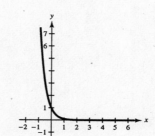

**6.** $f(x) = -6^{x-2}$

| $x$ | $-1$ | 0 | 1 | 2 | 3 | 4 | 5 |
|---|---|---|---|---|---|---|---|
| $f(x)$ | $-0.0046$ | $-0.0278$ | $-.1667$ | $-1$ | $-6$ | $-36$ | $-216$ |

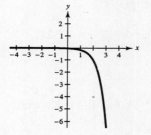

**7.** $f(x) = 1 - e^{2x}$

| $x$ | $-3$ | $-1$ | 0 | 1 | 2 |
|---|---|---|---|---|---|
| $f(x)$ | 0.9975 | 0.8647 | 0 | $-6.389$ | $-53.6$ |

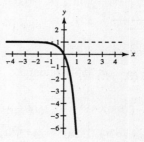

**8.** (a) $\log_7 7^{-0.89} = -0.89 \log_7 7 = -0.89$

(b) $4.6 \ln e^2 = 4.6(2)\ln e = 9.2$

**9.** $f(x) = -\log_{10} x - 6$

Domain: $x > 0$

Vertical asymptote: $x = 0$

$x$-intercept: $(10^{-6}, 0) \approx (0, 0)$

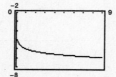

**10.** $f(x) = \ln(x - 4)$

Domain: $x > 4$

Vertical asymptote: $x = 4$

$x$-intercept: $(5, 0)$

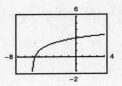

**11.** $f(x) = 1 + \ln(x + 6)$

Domain: $x > -6$

Vertical asymptote: $x = -6$

$x$-intercept: $(-5.632, 0)$

**12.** $\log_7 44 = \dfrac{\ln 44}{\ln 7} \approx 1.945$

**13.** $\log_{2/5}(0.9) = \dfrac{\ln(0.9)}{\ln\left(\frac{2}{5}\right)} \approx 0.115$

**14.** $\log_{24} 68 = \dfrac{\ln 68}{\ln 24} \approx 1.328$

**15.** $\log_2 3a^4 = \log_2 3 + \log_2 a^4 = \log_2 3 + 4 \log_2 a$

**16.** $\ln \dfrac{5\sqrt{x}}{6} = \ln 5 + \ln\sqrt{x} - \ln 6 = \ln 5 + \frac{1}{2}\ln x - \ln 6$

**17.** $\log_3 13 + \log_3 y = \log_3(13y)$

**18.** $4 \ln x - 4 \ln y = \ln x^4 - \ln y^4 = \ln\left(\dfrac{x^4}{y^4}\right) = \ln\left(\dfrac{x}{y}\right)^4$

**19.** $\dfrac{1025}{8 + e^{4x}} = 5$

$\qquad 1025 = 40 + 5e^{4x}$

$\qquad 985 = 5e^{4x}$

$\qquad e^{4x} = 197$

$\qquad 4x = \ln(197)$

$\qquad x = \dfrac{1}{4}\ln(197) \approx 1.321$

**20.** $\log_{10} - \log_{10}(8 - 5x) = 2$

$\qquad \log_{10}\left(\dfrac{x}{8 - 5x}\right) = 2$

$\qquad 10^2 = \dfrac{x}{8 - 5x}$

$\qquad 800 - 500x = x$

$\qquad 800 = 501x$

$\qquad x = \dfrac{800}{501} \approx 1.597$

**21.** $\frac{1}{2} = 1e^{k(22)}$ (half-life is 22 years)

$\ln \frac{1}{2} = 22k$

$\quad k = \frac{1}{22}\ln\frac{1}{2} = -\frac{1}{22}\ln 2 \approx -0.03151$

$A = e^{-0.03151(19)}$

$\quad = e^{-0.03151(19)}$

$\quad \approx 0.54953 \quad$ or $\quad 55\%$ remains

**22.** (a) $y_1 = -0.03095x^2 + 2.3667x + 41.3714$

$\qquad y_2 = 44.863(1.0328)^x = 44.8613\, e^{0.03228x}$

$\qquad y_3 = 35.06298\, x^{0.24661}$

(b)

 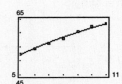

(c) $y_1$ fits best, although all are good fit

(d) For 2007, $x = 17$ and $y_1 \approx 72.7$ billion dollars

# Chapter 5    Chapter Test

**1.** $x - y = 6 \implies y = x - 6$. Then $3x + 5(x - 6) = 2 \implies 8x = 32 \implies x = 4, y = 4 - 6 = -2$.

*Answer:* $(4, -2)$

**2.** $y = x - 1 = (x - 1)^3 \implies x = 1$ or $1 = (x - 1)^2 = x^2 - 2x + 1 \implies x^2 - 2x = 0$.

Thus, $x = 1$ or $x(x - 2) = 0 \implies x = 0, 1, 2$.

*Answer:* $(0, -1), (1, 0), (2, 1)$

**3.** $x - y = 3 \implies y = x - 3 \implies$

$$4x - (x - 3)^2 = 7$$
$$4x - (x^2 - 6x + 9) = 7$$
$$x^2 - 10x + 16 = 0$$
$$(x - 2)(x - 8) = 0$$
$$x = 2, 8$$

*Answer:* $(2, -1), (8, 5)$

**4.** $\begin{cases} 2x + 5y = -11 & \text{Equation 1} \\ 5x - y = 19 & \text{Equation 2} \end{cases}$

$-\frac{5}{2}$ times Eq. 1 added to Eq. 2 produces

$-\frac{27}{2}y = \frac{93}{2} \implies y = -\frac{31}{9}$.

Then $2x + 5\left(-\frac{31}{9}\right) = -11 \implies x = \frac{28}{9}$.

*Answer:* $\left(\frac{28}{9}, -\frac{31}{9}\right)$

**5.** $\begin{cases} x - 2y + 3z = -5 & \text{Equation 1} \\ 2x \qquad - z = -4 & \text{Equation 2} \\ \qquad 3y + z = 17 & \text{Equation 3} \end{cases}$

$\begin{cases} x - 2y + 3z = -5 \\ \quad 4y - 7z = 6 \qquad (-2)\text{ Eq. 1} + \text{Eq. 2} \\ \quad 3y + z = 17 \end{cases}$

$\begin{cases} x - 2y + 3z = -5 \\ \quad 4y - 7z = 6 \\ \qquad \frac{25}{4}z = \frac{50}{4} \qquad \left(-\frac{3}{4}\right)\text{ Eq. 2} + \text{Eq. 3} \end{cases}$

$z = 2 \implies 4y - 7(2) = 6 \implies y = 5$

$z = 2, y = 5 \implies x - 2(5) + 3(2) = -5 \implies x = -1$

*Answer:* $(-1, 5, 2)$

**6.** $\begin{cases} 5x + 5y - z = 0 & \text{Equation 1} \\ 10x + 5y + 2z = 0 & \text{Equation 2} \\ 5x + 15y - 9z = 0 & \text{Equation 3} \end{cases}$

$\begin{cases} 5x + 5y - z = 0 \\ \quad - 5y + 4z = 0 \qquad (-2)\text{Eq. 1} + \text{Eq. 2} \\ \quad 10y - 8z = 0 \qquad (-1)\text{Eq. 1} + \text{Eq. 3} \end{cases}$

$\begin{cases} 5x + 5y - z = 0 \\ \quad - 5y + 4z = 0 \\ \qquad\qquad 0 = 0 \qquad (2)\text{Eq. 2} + \text{Eq. 3} \end{cases}$

Infinite number of solutions. They are all of the form $(-3a, 4a, 5a)$ or $\left(-\frac{3}{5}a, \frac{4}{5}a, a\right)$ where $a$ is any real number.

**7.** $6 = a(0)^2 + b(0) + c \implies c = 6$

$2 = a(-2)^2 + b(-2) + c$

$\frac{9}{2} = a(3)^2 + b(3) + c$

Hence, $\begin{cases} 4a - 2b + 6 = 2 \text{ or } 2a - b = -2 \\ 9a + 3b + 6 = \frac{9}{2} \text{ or } 9a + 3b = -\frac{3}{2} \end{cases}$

Solving this system for $a$ and $b$, you obtain

$a = -\frac{1}{2}, b = 1$. Thus, $y = -\frac{1}{2}x^2 + x + 6$.

**8.** $\dfrac{5x - 2}{(x - 1)^2} = \dfrac{A}{x - 1} + \dfrac{B}{(x - 1)^2}$

$5x - 2 = A(x - 1) + B = Ax + (-A + B)$

$\begin{cases} A = 5 \\ -A + B = -2 \implies B = 3 \end{cases}$

$\dfrac{5x - 2}{(x - 1)^2} = \dfrac{5}{x - 1} + \dfrac{3}{(x - 1)^2}$

**9.** $\begin{bmatrix} 2 & 1 & 2 & \vdots & 4 \\ 2 & 2 & 0 & \vdots & 5 \\ 2 & -1 & 6 & \vdots & 2 \end{bmatrix}$ row reduces to $\begin{bmatrix} 1 & 0 & 2 & \vdots & 1.5 \\ 0 & 1 & -2 & \vdots & 1 \\ 0 & 0 & 0 & \vdots & 0 \end{bmatrix}$

Infinite number of solutions. Let $z = a$, $y = 2a + 1$, $x = 1.5 - 2a$.

*Answer:* $(1.5 - 2a, 1 + 2a, a)$, where $a$ is any real number

**10.** $\begin{bmatrix} 2 & 3 & 1 & \vdots & 10 \\ 2 & -3 & -3 & \vdots & 22 \\ 4 & -2 & 3 & \vdots & -2 \end{bmatrix}$ row reduces to $\begin{bmatrix} 1 & 0 & 0 & \vdots & 5 \\ 0 & 1 & 0 & \vdots & 2 \\ 0 & 0 & 1 & \vdots & -6 \end{bmatrix}$

*Answer:* $(5, 2, -6)$

**11.** (a) $A - B = \begin{bmatrix} 1 & 0 & 4 \\ -7 & -6 & -1 \\ 0 & 4 & 0 \end{bmatrix}$    (b) $3A = \begin{bmatrix} 15 & 12 & 12 \\ -12 & -12 & 0 \\ 3 & 6 & 0 \end{bmatrix}$

(c) $3A - 2B = \begin{bmatrix} 7 & 4 & 12 \\ -18 & -16 & -2 \\ 1 & 10 & 0 \end{bmatrix}$    (d) $AB = \begin{bmatrix} 36 & 20 & 4 \\ -28 & -24 & -4 \\ 10 & 8 & 2 \end{bmatrix}$

**12.** $A^{-1} = \dfrac{1}{ad - bc} \begin{bmatrix} d & -b \\ -c & a \end{bmatrix} = \dfrac{1}{30 - 40} \begin{bmatrix} -5 & -4 \\ -10 & -6 \end{bmatrix} = \dfrac{1}{10} \begin{bmatrix} 5 & 4 \\ 10 & 6 \end{bmatrix} = \begin{bmatrix} \frac{1}{2} & \frac{2}{5} \\ 1 & \frac{3}{5} \end{bmatrix}$

$X = A^{-1}B = \begin{bmatrix} \frac{1}{2} & \frac{2}{5} \\ 1 & \frac{3}{5} \end{bmatrix} \begin{bmatrix} 10 \\ 20 \end{bmatrix} = \begin{bmatrix} 13 \\ 22 \end{bmatrix} \implies (x, y) = (13, 22)$

**13.** $\begin{vmatrix} -25 & 18 \\ 6 & -7 \end{vmatrix} = (-25)(-7) - 6(18) = 67$

**14.** $\det(A) = \begin{vmatrix} 4 & 0 & 3 \\ 1 & -8 & 2 \\ 3 & 2 & 2 \end{vmatrix} = 4(-16 - 4) - 0 + 3(2 + 24)$

$\qquad\qquad = -80 + 78 = -2$

**15.** $\begin{vmatrix} -5 & 0 & 1 \\ 3 & 2 & 1 \\ 4 & 4 & 1 \end{vmatrix} = -5(2 - 4) - 0 + 1(12 - 8) = 10 + 4 = 14$

Area $= \frac{1}{2}(14) = 7$

**16.**  $x = \dfrac{\begin{vmatrix} 11 & 8 \\ 21 & -24 \end{vmatrix}}{\begin{vmatrix} 20 & 8 \\ 12 & -24 \end{vmatrix}} = \dfrac{-432}{-576} = \dfrac{3}{4}$    $y = \dfrac{\begin{vmatrix} 20 & 11 \\ 12 & 21 \end{vmatrix}}{\begin{vmatrix} 20 & 8 \\ 12 & -24 \end{vmatrix}} = \dfrac{288}{-576} = -\dfrac{1}{2}$

*Answer:* $\left(\frac{3}{4}, -\frac{1}{2}\right)$

**17.**  Upper left:    $400 + x_2 = x_1$

Upper right:    $x_1 + x_3 = x_4 + 600$

Lower left:        $300 = x_2 + x_3 + x_5$

Lower right:    $x_5 + x_4 = 100$

$$\begin{cases} x_1 - x_2 & = 400 \\ x_1 \quad\;\; + x_3 - x_4 & = 600 \\ \quad\; x_2 + x_3 \quad\;\; + x_5 & = 300 \\ \quad\quad\quad\quad\quad x_4 + x_5 & = 100 \end{cases}$$

Solving this system,

$$\begin{bmatrix} 1 & -1 & 0 & 0 & 0 & \vdots & 400 \\ 1 & 0 & 1 & -1 & 0 & \vdots & 600 \\ 0 & 1 & 1 & 0 & 1 & \vdots & 300 \\ 0 & 0 & 0 & 1 & 1 & \vdots & 100 \end{bmatrix} \rightarrow \begin{bmatrix} 1 & 0 & 1 & 0 & 1 & \vdots & 700 \\ 0 & 1 & 1 & 0 & 1 & \vdots & 300 \\ 0 & 0 & 0 & 1 & 1 & \vdots & 100 \\ 0 & 0 & 0 & 0 & 0 & \vdots & 0 \end{bmatrix}$$

Letting $x_3 = a$ and $x_5 = b$ be real numbers, we have

$x_5 = b$

$x_4 = 100 - b$

$x_3 = a$

$x_2 = 300 - a - b$

$x_1 = 700 - b - a$

# Chapters 3–5    Cumulative Test

**1.** $f(x) = -\frac{1}{2}(x^2 + 4x)$

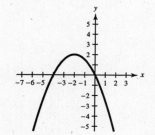

**2.** $f(x) = \frac{1}{4}x(x - 2)^2$

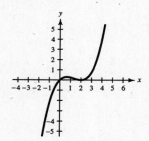

**3.** $x^3 + 2x^2 + 4x + 8 = (x + 2)(x^2 + 4)$

Zeros: $-2, \pm 2i$

**4.** Using a graphing utility, $x \approx 1.424$

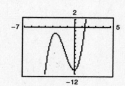

**5.**
$$\begin{array}{r} 4x + 2 \\ x + 3 \overline{)\, 4x^2 + 14x - 9\,} \\ \underline{4x^2 + 12x}\phantom{xx} \\ 2x - 9 \\ \underline{2x + 6}\phantom{x} \\ -15 \end{array}$$

$$\frac{4x^2 + 14x - 9}{x + 3} = 4x + 2 - \frac{15}{x + 3}$$

**6.**
$$\begin{array}{r|rrrr} 6 & 2 & -5 & 6 & -20 \\ & & 12 & 42 & 288 \\ \hline & 2 & 7 & 48 & 268 \end{array}$$

$$\frac{2x^3 - 5x^2 + 6x - 20}{x - 6} = 2x^2 + 7x + 48 + \frac{268}{x - 6}$$

**7.** $f(x) = (x - 0)(x + 3)\left[x - \left(1 + \sqrt{5}i\right)\right]\left[x - \left(1 - \sqrt{5}i\right)\right]$

$\quad = x(x + 3)[(x - 1)^2 + 5]$

$\quad = (x^2 + 3x)(x^2 - 2x + 6)$

$\quad = x^4 + x^3 + 18x$

**8.** $f(x) = \dfrac{2x}{x - 3}$

Vertical asymptote: $x = 3$

Horizontal asymptote: $y = 2$

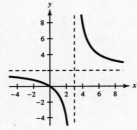

**9.** $f(x) = \dfrac{5x}{x^2 + x - 6} = \dfrac{5x}{(x + 3)(x - 2)}$

Vertical asymptotes: $x = -3, 2$

Horizontal asymptote: $y = 0$

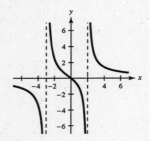

**10.** $f(x) = \dfrac{x^2 - 3x + 8}{x - 2} = x - 1 + \dfrac{6}{x - 2}$

Vertical asymptote: $x = 2$

Slant asymptote: $y = x - 1$

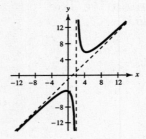

**11.** $(1.85)^{3.1} \approx 6.733$

**12.** $58^{\sqrt{5}} \approx 8772.934$

**13.** $e^{-20/11} \approx 0.162$

**14.** $4e^{2.56} \approx 51.743$

**15.** $f(x) = -3^{x+4} - 5$

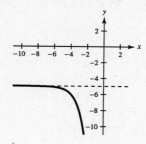

**16.** $f(x) = -\left(\frac{1}{2}\right)^{-x} - 3$

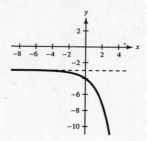

**17.** $f(x) = 4 + \log_{10}(x - 3)$

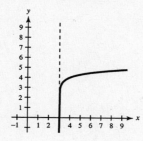

**18.** $f(x) = \ln(4 - x)$

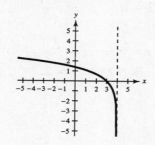

**19.** $\log_5 21 = \dfrac{\ln 21}{\ln 5} \approx 1.892$

**20.** $\log_9 6.8 = \dfrac{\ln 6.8}{\ln 9} \approx 0.872$

**21.** $\log_2\left(\dfrac{3}{2}\right) = \dfrac{\ln\left(\frac{3}{2}\right)}{\ln 2} \approx 0.585$

**22.** $\log_5\left(\dfrac{x^2 - 16}{x^4}\right) = \log_5[(x - 4)(x + 4)] - \log_5 x^4$

$\qquad = \log_5(x - 4) + \log_5(x - 4) - 4\log_5 x, \quad x > 4$

**23.** $2\ln x - \dfrac{1}{2}\ln(x + 5) = \ln x^2 - \ln\sqrt{x + 5}$

$\qquad\qquad\qquad\qquad = \ln\dfrac{x^2}{\sqrt{x + 5}}, \quad x > 0$

**24.** $6e^{2x} = 72$

$\qquad e^{2x} = 12$

$\qquad 2x = \ln 12$

$\qquad x = \tfrac{1}{2}\ln 12 \approx 1.242$

**25.** $4^{x-5} + 21 = 30$

$\qquad 4^{x-5} = 9$

$\qquad (x - 5)\ln 4 = \ln 9$

$\qquad x - 5 = \ln 9/\ln 4$

$\qquad x = 5 + \dfrac{\ln 9}{\ln 4} \approx 6.585$

**26.** $\log_2 x + \log_2 5 = 6$

$\qquad \log_2 5x = 6$

$\qquad 5x = 2^6 = 64$

$\qquad x = \dfrac{64}{5} = 12.8$

**27.** $\begin{cases} 4x - 5y = 29 \\ -x + 9y = 16 \end{cases}$

From Eq. 2, $x = 9y - 16$.

Then $4(9y - 16) - 5y = 29 \Rightarrow 31y = 93 \Rightarrow y = 3$, and $x = 9(3) - 16 = 11$.

*Answer:* $(11, 3)$

**28.** $\begin{cases} 2x - y^2 = 0 \\ x - y = 4 \end{cases} \Rightarrow \quad x = y + 4$

$\qquad\qquad 2(y + 4) - y^2 = 0$

$\qquad\qquad\quad y^2 - 2y - 8 = 0$

$\qquad\qquad\quad (y - 4)(y + 2) = 0$

$\qquad\qquad\qquad\qquad y = 4 \Rightarrow x = 8$

$\qquad\qquad\qquad\qquad y = -2 \Rightarrow x = 2$

$\quad (2, -2), (8, 4)$

**29.** $y_1 = \log_3 x = \dfrac{\ln x}{\ln 3}$ and $y_2 = -\dfrac{1}{3}x + 2$

Intersect at $(3, 1)$

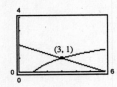

**30.** $\begin{cases} x - y + 3z = -1 \\ 2x + 4y + z = 2 \end{cases}$

$\begin{cases} x - y + 3z = -1 \\ \quad\ 6y - 5z = 4 \end{cases}$

Infinite number of solutions. Let $z = a$.

$6y - 5a = 4 \Rightarrow y = \frac{1}{6}(4 + 5a)$

$x - \frac{1}{6}(4 + 5a) + 3a = -1 \Rightarrow x = \frac{1}{6}(-2 - 13a)$

*Answer:* $\left(\frac{1}{6}(-2 - 13a), \frac{1}{6}(4 + 5a), a\right)$

**31.** $\begin{bmatrix} 2 & -3 & 1 & \vdots & 13 \\ -4 & 1 & -2 & \vdots & -6 \\ 1 & -3 & 3 & \vdots & 12 \end{bmatrix}$ row reduces to $\begin{bmatrix} 1 & 0 & 0 & \vdots & 0.6 \\ 0 & 1 & 0 & \vdots & -4 \\ 0 & 0 & 1 & \vdots & -0.2 \end{bmatrix}$

*Answers:* $(0.6, -4.0, -0.2)$

**32.** $\begin{bmatrix} 1 & -4 & 3 & \vdots & 5 \\ 5 & 2 & -1 & \vdots & 1 \\ -2 & -8 & 0 & \vdots & 30 \end{bmatrix}$ row reduces to $\begin{bmatrix} 1 & 0 & 0 & \vdots & 1 \\ 0 & 1 & 0 & \vdots & -4 \\ 0 & 0 & 1 & \vdots & -4 \end{bmatrix}$

*Answer:* $(1, -4, -4)$

**33.** $3A - 2B = \begin{bmatrix} -7 & -10 & -16 \\ -6 & 18 & 9 \\ -12 & 16 & 7 \end{bmatrix}$

**34.** $5A + 3B = \begin{bmatrix} -18 & 15 & -14 \\ 28 & 11 & 34 \\ -20 & 52 & -1 \end{bmatrix}$

**35.** $AB = \begin{bmatrix} 3 & -31 & 2 \\ 22 & 18 & 6 \\ 52 & -40 & 14 \end{bmatrix}$

**36.** $BA = \begin{bmatrix} 5 & 36 & 31 \\ -36 & 12 & -36 \\ 16 & 0 & 18 \end{bmatrix}$

**37.** Determinant $= (-10)(2)(5)(-3) = 300$

(upper triangle)

**38.** $\begin{vmatrix} -2 & -8 & 1 \\ 3 & 7 & 1 \\ 7 & 19 & 1 \end{vmatrix} = 0 \implies$ the points are collinear.

**39.** $A = Pe^{rt} = 2500e^{(0.075)(25)} \approx \$16,302.05$

**40.**  $P = 228e^{kt}$

$166 = 228e^{k(-30)} \implies k \approx 0.01058$

For 2010, $P = 228e^{k(10)} \approx 253.4$ or 253,400 people

**41.** (a)  $y = 0.248x^2 - 0.28x + 50.3$

$y = 42.62(1.05)^x = 42.62e^{0.0515x}$

$y = 32.45x^{0.3329}$

(b)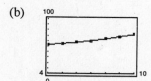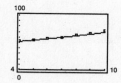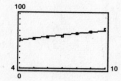

(c)  The quadratic model fits best.

(d)  In 2006, $x = 16$ and $y \approx 109$ or 109,000 pilots and copilots.

**42.** (a)  $C = 59.95x + 150,000$

$R = 200x$

(b)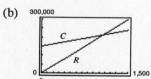

(c)  $59.95x + 150,000 = 200x$

$150,000 = 140.05x$

$x \approx 1071$ units

Must sell 1071 units to break even.

**43.** $\begin{cases} 2l + 2w = 76 \\ lw = 352 \implies l = \dfrac{352}{w} \end{cases}$

$2\left(\dfrac{352}{w}\right) + 2w = 76$

$704 + 2w^2 = 76w$

$w^2 - 38w + 352 = 0$

$(w - 22)(w - 16) = 0$

$w = 16, 22$

Dimensions: $16 \times 22$ meters

# Chapter 6    Chapter Test

**1.** $a_n = \left(-\frac{2}{3}\right)^{n-1}$    $a_1 = \left(-\frac{2}{3}\right)^{1-1} = \left(-\frac{2}{3}\right)^0 = 1$

$a_2 = -\frac{2}{3}$

$a_3 = \left(-\frac{2}{3}\right)^2 = \frac{4}{9}$

$a_4 = \left(-\frac{2}{3}\right)^3 = -\frac{8}{27}$

$a_5 = \left(-\frac{2}{3}\right)^4 = \frac{16}{81}$

**2.** $a_1 = 12, \ a_{k+1} = a_k + 4$

$a_2 = 12 + 4 = 16$

$a_3 = 16 + 4 = 20$

$a_4 = 20 + 4 = 24$

$a_5 = 24 + 4 = 28$

**3.** $\dfrac{11!4!}{4!7!} = \dfrac{11!}{7!} = \dfrac{11 \cdot 10 \cdot 9 \cdot 8 \cdot 7!}{7!} = 11 \cdot 10 \cdot 9 \cdot 8 = 7920$

**4.** $a_n = dn + c, \; c = a_1 - d = 5000 - (-100) = 5100$

$\Rightarrow a_n = -100n + 5100 = 5000 - 100(n - 1)$

**5.** $a_n = a_1 r^{n-1}, \; a_1 = 4, \; r = \dfrac{1}{2} \Rightarrow a_n = 4\left(\dfrac{1}{2}\right)^{n-1}$ 　　　　**6.** $\displaystyle\sum_{n=1}^{12} \dfrac{2}{3n + 1}$

**7.** $\displaystyle\sum_{n=1}^{7} (8n - 5) = 8\left(\dfrac{7(8)}{2}\right) - 5(7) = 224 - 35 = 189$

**8.** $\displaystyle\sum_{n=1}^{8} 24\left(\dfrac{1}{6}\right)^{n-1} = 24\left(\dfrac{1 - \left(\dfrac{1}{6}\right)^8}{1 - \dfrac{1}{6}}\right) = 24\left(\dfrac{6}{5}\right)\left(1 - \left(\dfrac{1}{6}\right)^8\right) \approx 28.79998 \approx 28.80$

**9.** $\displaystyle\sum_{n=1}^{\infty} 5\left(\dfrac{1}{10}\right)^{n-1} = 5[1 + 0.1 + 0.01 + \ldots] = 5[1.1111\ldots] = 5.555\ldots = \dfrac{50}{9}$

**10.** (1) For $n = 1, 3 = \dfrac{3(1)(1 + 1)}{2}$

(2) Assume $S_k = 3 + 6 + \cdots + 3k = \dfrac{3k(k + 1)}{2}$.

Then $S_{k+1} = 3 + 6 + \cdots + 3k + 3(k + 1)$

$= S_k + 3(k + 1)$

$= \dfrac{3k(k + 1)}{2} + 3(k + 1)$

$= \dfrac{k + 1}{2}[3k + 6]$

$= \dfrac{3(k + 1)(k + 2)}{2}$

Therefore, the formula is true for all positive integers $n$.

**11.** $(2a - 5b)^4 = (2a)^4 - 4(2a)^3(5b) + 6(2a)^2(5b)^2 - 4(2a)(5b)^3 + (5b)^4$

$= 16a^4 - 160a^3b + 600a^2b^2 - 1000ab^3 + 625b^4$

**12.** $_8C_3(3)^3(2)^5 = 56(27)(32) = 48{,}384$ 　　　　　　　**13.** $_9C_3 = 84$

**14.** $_{20}C_3 = 1140$

**15.** $_9P_2 = \dfrac{9!}{7!} = 9 \cdot 8 = 72$ 　　　　　**16.** $_{70}P_3 = \dfrac{70!}{67!} = 70 \cdot 69 \cdot 68$

$= 328{,}440$

**17.** $26 \cdot 10 \cdot 10 \cdot 10 = 26{,}000$ ways

**18.** $_{25}C_4 = \dfrac{25!}{21! \, 4!} = \dfrac{25 \cdot 24 \cdot 23 \cdot 22}{24} = 12{,}650$ ways

**19.** There are 6 red face cards $\Rightarrow$ probability $= \frac{6}{52} = \frac{3}{26}$.

**20.** There are $_4C_2 = 6$ ways to select 2 spark plugs. Only one way corresponds to both being selected. Probability $= \frac{1}{6}$.

**21.** (a) $\left(\frac{30}{60}\right)\left(\frac{30}{60}\right) = \frac{1}{2} \cdot \frac{1}{2} = \frac{1}{4}$

(b) $\frac{11}{60} \cdot \frac{11}{60} = \frac{121}{3600} \approx 0.0336$

(c) $\frac{1}{60} \approx 0.0167$

**22.** $1 - 0.75 = 0.25 = 25\%$

# Chapter 7    Chapter Test

**1.** $y^2 = 8x = 4(2)x$

Vertex: $(0, 0)$

Focus: $(2, 0)$

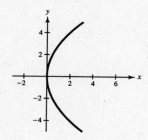

**2.** $4x^2 + y^2 = 4$

$x^2 + \dfrac{y^2}{4} = 1$

Vertices: $(0, \pm 2)$

$c^2 = a^2 - b^2 = 4 - 1 = 3$

Foci: $\left(0, \pm \sqrt{3}\right)$

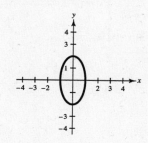

**3.**　$x^2 - 4x - 4y^2 = 0$

$x^2 - 4x + 4 - 4y^2 = 4$

$(x - 2)^2 - 4y^2 = 4$

$\dfrac{(x - 2)^2}{4} - y^2 = 1$

Hyperbola

Center: $(2, 0)$

Vertices: $(4, 0), (0, 0)$

$c^2 = a^2 + b^2 = 2^2 + 1^2 = 5$

Foci: $\left(2 \pm \sqrt{5}, 0\right)$

**4.** Vertex: $(0, 7)$

$(y - 7)^2 = 4p(x - 0)$

$(0 - 7)^2 = 4p(-8 - 0)$

$49 = -32p \implies p = \dfrac{-49}{32}$

$(y - 7)^2 = \dfrac{-49}{8}x$

$(y - 7)^2 = 4\left(-\frac{49}{32}\right)x$

**5.** Vertex: $(6, -2), p = 2$

$(y + 2)^2 = 4(2)(x - 6)$

$(y + 2)^2 = 8(x - 6)$

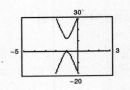

**6.** Center: $(4, 2), a = 4, b = 2$

$$\frac{(x - 4)^2}{16} + \frac{(y - 2)^2}{4} = 1$$

**7.** Center: $(-6, 3), a = 7, b = 4$

$$\frac{(x + 6)^2}{16} + \frac{(y - 3)^2}{49} = 1$$

**8.** Vertical transverse axis, $a = 3$

Asymptotes: $y = \pm\frac{a}{b}x = \pm\frac{3}{2}x \implies b = 2$

$$\frac{y^2}{9} - \frac{x^2}{4} = 1$$

**9.** Solve for $y$: $(y - 6)^2 = 36\left[\dfrac{(x + 1)^2}{\frac{1}{16}} + 1\right] = 36[16(x + 1)^2 + 1]$

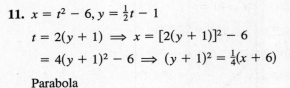

$y - 6 = \pm 6\sqrt{16(x + 1)^2 + 1}$

$\quad y = 6 \pm 6\sqrt{16(x + 1)^2 + 1}$

Center: $(-1, 6)$

**10.** $(x^2 - 10x + 25) + (y^2 + 4y + 4) = -4 + 25 + 4$

$(x - 5)^2 + (y + 2)^2 = 25$

$(y + 2)^2 = 25 - (x - 5)^2 = 10x - x^2$

$\quad y = -2 \pm \sqrt{10x - x^2}$

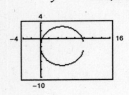

**11.** $x = t^2 - 6, y = \frac{1}{2}t - 1$

$t = 2(y + 1) \implies x = [2(y + 1)]^2 - 6$

$\quad = 4(y + 1)^2 - 6 \implies (y + 1)^2 = \frac{1}{4}(x + 6)$

Parabola

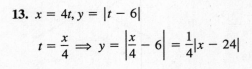

**12.** $x = \sqrt{t^2 + 2}, y = \dfrac{t}{4}$

$t = 4y \implies x = \sqrt{16y^2 + 2} \implies x^2 = 16y^2 + 2$

$\implies x^2 - 16y^2 = 2$

Right portion
of hyperbola

$\left(x \geq \sqrt{2}\right)$

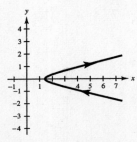

**13.** $x = 4t, y = |t - 6|$

$t = \dfrac{x}{4} \implies y = \left|\dfrac{x}{4} - 6\right| = \dfrac{1}{4}|x - 24|$

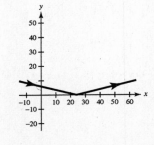

**14.**

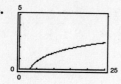

**15.** $x = t, y = 7t + 6$

$x = -t, y = -7t + 6$

(other answers possible)

**16.** $x = t, y = t^2 + 10$

$x = t + 1, y = (t + 1)^2 + 10$

(other answers possible)

**17.** Vertex: $(0, 16) \Rightarrow x^2 = 4p(y - 16)$

$(6, 14)$ on parabola $\Rightarrow 36 = 4p(14 - 16) = -8p \Rightarrow p = \dfrac{-9}{2}$

$x^2 = 4\left(-\dfrac{9}{2}\right)(y - 16)$

$x^2 = -18(y - 16)$

let $y = 0 \Rightarrow x^2 = -18(0 - 16) = 288 \Rightarrow x = \pm 12\sqrt{2}$

width $= 2\left(12\sqrt{2}\right) = 24\sqrt{2} \approx 33.94$ meters

**18.** $2a = 768,800 \Rightarrow a = 384,400$

$2b = 767,740 \Rightarrow b = 383,870$

$c = \sqrt{a^2 - b^2} \approx 20,179$

perihelion $= a - c \approx 364,221$ km

aphelion $= a + c \approx 404,579$ km

## Chapters 6–7   Cumulative Test

**1.** (a) $a_1 = \dfrac{(-1)^{1+1}}{2(1) + 3} = \dfrac{1}{5}$

$a_2 = \dfrac{-1}{7}$

$a_3 = \dfrac{1}{9}$

$a_4 = \dfrac{-1}{11}$

$a_5 = \dfrac{1}{13}$

(b) $a_1 = 3(2)^{1-1} = 3$

$a_2 = 6$

$a_3 = 12$

$a_4 = 24$

$a_5 = 48$

**2.** $\dfrac{49!}{46!} = \dfrac{49 \cdot 48 \cdot 47 \cdot 46!}{46!} = 49 \cdot 48 \cdot 47 = 110,544$

**3.** $\displaystyle\sum_{k=1}^{6} (7k - 2) = 7\frac{6(7)}{2} - 2(6) = 7(21) - 12 = 135$

**4.** $\displaystyle\sum_{k=1}^{4} \frac{2}{k^2 + 4} = \frac{2}{1 + 4} + \frac{2}{4 + 4} + \frac{2}{9 + 4} + \frac{2}{16 + 4}$
$$\approx 0.9038$$

**5.** $\displaystyle\sum_{k=1}^{5} 10 = 5(10) = 50$

**6.** $\displaystyle\sum_{k=3}^{6} (k - 1)(k + 2) = \sum_{k=3}^{6} (k^2 + k - 2)$
$$= 10 + 18 + 28 + 40 = 96$$

**7.** $\displaystyle\sum_{n=0}^{10} 9\left(\frac{3}{4}\right)^n = \sum_{n=1}^{11} 9\left(\frac{3}{4}\right)^{n-1} = 9\frac{1 - \left(\frac{3}{4}\right)^{11}}{1 - \frac{3}{4}} = 36\left(1 - \left(\frac{3}{4}\right)^{11}\right)$
$$\approx 34.4795$$

**8.** $\displaystyle\sum_{n=1}^{\infty} 8(0.9)^{n-1} = \sum_{n=0}^{\infty} 8(0.9)^n = \frac{8}{1 - 0.9} = \frac{8}{0.1} = 80$

**9.** For $n = 1$, $3 = 1(2(1) + 1) = 3$. Assume true for $n = k$. Then,
$$3 + 7 + \cdots + (4k - 1) + [4(k + 1) - 1] = k(2k + 1) + [4k + 3]$$
$$= 2k^2 + 5k + 3 = (2k + 3)(k + 1)$$
$$= (k + 1)[2(k + 1) + 1].$$

**10.** $(x + 5)^4 = x^4 + 20x^3 + 150x^2 + 500x + 625$

**11.** $(2x + y^2)^5 = 32x^5 + 80x^4y^2 + 80x^3y^4 + 40x^2y^6 + 10xy^8 + y^{10}$

**12.** $(x - 2y)^6 = x^6 - 12x^5y + 60x^4y^2 - 160x^3y^3 + 240x^2y^4 - 192xy^5 + 64y^6$

**13.** $(2x - 1)^8 = 256x^8 - 1024x^7 + 1792x^6 - 1792x^5 + 1120x^4 - 448x^3 + 112x^2 - 16x + 1$

**14.** $\dfrac{5!}{2!2!1!} = \dfrac{120}{4} = 30$   **15.** $\dfrac{6!}{3!} = 6 \cdot 5 \cdot 4 = 120$   **16.** $\dfrac{10!}{2!2!2!} = 453{,}600$   **17.** $\dfrac{10!}{3!2!2!} = 151{,}200$

**18.** Hyperbola

Center: $(-5, -3)$

Vertical transverse axis

Vertices: $(-5, 3), (-5, -9)$

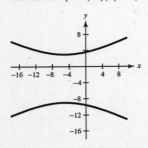

**19.** Ellipse with center $(2, -1)$:

Vertices: $(2, 2), (2, -4)$

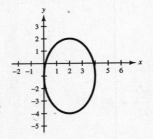

**20.** Hyperbola: $\dfrac{y^2}{16} - \dfrac{x^2}{16} = 1$

Center: $(0, 0)$

Vertices: $(0, \pm 4)$

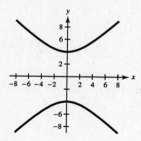

**21.** $(x^2 - 2x + 1) + (y^2 - 4y + 4) = -5 + 1 + 4 = 0$

$$(x - 1)^2 + (y - 2)^2 = 0$$

Point: $(1, 2)$

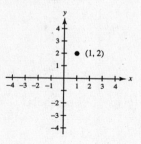

**22.** $y^2 - 6y + 9 = 6x - 42 + 9$

$$(y - 3)^2 = 6x - 33$$

Parabola

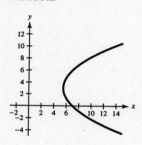

**23.** $9(x^2 + 6x + 9) + 25y^2 = 144 + 81$

$$9(x + 3)^2 + 25y^2 = 225$$

$$\dfrac{(x + 3)^2}{25} + \dfrac{y^2}{9} = 1$$

Ellipse

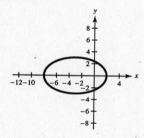

**24.** Vertex: $(2, 3) \implies (x - 2)^2 = 4p(y - 3)$

$(0, 0)$ on parabola: $(0 - 2)^2 = 4p(0 - 3)$

$$4 = 4p(-3) \implies p = -\dfrac{1}{3}$$

$$(x - 2)^2 = 4\left(-\dfrac{1}{3}\right)(y - 3)$$

$$(x - 2)^2 = -\dfrac{4}{3}(y - 3)$$

**25.** Center: $(1, 4)$, $a = 5$, $b = 2$

$$\dfrac{(x - 1)^2}{25} + \dfrac{(y - 4)^2}{4} = 1$$

**26.** Center: $(0, -4)$, $a = 2$

$$\dfrac{(y + 4)^2}{4} - \dfrac{x^2}{b^2} = 1$$

$(4, 0)$ on hyperbola: $\dfrac{(0 + 4)^2}{4} - \dfrac{4^2}{b^2} = 1 \implies \dfrac{16}{b^2} = 4 - 1 = 3 \implies b^2 = \dfrac{16}{3}$

$$\dfrac{(y + 4)^2}{4} - \dfrac{x^2}{\dfrac{16}{3}} = 1$$

**27.** The center is $(0, 2)$ and $c = 2$. The transverse axis is vertical and $\dfrac{a}{b} = \dfrac{1}{2}$.

$$b^2 = c^2 - a^2 = 4 - \left(\frac{b}{2}\right)^2 \implies \frac{5b^2}{4} = 4 \implies b^2 = \frac{16}{5} \text{ and } a^2 = \frac{4}{5}.$$

Thus, $\dfrac{(y-2)^2}{4/5} - \dfrac{x^2}{16/5} = 1.$

**28.** $x = 2t + 1, y = t^2$

$t = \frac{1}{2}(x - 1) \implies y = \left(\frac{1}{2}(x-1)\right)^2$ parabola

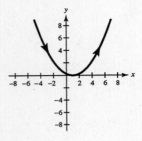

**29.** $x = 8 + 3t, y = 4 - t$

$t = \frac{1}{3}(x - 8) \implies y = 4 - \frac{1}{3}(x - 8) = -\frac{1}{3}x + \frac{20}{3}$
line

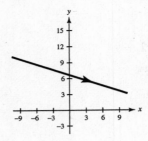

**30.** $x = 4 \ln t, y = \frac{1}{2}t^2$

$t = e^{x/4} \implies y = \frac{1}{2}e^{x/2}$

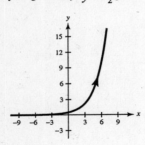

**31.** $y = 3x - 2$. Sample answers:

$x = t$            $x = -t$

$y = 3t - 2$       $y = -3t - 2$

**32.** $y = \dfrac{2}{x}$. Sample answers:

$x = t$            $x = -t$

$y = \dfrac{2}{t}$        $y = -\dfrac{2}{t}$

**33.** $y = x^2 - 1$. Sample answers:

$x = t$            $x = -t$

$y = t^2 - 1$       $y = t^2 - 1$

**34.** $x = \phantom{x} 2 + (6 - 2)t \implies x = 4t + 2$

$y = -3 + (4 + 3)t \qquad y = 7t - 3$

**35.** $\displaystyle\sum_{n=7}^{11} (13.16n^2 - 177.7n + 758) = \$1254.9$ million

**36.** $28{,}000 + 28{,}000(1.05) + \cdots + 28{,}000(1.05)^{14}$

$= 28{,}000 \displaystyle\sum_{n=1}^{15} (1.05)^{n-1} = 28{,}000 \dfrac{1 - 1.05^{15}}{1 - 1.05}$

$\approx \$604{,}199.78$

**37.** There are 2 choices for the first digit (4 or 5). Then there remains 2 choices for the second digit, and 1 for the third. Thus, probability $= \frac{1}{4}$.

**38.** Placing the center of the ellipse at the origin, $a = \dfrac{97}{2} = 48.5$ and $b = \dfrac{46}{2} = 23$. Thus,

$$\frac{x^2}{48.5^2} + \frac{y^2}{23^2} = 1 \text{ or } \frac{x^2}{23^2} + \frac{y^2}{48.5^2} = 1$$

**39.** $x = 99.6t$

$y = 3 + 57.5t - 16t^2$

When $x = 375$, $t \approx 3.765$.

Then $y \approx -7.32$ which implies that the ball does not go over the fence.